NANO:
The Essentials

Understanding Nanoscience
and Nanotechnology

About the Author

T. Pradeep, Ph.D., is a Professor in the Department of Chemistry and Sophisticated Analytical Instrument Facility (SAIF) at the Indian Institute of Technology, Madras. He is the editor of *Advances in Physical Chemistry* and co-author of *Nanofluids: Science and Technology*. Dr. Pradeep also has written over 150 research papers and holds 5 patents.

NANO:
The Essentials

Understanding Nanoscience
and Nanotechnology

T. PRADEEP
Professor
Indian Institute of Technology, Madras
Chennai, India

McGraw-Hill
New York   Chicago   San Francisco   Lisbon   London
Madrid   Mexico City   Milan   New Delhi   San Juan
Seoul   Singapore   Sydney   Toronto

The **McGraw·Hill** Companies

Copyright © 2008 by The McGraw-Hill Companies, Inc. All rights reserved. Printed in the United States of America. Except as permitted under the United States Copyright Act of 1976, no part of this publication may be reproduced or distributed in any form or by any means, or stored in a data base or retrieval system, without the prior written permission of the publisher.

1 2 3 4 5 6 7 8 9 0   DOC/DOC   0 1 4 3 2 1 0 9 8

ISBN 978-0-07-154829-8
MHID 0-07-154829-7

The sponsoring editor for this book was Stephen S. Chapman and the production supervisor was Richard C. Ruzycka. The art director for the cover was Jeff Weeks.

Printed and bound by RR Donnelley.

This book was previously published by Tata McGraw-Hill Publishing Company Limited, New Delhi, India, copyright © 2007.

McGraw-Hill books are available at special quantity discounts to use as premiums and sales promotions, or for use in corporate training programs. To contact a special sales representative, please visit the Contact Us page at www.mhprofessional.com.

This book is printed on acid-free paper.

Information contained in this work has been obtained by The McGraw-Hill Companies, Inc. ("McGraw-Hill") from sources believed to be reliable. However, neither McGraw-Hill nor its authors guarantee the accuracy or completeness of any information published herein, and neither McGraw-Hill nor its authors shall be responsible for any errors, omissions, or damages arising out of use of this information. This work is published with the understanding that McGraw-Hill and its authors are supplying information but are not attempting to render engineering or other professional services. If such services are required, the assistance of an appropriate professional should be sought.

To
My Teacher

CNR Rao

For he made me realize that there is an ocean in every drop.
With his fathomless patience, unfailing support, critical suggestions and iron will, nothing was ever a hurdle!

Contents

Preface   *xv*
Acknowledgments   *xix*

Part One
Introduction

1. Introduction—The Canvas of Nano   3

1.1   Nano and Nature   3
1.2   Our Technologies and the World We Live in   5
1.3   Nano—The Beginning   9
     Review Questions   12
     References   12

Part Two
Experimental Methods

2. Investigating and Manipulating Materials in the Nanoscale   15

2.1   Introduction   15
2.2   Electron Microscopies   20
2.3   Scanning Probe Microscopies   43
2.4   Optical Microscopies for Nanoscience and Technology   54
2.5   Other Kinds of Microscopies   59
2.6   X-Ray Diffraction   75
2.7   Associated Techniques   81
     Review Questions   81
     References   82
     Additional Reading   84

PART THREE
Diversity in Nanosystems

3. Fullerenes — 89

    3.1 Introduction — 89
    3.2 Discovery and Early Years — 91
    3.3 Synthesis and Purification of Fullerenes — 94
    3.4 Mass Spectrometry and Ion/Molecule Reactions — 95
    3.5 Chemistry of Fullerenes in the Condensed Phase — 96
    3.6 Endohedral Chemistry of Fullerenes — 99
    3.7 Orientational Ordering — 100
    3.8 Pressure Effects — 101
    3.9 Conductivity and Superconductivity in Doped Fullerenes — 102
    3.10 Ferromagnetism in C_{60}.TDAE — 103
    3.11 Optical Properties — 103
    3.12 Some Unusual Properties — 104
          Review Questions — 105
          References — 106
          Additional Reading — 112

4. Carbon Nanotubes — 114

    4.1 Introduction — 114
    4.2 Synthesis and Purification — 117
    4.3 Filling of Nanotubes — 119
    4.4 Mechanism of Growth — 120
    4.5 Electronic Structure — 120
    4.6 Transport Properties — 122
    4.7 Mechanical Properties — 122
    4.8 Physical Properties — 123
    4.9 Applications — 123
    4.10 Nanotubes of Other Materials — 124
          Review Questions — 125
          References — 126
          Additional Reading — 127

5. Self-assembled Monolayers — 128

5.1	Introduction	128
5.2	Monolayers on Gold	129
5.3	Growth Process	138
5.4	Phase Transitions	141
5.5	Patterning Monolayers	142
5.6	Mixed Monolayers	144
5.7	SAMS and Applications	144
	Review Questions	153
	References	154
	Additional Reading	155

6. Gas Phase Clusters — 156

6.1	Introduction	156
6.2	History of Cluster Science	158
6.3	Cluster Formation	158
6.4	Cluster Growth	161
6.5	Detection and Analysis of Gas Phase Clusters	163
6.6	Types of Clusters	166
6.7	Properties of Clusters	172
6.8	Bonding in Clusters	176
	Review Questions	177
	References	177
	Additional Reading	178

7. Semiconductor Quantum Dots — 179

7.1	Introduction	179
7.2	Synthesis of Quantum Dots	182
7.3	Electronic Structure of Nanocrystals	187
7.4	How Do We Study Quantum Dots?	189
7.5	Correlation of Properties with Size	194
7.6	Uses	195
	Review Questions	197
	References	198
	Additional Reading	198

8. Monolayer-protected Metal Nanoparticles — 199

- 8.1 Introduction — 199
- 8.2 Method of Preparation — 200
- 8.3 Characterization — 200
- 8.4 Functionalized Metal Nanoparticles — 204
- 8.5 Applications — 206
- 8.6 Superlattices — 208
  - *Review Questions* — 212
  - *References* — 213
  - *Additional Reading* — 214

9. Core-shell Nanoparticles — 215

- 9.1 Introduction — 215
- 9.2 Types of Systems — 216
- 9.3 Characterization — 225
- 9.4 Properties — 227
- 9.5 Applications — 234
  - *Review Questions* — 240
  - *References* — 241
  - *Additional Reading* — 243

10. Nanoshells — 244

- 10.1 Introduction — 244
- 10.2 Types of Nanoshells — 245
- 10.3 Properties — 252
- 10.4 Characterization — 255
- 10.5 Applications — 257
  - *Review Questions* — 259
  - *References* — 259
  - *Additional Reading* — 260

PART FOUR
Evolving Interfaces of Nano

11. Nanobiology — 263

- 11.1 Introduction — 263
- 11.2 Interaction Between Biomolecules and Nanoparticle Surfaces — 264

11.3	Different Types of Inorganic Materials Used for the Synthesis of Hybrid Nano-bio Assemblies	269
11.4	Applications of Nano in Biology	271
11.5	Nanoprobes for Analytical Applications—A New Methodology in Medical Diagnosis and Biotechnology	276
11.6	Current Status of Nanobiotechnology	278
11.7	Future Perspectives of Nanobiology	280
	Review Questions	280
	References	281
	Additional Reading	282

12. Nanosensors — 283

12.1	Introduction	283
12.2	What is a Sensor?	284
12.3	Nanosensors—What Makes Them Possible?	285
12.4	Order from Chaos—Nanoscale Organization for Sensors	285
12.5	Characterization—To Know What has been Put In	288
12.6	Perception—Nanosensors Based on Optical Properties	289
12.7	Nanosensors Based on Quantum Size Effects	291
12.8	Electrochemical Sensors	293
12.9	Sensors Based on Physical Properties	294
12.10	Nanobiosensors—A Step towards Real-time Imaging and Understanding of Biological Events	296
12.11	Smart Dust—Sensors of the Future	298
	Review Questions	299
	References	299
	Additional Reading	300

13. Nanomedicines — 301

13.1	Introduction	301
13.2	Approach to Developing Nanomedicines	302
13.3	Various Kinds of Nanosystems in Use	303
13.4	Protocols for Nanodrug Administration	305
13.5	Nanotechnology in Diagnostic Applications	307
13.6	Materials for Use in Diagnostic and Therapeutic Applications	310
13.7	Future Directions	312
	Review Questions	313

	References	313
	Additional Reading	315

14. Molecular Nanomachines — 316

- 14.1 Introduction — 316
- 14.2 Covalent and Non-covalent Approaches — 317
- 14.3 Molecular Motors and Machines — 318
- 14.4 Molecular Devices — 319
- 14.5 Single Molecule Devices — 320
- 14.6 Practical Problems with Molecular Devices — 327
  - *Review Questions* — 328
  - *References* — 328
  - *Additional Reading* — 329

15. Nanotribology — 330

- 15.1 Introduction — 330
- 15.2 Studying Tribology at the Nanoscale — 331
- 15.3 Nanotribology Applications — 337
- 15.4 Outstanding Issues — 341
  - *Review Questions* — 342
  - *References* — 342
  - *Additional Reading* — 343

PART FIVE
Society and Nano

16. Societal Implications of Nanoscience and Nanotechnology (in Developing Countries) — 347

- 16.1 Introduction — 348
- 16.2 From the First Industrial Revolution to the Nano Revolution — 349
- 16.3 Implications of Nanoscience and Nanotechnology on Society — 351
- 16.4 Issues—An Outlook — 353
- 16.5 Nano Policies and Institutions — 358
- 16.6 Nanotech and War—Nano Arms Race — 360
- 16.7 Public Perception and Public Involvement in the Nano Discourse — 360
- 16.8 Harnessing Nanotechnology for Economic and Social Development — 362

16.9	Conclusions	368
	Review Questions	368
	References	369
	Additional Reading	371
Appendix		372
Glossary		413
Index		427

Preface

The world of materials science is witnessing a revolution in the exploration of matter at the small scale. Sub-atomic particles have been a fascination since the first half of the 20th century. High-energy accelerators allow us now to penetrate the constituents of sub-atomic particles. This is an ongoing quest. New and improved properties of materials whose constituting units are nanosized objects make one explore these objects in further detail. When matter at nanoscales can perform functions hitherto done by bulk materials and their assemblies, many things inconceivable in the past can be achieved. Imagine devices such as moving bodies 1000 times smaller than a bacterium. Imagine complex machines as small as a virus. In fact the virus itself is such a machine, created by nature.

Obviously reducing dimension and consequent exploration of properties has no limit. This takes one to the famous, oft-repeated statement of Feynman, "there is plenty of room at the bottom." One can start arranging molecules to achieve the functions of a complex machine. A future with controlled molecular assemblies of this kind, the molecular nanotechnology, will revolutionize everything—from food to thought will change with this newly acquired power.

Atoms and typical molecules are a few angstroms long. That is 10^{-10} m or 10 billionth of a meter. Numbers of this kind are very hard to comprehend as they are not in everyday use. This length is as small as a millimeter if one were to take a wire stretched between Chennai and Kanyakumari, the southern tip of India. In order to understand 10^{-10}, it is useful to imagine 10^{10}, a number which is astronomically big. The distance of 10^{10} m is 10 million kilometers or it is 26 times the distance between the earth and the moon. Obviously no one, except astronauts, travel that kind of distances. In day-to-day life we feel distances of the order of meters and centimeters, the smallest distance one can see without instruments is 0.1 mm, or the thickness of a cotton fiber. This is 10^{-4} m. The distance of 10^{-10} m is a million times smaller, or it is about 1 mm if the cotton fiber were to be expanded to appear like a 100 m wide highway. We are talking about very small things, and consequently we need the right tools to see these objects.

In order to get a nano object to function, it is necessary to assemble the constituent atoms or molecules, perhaps into a large single molecule such as a protein. These objects are of the size of a nanometer (10^{-9} m). The science of nanometer scale objects is nanoscience. The resulting technology is called nanotechnology. Nanotechnology involves achieving the capability to manipulate matter in a desired fashion, atom by atom. At this scale, the constituents of matter do functions, which are different from those of the constituents or bulk materials. While molecular properties bridge material functions at this interface, a wide gap opens up in our understanding of properties in this size domain. This makes it necessary to do additional investigation. Obviously there are many surprises in such studies which make this area scientifically fascinating.

The advances in this area will result in newer technologies: nanoscience and nanotechnology market in 2015 is predicted to be worth 350 billion dollars. That consequently calls for new investments in human resource development. These people must have strong foundation in nanoscience and technology with a fair background in chemistry, physics, mathematics and biology as well as in electrical, mechanical and chemical engineering. While the basic science and engineering degrees are absolutely essential, training in nano related areas with a desire to constantly update their knowledge will be necessary to launch them into the demands of the world. It is not possible to bridge the gap within the undergraduate curriculum alone by offering a few additional courses. Postgraduate programmes may have to be thought of in nanoscience and technology.

Such programmes must provide sufficient information on experimental methodologies with necessary theoretical background. Current undergraduate curriculum does not provide an introduction to the modern experimental tools, at least in India, where the author has first hand knowledge. It is also impossible to expect it at this level, considering the time available and the advances in the respective fields. Therefore, any programme directed towards nanoscience and technology has to be done at the postgraduate or graduate level. While a devoted postgraduate course in the area may be useful, this has to be done with sufficient practical exposure to all the areas and experimental methodologies. This calls for large scale investment on infrastructure, especially in countries such as India, as most universities are ill equipped to meet the demands in instrumentation intensive areas. The few who have the necessary competence both in terms of the infrastructure and human resource may initiate such programmes. Courses at the graduate level are certainly useful, but considering the selected few that one takes during the graduate programme, this may not provide an overall perspective of the area, unless one is pursuing research in it.

This book has been written to provide an overall appreciation of the area, starting from the basics. The subject matter can be easily comprehended by an undergraduate student in science or engineering. Certain details are omitted deliberately, but sufficient directions are given to both students and teachers who require detailed understanding in any specific area. For the benefit of those interested in further exploration, a brief summary of the discoveries in the area is given at the end of the book with original references. This may be useful for a person completing a master's degree yet unsure of the area in which he would like to specialize. A glossary of nano terms is also included. The entire material can be structured as a one semester course, with about 36 lectures, the suggested number of lecture hours is only a guideline. Lectures may also be added to bring more quantitative aspects into the classroom, especially in specific areas such as self assembled monolayers (e.g. experimental structure of monolayers), metal nanoparticles (e.g. electronic absorption spectrum) and quantum dots (e.g. size effects). This may be done in several different ways. An example would be 20–25 classroom lectures followed by student presentations, which can include additional material from current research or cited references. The summary of discoveries in the area may be used to pick topics for presentations. It is also possible to cover selected areas using this book while a course on a more focused subject is presented.

Subject matter of the kind discussed in this book is rapidly evolving. Therefore, any book of this kind has to be updated at periodic intervals. Science at the nano-bio-mechanical-electrical interfaces is opening up newer disciplines. Advanced technology is rapidly changing instrumentation, which in turn makes it possible to explore newer things. The problems and the way they are looked at are changing. All of these would suggest a more frequent revision of the book than would be necessary in more traditional areas.

This book has also been written in view of specific demands of our times. There is a realization that nano is the direction to go in the years to come. This has come to the minds of educationists, planners and administrators alike—thanks to the media. As a result of this, several institutions are planning to establish courses and programmes in the area. This book may be used at least as a pointer while designing the course content and structure of such programmes. The present book along with the suggested references will be adequate for an elective course. A more detailed textbook with lot more material covering additional subject areas such as nanotechnology will be necessary for a specialized degree programme on nano.

The author is always receptive to constructive criticism and can be contacted at *pradeep@iitm.ac.in*.

T. PRADEEP

Acknowledgments

This book became a reality due to the continuous support of several people. First of all I should thank my students in the undergraduate, masters and graduate levels at the Indian Institute of Technology Madras, who were instrumental in making me organize bits and pieces necessary for this writing. The courses, CY305-Molecular architecture and evolution of functions, CY653-Electron spectroscopy and CY722-Novel materials, which I have been handling for the past several years contributed to the evolution of the contents of this book. My graduate students have contributed heavily to the writing, in fact several of the chapters are largely due to them. I thank the following students for their contributions in writing the chapters: D.M. David Jeba Singh (Gas Phase Clusters), V.R. Rajeev Kumar (Self-assembled Monolayers and Monolayerprotected Metal Nanoparticles), A. Sreekumaran Nair (Core-shell Particles), M.J. Rosemary (Nanoshells), Renjis T. Tom (Nanobiology) and C. Subramaniam (Nanosensors). Three of my undergraduate students, Anshup (Nanomedicines), Sahil Sahni and Richie Khandelwal (Nanotribology) also contributed. Although their contributions have been substantial, the errors are mine. Dr Birgit Bürgi, who spent a few months in our lab, with whom I wrote an article on societal implications of nanoscience and technology was kind to permit the use of the article in this volume. The article was published originally in Current Science. Similarly, another article of mine, on fullerenes, published in the same journal also finds place in this book, although with some modifications. Support of C. Subramaniam and V. Sajini for preparing the history and index, respectively, are acknowledged. My former assistants, K.A. Arun and K.R. Antony helped me at various stages of manuscript preparation.

I thank all my present and previous students whose contributions enriched my understanding of the subject.

All the publishers and authors readily agreed to reproduce their work, for which I am grateful. There have been a few websites from which we have used figures and I thank them for their kind permission. Several textbooks, monographs, review articles and websites have helped me in gathering information and it is my earnest hope that all the intellectual property has been adequately acknowledged. In spite of the best intention, if any material has not been referenced adequately or appropriately I request the reader to point out such lapses, which will be corrected at the earliest available opportunity.

This work reaches you with the hard and sustained work of an excellent team at McGraw-Hill Education (India) Ltd. I thank the support of each one of them.

I thank the Indian Institute of Technology Madras for encouragement. Financial support of the Curriculum Development Cell of the Centre for Continuing Education is acknowledged.

Finally, I thank my wife and children for their patience.

T. Pradeep

NANO:
The Essentials

Understanding Nanoscience
and Nanotechnology

PART ONE

Introduction

Contents:

- Introduction—The Canvas of Nano

Chapter 1

Introduction—The Canvas of Nano

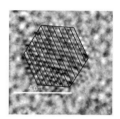

Nanoscience and nanotechnology refer to the control and manipulation of matter at nanometer dimensions. This control has made it possible to have life, which is a collection of most efficient nanoscale processes. The best eco-friendly and efficient processes must learn from nature. When we explore life around us, it is found that organization of nanomaterials is central to biology. Architectures made by organisms are all based on nanoassemblies. Today we know that it is possible to use biological processes to make artificial nanostructures. Chemically synthesized nanostructures have been used at various stages of civilization.

Learning Objectives

- Why nanotechnology?
- What are the connections between nanotechnology and biology?
- What are wet and dry nanotechnologies?
- What are the historical landmarks in this area?

1.1 Nano and Nature

This chapter should begin with an apology. In it, you will find a popular science introduction to nano, not the one commonly found in textbooks. Yet, this needs to be done as science at the nanoscale has larger implications and needs to be understood from diverse viewpoints.

Man has learnt a lot from nature. Yet his manufacturing practices are primitive. Everyone knows that a lot more needs to be done to get closer to nature. For example, no one has reached the efficiency of photosynthesis in storing energy. No one can facilitate energy transfer (or electron transfer) as efficiently as biomolecules. No factory does water purification and storage as efficiently as coconut trees or water melons. The brain of one person can, in principle, store and process more information than today's computer. It is unlikely for any movie camera to capture visuals more vividly than the human eye. The olfactory receptors of the dog are much more sensitive than the sensors we have developed, though single molecule detectors have been reported. Most early warning systems are primitive when compared to the sixth sense of animals. Well, all these functions are performed in nature without any fanfare; this has been happening since time immemorial and with precision each time.

Conventional wisdom says that what happens in a factory is high-tech. Technology converts primitive, unusable materials into modern, useful materials. But technology has a much greater impact on nature especially as the complexity of the technology increases. The impact of the wheel is not as significant as that of the automobile. When spaced in time, the impact of technology increases along with the progress of civilization. The chisel symbolized the highest technology of the Neolithic era. The man who could make his chisels better would get a greater share of food. The best or most high-tech product today would be the super chips used in the fastest computers. These, in the course of production from sand to wafers and then to integrated circuits, have caused severe damage to the environment, even as they contribute to the information explosion. The impact of modern technology is evident on all natural resources—water, air and everything around us. Of course, what we have developed is not high-tech in totality.

The use of conventional technology has not ensured optimum efficiency in energy conversion. Our best photovoltaic devices convert light with only 16 per cent efficiency. Our best internal combustion engines work at around 52 per cent efficiency. While cooking, we use 38 per cent (at best) of the thermal energy produced by gas. But our body utilizes almost the entire chemical energy it produces. Plants utilize this energy much better, as do bacteria. If we were to be as inefficient as an electric motor we would be consuming several times more food than we do today and there would not be enough food for all of us! It is therefore clear that ultimate efficiency or value for money is achieved only if we traverse nature's way.

Nature as a whole fixes about 110–120 billion tonnes of carbon per annum through photosynthesis. We humans emit only 0.65 billion tonnes of carbon dioxide through respiration. But carbon emissions due to human activity constitute about 8 billion tonnes, 77.5 per cent of which is due to the burning of fossil fuels alone. During this process, we produce a lot of other wastes such as smoke, complex organic compounds and oxides of nitrogen. Obviously, the technologies we have developed are much less efficient than those operating in nature. But most importantly, the benefits accruing from the processes and machines developed by us are incongruent with the huge amounts of resources we utilize for these developments.

Eric Drexler (Ref. 1) has suggested an alternate way of producing things, by assembling things from the bottom, which can be called molecular nanotechnology. This is akin to the humble way in which plants take carbon dioxide and water from the environment to produce organic compounds like carbohydrates in the presence of sunlight. A vast majority of living beings on this planet subsist on these carbohydrates, excepting a few organisms, which abstract other forms of chemical energy. In fact, one carbon CO_2 is assembled by a series of chemical processes to yield complex structures. This one-by-one assembly has facilitated functions with single molecules. Examples of this include molecular motors, muscle fibres, enzymes, etc., each of which is designed to perform a specific activity. The complexity of this molecular architecture is such that one molecule can communicate precisely with another so that the structure, as a whole, achieves unusually complex functions, that are necessary to sustain life. Nature has taken a long time to master this complexity. Maybe, that is the path one must pursue if one has to look toward the future.

Any production achieved through biological processes is extremely complex, but very cheap in real terms. The constitution of a water melon is more complex than the most complex integrated circuit, yet it costs far less. On the other hand, the power to manipulate atoms and arrange them in the way we please can facilitate the creation of complex inorganic structures merely at the price of vegetables. This power can, in fact, facilitate the creation of all man-made products. That is nanotechnology. In many ways, this is the wet side of nanotechnology (Ref. 2). There is a corresponding dry side wherein the ability to organize things atom by atom would make it possible to have structures and devices with functions. That would not only make computers smaller, and surgical procedures feasible without blood loss, but also help harness solar energy efficiently so that we can avoid climate changes. The organisation of the molecules is such that they can communicate with one another. Such an arrangement enables the execution of usually complex functions necessary to sustain life.

1.2 Our Technologies and the World We Live in

Implementing Nature's ways would imply a thorough understanding of molecular machinery. This knowledge, if applied to inorganic matter, results in functional materials. Through these, a superstructure may be built, which has functions similar to those encountered in biology. Think of molecules transferring matter from one end to the other. Think of molecules, which bend, stretch or curl in response to external stimulus such as temperature and come back to their original shape when the stimulus is reversed. Consider chemical reactions which can be turned on and off by light. Think of molecules converting one chemical into another without using anything else in the medium and assume that such transformation occurs with precision and within the shortest time. These kinds of functions and indeed many more have already been achieved and there is much greater scope for such achievements in the future.

In the course of the evolution of mankind, technologies have come and gone. One large difference between today's technology and that of the past lies in the time taken to perfect technology. The agrarian era, driven by the associated technologies of irrigation, tools, fertilizers, etc., took a few thousand years to evolve, with the time period varying depending on the location. The Industrial Age, which came just after the agrarian era, starting around the 1800s, took about 150 years in evolution. Then came the Information Age, starting from the 1950s and in many projections, it is said to have evolved, to a great extent. The impact of that evolution may not have been felt in many societies, as technologies are absorbed differently (see Chapter 16), but in the developed countries of the West can be said to be technologically fully evolved. The next age, at the threshold of which we stand today, is expected to evolve within a generation. But the way in which that technology is absorbed will be vastly different from the absorption of earlier technologies.

A partial list of technologies developed in the 1900s, along with the year of invention and names of the people who invented them, is given in Table 1.1.

It is important to note that many technologies took a long time to reach the marketplace, but the most recent technologies are already in the marketplace (see those of the 1990s).

Table 1.1: *Details of technological inventions of the 20th Century*

Year	Technology	Inventor
1920s:		
1924	Frozen foods	Clarence Birdseye
1926	Rocket engine	Robert Goddard
1926	Television	John Logie Baird
1928	Penicillin	Alexander Fleming
1930	Synthetic rubber	Julius Nieuwland
1930s:		
1930	Jet engine	Frank Whittle and Hans Von Ohain
1932	Automatic transmission	Richard Spikes
1934	Nylon	Wallace Hume Carothers
1937	Pulse code modulation to convert voice signals into electronic pulses	Alec Reeves
1937	Xerography or Xerox machines	Chester Floyd Carlson
1940s:		
1940	Radar	Robert Watson–Watt
1946	Microwave oven	Percy Spencer
1947	Cellular phone (conceptually)	D.H. Ring
1947	Transistor	Nillian Shockley, John Bardeen, and Walter Brattain
1949	Magnetic core memory	An Wang and then Jay Forrester
1950s:		
1951	The pill	Gregory Pincus
1952	Thorazine	Henri Laborit
1954	Fortran, the first high-level programming language	Griffith John Backus
1955	Polio vaccine	Jonas Salk
1956	Disk drive	Reynold B. Johnson
1958	Implantable pacemaker	Wilson Greatbatch
1958	Lasers	Schnwlow, Towens Basov, and Prokhorov
1959	Integrated circuit	Robert Noyce, Jack Kilby

Contd.

Table 1.1 Contd.

Year	Technology	Inventor
1960s:		
1962	Modem	US Airforce, AT&T
1968	Automated teller machines (ATMs)	Don Wetzel
1968	Mouse	Douglas Engelbart
1969	Charge-coupled devices	George E. Smith, Williard S. Boyle
1969	The Internet	UCLA, Stanford, among others
1970s:		
1970	Compact disc (CD)	James T. Russell
1970	Liquid crystal displays	James Fergason
1971	Microprocessor	Intel, Busicom
1972	Computed tomography imaging	Godfrey Housnfield, Allan Cormack
1972	Ethernet	Robert Metcalfe
1972	E-entertainment and precursor to video games	Nolan Bushnell
1974	Catalytic converter	Rodney Bagley, Irwin Lachman, Ronald Lewis
1975	Recombinant DNA	Herbert Boyer and Stanely Cohen
1979	Spreadsheet	Daniel Bricklin, Bob Frankston
1980s:		
1986	Automated DNA sequencing machines	Leroy Hood, Llyod Smith and Mike Hunkapiller
1987	Mevacor to reduce cholesterol	Merck
1987	Prozac to reduce depression	Ray Fuller of Eli Lilly Company
1989	World wide web	Tim Berners-Lee
1990s:		
1994	Viagra	Albert Wood, Peter Dunn, Nicholas Terrett, Andrew Bell, Peter Ellis
1996	Protease inhibitors for patients suffering from HIV	S. Oroszlan, T.D. Copeland

Although technologies have come and gone, it is important to assess what they have given us. The industrial age of the 1900s gave us advanced agricultural practices like the use of chemical fertilizers, radio, TV, air conditioning, car, jet planes, modern medicines, fabrics, etc. which can help the rich live like the kings of the past, and also enable paupers to become kings if they have an appropriate understanding of the markets. The Industrial Age removed the distinction between the king and the commonman. The Information Age has given us mobile phones, Internet, cable TV, email, ATMs, administrative reforms, and reduced distances, among other things, which have transformed our neighbourhood completely. It has eliminated distances (and distinctions) of all sorts. The next era may remove the barrier between humans and their surroundings in every possible way; life may acquire a seamless link with nature. A prophetic statement indeed!

This change in our lives has occurred due to science. Chemistry has been the driving force in the front, which made major changes possible in the 19th and 20th centuries. The large production of ammonia, sulphuric acid, cement, iron, aluminum, drugs, fibres, dyes, polymers, plastics, petroleum products, etc. has changed the world. What the world of chemistry has produced drives society. Chemistry contributes to more than half the global production, including that of computers. Chemicals drive several economies. While this growth of chemistry was continuing, a major change occurred in 1947 with the discovery of the transistor. In the 1950s and beyond, the advent of semiconductor devices facilitated the development of consumer electronics. Everything that one purchases today has an integrated circuit in it. Gadgets from toys to cars function with these circuits. This, along with computers, helped build an era of physics. Many predict that the next era would be that of biology and materials. In that era, it is highly likely that the interfaces between disciplines rather than the disciplines themselves would contribute largely to developments.

The manner in which this change has occurred has left many distinct marks on society. The typewriter, which was a prized possession till recently, has disappeared completely. Electronic typewriters—manufactured in the era between typewriters and word processors—have also disappeared. The 'typist' has become non-existent in many institutions and there is no more recruitment under this category. Factories producing goods related to the typewriter have vanished before our eyes, in the recent past. This may be contrasted with the disappearance of the blacksmith from Indian villages. The blacksmith, a reminder of the agrarian era disappeared slowly as a result of the advent of new factory-made agricultural implements. While the former development took place over a few decades, the latter has taken centuries. As a result of the changing times, it became necessary to develop several other skills. Programming skills, which were expected to be specialized, became a part of the established set of skills. Today, computer knowledge is no longer considered special, but is expected to be part of the school training.

In spite of the large economic boom that has resulted from the developments in various fields, many problems remain. Poverty is widespread and several communities in the world still suffer from starvation. Clean drinking water is inadequate and many still have no access to it. In India, around 15 per cent of the population has no access to clean drinking water. The global estimate of power requirement for 2050 is 30PWh (P = peta, 10^{15}). Radically new forms of technology will be needed to harness that power. Maintaining a clean environment will be the biggest challenge that humanity will face in the years to come. The alarming depletion of non-renewable resources, forest areas and wetlands, the extinction of animal and plant species, and the deterioration in air and water quality are gigantic problems. These issues are of larger importance to marginalized societies where life is inexorably interlinked with nature. These and the associated problems of healthcare, education and housing are astronomical even with a nominal population growth rate of 1.14 per cent (2004 estimate), which would make us a country of 10 billion people by 2050. Note that it was just 1 billion in 1820, 2 billion in 1930, 3 billion in 1960, 4 billion in 1974, 5 billion in 1988, and 6 billion in 2000; in other words we have grown six times in just 180 years. The way in which we manage our lives has created widespread problems of over-population, industrial disasters, pollution (air, water, acid rain, toxic substances), loss of vegetation (over-grazing, deforestation, desertification), loss of wildlife, and degradation, depletion and erosion of the soil. Obviously with national boundaries and with each nation having its own problems and priorities, these issues will never be solved to the satisfaction of all. There is a larger problem of the distribution of wealth. The per capita gross domestic product (GDP) of the United States is $37800, 13 times higher than that of India on a purchasing power parity basis (2004 estimate). What this obviously means is that a lot needs to be done to achieve a comparable quality of life for everyone. Achieving this goal would necessitate a completely new kind of technological initiative. Agriculture, which currently contributes 23.6 per cent to the GDP in India, will not be in a position to contribute further. Globally, it contributes only 4 per cent to the GDP. In the US, its contribution is only 1.4 per cent. Obviously, people don't live on food alone these days. Money is made largely by industries and services, but the new industry must not only be kind towards nature but also as efficient as the latter. Where can one look for solutions? Nanotechnology does offer solutions to some of these issues though it is not a savior for all.

1.3 Nano—The Beginning

What are the historical milestones in the saga of nano? Many nano forms of matter exist around us. One of the earliest nano-sized objects known to us was made of gold. Faraday prepared colloidal gold in 1856 and called it 'divided metals'. In his diary dated 2 April 1856, Faraday called the particles he made the 'divided state of gold' (*http://personal.bgsu.edu/~nberg/faraday/diary2.htm*). The solutions he prepared are preserved in the Royal Institution, see Fig. 1.1 (Plate 1).

Metallic gold, when divided into fine particles ranging from sizes of 10–500 nm particles, can be suspended in water. In 1890, the German bacteriologist Robert Koch found that compounds made with gold inhibited the growth of bacteria. He won the Nobel prize for medicine in 1905. The use of gold in medicinal preparations is not new. In the Indian medical system called *Ayurveda*, gold is used in several preparations. One popular preparation is called '*Saraswatharishtam*', prescribed for memory enhancement. Gold is also added in certain medicinal preparations for babies, in order to enhance their mental capability. All these preparations use finely ground gold. The metal was also used for medical purposes in ancient Egypt. Over 5,000 years ago, the Egyptians used gold in dentistry. In Alexandria, alchemists developed a powerful colloidal elixir known as 'liquid gold', a preparation that was meant to restore youth. The great alchemist and founder of modern medicine, Paracelsus, developed many highly successful treatments from metallic minerals including gold. In China, people cook their rice with a gold coin in order to help replenish gold in their bodies. Colloidal gold has been incorporated in glasses and vases to give them colour. The oldest of these is the fourth Century AD Lycurgus cup made by the Romans, see Fig. 1.2 (Plate 1). The cup appears red in transmitted light (if a light source is kept within the cup) and appears green in reflected light (if the light source is outside). Modern chemical analysis shows that the glass is not much different from that used today. The compositions are given in Table 1.2.

Table 1.2: *Compositions of Lycurgus cup and modern glass*

Constituent	Lycurgus Cup	Modern Glass
Silicon dioxide	73%	70%
Sodium oxide	14%	15%
Calcium oxide	7%	10%

So what helps to impart colour to the glass? It contains very small amounts of gold (about 40 parts per million) and silver (about 300 parts per million) in the form of nanoparticles. A review of the historical developments in the area of gold colloids can be found in Ref. 3.

Nature makes nano objects of varying kind. Magnetite (Fe_3O_4) particles of nanometer size are made by the bacteria, *Magnetosperillum magnetotacticum*. These bacteria make particles of specific morphology. For a bacterium, the magnetism caused by the particles helps in finding a direction favourable for its growth. There are several bacteria like the familiar *Lactobacillus* which can take up metal ions added into buttermilk, and reduce them inside the cell and make nanoparticles. In Fig. 1.3, we see the transmission electron microscopic picture of a single *Lactobacillus* bacterium after incubation with gold ions for several hours. Fungi and viruses are known to make nanoparticles.

However, the science of nanometer scale objects was not discussed until much later. On December 29, 1959, the Nobel prize winning physicist, Richard Feynman gave a talk at the annual meeting of the American Physical Society entitled "There's plenty of room at the bottom'. In this talk, he stated, "The principles of physics, as far as I can see, do not speak against the possibility of maneuvering things atom by atom." He, in a way, suggested the bottom up approach, "…it is interesting that it would be, in principle, possible (I think) for a physicist to synthesize any chemical substance that the chemist writes down. Give the orders and the physicist synthesizes it. How? Put the atoms down where the chemist says, and so you make the substance. The problems of chemistry and biology can be greatly helped if our ability to see what we are doing, and to do things on an atomic level, is ultimately developed—a development which I think cannot be avoided" (Ref. 4). However, the world had to wait a long time to put down atoms at the required place. In 1981, the

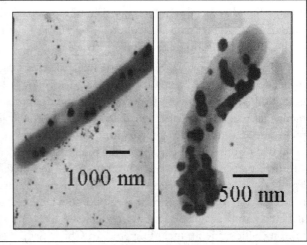

Fig. 1.3: *Gold nanoparticles within the Lactobacillus contour. This transmission electron microscopic image shows large particles of more than 200 nm diameter. However, smaller particles are also made (from the Author's work).*

scanning tunneling microscope was made and later a number of tools collectively called scanning probe microscopes were developed. The team associated with these developments got the 1986 Nobel prize for physics. The tools they developed can help see and place atoms and molecules wherever needed. An exhaustive summary of the historical development in the area of nanoscience and technology is listed separately.

The current growth of technology suggests that reductions are needed in the dimensions of devices and active materials. This is evident in the case of computer technology. The number of transistors used in an integrated circuit has increased phenomenally in the past 40 years. In 1965, Gordon Moore, the co-founder of Intel, observed that the number of transistors per square inch on integrated circuits doubled every year since the integrated circuit was invented. Moore predicted that this trend would continue in the foreseeable future. In the subsequent years, this pace slowed down, but the data density doubled approximately every 18 months. This is the current definition of Moore's Law. Most experts, including Moore himself, expect Moore's Law to hold for some more time. For this to happen the device dimension must shrink, touching the nanometer regime very soon. The Pentium 4 of 2000 (see Table 1.3), used a 130 nm technology, i.e. the device structure drawn on silicon was as small as this dimension. In 2004, the technology graduated to 90 nm, well into the nanotechnology domain (under 100 nm) and 45 nm technology is being discussed currently.

Table 1.3: *The complexity of integrated circuits as seen in the evolution of Intel microprocessors*

Name	Year	Transistors	Microns	Clock speed
8080	1974	6,000	6	2 MHz
8088	1979	29,000	3	5 MHz
80286	1982	134,000	1.5	6 MHz
80386	1985	275,000	1.5	16 MHz
80486	1989	1,200,000	1	25 MHz
Pentium	1993	3,100,000	0.8	60 MHz
Pentium II	1997	7,500,000	0.35	233 MHz
Pentium III	1999	9,500,000	0.25	450 MHz
Pentium 4	2000	42,000,000	0.18	1.5 GHz
Pentium 4 "Prescott"	2004	125,000,000	0.09	3.6 GHz

Obviously, with all these developments, new nanotech products will indeed reach the marketplace in the immediate future. However, the answer to when this need will be felt by the people varies. Many believe that there will be nanotech laws in the near future as there can be economic, social, health and security implications related to nanotechnology which would be of concern to many nations. The implications of nanotechnology for society may be significant enough for nations to discuss it as part of their election campaigns. A detailed discussion of such aspects is given in the last chapter. In between the first and the last chapters, we have highlighted the various aspects of this rapidly emerging and fascinating science.

Review Questions

1. What is nanoscience?
2. What is nanotechnology?
3. What are nanomaterials?
4. Why nanotechnology now? Why we did not hear about it in the past?
5. Is there a systematic evolution of nanotechnology from microtechnology? Will there be picotechnology?
6. Are there nano objects around you? Are there such objects in your body? Name a few.
7. Have a look at natural objects such as sea shells, wood, bone, etc. Is there any nanotechnology in them?
8. If nature is full of nano, what limits us from making nanomaterials or nanodevices?
9. What are the likely impacts of nanotechnology?

References

1. Drexler, K.E., (1986), *Engines of Creation*, Garden City, Anchor Press/Doubleday, New York; Drexler, K.E., and C. Peterson, and G. Pergamit, (1991), *Unbounding the Future: The Nanotechnology Revolution*, William Morrow and Company, Inc. New York.
2. The term, 'wet nanotechnology' was first introduced by R.E. Smalley.
3. M–C. Daniel and D. Astruc, *Chem. Rev.*, **104** (2004), 293–346.
4. http://www.zyvex.com/nanotech/feynman.html.

PART TWO

Experimental Methods

Contents:

- Investigating and Manipulating Materials in the Nanoscale

Chapter 2

INVESTIGATING AND MANIPULATING MATERIALS IN THE NANOSCALE

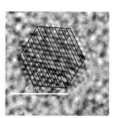

The observation of materials in the nanoscale can be done using electrons, photons, scanning probes, ions, atoms, etc. A wide range of techniques is available in each of these areas and a systematic application of several tools leads to a complete understanding of the system. In addition, <u>in-situ</u> nano measurements become a reality with these tools. The properties of individual nano objects can be studied with precision and some examples of this are illustrated. It is also possible to adapt the techniques mentioned for nanomanipulation, which becomes the basis of nanotechnology.

Learning Objectives

- What are the principal properties used to explore nanomaterials?
- What are the differences between photon, electron, and scanning probe techniques?
- What are the modern advances in these techniques?
- How do we manipulate objects in the nano dimension?

2.1 Introduction

Observation is the key to making new discoveries, and this is especially true in the nanoscale. In fact, as far as nano objects are concerned, one cannot proceed further with the investigations without observing these objects. Observation is done with a probe which may consist of photons, electrons, neutrons, atoms, ions or even an atomically sharp pin. For nanomaterials, the probing light or particle often has varying frequencies, ranging from gamma to infrared rays or beyond in the case of photons or hyper thermal (<100 eV) to relativistic energies in the case of particles. The resulting information can be processed to yield images or spectra which reveal the topographic, geometric, structural, chemical or physical details of the material. Several techniques are available under the broad umbrella of characterization of materials, which may be used to study nanomaterials in one way or the other. A partial list of these techniques is given in Table 2.1. Some of these techniques may be used in a spatially resolved fashion. In this chapter, we look at some of the more important tools used in the context of nanoscience and technology.

Table 2.1: *Common analytical tools used for characterization of materials*

AES	Aüger Electron Spectroscopy★
AFM	Atomic Force Microscopy★
APECS	Aüger Photoelectron Coincidence Spectroscopy
APFIM	Atom Probe Field Ion Microscopy★
APS	Appearance Potential Spectroscopy
ARPES	Angle Resolved Photoelectron Spectroscopy★
ARUPS	Angle Resolved Ultraviolet Photoelectron Spectroscopy★
ATR	Attenuated Total Reflection
BEEM	Ballistic Electron Emission Microscopy★
BIS	Bremsstrahlung Isochromat Spectroscopy
CFM	Chemical Force Microscopy★
CM	Confocal Microscopy (especially with fluorescence and Raman detection)★
DRIFTS	Diffuse Reflectance Infra-Red Fourier Transform Spectroscopy
EDX	Energy Dispersive X-ray Analysis
EELS	Electron Energy Loss Spectroscopy★
	Ellipsometry, see RDS
EMS	Electron Momentum Spectroscopy
EPMA	Electron Probe Micro-analysis★
ESCA	Electron Spectroscopy for Chemical Analysis, also XPS
	(X-ray photoemission spectroscopy)★
ESD	Electron Stimulated Desorption
ESDIAD	Electron Stimulated Desorption Ion Angle Distributions
EXAFS	Extended X-ray Absorption Fine Structure
FEM	Field Emission Microscopy★
FIM	Field Ion Microscopy★
FRET	Fluorescence Resonance Energy Transfer★
FTIR	Fourier Transform Infra-red Spectroscopy★
FT RA-IR	Fourier Transform Reflectance-Absorption Infra-red

Contd.

Table 2.1 Contd.

HAS	Helium Atom Scattering
HEIS	High Energy Ion Scattering
HREELS	High Resolution Electron Energy Loss Spectroscopy
IETS	Inelastic Electron Tunneling Spectroscopy
KRIPES	k-Resolved Inverse Photoemission Spectroscopy
ILS	Ionization Loss Spectroscopy
INS	Ion Neutralization Spectroscopy
IPES	Inverse Photoemission Spectroscopy
IRAS	Infra-red Absorption Spectroscopy
ISS	Ion Scattering Spectroscopy
LEED	Low Energy Electron Diffraction★
LEEM	Low Energy Electron Microscopy★
LEIS	Low Energy Ion Scattering
LFM	Lateral Force Microscopy★
MBS	Molecular Beam Scattering
MCXD	Magnetic Circular X-ray Dichroism
MEIS	Medium Energy Ion Scattering
MFM	Magnetic Force Microscopy★
MIES	Metastable Impact Electron Spectroscopy
MIR	Multiple Internal Reflection
NEXAFS	Near-Edge X-ray Absorption Fine Structure
NSOM	Near Field Scanning Optical Microscopy★
PAES	Positron Annihilation Aüger Electron Spectroscopy
PEEM	Photoemission Electron Microscopy★
PED	Photoelectron Diffraction
PIXE	Proton Induced X-ray Emission
PSD	Photon Stimulated Desorption
RAIRS	Reflection Absorption Infra-red Spectroscopy
RAS	Reflectance Anisotropy Spectroscopy
RBS	Rutherford Back Scattering
RDS	Reflectance Difference Spectroscopy

Contd.

Table 2.1 Contd.

REFLEXAFS	Reflection Extended X-ray Absorption Fine Structure
RHEED	Reflection High Energy Electron Diffraction★
RIfS	Reflectometric Interference Spectroscopy
SAM	Scanning Aüger Microscopy★ also Scanning Acoustical Microscope★
SCM	Scanning Confocal Microscope★ also CM
SEM	Scanning Electron Microscopy★
SEMPA	Scanning Electron Microscopy with Polarization Analysis★
SERS	Surface Enhanced Raman Scattering★
SEXAFS	Surface Extended X-ray Absorption Spectroscopy
SFS	Sum Frequency Spectroscopy
SHG	Second Harmonic Generation
SH-MOKE	Second Harmonic Magneto-optic Kerr Effect
SIM	Scanning Ion Microscope★
SIMS	Secondary Ion Mass Spectrometry★
SKS	Scanning Kinetic Spectroscopy
SLM	Scanning Light Microscope★
SMOKE	Surface Magneto-optic Kerr Effect
SNMS	Sputtered Neutral Mass Spectrometry
SNOM	Scanning Near Field Optical Microscopy★
SPIPES	Spin Polarized Inverse Photoemission Spectroscopy★
SPEELS	Spin Polarized Electron Energy Loss Spectroscopy
SPLEED	Spin Polarized Low Energy Electron Diffraction★
SPM	Scanning Probe Microscopy★
SPR	Surface Plasmon Resonance
SPUPS	Spin Polarized Ultraviolet Photoelectron Spectroscopy
SPXPS	Spin Polarized X-ray Photoelectron Spectroscopy
STM	Scanning Tunneling Microscopy★
SXAPS	Soft X-ray Appearance Potential Spectroscopy
SXRD	Surface X-ray Diffraction
TDS	Thermal Desorption Spectroscopy
TEAS	Thermal Energy Atom Scattering
TIRF	Total Internal Reflectance Fluorescence

Contd.

Table 2.1 Contd.

TPD	Temperature Programmed Desorption
TPRS	Temperature Programmed Reaction Spectroscopy
TXRF	Total Reflection X-ray Fluorescence
UPS	Ultraviolet Photoemission Spectroscopy★
XANES	X-ray Absorption Near-Edge Structure
XPD	X-ray Photoelectron Diffraction★
XPS	X-ray Photoemission Spectroscopy★
XRR	X-ray Reflectometry
XSW	X-ray Standing Wave

★*Tools which are either microscopy or with which microscopy is possible.*

We use microscopy in order to see objects in more detail. The best distance that one can resolve with optical instruments, disregarding all aberrations, is about 0.5 λ, or of the order of 250 nm with visible radiation. All forms of microscopy are aimed at improving our capacity to see. Under ideal conditions, the smallest object that the eye can resolve is about 0.07 mm. This limit is related to the size of the receptors in the retina of the eye. Any microscope is designed to magnify the image falling on the retina. The advantage of a microscope is that it effectively brings the object closer to the eye. This allows us to see a magnified image with greater details.

Several forms of microscopy are available for studying nanomaterials. These can be broadly grouped under the following categories:

1. Optical microscopes
2. Electron microscopes
3. Scanning probe microscopes
4. Others

While the first three categories are more common, others are also used for nano measurements. A few of the techniques which are also useful as spectroscopies, are also discussed. In this chapter, we review all these forms of microscopic tools in detail. The discussion, however, does not include the theoretical aspects in such detail. The texts listed in the bibliography contain in-depth discussions of each technique. Every topic, however, may not be discussed to the same extent.

A microscope is an instrument used to form enlarged images. The word 'microscope' is derived from two Greek words, "*micros*" meaning 'small'; and "*skopos*" meaning 'to look at'. Microscopes developed by Antoni van Leeuwenhoek (1632–1723) were the state-of-the-art for about 200 years. These single lens microscopes had to be held against the eyeball because of their short focal lengths. They helped in the discovery of bacteria. As his research was not appreciated, van Leeuwenhoek destroyed most of his 500 odd microscopes before his death at the age of 91. Only two or three microscopes developed by Leeuwenhoek are known to exist today.

The following definitions must be listed before we discuss microscopies.

Resolution: A measure of the capacity of the instrument to distinguish two closely spaced points as separate points, given in terms of distance.

Resolving power: The resolution achieved by a particular instrument under optimum conditions. While resolving power is a property of the instrument and is a quantity that may be estimated, resolution is equal to or poorer than the resolving power and has to be determined for the instrument.

The following two kinds of microscopes exist:

1. *Transmitting*—The probe beam passed through the specimen is differentially refracted and absorbed.
2. *Scanning*—The probe beam is scanned over the surface. The image is created point-by-point.

There are several kinds of scanning microscopes. These are listed below.

1. *Scanning Electron Microscope (SEM)*—In this, a monochromatic electron beam is passed over the surface of the specimen which induces various changes in the sample. The resulting particles from the sample are used to create an image of the specimen. The information is derived from the surface of the sample. The most important advantage of SEM is its large depth of field. Although the images appear to be three-dimensional, a true three-dimensional image is obtained only by using a combination of two pictures.
2. *Scanning Ion Microscope (SIM)*—In this, charged ions are used to obtain the image and the process etches away the top surface.
3. *Scanning Acoustical Microscope (SAM)*—This uses ultrasonic waves to form images. The best resolution achieved is of the order of 2.5 microns, which is limited by the wavelength of sound. Its advantage is that it allows one to look at live biological materials.
4. *Scanning Light Microscope (SLM)*—In this, a fine beam of visible light is passed over the surface to build up the image point-by-point. It facilitates increased depth of field and colour enhancement.
5. *Scanning Confocal Microscope (SCM)*—In this, a finely focused beam of white or monochromatic light is used to scan a specimen. It allows one to optically section through a sample. This technique is more commonly referred to as confocal microscopy.

2.2 Electron Microscopies

Microscopes consist of an illumination source, a condenser lens to converge the beam on to the sample, an objective lens to magnify the image, and a projector lens to project the image onto an image plane which can be photographed or stored. In electron microscopes, the wave nature of the electron is used to obtain an image. There are two important forms of electron microscopy, namely scanning electron microscopy and transmission electron microscopy. Both these utilize electrons as the source for illuminating the sample. The use of optical analogy makes it easier to understand the functioning of a microscope. Both the tools use similar illumination sources, but they differ in a number of other aspects. The lenses used in electron microscopes are electromagnetic lenses, which are widely different from glass lenses, though similar principles apply in both cases.

Due to its simplicity, scanning electron microscopy (SEM) will be discussed before transmission electron microscopy (TEM). The discussion of TEM will be brief because even though TEM is more complex in usage, an understanding of SEM makes it easier to use TEM. Several other modifications to each of these techniques are possible and we will highlight these advanced tools at the appropriate places.

2.2.1 Scanning Electron Microscopy

Basics

The de Broglie wave equation relates the velocity of the electron with its wavelength, $\lambda = h/mv$ (h is Planck's constant, m is the rest mass of electron and v is its velocity). An electron of charge e (1.6×10^{-19} coulomb), and mass m (9.11×10^{-28} gm), when passing through a potential difference of V volts (expressed in joules/coulomb), acquires a kinetic energy of $1/2 mv^2 = eV$. This will give: $v = \sqrt{(2eV/m)}$ and $\lambda = \sqrt{(h^2/2meV)}$. Substituting (with conversion factors, 1 joule = 10^7 dyne.cm = 10^7 cm^2.gm/sec^2), λ (in nm) = $1.23/\sqrt{V}$. When $V = 60,000$ volts, $\lambda = 0.005$ nm. This shows that the velocity of electrons will reach the speed of light in vacuum ($c = 3.10^{10}$ cm/sec) at high extraction potentials. The electron velocities at various acceleration voltages are given in Table 2.2.

Table 2.2: *Electron velocities at different acceleration voltages*

V	λ(nm)	$v(\times 10^{10}$ cm/sec)	v/c
50,000	0.0055	1.326	0.442
100,000	0.0039	1.875	0.625
1,000,000	0.0012	5.930	1.977

However, the equation breaks down when the electron velocity approaches the speed of light as mass increases. At such velocities, one needs to do relativistic correction to the mass so that it becomes $m = m_o/\sqrt{[1 - (v^2/c^2)]}$. This makes $\lambda = 1.23/(V + 10^{-6} V^2)$ nm.

After including the relativity effects, the velocities and the wavelengths achieved are as shown in Table. 2.3.

Table 2.3: *Voltage and velocity after inclusion of relativity effects*

V	v(× 10¹⁰ cm/sec)	v/c
50,000	1.283	0.414
100,000	1.699	0.548
1,000,000	2.917	0.941

Resolving Power

1. Abbe criterion: In 1893, Abbe showed that the smallest resolvable distance between objects is about half the wavelength of the light used. What does this mean for magnification? The maximum magnification that can be used is equal to the resolving power of the eye divided by the resolving power of the microscope. In the case of light microscopes, the resolving power is about 250 nm. What we can see with our naked eye without difficulty is about 250 μm. So the useful magnification is 250/0.25 = 1000X (X means diameter). A magnification higher than this has no value as it represents empty magnification, the effect of which is only a magnified blur. The resolving power of an electron microscope would be 0.0027 nm for an electron energy of 50,000 V. This means that one can obtain a magnification of the order of 100,000,000. However, this is not achieved due to aberrations of the electron lenses and the complex nature of electron–sample interactions. It is important to know that we have achieved the theoretical limit of resolution as far as optical microscopes are concerned as the various aberrations have been resolved.

2. Rayleigh criterion: An ideal lens projects a point on the object as a point on the image. But a real object presents a point as a disk in the image plane. This disk is called the Airy disk named after George Airy (Fig. 2.1). The diameter of the disk depends on the angular aperture of the lens. If two points are placed close to each other, the closest distance at which they appear to be separated in the image is about half the width of the disks. This separation (shown in the arrow) can be given as: $d = 0.61\ \lambda/n \cdot \sin\theta$, where n is the refractive index of the medium and θ is the semi-angular aperture of the lens. $n \cdot \sin\theta$ is the numerical aperture (NA) of the lens.

In order to maximize resolving power, λ must be decreased, and n or θ increased. Recall that at the moment, we are concerned with an **aberration-free** optical system. In the case of light microscopes, with oil immersion optics ($n = 1.5$), $\sin\theta = 0.87$ and $\lambda = 400$ nm (for the blue end of the visible spectrum), the limit is $d = 0.2\ \mu$m. In the case of TEM, $n = 1$ (vacuum), $\sin\theta = 10^{-2}$ and λ is of the order of 0.005 nm and $d = 0.3$ nm.

Classical vs. Electron Optics

Light is refracted by lenses. This is the property used for magnifying or demagnifying an object. At the lens surface, the refractive index changes abruptly and remains constant between the surfaces. Imaging lenses are constructed by using the principles of refraction. Glass surfaces of the lenses are shaped in order to obtain the desired results. In electron optics, the trajectory of the electrons is changed by applying electric or magnetic fields. This change is continuous depending on the electric and magnetic fields. Electromagnetic lenses, which have fewer aberrations than electrostatic lenses, are used in electron microscopy.

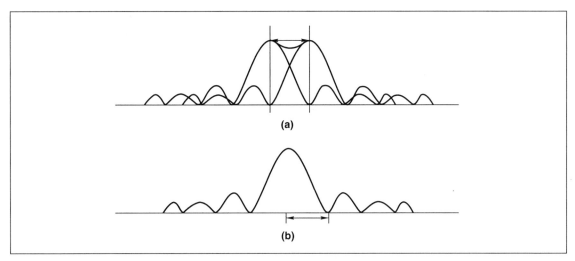

Fig. 2.1: (a) Intensity distribution at the image plane forming an Airy disk. The points are considered resolved if the intensity maxima are separated at half the height. (b) Rayleigh criterion of resolution.

What does an SEM contain?

The principal elements of an SEM are discussed below.

1. Electron gun: The electron gun provides a stable beam of electrons. The most common form of an electron gun is a thermionic emitter, wherein the work function of the metal is overcome by the surface temperature of the filament. Typically a hairpin filament made of a tungsten wire of 100 μm, with a tip radius of 100 μm, is used to generate electrons. The filament is held at a negative potential and heated to about 2000–2700 K by resistive heating. The electrons are confined and focused by a grid cap (Wehnelt cap) held at a slightly higher negative potential than the filament. The confined beam is accelerated to the anode held at the ground potential and a portion of the beam is passed through a hole. The beam current measured at the anode is used to regulate the power supply which drives the electron gun for the desired current. The tungsten wire gradually evaporates due to the high operating temperature, its resistance decreases and finally it becomes so thin that it breaks. Higher temperature (higher operating current) increases the evaporation rate and reduces the filament life.

The filament characteristic is given in terms of brightness which is given as current/area.solid angle. The brightness is of the order of 10^5 (A/cm^2 sr) for the tungsten filament. An order of magnitude increased brightness is possible in the case of lanthanum hexaboride (LaB$_6$)-based guns and a still larger brightness is possible in the case of field emission guns. This is achieved by reducing the emission area, which results in significant reduction in the demagnification required in the operation.

LaB$_6$ has greater brightness and a longer lifetime as compared to tungsten. The emitter is a single crystal of LaB$_6$ of 10 μm diameter with 500 μm length. The tip is ground to a sharp point of 10 μm and the crystal is mounted on a rhenium or graphitic carbon base. The base is resistively heated so that the tip emits electrons. The gun requires a better vacuum than that needed for a tungsten filament as the tip can be contaminated easily. The filament lifetime is of the order of 1000 h and the increased brightness justifies an order of magnitude increase in cost.

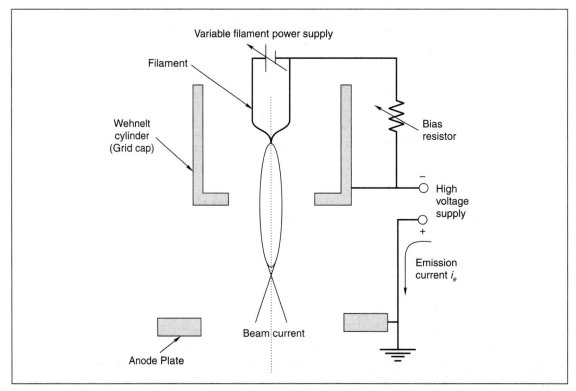

Fig. 2.2: Schematic of an electron gun. The gun is normally operated at a constant emission current, which implies regulation of the current supplied to the filament. Adapted from J.I. Goldstein, et al. (2003) (Ref. 1).

Field emission is another way of making electrons. This kind of source has a high brightness. In field emission, the field at the tip of the emitter reaches a magnitude of 10 V/nm, and the potential barrier for electron emission gets reduced and becomes narrower so that electron can tunnel and escape the cathode. Two kinds of field emission sources are commonly used, i.e. the cold field emitter (CFE) and the Schottky field emitter (SFE). In CFE, a sharp single crystal of <310> tungsten tip is spot-welded on a tungsten wire. The tip is made of tungsten as it has the strength required to withstand the mechanical stress produced at the tip. The tip itself is mounted in a triode configuration so that the potential difference between the tip and the first electrode is of the order of 3–5 kV to produce 10 μA emission. The potential difference between the second electrode and the tip determines the beam energy. Field emission is sensitive to adsorbed gases and the surface can be cleaned by flashing the tip to a temperature of 2500 K. After flashing, the emission is high but soon it reduces and stabilizes as the surface is covered. The tip has to be re-flashed to get a clean surface again after several hours of operation. In each flashing, the tip gets more blunt and after several thousands of flashes, the tip is rendered unfit for use and it has to be changed. This occurs only after several years as one flashing is enough for a day.

SFEs are operated at high temperatures. They are self-cleaning and their tips sharpen in the extraction field. In SFEs, the field at the tip is used to reduce the work function barrier. The tip is held at a high temperature. In order to achieve better reduction of the work function, the surface is coated with ZrO_2 from a dispenser. The SFE gun is run continuously even when no current is drawn. This keeps the system clean. However, as the ZrO_2 reservoir is finite, the lifetime is finite and a replacement is needed every year.

2. Electron lenses: An electromagnetic lens used in electron microscopy is shown in Fig. 2.3(a). This consists of a coil of wire generating a magnetic field enclosed in an iron casing. A magnetic field is generated between the polepieces by applying a current through the coils. The electron in a magnetic field undergoes rotation. As the radial component of the magnetic field reverses after the center of the lens, the rotation in the first half of the lens is reversed. The electron leaves the lens without any net change in the angular momentum, but it undergoes deflection towards the axis. Since the radial force is directed toward the axis, the magnetic lens is always convergent. An electron travelling off-axis to the beam path will spiral through the lens towards the optic axis. The point from which the beam starts bending to the point where it crosses the lens axis is called the focal length. The focal length can be varied by changing the current in the coils and this is a difference from the optical lenses where the focal length is fixed. Referring to Fig. 2.3(b), the focal length can be given as $f = 1/p + 1/q$. From this, one can calculate the magnification, $M = q/p$ and demagnification, $m = p/q$, where p is the distance from the object to the centre of the lens and q is the distance from the image to the centre of the lens.

An electron microscope contains two kinds of lenses. The condenser lens has a large bore giving a long focal length, while the objective lens has a strong field of short axial extent giving a short focal length, resulting in high magnification. In an electromagnetic lens, an electric current passing through a conductor gives rise to a magnetic field. If the conducting wire forms a solenoid (several turns around a cylinder), each turn contributes to the induced magnetic field at the centre. The flux density at the center is: $B = \mu(NI/l)$, where μ is the permeability of the surrounding material, N is the number of turns, I is the current flowing through the coil and l is the length of the solenoid. $H = NI/l$ and so by substituting, $B = \mu H$. So, μ is the flux density per unit field. For air and non-magnetic materials, $\mu = 1.0$ and $B = H$. The permeability of iron is field-dependent and it is unity at high field and so, $B = H$. At high magnetic fields, iron reaches magnetic saturation and that is the reason why $\mu = 1$. Due to the hysteresis of iron, the extent of current used to energize a magnetic lens cannot be used to get the lens characteristics such as focal length. Hysteresis has to be kept small and soft iron is used for this reason. The current results in the heating and cooling by water is needed to keep the system cool. The system is shielded from external magnetic fields. For this, a high permeability material is used (μ metal is one such material).

When the solenoid is encased in a case of soft iron, the magnetic field along the axis is increased. If the entire coil except at a narrow annular gap is made of soft iron, a greater concentration of the magnetic field along a short axial distance occurs. The polepieces are these annular rings which focus the magnetic field at a given location. The field strength used is below 20,000 gauss in typical instruments.

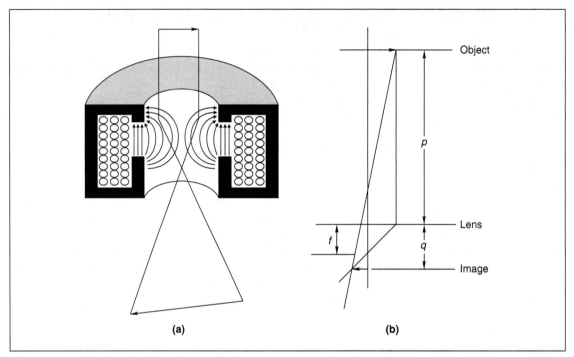

Fig. 2.3: (a) Schematic of an electromagnetic electron lens. The magnetic field lines are shown. (b) Illustration of the concept of magnification.

The focal length is given by $f = Kv/(NI)^2$ where K is a constant; v, the relativistically corrected accelerating voltage; and NI, the number of turns of the coils. Focusing is achieved by varying the current through the coils. It is important to note that the focal length is not linearly related to current. The focal length of the lens is directly proportional to the accelerating voltage. So the image quality is affected by varying the electron velocity.

Lens aberrations: Electron optics suffers from aberrations. But unlike in light optics, there is no way to solve these aberrations. All that can be done is to reduce them through proper design considerations.

Spherical aberration: Electrons in far away trajectories from the optic axis are bent more strongly. As a result, the electron beam entering the lens near the edge of the lens, is brought to focus at a different spot than the spots closer to the centre. This is schematically illustrated in Fig. 2.4(a). The error in the image due to this becomes more pronounced as the beam is moved further away from the optical axis of the lens.

Differential focusing causes the image at the perimeter to get smeared instead of at the centre. As a result of this, the image appears as a disk and not as a point. The smallest disk is called the spherical aberration disk of least confusion, $d_s = 1/2\, C_s \alpha^3$ where C_s is the spherical aberration coefficient and α is the angle of the outer ray through the lens. C_s is typically a few mm for lenses with short focal lengths. The spherical aberration can be minimized by removing the outer edge of the beam. This is achieved by placing a small-holed aperture at the centre of the magnetic field or immediately below it. However, a smaller diameter reduces the beam current and also leads to aperture diffraction.

Chromatic aberration: This aberration is due to the energy spread of the electrons. When light of different energies enters a converging lens at the same point, the extent of deflection will depend on the energy. In light optics, radiation of a shorter wavelength is deflected more strongly than that of a longer wavelength. In electron optics, the reverse happens, i.e. a shorter wavelength is deflected less strongly. This is due to the fact that electrons are subjected to lesser deflection when the beam energy is high. As a result, the beams of two different energies form images at different points as illustrated in Fig. 2.4(b). Due to this aberration, instead of a point, a disk results and the diameter of the disk of least confusion can be given as, $d_c = C_c\, \alpha\, (\Delta E / E_o)$, where C_c is the chromic aberration coefficient, E_o is the beam energy and ΔE is the energy spread. The fractional variation in beam energy is the significant factor.

Chromatic aberration can be reduced by stabilizing the energy of the electron beam. Stabilized acceleration voltage and improved gun design ensure the stability of the beam. The effect of chromatic aberration is pronounced near the perimeter of a converging lens and an aperture can be used to eliminate these electrons.

Aperture diffraction: The wave nature of electrons causes the beam to diffract upon passing through a narrow slit. Each beam passing through the slit sets up its own waves. These will interact to give a bright spot in the middle and a set of concentric rings in the image plane called the 'Airy disk'. If the intensity distribution is plotted in one dimension, it looks like that one shown in Fig. 2.4(c). The contribution to the spot size due to diffraction is given as the half diameter of the Airy disk, $d_d = 0.61\, \lambda / \alpha$, where λ is the wavelength of the beam and α is the aperture half angle. In order to reduce the effects of diffraction, it is necessary to have as great an angle as possible between the optical axis and the lens perimeter. That would amount to having no aperture at all. But a smaller aperture is needed to reduce the effects of spherical aberration and chromatic aberration, which would cause diffraction problems. Thus an optimum aperture size must be chosen.

Astigmatism: Astigmatism refers to the improper shape of the beam. A point object is focused to two-line foci at the image plane and instead of a point, an ellipse appears. The two-line foci may be forced to coincide for correcting this defect. Astigmatism occurs due to defects in the focusing fields, which could be due to several aspects related to electromagnetic lenses, apertures and other column components. Imperfection in machining can cause astigmatism. Astigmatism is corrected by a set of magnets called 'stigmators', which are placed around the circumference of the column. These are adjusted according to strength and position in an effort to induce an equal and opposite effect on the beam.

The effect of lens aberrations is important for the objective lens as the effects caused to the beam would be small in comparison to the diameter at other lenses. Typical spherical aberration can be corrected

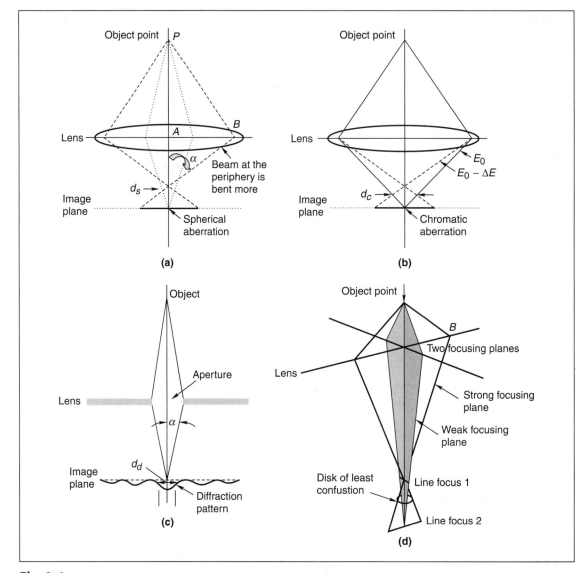

Fig. 2.4: *Various aberrations of electromagnetic lenses. (a) Spherical aberration, (b) chromic aberration, (c) aperture diffraction, and (d) astigmatism. Adapted from Goldstein, et al. 2003 (Ref. 1).*

completely. But the correction leads to aperture diffraction and therefore these two should be controlled properly. Chromatic aberration is significant below an accelerating voltage of 10 kV.

3. Scan coils: The next part of the SEM consists of the scan coils. In SEM, the scanned image is formed point by point and the scan is achieved by the scan coils. There are two pairs of coils, one each for the X

and Y axes (Fig. 2.5). The scan coils lie within the column and move the electron beam as per the requirement across the specimen. They are electromagnetic coils and are energized by the scan generator. The scan generator is connected to other components such as the cathode ray tube (CRT) and the magnification module.

The scan is made as follows. The electron beam is swept across the sample. The pattern over the sample is synchronous with that observed in the CRT. The secondary electrons produced by the sample are detected. The intensity of the signal at the CRT is proportional to the secondary electrons. An intense signal can illuminate several dots on the screen, while a weak signal would mean that no dots will be illuminated by the electron gun. The detector therefore gives the intensity of the signal, while the raster pattern gives the location of the signal. In this way, the image on the CRT is built up point by point to match what is happening on the surface of the sample.

This manner in which an image is formed is the essential difference between the transmission and scanning types of microscopes. A couple of important observations need to be made in this type of image formation. Firstly, the focus is dependent upon the size of the electron beam spot. The smaller the spot on the sample, the better is the focus. Secondly, magnification is not produced by a magnification or enlarging lens but rather by taking advantage of the differential between the size of the scan pattern on the sample and the size of the CRT.

The size of the CRT is fixed. The size of the scan pattern on the sample is variable and is determined by the magnification module. By narrowing the size of the area which is scanned and conveying that to the CRT, we can increase the magnification of the image. The smaller the area scanned, the lesser is the distance between the raster points, and the smaller is the amount of current needed to shift the beam from point to point. The greater the area scanned, the lower is the magnification, while the greater the distance between the raster points, the greater is the amount of current needed to shift the beam from point to point. In this way, when we operate the SEM at relatively low magnifications, we actually push the scan coils to their extremes.

The scan generator changes the step current to the scan coils. This current is then multiplied by a constant by the magnification module and sent to the scan coils. The higher the total magnification, the lower is the multiplier constant.

4. *Electron detector*: There are two kinds of electrons coming out from the sample in an SEM; the backscattered electrons, with high energies being larger or lower than the primary beam energy, and the secondary electrons which have energies of a few eV, or less than a few tens of eV. The most common kind of detector used in SEM is the Everhart-Thornley (*E-T*) detector (Fig. 2.5). In this, a scintillator material is exposed to the electrons. Energetic electrons, upon impact with the material result in photon emission. The photons are transmitted through a light pipe. This is a glass rod or a piece of plastic. At the other end of the light pipe, a photomultiplier is placed, which converts the photons to electrons through photoemission. The electrons can be made to undergo a series of collisions on surfaces producing a cascade of electrons and a gain of 10^6 can be achieved this way. The detector has a high bandwidth, which means it responds to a rapidly varying signal. This would be the case when the primary beam is scanned over the sample at high speeds.

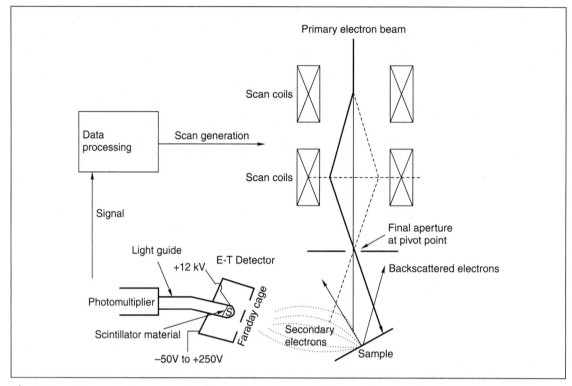

Fig. 2.5: *Schematic of an SEM. The electron beam is scanned by a set of scan coils and the secondary electrons are detected by the detector. By applying a negative potential to the Faraday cage, the secondary electrons are rejected completely. Adapted from Goldstein, et al. 2003 (Ref. 1).*

In the practical operation of the detector, a thin metal coating is applied on the surface of the scintillator. A high positive potential is applied to the metal surface so that all electrons, including the low energy secondary electrons, are accelerated to it so as to generate photons. The high voltage should not affect the primary beam and for this reason, a Faraday cage is kept over the scintillator, which is electrically insulated. By applying a desired potential in the range of –50 to +250 V to the Faraday cage, a complete rejection or collection of secondary electrons becomes possible.

For collecting backscattered electrons, a separate detector can be employed. Several kinds of detectors are used and one of the more popular ones is the solid state diode detector. This works on the principle that in a semiconductor, the electron impact can produce electron-hole pairs if the electron energy is above a threshold value, which is of the order of a few eV. Thus, when a high energy electron falls on it, several thousands of electron-hole pairs are formed. The charges can be swept in opposite directions if a bias is applied. The charge collected can be amplified and detected.

In low voltage SEM, the common electron detector is the channel plate. Here primary electrons make secondaries in a channel, and a cascade of secondary electrons is produced. Several such channels are made on a disk which is a few cm in diameter, and the electrons coming out at the other side are collected and amplified. As the secondary electron yield increases in the low keV range, channel plate works well for low energy operation. Low energy secondary electrons require acceleration for detection by a channel plate.

SEM: Modern Advances

SEM is the most widely used electron microscopic technique. This is largely because of its versatility, its various modes of imaging, ease of sample preparation, possibility of spectroscopy and diffraction, as well as easy interpretation of the images. The method has high throughput making it an accessible facility. A very wide range of magnification is available which facilitates the visualization of virtually every detail. Best SEMs can obtain image resolutions in the range of 0.5 nm and for this, the sample need not be specially prepared. Sample size is not a limitation in SEM and samples as large 6" silicon wafers can be put directly in a modern machine.

Modern advanced SEMs utilize field emission sources. There have been numerous advances in various aspects of the hardware such as lenses, detectors and digital image acquisition. These advances in hardware, coupled with advances in other areas of instrumentation such as power supplies, high vacuum instrumentation have now made it possible to acquire SEM images from almost anything, including wet biological samples.

Low voltage SEM is another new development. The extreme surface sensitivity of this technique is a result of the reduced interaction volume. This allows the measurement of images with nanometer scale resolution with under 1 kV acceleration. At these low energies, charging is not an issue and it is possible to measure images without conductive coatings. The electron energy can be reduced further by applying a negative potential on the sample and in this way, the beam energy can be reduced to below 100 eV. It also makes ultra low voltage SEM (ULV-SEM) possible, which is extremely surface-sensitive and avoids beam-induced damage at surfaces.

When an electron beam interacts with matter, several processes occur. These result in particle or photon emission processes which are summarized in Fig. 2.6. The electron emission processes include elastic and inelastic scattering, and the emission of secondary electrons and Aüger electrons. Inelastic scattering occurs as the beam interacts with the sample and electronic excitations of the constituent atoms can occur. These excitations can lead to valence and core electron excitations and emission. The core hole thus created may get filled by electron de-excitation resulting in X-rays. The de-excitation can also result in electron ejection, called Aüger emission. In addition, collision of the primary beam can also lead to excitations of lattice vibrations. All these electrons can be used to gather microscopic information of the sample. In addition, they can also be used to obtain chemical or compositional information as in the case of Aüger electrons or structural information as in the case of backscattered electrons. In a normal SEM image, only secondary electrons are detected. Emissions of characteristic X-rays as well as continuous X-rays occur. The characteristic X-rays are used for qualitative and quantitative information. In addition to these, electron-induced desorption, ion ejection, etc., also occur.

High resolution SEM is now a routine analytical tool. In nanoscience and technology, this becomes an important high throughput characterization tool. It is important to know the dimensions of the structures fabricated and the materials prepared when characterizing device structures. Thus SEM becomes an indispensable tool in nanometrology (a branch involving nanoscale calibration).

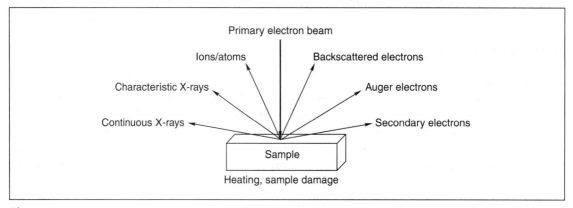

Fig. 2.6: *Electron beam-induced processes in the sample.*

Microanalysis

In both SEM and TEM, high spatial resolution microanalysis of materials is possible. The spatial resolution of the analysis is made possible by the small dimension of the excitation beam, which is of the order of a few nanometers in state-of-the-art instruments. The electron beam causes various excitations in the sample, which are characteristic of the elements present in the material. Characteristic X-rays emitted by the sample as a result of core hole decay can be used for elemental identification. The intensity of the signal can be used for quantitative analysis.

Microanalysis is done in two ways. One corresponds to the energy analysis called 'energy dispersive spectrometry' (EDS), while the other corresponds to wavelength analysis called 'wavelength dispersive spectrometry' (WDS). While improved energy resolution is possible in WDS, it is more cumbersome and time-consuming than EDS. In EDS, a signal from the detector is proportional to both the energy and intensity of the X-rays. In WDS, the wavelength and intensity of the X-ray are determined separately.

When an electron approaches the atom, it gets decelerated due to the coulombic field. This results in a loss of energy for the electron and that energy appears as photon, referred to as bremsstrahlung or 'breaking radiation'. This radiation contains photons of all energies till the energy of the original electron, as an electron can lose any energy, from zero to the energy of the primary electrons. The characteristic X-rays emitted by the atoms will appear as spikes on this large, smoothly varying photon intensity. There are several characteristic X-ray lines with which an atom can be identified. The intensities of these lines can be related to the concentrations of the emitting species in the sample, though various parameters determine the intensities. The important aspects related to X-ray emission intensity are, inner shell ionization cross section, X-ray absorption cross section of materials and X-ray production range.

In EDS, all photons emitted by the sample are collected and measured simultaneously by a solid state X-ray detector. The common EDS detector is lithium-drifted silicon, Si(Li). Intrinsic or high purity Ge (HPGe) is also used, which is more common in the case of TEM instruments due to their higher collection efficiency. When an X-ray photon falls on an intrinsic semiconductor (having no charge carriers), due to photoabsorption, charge carriers (electrons and holes) are created. These are swept by an applied bias forming a charge pulse. This charge pulse is then converted into a voltage signal. Intrinsic condition is hard to achieve and detector crystals are made to behave like intrinsic silicon. This is made by applying Li on p-type Si, thereby forming a p-n junction. The junction region will behave like an intrinsic semiconductor. The width of this region can be expanded by applying an electric field at elevated temperatures. The removal of most of the p-type Si region makes a detector (which is the Si left behind). However, the Li present is mobile at room temperature and it is necessary to cool down the detector to 77 K in order to operate the detector under bias.

The intrinsic silicon active layer is covered on the front side with p-type Si and in the back with n-type Si. The front is also coated with a thin gold layer for electrical contact. The device is kept behind a Be window which blocks visible light. Also, it mechanically seals the detector assembly from the vacuum chamber, thereby avoiding contamination. When an energetic photon strikes the detector, charge carriers are created. The detector is reverse-biased, i.e. p-type is connected to negative potential. This means that when the charge carriers are created, the holes move to the p-side while the electrons move to the n-side, thus creating a pulse of electrons at the n-side, which can be amplified. In order to reduce noise and also Li mobility, to maintain the intrinsic condition, the detector is cooled. When photons of high energy above the Si K shell binding energy (1.84 keV) fall on it, photoelectron emission takes place. The photoelectron loses energy inelastically, thus exciting electron-hole pairs. The core hole produced may be filled by photoemission or Aüger emission. The photon may be reabsorbed resulting in photoemission, or it may undergo inelastic scattering. Thus, all the energy of the initial photon is used in generating electrons. In favourable conditions, the number of charges created in the detector per photon is equal to photon energy/energy required for the creation of an electron-hole pair (ε). For a 5000 eV photon, this works out to 5000/3.86 or 1295 electrons (ε at 77 K is 3.86 eV for Si). However, this number is very small, and generates a very small charge pulse of 2×10^{-16} C. Each charge pulse, converted to voltage, represents the energy of the photons. The resolution of the photon energy, distinguishable by solid state detectors of this kind, is 130 eV at MnKα (5890 eV). The detector resolution is a function of X-ray energy and it is about 65 eV at 282 eV. There are several advanced detectors such as the silicon drift dectector, which works at 250 K, having the same resolution but an increased sampling rate. A microcalorimeter EDS works by measuring the temperature difference caused by the absorption of X-rays. With this, an increased energy resolution is possible, which is about 4.5 eV at MnKα.

Wavelength dispersive spectrometers work on the principle of Bragg diffraction, $n\lambda = 2d\sin\theta$, where λ is the wavelength of the X-ray and θ is the angle of incidence. Therefore, at various angles of incidence, it is possible to scatter rays of different wavelengths to the detector. As $\sin\theta$ cannot be larger than one, the longest wavelength diffracted is 2d, therefore the crystal puts a limit on the range of elements covered. There are also other limitations of the hardware. When θ is 90°, the detector has to sit at the X-ray source; when $\sin\theta$ is close to zero, the movement of the crystal near the specimen is a problem. In order to diffract the low energy X-rays emitted by Be, B, C and N, it is necessary to use crystals of large d values.

This is made possible by Langmuir–Blodgett films (see the chapter on Self-assembled Monolayers) whose interlayer spacing can be tuned by varying the chain length of the monolayer. Most of the measurements of long wavelength X-rays are done by layered synthetic microstructures. This contains evaporated alternate layers of heavy and light elements. The thickness of the layers can be easily controlled.

The detector commonly employed is a gas proportional counter. This has a gas mixture (typically 90 per cent argon, 10 per cent methane). The detector assembly has a thin window through which the X-ray photon falls in. It causes the ejection of a photoelectron, which is accelerated by the high voltage applied on the collector. The electron causes subsequent ionization in the gas and as a result, a gain factor of the order of 10^5 is achieved. There have been numerous improvements in gas-filled detectors involving the use of improved windows, other gases, etc.

The spatial resolution of SEM is affected by the beam size as mentioned before. But the beam causes interaction with the sample and the region of interaction exceeds the physical dimension of the beam itself. The spatial extent increases laterally as well as depth-wise, and this interaction volume depends on the material and also on the beam energy. At large energies, the beam interacts to a greater extent, while the depth of interaction decreases with a decrease in electron energy till about 100 eV and at very low energies of the order of a few eV, it increases again. The range of interaction is decided by the electron attenuation length. The universal electron attenuation length of materials is given in Fig. 2.7. As seen in the figure, the attenuation length increases with electron energy and has a minimum at a few hundred eV. This makes it possible to design highly surface-sensitive SEMs by reducing the beam energy. This development is discussed in low energy SEM.

2.2.2 Transmission Electron Microscopy

In TEM, the transmitted electrons are used to create an image of the sample. Scattering occurs when the electron beam interacts with matter. Scattering can be elastic (no energy change) or inelastic (energy change). Elastic scattering can be both coherent and incoherent (with and without phase relationship). Elastic scattering occurring from well-ordered arrangements of atoms as in a crystal, results in coherent scattering, giving spot patterns. This can be in the form of rings in the case of a polycrystalline material. However, inelastic scattering also occurs, which also gives regular patterns as in the case of Kikuchi patterns. Inelastic processes give characteristic absorption or emission, specific to the compound or element or chemical structure. Because of all these diverse processes, a transmission electron microscope is akin to a complete laboratory.

There are two main mechanisms of contrast in an image. The transmitted and scattered beams can be recombined at the image plane, thus preserving their amplitudes and phases. This results in the phase contrast image of the object. An amplitude contrast image can be obtained by eliminating the diffracted beams. This is achieved by placing suitable apertures below the back focal plane of the objective lens. This image is called the bright field image. One can also exclude all other beams except the particular diffracted beam of interest. The image using this is called the dark field image (Fig. 2.8).

TEMs with resolving powers in the vicinity of 1Å are now common. As a result, HRTEM is one of the most essential tools of nanoscience. Interaction of the electrons with the sample produces elastic and

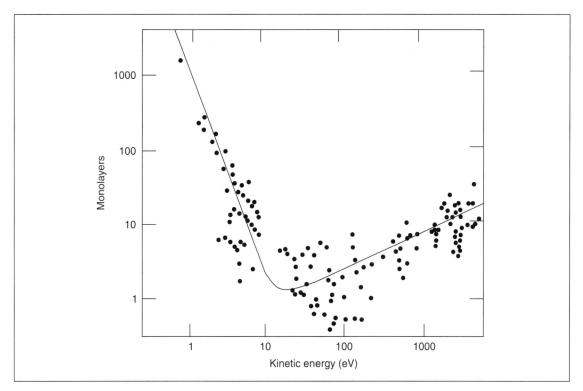

Fig. 2.7: *Universal attenuation length data of materials as a function of electron kinetic energy. At low kinetic energies, the electrons are sensitive to only a few monolayers of the material. The data points correspond to the experimental data from a number of materials and the line is an approximate fit.*

inelastic scattering. Most of the studies are done with the elastically scattered electrons which form the bright field image. The inelastically scattered electrons are used for electron energy loss spectroscopy as well as for energy filtered imaging. Electrons emerging from the sample, after a series of interactions with the atoms of the target material, have to be transferred to the viewing screen to form an image. This is greatly influenced by the transfer characteristics of the objective lens. A parameter which can be used to understand the transfer properties is the phase contrast transfer function (CTF). This function modulates the amplitudes and phases of the electron diffraction pattern formed in the back focal plane of the objective lens. The function is given as:

$$T(k) = -\sin\left[\frac{\pi}{2} C_s \lambda^3 k^4 + \pi \Delta f \lambda k^2\right],$$

where, C_s is the spherical aberration coefficient, λ, the wavelength of the beam, Δf, the defocus value and k, is the spatial frequency. Defocus setting (with reference to the Gaussian focus) in the electron microscope determines the shape of the CTF. This is given in Angstrom units. In the electron optical

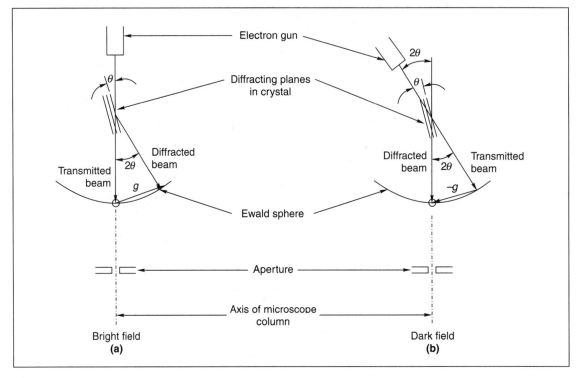

Fig. 2.8: *Two kinds of image collection: (a) bright field, and (b) dark field. In bright field, the transmitted beam is used for imaging. In dark field, the diffracted beam is used for imaging. The diffracted beam can be used for imaging either by using a movable aperture or by shifting the incident beam, keeping the aperture constant. Adapted from Thomas and Goringe, 1979 (Ref. 2).*

convention, over-focus implies negative defocus values, while under-focus (the only useful range) implies positive defocus values. The CTF of a 200 keV microscope will look like the one shown in Fig. 2.9.

The general characteristic to be noted is its oscillatory nature. When it is negative, we have a positive phase contrast and atoms will appear dark on a bright background. When it is positive, a negative phase contrast occurs, and atoms will appear bright on a dark background. When it is equal to zero, there is no contrast or information transfer. CTF can continue forever but it is modified by functions known as 'envelope functions' and eventually dies off.

In any instrument, the most important aspect is the resolution. This is the ability of the instrument to discriminate between two closely lying objects. As the wavelength of electrons is in the picometer range, it is only natural to expect a resolution of that order. However, this is not the case. This is because of the fact that electron lenses have aberrations and also due to the fact that it is impossible to get a coherent electron beam. We can express the resolution as, $d \alpha C_s^{1/4} \lambda^{3/4}$.

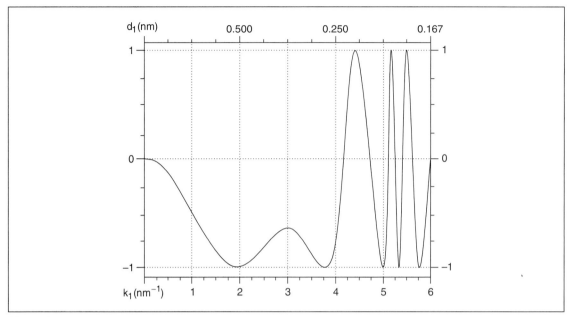

Fig. 2.9: *A contrast transfer function.*

The interpretable resolution, or the structural or point resolution is the first zero crossover of the CTF at the optimum defocus. This can be given as $\delta \sim 0.66(C_s \lambda^3)^{1/4}$. C_s increases with electron beam energy, but the point resolution improves with acceleration voltage as λ is reduced. C_s ranges from 0.3 mm for 100 keV to 1.5 mm for 1 MeV. The resolution changes from 0.25 to 0.12 nm. In view of the increasing cost of the instrument at larger acceleration voltages and irradiation damages at larger acceleration voltages, it is more advisable to use intermediate energy ranges. The instrumental resolution is lesser than this value, but if one knows the defocus and C_s accurately, one can obtain improved images by subsequent processing. The lattice resolution refers to the finest lattice spacings observable in images. The lattice images may not provide better information about the sample in several cases as far as the local atomic environment is concerned. Interpretable and instrumental resolutions are important in assessing the performance of the instrument.

2.2.3 STEM

STEM is a hybrid instrument with the features of both SEM and TEM. The same probe beam in TEM can be demagnified and used as a probe beam, which can be scanned over the sample. The probe beam has to be small and bright and therefore, field emission sources are needed to obtain beam dimensions that are smaller than a nanometer. The principal components of STEM are shown in Fig. 2.10. The beam coming off the gun is demagnified by the objective lens. A set of condenser lenses is placed above the objective lens to add flexibility to the beam parameters. The scanning coils are incorporated into the objective lens itself. The beam falling on the sample produces a diffraction pattern. The pattern is a convergent beam electron diffraction and is also called electron nano diffraction. The beam size is close to the resolution limit of the instrument. The observation screen may have an aperture and some part of the transmitted beam is used for electron energy loss spectroscopy (EELS, see below). This analysis is usually done with the central part of the beam (Fig. 2.10(a), bright field), but can also be done with any of the diffracted spots (Figs 2.10(b) and (c)). That is, the dark field images may be formed either by the deflection of the diffracted beam into a centrally placed objective aperture, or by the displacement of the aperture which allows only a specific diffracted beam to pass through. The measurement may correspond to either the EEL spectrum or an image using electrons with characteristic energy loss. The normal bright field image is made with no-loss electrons from the central beam. Dark field images are made with electrons deflected from the beam axis with or without energy loss. Deflection coils placed after the sample are used to direct the chosen part of the diffraction pattern to the entrance aperture of the spectrometer. Common dark field STEM images are collected by imaging part of the diffraction pattern outside the central beam spot.

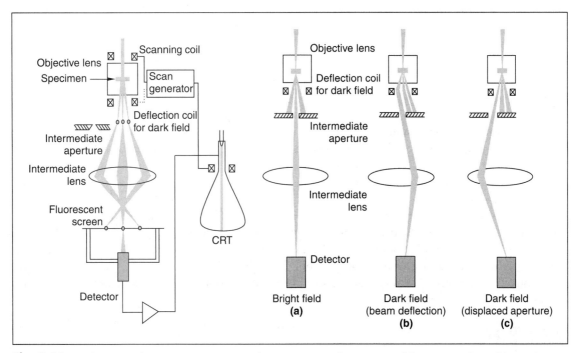

Fig. 2.10: *Schematic of a STEM instrument and various ways of operation of the same. Adapted from Thomas and Goringe, 1979 (Ref. 2).*

Detectors can be placed around the sample to detect the secondary radiations coming out from the sample. X-rays, secondary electrons and Aüger electrons can be analyzed, with the last two normally being used for imaging purposes and the first one for microanalysis. In microanalysis, the important aspect to be noted is that since the sample thickness is small in STEM, the interaction volume is very small. The chemical analysis of small volumes is limited only by the quality of the signal obtained as far as an X-ray is concerned. A combination of ELS and X-ray analysis may be used in most cases.

There are numerous areas where STEM can be of use in nanotechnology. With the current instrumental resolution of the order of 0.1 nm, complete characterization of materials up to 1 nm is possible. Diffraction of objects at 1 nm is possible. All kinds of materials such as crystalline, amorphous and biological materials have been investigated. With a growing need to know everything about a nanostructure, applications of STEM are also likely to increase.

2.2.4 Image Collection in Electron Microscopes

In SEM, the image is collected in a video monitor (CRT), which is captured by a camera or a computer. In TEM, the image is formed when transmitted electrons fall on a phosphor screen. The image formed can be photographed. Photography is a result of a series of chemical reactions. This uses the light sensitivity of silver halides. While silver bromide is commonly used, iodide and chloride are also used. When the grains of silver salt present in a gelatinous film are impinged with photons, they reach an activated state, which get reduced by a reducing agent (developer). The reduction will ultimately happen on all the silver halide grains, including those which are not exposed to photons. As a result, one needs to stop the process of reduction. This is done by putting the film in water. The remaining silver grains can still get reduced later on. In order to avoid this, these grains have to be removed by a process called fixation for which thiosulphate is used. After the process of development and fixation, the film is dried and stored.

Basically two types of emulsions are used in photographic films. The classification is based on light energy. Panchromatic emulsions are sensitive to all wavelengths and are therefore handled in total darkness till fixation. Orthochromatic emulsions are sensitive to certain wavelengths only. There is a safe light in which it can be handled. The film has two characteristics. Light sensitivity, which decides the length of the exposure time and grain size, which determines the extent to which one can enlarge the image. The extent to which the film is darkened depends on the light intensity and duration of exposure. These parameters have to be optimized for a given film.

Digital Imaging in TEM

In digital TEM imaging, the image formed by the electron beam is captured by a charge-coupled device (CCD). CCD is an image sensing device, in which an array of light sensitive capacitors is connected on an integrated circuit. By external control, the charge at one of the capacitors can be transferred to another. Depending on how the CCD is implemented, there are three common kinds of architectures. In the 'full-frame' CCD, the charge is collected on all the available area of the CCD and the information is read out by placing a mechanical shutter over the sensor. In the 'frame-transfer' CCD, half of the image area is covered with an opaque mask and the image collected is transferred immediately to the opaque area and read out slowly. As double the area (sensor and intermediate storage) is needed, this adds to cost. But quick accumulation of image becomes possible by this way. In the 'interline' architecture, every other column of the sensor is masked and the image is shifted only one pixel after collection. This adds to a fast shutter speed and smearing of the data is less, but only half the area is effectively used and so the fill-factor is only half. There are ways to improve this, however.

In the CCD detectors used in TEM, the photons have to be generated first. For this a scintillator plate is used. The electrons fall on this and photons are generated. The photons are transferred to the CCD by fiber-optics. At the CCD photons generate charges and they are read out, pixel by pixel. CCDs are generally cooled to reduce the noise (dark current).

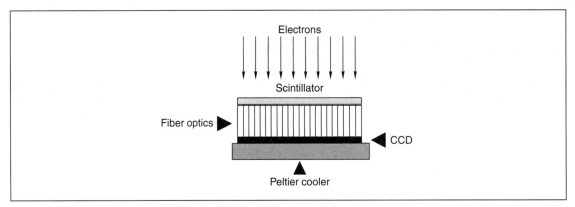

Scheme 2.1: *Schematic of a CCD detector used for TEM.*

2.2.5 Environmental Transmission Electron Microscopy (ETEM)

Science at the nanoscale is concerned with the variation of properties with size and shape. Structural and morphological changes occur in these materials with minor changes in temperature and other parameters such as atmosphere. Because of these reasons, nanomaterials have to be studied at higher pressures and temperatures. Such studies have been important in areas such as catalysis. Gas–solid interactions, in general, have been of interest in areas such as corrosion, oxidation, etc. Biological materials need to be investigated close to their natural environments. All these requirements have necessitated the modification of conventional TEMs to work at modified conditions. These microscopes are called environmental transmission electron microscopes. They facilitate the investigation of structural and chemical changes on materials upon gas–solid and liquid–solid interactions at varying temperatures.

In ETEM, the sample is confined in a high pressure region, as high as 150 torr, in such a way that the column vacuum does not deteriorate significantly. The key instrumental feature is an environmental cell in which the pressure–temperature conditions are regulated. This can be done by using two kinds of approaches: 1. using electron transparent windows to confine the system under investigation, and 2. using differential pumping to confine the system. The window method has an advantage in that it makes higher pressures possible, but it is difficult to get high resolution due to the additional scattering as a result of the window material. Due to the space requirements of such a cell, these kinds of cells have been used only in high energy instruments because of their larger pole-piece gaps. The medium energy TEMs today have pole-piece gaps of 7–9 mm, large enough to accommodate a cell. Differentially pumped cells are used nowadays.

Such instruments are used for the *in-situ* synthesis of nanoparticles, the study of carbon nanotube growth, *in-situ* chemical transformations of nanoparticles, etc. Combining diffraction, microscopy and elemental analysis in ETEM facilitates complete investigations as a process occurs. It is possible to make nanolithographic structures in such an instrument.

2.2.6 Electron Energy Loss Spectroscopy at the Nanometer Scale

The characterization of materials at the nanometer scale is central to nanoscience and technology. With improvements in electron sources and optics, it is possible to generate beams of sub-nanometer spatial spread, which can probe into the chemical and structural aspects of the sample. In electron energy loss spectroscopy, one is interested in the inelastic energy losses suffered by the primary beam as a result of scattering. The energy changes may be due to the collective electron oscillations or inter-band transitions of the sample or the characteristic energy losses occurring due to core electron excitations resulting in characteristic edges superposed on a monotonically decaying background. While the former occur in the range of 5–50 eV, the other occurs within about a few thousand eV. The collective electron oscillations or inter-band transitions may not have much elemental or material significance as the energies of these transitions change with size at the nanometer level. At extremely low energies, these transitions may disappear altogether as well. The core electron excitations are characteristic of the element. Thus a qualitative analysis of the material is possible in terms of the characteristic energy losses obtained. An energy loss at 284 eV is characteristic of carbon and that at 530 eV is due to oxygen. The intensity at this energy loss after subtracting the background suitably is related to the concentration of the atom. Thus quantitative analysis is possible. The position and shape of the energy loss are characteristic of the chemical environment of the sample. Because of all these, electron energy loss spectroscopy at the nanometer scale is a unique probe.

We can calculate core level excitation spectra. In this, the approach will be to calculate the inelastic scattering cross section. The essential aspect of these computations will be to evaluate the probability of electronic transitions between two orbitals as a result of a perturbing electric field. Atomic orbitals of the initial states and molecular orbitals of the final states can be used for such computations. By performing refined calculations, it is possible to obtain a positive agreement between theory and the fine structures seen experimentally.

The energy loss information can be collected as a function of space. This can be done in two different modes, either as a spectrum image wherein at each point defined on the sample, one collects a complete spectrum or the spectrum in an energy window ΔE is collected for a two-dimensional space called 'image-spectrum'. Both the methods have their own advantages and disadvantages. The former is collected in the STEM and is superior in the sense that all spectral channels are collected parallely as a function of space. A post-column energy analyzer does the energy analysis of the electrons. In the image-spectrum mode, an energy filter is used. An imaging stage which converts the spectrum into an image is put after the filter. The filter fixes the energy of the electrons to be transmitted and the slit at the exit plane of the filter decides the energy width.

EELS has been used for elemental mapping of materials at the nanometer range. Elemental mapping of doped elements in singe-walled carbon nanotubes, nanobubbles in alloy matrices, segregation of atoms in ultra-thin films, etc. have been performed with EELS. Mapping of individual atoms in nanostructures has been demonstrated. Apart from elemental mapping, the distribution of the bonding characteristics of an element within a material at the nanometer level can be investigated. Measuring atomic properties of nanometer sized objects is a distinct reality with EELS.

2.2.7 *In-situ* Nano Measurements

Properties of individual nano objects have to be understood in a number of cases. This has a bearing on the properties of the bulk systems. For example, how would one measure Young's modulus of a single carbon nanotube? There are several methods one can use for this, but it is important to make the measurement with an instrument wherein the nano object is seen while the measurement is performed. There are several instances of this kind such as the measurement of electrical conductivity, thermal conductivity, temperature stability, melting point, chemical reactivity, etc. of single nano objects. All these are possible in the transmission electron microscope. Of course for each of these measurements, the TEM has to be modified significantly. It is also necessary to invent new methods and appropriate theory for the nano measurements.

One of the easier kinds of *in-situ* measurements in TEM (as well as in other cases) is temperature dependence. This measurement entails varying the sample temperature in the rage of liquid helium (or more generally liquid nitrogen) to 1200 K. The phase or morphology change of the material is investigated with specific reference to a single nano object. This kind of investigations reveal that different morphologies have different stability. For example, among a group of particles with cubic and tetrahedral shapes, the cubic one is found to change into a spherical structure at a lower temperature than the tetrahedral one. This is attributed to the lower surface energy of the [111] planes present in a tetrahedral structure than the [100] present on cubes.

One of the ways of measuring Young's modulus of a nanotube is to investigate the blurring contrast due to the thermal vibration of the tip of a carbon nanotube image. This is shown in Fig. 2.11. Some of the nanotubes, shown with arrows, show a blurring contrast. The blurring amplitude, u can be related to the vibration energy, which is related to Young's modulus. By measuring the blurring contrast as a function of temperature, Young's modulus of a multi-walled carbon nanotube is calculated to be in the range of 1 TPa (Ref. 3).

Numerous different kinds of measurements have been demonstrated which include *in-situ* transport measurements in a nanotube, mapping the electrostatic potential at the tips, work function measurements of individual nano objects, field emission from nanotubes and applications of single nanotubes as balances. In several cases, it is necessary to develop new methods and appropriate theory. It is very clear that numerous kinds of nano measurements are possible in TEM with appropriate modifications of the equipment.

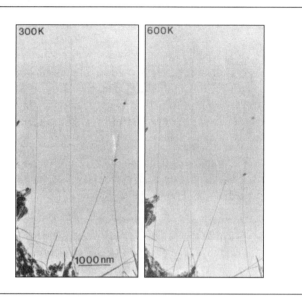

Fig. 2.11: *Low magnification TEM images of carbon nanotubes grown by the arc-discharge method. The tips of the nanotubes are blurred due to thermal vibrations. The amplitude of the vibrations increase with temperature. From Treacy, et al. 1996. Copyright Nature Publishing Group, used with permission from the author.*

2.3 Scanning Probe Microscopies

There are broadly two kinds of SPM techniques, namely scanning tunneling microscopy (STM) and atomic force microscopy (AFM). Numerous variations of these techniques exist. The objective of this section is to review some of the most prominent of these techniques so as to give a flavour of them. Those interested in additional details may consult original papers and books, some of which are listed in the references.

In a scanning probe technique, a probe of nanometer dimensions is used to investigate a material. The investigation is conducted on the surface of the material. This is done by keeping the tip stationary and moving the sample or vice versa. The information that is collected by moving the sample can be of several kinds which differ from technique to technique. The collected data and its variation across the sample are used to create an image of the sample. The resolution of such an image depends on the sample, the control that one has on the movement on the tip/sample, and the inherent nature of the data.

2.3.1 Scanning Tunneling Microscopy

STM was developed (Ref. 4) in 1982 and the inventors were awarded the Nobel Prize for Physics in 1986. In STM, the phenomenon of electron tunneling is used to obtain an image of the topography of the surface. This utilizes the principle of vacuum tunneling. Here two surfaces, a tunneling probe and a surface are brought near contact, at a small bias voltage. If two conductors are held close together, their wavefunctions can overlap. The electron wavefunctions at the Femi level have a characteristic exponential inverse decay length, K which can be given as, $K = \sqrt{(8m\varphi)/h}$, where m is mass of electron, and φ is the local tunneling barrier height or the average work function of the tip and sample. When a small bias voltage, V is applied between the tip and the sample, the overlapped electron wavefunction permits quantum mechanical tunneling and a current, I to flow through. The tunneling current, I decays exponentially with a distance of separation as $I \alpha V e^{-\sqrt{(8m\varphi)/h} 2d}$, where d is the distance between the tip and the sample and φ is the work function of the tip. The tunneling current is a result of the overlap of electronic wave functions of the tip and the sample. By considering the actual values of the electron work functions of most materials (typically about 4 eV), we find that the tunneling current drops by an order of magnitude for every 1Å of distance. The important aspect is that the tunneling current itself is very small and in addition, it is strongly distance-dependent. As a result, the direct measurement of vacuum tunneling was not observed till 1970s. Tunneling, however, was observed and was limited to tunneling through a barrier.

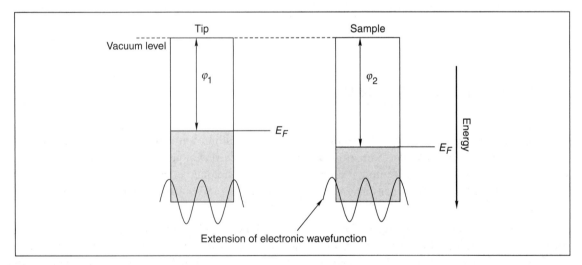

Fig. 2.12: *Electron energy states of the sample and the tip before the bias. When a sample and a tip are held close together there is a possibility of their electronic wave functions overlapping. φ_1 and φ_2 are the workfunctions of the tip and the sample, respectively.*

The basic principles of such a device are simple and are illustrated in Figures 2.12 and 2.13. In this, a tip is brought close to the sample so that the electrons can tunnel through the vacuum barrier. The position of the tip is adjusted by two piezoelectric scanners, with x and y control. The z-axis position is continuously adjusted, taking feedback from the tunneling current so that a constant tunneling current is maintained. The position of the z-axis piezo therefore reproduces the surface of the material. The position of the piezo is directly related to the voltage supplied to the piezoelectric drives. Scanning is possible in the constant height mode as well, but this is done only on extremely flat surfaces so that the tip does not crash by accident. The other mode of imaging is by modulating the tip at some frequency and measuring the resulting current modulation. This helps in understanding the compositional variation across the sample.

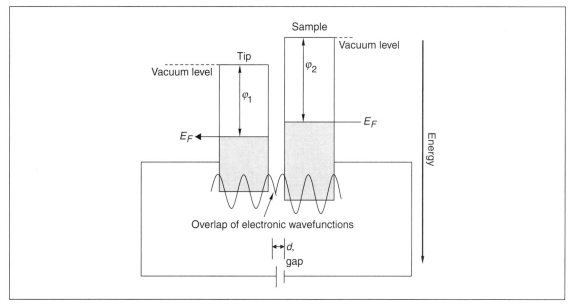

Fig. 2.13: *Electronic wavefunctions overlap at lower separations and a tunneling current is observed.*

Tunneling happens through a few atoms and it is believed that the STM images occur as a result of tunneling to a single atom or to a few atoms. The tip manufacturing process does not make one atomically sharp tip alone. However, what is apparently happening is that electron tunneling takes place to whichever atomically sharp tip which is closer to the surface.

STM has been used to understand numerous processes but a review of these is out of place here. Most of these studies relate to understanding various surface processes. The techniques available earlier for these studies before the arrival of STM, are low energy electron diffraction (LEED), reflected high energy electron diffraction (RHEED), X-ray diffraction and variations of these techniques. Instead of probing the average structure of the surface, as is possible with these techniques, STM allowed the investigation of local structures. These studies focused on surface reconstruction, adsorption, chemical transformations, etc. on metal, semiconductor and even on insulating surfaces (under appropriate conditions to observe tunneling current). In the brief discussion below, we shall illustrate a few of the experiments of relevance to nanoscience and technology.

STM provides information on the local density of states. The Density of States (DoS) represents the quantity of electrons existing at specific values of energy in a material. Keeping the distance between the sample and the tip constant, a measure of the current change with respect to the bias voltage can probe the local DoS (LDoS) of the sample. A plot of dI/dV as a function of V represents the LDoS. This is called scanning tunneling spectroscopy (STS). An average of the density of states mapped by using STM is comparable to the results from ultraviolet photoelectron spectroscopy and inverse photoemission spectroscopy (as discussed below). Such comparisons have been done in a few cases and the results indicate that the tip effects are unimportant. The important aspect is that such mapping of density of states is possible with spatial specificity.

Electrons can only tunnel to states which are present in the sample or tip. When the tip is negatively biased, electrons from the tip tunnel from the occupied states of the tip to the unoccupied states of the sample. If it is the other way, electrons tunnel from the occupied states of the sample to the unoccupied states of the tip. Thus a change in polarity makes it possible to probe the occupied or unoccupied states of the sample as illustrated in Fig. 2.14.

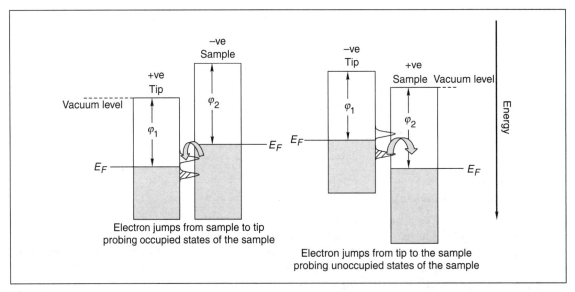

Fig. 2.14: *Depending on the sample bias, one can probe various kinds of states of the sample.*

STM has revolutionized a number of areas of fundamental science. Among the advantages of STM are its capability to analyze samples with atomic resolution, to see processes as they are occurring and to perform all these while the sample is in atmospheric conditions. In nanoscience and technology the uses of STM are manifold. An example is given here for the benefit of the reader. STM has been used in different ways to understand the electronic structure and properties of carbon nanotubes. An atomically resolved image (Ref. 5) of a single-walled carbon nanotube is shown in the Fig. 2.15. The arrangement of the hexagons is seen. One has to remember that the graphitic sheet has been folded so that the hexagons are not planar. From this, we can determine the tube axis and the chiral angle. The chiral angle is 7° for this tube and the tube diameter is 1.3 nm. The STS measurements give the band gap and local electronic structure of the tubes.

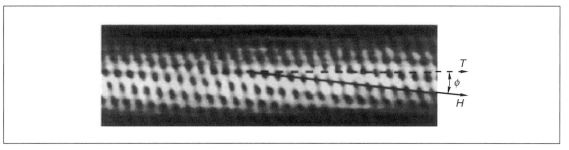

Fig. 2.15: *Atomically resolved STM images of an individual single-walled carbon nanotube. The lattice on the surface of the cylinders allows a clear identification of the tube chirality. Dashed arrow represents the tube axis T and the solid arrow indicates the direction of nearest-neighbour hexagon rows H. The tube has a chiral angle $\phi = 7°$ and a diameter d = 1.3 nm, which corresponds to the (11,7) type tube. See the chapter on carbon nanotubes. From Wildoer, et al., 1998 (Ref. 5). Copyright Nature Publishing Group.*

2.3.2 STM-based Atomic Manipulations

In a normal STM imaging process, the tip-sample interaction is kept small. This makes the analysis non-destructive. However, if the interaction is made big and as a result, the tip can move atoms on the substrate, there is a distinct possibility of writing atomic structures by using STM. This was done in 1990 by Eigler. When he manipulated Xe atoms on the surface of Ni(110) in a low temperature ultra high vacuum (UHV) STM instrument, a new branch of science was born (Ref. 6). The process of manipulating atoms is technically simple. This is schematically illustrated in Fig. 2.16. The STM scanning process is stopped and the tip is brought just above the sample atom. The tip is then lowered to increase the interaction between the tip and the atom. This is done by increasing the tunneling current (to the tune of ~30 nA). Note that the typical tunneling current used for imaging is of the order of 1 nA. As a result of this, the tip-atom interaction is strengthened and the tip can now be moved to the desired location. The interaction potential between the tip and the atom is strong to enable it to overcome the energy barrier so that the atom can slide on the surface. It is, however, not transferred to the tip from the substrate. After moving to the desired location, the tip is withdrawn by reducing the tunneling current. This process can be repeated to obtain the desired structure, atom by atom. A quantum corral built by arranging 48 Fe atoms on a Cu(111) surface (Ref. 7) is shown in Fig. 2.17 (Plate 2). The electron waves confined in the corral are seen in this picture. The confinement of electrons comes about as a result of the nano structure constructed. At such length scales, quantum mechanical phenomena can be observed. By having an elliptical corral of Co atoms and placing another Co atom at one of its foci, the other focus manifests some of the features of the atom, wherein no atom exists. When the atom is moved from the focus, the effect disappears (Ref. 8). This quantum mirage effect suggests the transportation of data in the quantum mechanical size limit.

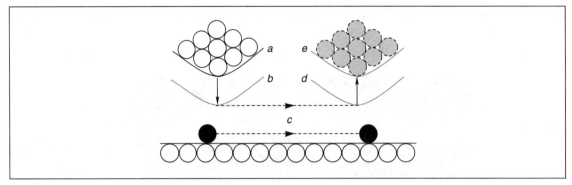

Fig. 2.16: *Process of manipulating atoms on the surface. An STM tip at position **a** is brought to position **b** by vertical movement at which point the tunneling current is large. Then the tip is slid over the surface to the desired location **d**, and subsequently the tip is brought back to position **e**.*

Numerous other manipulation strategies have been demonstrated. Tunneling current has been used to break chemical bonds. This has been shown in the case of oxygen and organic molecules. In the case of organic molecules, the detached fragments have been moved and further recombined in the desired fashion. Feynman's prediction of atomically constructed matter has come true.

Numerous modifications of STM are available. A list of variants is reproduced in Reference 9 in the Reference list. The reader may consult the references and additional reading material listed at the end of the chapter. The most current developments are in the areas of fast scanning STM, ultra low temperature STM and spin polarized STM. In the first, the dynamic processes taking place at the surface such as a chemical reaction are monitored and the images are captured so as to construct a movie. This can be combined with a solution phase STM so that the reactions in solutions can be investigated. In spin polarized STM, a magnetic tip is used so that the tunneling current is sensitive to the spin. In ultra low temperature STM, the measurements are done at the milliKelvin temperature range so that the phenomena at low temperatures can be probed.

2.3.3 Atomic Force Microscopy

In this technique, the interactions between a sharp probe and a sample are used for imaging. The cantilever which probes the surface has an atomically sharp tip which is brought into contact with the surface. The large scale use of AFM today is because of the application of microfabricated tips of Si or Si_3N_4. The spring constant of the tip is of the order of 1 N/m and the shortest vertical displacement, d measurable can be obtained from, $<1/2\ kd^2> \sim \frac{1}{2} k_B T$. With $k_B T$ of the order of 4×10^{-21} J at 298 K, the smallest vertical displacement observable is 0.5 nm. The extent of interaction between the cantilever and the tip is measured by cantilever displacements. The interaction between the tip and the sample is of the order of a nano Newton, which is not directly measured in AFM. The displacement of the cantilever is monitored by the reflection of a laser from the back of the cantilever, detected on a segmented photodetector. A four-segment photodiode is used for this purpose. In the very first AFM, the interaction was measured by the difference in tunneling current, with the tip being fixed on the back of the cantilever. This allows the detection of normal and lateral displacements of the cantilever. Optical detection is far superior to other forms of detection, though there are problems associated with the laser such as the heating of the cantilever and the sample. The image is generated from the interaction force. In the scan, the interaction force is kept constant by a feedback control. The increase in the interaction force when the tip approaches an elevated part is related to the vertical displacement of the scanner needed to eliminate this increase in signal. This is converted to height. Thus the basic components of the microscope are the cantilever, the detection system, scanners and the electronics. These components are schematically represented in Fig. 2.18. This also suggests that depending on the kind of interactions between the cantilever and the surface, various kinds of microscopies are possible. The probe can be made magnetic to investigate the magnetic interactions with materials. This results in magnetic force microscopy. The tip can have specific temperature probes or the tip itself can be made of a thermocouple. This facilitates scanning thermal microscopy (SThM). The tip may be attached with molecules which are designed to have specific molecular interactions with the surface. This results in chemical force microscopy. There are several such variations, some of which are listed under Ref. 9.

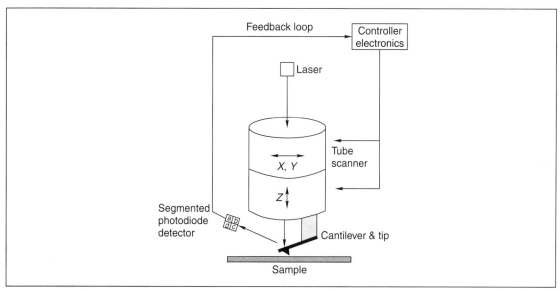

Fig. 2.18: *Schematic representation of an atomic force microscope. The sample surface is scanned by the cantilever, connected to a tubular scanner. The principal functional units in it are three piezoelectric scanners. The deflections of the cantilever are monitored by the segmented photodiode detector.*

Resolution in scanning probe microscopy cannot be defined in the same way as for optical methods, wherein the diffraction limit determines the resolution that is practically achievable. SPM is a three-dimensional imaging technique and the resolution is affected by the tip geometry. As would be seen in Fig. 2.19, improved resolution can be obtained for sharper tips. In practical description of resolution, especially in the biological context, the width of DNA measured is considered as a measure of resolution.

DNA in its β form is known to have a diameter of 2 nm. Width alone is not enough to describe the resolution as SPM is a three-dimensional technique and height is important.

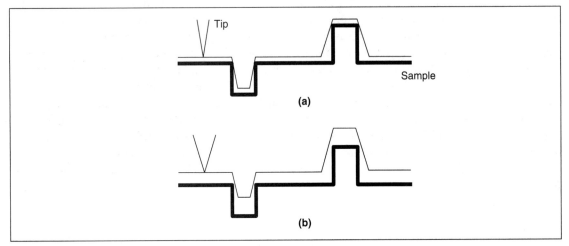

Fig. 2.19: *Resolution in SPM depends on the tip details; (a) gives better resolution in comparison to (b).*

AFM is commonly operated in two modes, the contact mode and the non-contact or tapping or intermittent contact mode. In the contact mode, the tip comes into contact with the surface. The force between the sample and the tip is the product of the displacement of the tip and the force constant of the cantilever ($f = -kx$). The contact with the surface allows an evaluation of the surface friction. When the interaction is strong, the surface damage can be significant, which makes the contact mode difficult to use for soft materials.

In the non-contact mode of operation, the tip is oscillated at its resonant frequency by an actuator. The decrease in the amplitude of the motion when the cantilever comes close to the sample is used to measure the tip-sample interaction. The drop in the amplitude is set to a pre-determined value. The intermittent contact that the tip makes is gentle and does not damage the material, though the probes are generally harder. Since it is a gentle mode of scanning, the non-contact mode is the most often used, especially in the case of materials with surfaces delicate such as a polished silicon wafer.

Typical AFM images have a resolution of the order of 5 nm. Atomic features have been observed, but this is not a routine development. In the case of specially fabricated tips, a resolution of 1 nm can be observed. True atomic features have been demonstrated in specific cases. The best known examples of nanoscale structures are DNA strands. Images of DNA spread on mica are shown in Fig. 2.20 (Ref. 10). These images show variations in the shape and width of the curved structures depending on the type of imaging. The width of the molecule seen in AFM images need not be of the actual width due to several factors. One of the factors is that it corresponds to the relaxation of the molecule on the substrate on which it is held for imaging. The other factors has to do with the tip-induced deformation in the sample. On the contrary, the contour length of the macromolecule is a measure of the molecular weight of the material.

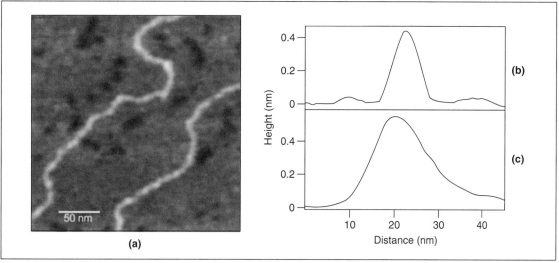

Fig. 2.20: *(a) Height image obtained with a SWNT tip of double-stranded DNA adsorbed on mica. (b) Typical height cross section from the image in (a). The FWHM is 5.6 nm. (c) Typical height cross section from an image of the lambda-DNA obtained with a conventional Si tip. The FWHM is 14.4 nm. Reused with permission from Wong et al. (Ref. 10). Copyright 1998, American Institute of Physics.*

Mechanical properties are measured by using AFM. It is important to correlate these properties to the chemical composition and structure in order to facilitate a complete understanding of the material. This can be done by combining spectroscopy with imaging. Although a few such tools such as confocal Raman microscopy and infrared microscopy are available, the spatial resolution is of the order of microns or hundreds of nanometers. A combination of AFM with spectroscopy will be immensely useful. Scanning near-field optical microscopy with Raman microscopy could prove to be useful in this regard, but the current resolution of this is only of the order of 50 nm.

2.3.4 Scanning Probe Lithography (SPL)

Manipulating objects and the tools associated with such manipulation are the central aspects of the development of civilizations. The names of various civilizations are coined on the basis of the tools used, such as the Stone Age, Iron Age, etc. In each of these ages, different kinds of tools were used to manipulate objects, which were mostly of large or macroscopic dimension. In the nano era, atoms are manipulated.

SPL refers to the use of SPM-based techniques for modifying substrates by the application of various actions such as scratching, writing, chemistry, photo-irradiation, etc. in a spatially confined manner. The structures that one can make as a result of these manipulations are of the order of 10–100 nm. These techniques are summarized in Table 2.4 (Ref. 11).

There are basically three kinds of probes with which SPL can be done. These are STM, AFM and SNOM, each of which is discussed separately elsewhere in this book. The motivation to use these

Table 2.4: Scanning Probe Lithographic Techniques. Adapted from S. Hong, et al., 2005 (Ref. 11).

	SPL Method	Instruments	Environment	Key Mechanism	Typical Resolution	Patterning Materials	Possible Applications
Nanoscale Pen Writing	Dip-Pen Nano-lithography	AFM	Ambient	Thermal Diffusion of Soft Solid	~10	SAM, Biomolecules Sol-Gel, Metal, etc.	Biochip, Nanodevice, Mask Repair, etc.
	Nanoscale Printing of Liquid Ink	NSOM	Ambient	Liquid Flow	~100 nm	Etching Solution, Liquid	Mask Repair, etc.
Nanoscale Scratching	Nanoscale Indentation	AFM	Ambient	Mechanical Force	~10 nm	Solid	Mask Repair, etc.
	Nanografting	AFM	Liquid Cell	Mechanical Force	~10 nm	SAM	Biochip, etc.
	Nanoscale Melting	AFM	Ambient	Mechanical Force and Heat	~10 nm	Low Melting Point Materials	Memory, etc.
Nanoscale Manipulation	Atomic and Molecular Manipulation	STM	Ultrahigh Vacuum (Often Low Temperature)	van der Waals or Electrostatic Forces	~0.1 nm	Metals, Organic Molecules, etc.	Molecular Electronics, etc.
	Manipulation of Nano-structures	AFM	Ambient	van der Waals or Electrostatic Forces	~10 nm	Nanostructure, Biomolecules, etc.	Mask Repair, Nanodevices, etc.
	Nanoscale Tweezers	Possibly AFM	Ambient	van der Waals or Mechanical Forces	~100 nm	Nanostructures	Electrical Measurement, etc.

Contd...

Investigating and Manipulating Materials in the Nanoscale 53

Table 2.4 Contd...

	SPL Method	Instruments	Environment	Key Mechanism	Typical Resolution	Patterning Materials	Possible Applications
Nanoscale Chemistry	Nanoscale Oxidation	STM or AFM	Humid Air	Electrochemical Reaction in a Water Meniscus	~10 nm	Si, Ti, etc.	Nanodevices, etc.
	Nanoscale Desorption of SAM	STM or AFM	Humid Air	Electrochemical Reaction in a Water Meniscus	~10 nm	SAM	Nanodevices, etc.
	Nanoscale Chemical Vapour Deposition	STM	Ultrahigh Vacuum with Precursor Gas	Nanoscale Chemical Vapor Deposition	~10 nm	Fe, W, etc.	Magnetic Array, etc.
Nanoscale Light Exposure	Nanoscale Light Exposure	NSOM	Ambient	Photoreaction	~100 nm	Photosensitive Materials	Nanodevices, etc.

lithographic techniques is to overcome the limit of current technology. The semiconductor industry is dependent on ultraviolet lithography, which uses ultraviolet rays to pattern surfaces, which are pre-coated with a photoresist. The chemical reactions on the resist will eventually strengthen or weaken the molecular bonding in the resist and a pattern can be made by subjecting the modified resist-coated material to a solvent wash. The pattern thus created can be used in an etching process. The process can be repeated and a complicated structure can be created on the surface. This process can be undertaken on large sizes of device structures and a huge number of devices can be made on small areas, which is the basis of semiconductor technology. The smallest structures that can be created today are of the order of ~90 nm, using ultraviolet light of ~190 nm wavelength. The lithography technique meets its natural limits as the size approaches the resolution limit of the optical techniques. This is given by the Rayleigh equation, resolution = $k\lambda/NA$ where k is a constant and NA is the numerical aperture of the lens system ($n\sin\theta$, where θ is the angle of incidence and n is the refractive index of the medium). The resolution is normally taken to be approximately $\lambda/2$. As a result, reduction in the size of structures possible using X-ray lithography, e-beam lithography and stamping-based methods is being investigated.

Although SPL methods are important tools, they are serial techniques as the probe makes the transformation in steps. In this they have a distinct disadvantage in comparison to the traditional methods. However, it is important to note that the transformations carried out by lithographic methods generally involve higher temperatures of the order of 100°C and at such temperatures, biological materials lose activity. SPL does not require higher temperature and the methods used are generally delicate. This allows the use of such techniques for manipulating biological or soft materials. With the bio-nano interface growing significantly, SPL-based lithography is bound to find newer applications.

2.4 Optical Microscopies for Nanoscience and Technology

The minimum distance that an optical microscope can resolve is: $\Delta x = 0.61\, \lambda/n\sin\theta$, where λ is the wavelength of the light in vacuum, θ is the collection angle and n is the index of diffraction. One can improve the resolution by decreasing the wavelength of illumination, by decreasing n or by increasing θ. However, irrespective of the various improvements, the fundamental limits imposed by the methodology cannot be overcome. When it comes to particle beams such as electrons, the image resolution can be increased as electrons at high energies have very short wavelengths.

Light, especially visible light, enjoys a lot of advantages in the investigation of matter in spite of its limitations. One advantage is that light at this energy does not modify matter as the energy involved is small. Light also results in excitations in matter which leads to phenomena such as fluorescence, that can be used for studying materials with chemical specificity. Light also leads to absorption and inelastic scattering, both of which can be used for imaging purposes. These are also molecule-specific.

Improved resolution in optical microscopies can broadly be brought about in two different ways. The first belongs to far field imaging and the second to near field imaging. In the former, the illumination occurs at a distance several microns away from the object to be imaged and in the latter, it occurs within distance of a few nanometers from the surface. While the former looks at the bulk features of the sample, the latter looks at the surface features. The diffraction limit is valid in the former and therefore the resolution is limited. In the latter, a resolution of the order of 20 nm has been demonstrated. Obviously, one may then come to the conclusion that nanoscale objects cannot be imaged by far field techniques. But this is not true. The way in which imaging nanoscale objects is done is through confocal microscopy.

2.4.1 Confocal Microscopy

Confocal means 'having the same focus'. In such a microscope, a point-like light source, generally a laser, is used. This point source is derived by passing the light through a pinhole, which can be conveniently achieved by using a fiber-optic connector. This is directed to the specimen through a beam splitter and an objective, which illuminates a spot. The point of illumination can be moved across the sample by a scanner and there are several ways by which this can be done. The emitted light from the sample, generally fluorescence (or less common scattered light, as in the case of Raman spectroscopy), passes through the detecting pinhole and forms a point-like image on the detector. The light is scanned across the sample to obtain a two-dimensional image or the depth from which the light is collected is varied by moving either the objective or the sample thus giving a three-dimensional image of the sample. All these three points, namely the illumination pinhole, sample spot and the detector pinhole, are optically conjugated together, thus giving the confocal microscope (Fig. 2.21). The confocal microscope is therefore a confocal scanning optical microscope. The optical sectioning aspect is the most important advantage of confocal microscope. The sections can be as thin as the wavelength of light and its spatial resolution of the microscope is the best that can be achieved by using optical microscopy.

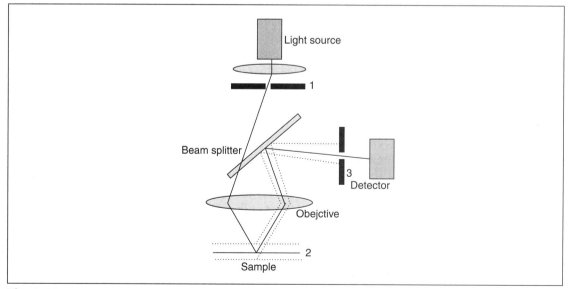

Fig. 2.21: *Schematic of a confocal microscope. 1 and 3 are confocal pinholes. Light emanating from another plane of the sample, indicated in dotted line, is not detected. 1, 2 and 3 are optically conjugated in the arrangement.*

Collecting the imaging involves scanning the light. The simplest approach would be to move the sample as the optical system is optimized. This process is slow as the piezoelectric scanners take time and real-time observation of processes is impossible. The other approach is to have a set of two mirrors to scan the laser in the xy plane, by first doing an x-scan and then making a y-shift, then an x-scan again and so on. The laser beam itself can be split into several smaller beams and all the beams may be simultaneously used for imaging. In this way, each beam needs to be moved only for a short distance for imaging the entire sample. This methodology uses a Nipkow disk, in which thousands of microlenses are mounted on a disk and the light is focused into thousands of pinholes created on another disk. All the beams are focused by the objective simultaneously and the light coming out from the sample is collected through the pinholes and microlenses and detected parallely. The holes can be arranged in a spiral fashion so that the entire space can be scanned by rotating the disk. Increasing the speed of rotation will increase the speed of imaging.

As can be seen in Fig. 2.21, the signal collected from the sample is confined to specific illumination volume by the use of an aperture. Thus the aperture sits at the same focal point of the objective, rejecting all light that comes from other regions. This facilitates localization of the illumination volume. Thus an object whose spatial dimension is smaller than the wavelength of light can be studied by localizing the illumination volume. In this illumination volume, one can look at the fluorescence of a molecule or a quantum dot. These can be part of a living cell or a polymeric composite. Thus direct localization of the illumination volume smaller than the resolution of light microscopy, is possible in the confocal technique.

The principal advantage of confocal microscopy is the image contrast, which is achieved by rejecting light that comes from other focal planes of the sample. The smaller the slit, larger is the rejection, but the overall signal quality decreases in this way.

The most important recent development in confocal is 4pi confocal microscopy. In this technique, two objectives are used to illuminate or collect light from the sample. An interference pattern is created by using the light from the objectives. This pattern has one major central peak and several side lobes. The central peak has a reduced axial spread, which is less than the peak that one would get from a pinhole. Depending on how the interference pattern is created, it is possible to get even an FWHM, which is one-sixth of the wavelength of illumination. This results in increased resolution images by detecting only the central peak, and avoiding the side lobes.

The most important application of confocal microscopy in nanoscience is in the investigation of the interactions of nanosystems with biological components. There are numerous examples of this kind wherein nanoparticles, nanoshells, nanotubes and such other objects are made to interact with cells, bacteria, viruses, etc. The interaction takes place within the cell in most cases and confocal microscopy is used to monitor the processes. For this purpose, a fluorescent tag is often attached to the nanosystem or the nanosystem itself is fluorescent as in the case of a semiconductor quantum dot. A typical example is shown in Fig. 2.22 (Plate 2) (Ref. 12).

2.4.2 Scanning Nearfield Optical Microscopy

The finite resolution of conventional optical microscopy shows that the limit of resolution is approximately $\lambda/2$ where λ is the wavelength of light used for illumination. In his three papers presented during the period 1928–32, E.H. Synge (Ref. 13) has shown that this limit can be overcome if the illumination volume is reduced to a dimension that is smaller than the wavelength of light. The fundamental aspect is that the nearfield light intensity decreases rapidly when the aperture dimension is small. This was realized experimentally in 1983 and 1984 by two independent groups soon after the discovery of STM. In this method, the light source dimension is reduced by using one of the several tools discussed below. The probe is a tip which interacts with the sample at a close distance ($d \ll \lambda$). The sample–light interaction is detected by the photodetector placed at a distance away from the sample. The distance between the sample and the tip is kept very small, by the use of a feedback mechanism which is typically based on the modulation of the quartz tuning fork. The sample or the tip is scanned so as to construct an image of the sample. This process is illustrated schematically in Fig. 2.23.

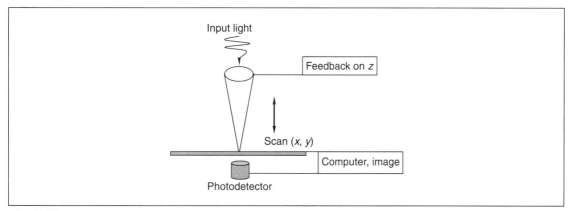

Fig. 2.23: *Schematic of a scanning nearfield optical microscope.*

There are numerous ways in which probes suitable for SNOM are made. One approach is to make probes from fibres, with one end tapered. The outer surface of the fibre is coated with a metal film and the very end of the fibre is exposed. This allows the light to pass through the tip. The standard approach to make a tapered fibre is the pulling technique in which a fibre is held in a stretched position and the middle of the stretched region is melted by a CO_2 laser or a heating filament. This allows both the ends to get stretched, which creates tapered ends. The other approach is to acid-etch the glass to produce a sharp tip. The fibre is then coated with a metal film such as aluminium or gold and the coating at the tip is removed. There are several approaches for making such an opening. The more common approach is the shadowed evaporation method. Here the metal is evaporated at a tilted direction in comparison to the probe axis. This facilitates evaporation of the metal all over, except at the tip. A uniform coating on all sides is achieved by rotation. This method makes it possible to batch process tips, but the uniformity of the tips is a problem, which depends upon the angle of evaporation, rotation, etc. The other methods used are electro-erosion and ion beam milling. Tips with physical holes are also possible. In this approach, tips are micro-machined in the same way as AFM tips. Holes are then made on tips by reactive ion etching or electron beam lithography. With this technique, uniform 50 nm diameter holes have been made on Si_3N_4 cantilevers.

There are several ways in which SNOM measurements are done. Illuminating the object at farfield and collecting light at nearfield, nearfield illumination and farfield collection, and illumination and collection at nearfield are the common approaches. The other approach is by illuminating the object with evanescent waves and collecting the light with a nearfield probe. In this approach, the sample is transparent and total internal reflection of the light occurs. The evanescent waves penetrate a few hundred nanometers above the interface and they propagate into the aperture when it is brought close to the surface. This is sometimes referred to as scanning tunneling optical microscopy (STOM) or photon scanning tunneling microscopy (PSTM). While all the above methods use tips with an aperture, apertureless tips are also employed which utilize the interaction between the tip and the sample to enhance or scatter the nearfield signal. This depends on the tip–sample interaction. The sample is illuminated at farfield and the collection is also done at farfield.

The most important application of SNOM is in the study of single molecules. Here the signal, usually fluorescence, of a single molecule is detected. The dynamics of the molecules, the local environment of the molecule, the polarization dependence of emission and thereby the orientation of the dipole moment of the molecule, photobleaching, etc. have been investigated.

The lateral resolution obtained by the technique is given approximately by the diameter of the aperture. There are practical limitations to the aperture diameter. Penetration of light to the metal coating of the tip and therefore the thickness of the coating determines the illumination volume. The amount of light that can be transmitted at farfield is reduced drastically at very small aperture sizes. Increasing the light output by the sample using the field enhancement of fluorescence, for example, is a way in which one can increase the signal to noise. Nanoparticles-attached probes can be used to enhance local fluorescence intensity. Among other approaches, fluorescence resonance energy transfer (FRET) is a promising method in which the excitation energy of a probe is non-radiatively transferred to an acceptor at a very short distance. The energy transfer efficiency decreases rapidly with distance and also requires an overlap of the spectra of the molecules concerned. If one of the molecules involved in FRET is attached to the probe and the other is in the sample, probing molecular distances becomes feasible. Currently, lateral resolution of the order of 15 nm has been reported.

Light can be used for vibrational excitations as well. Of all the techniques, Raman microscopy has been used well with SNOM. However, the main problem in this is the poor Raman intensity that one gets for normal molecules and materials. The enhancement of Raman by attaching molecules on nanoparticles of noble metals or rough surfaces, called the surface enhanced Raman effect, has been used for the detection of Raman signals from single molecules.

A SNOM image of human cells (SiHa) is shown in Fig. 2.24 (Plate 3) (Ref. 14). This image was taken in the author's laboratory after the cells were incubated with Au^{3+} ions for 96 hours. The cells were removed and were embedded in a polymer matrix. Thin slices of 70 nm thickness of the processed material were taken with an ultramicrotome. The image shows the growth of gold nanoparticles in the cytoplasm of the cells. There are small nanoparticles in the nucleus of the cells too (as shown by TEM), but these are not revealed here as the resolution is poor.

2.5 Other Kinds of Microscopies

2.5.1 Secondary Ion Mass Spectrometry (SIMS)

Ion-based tools are important in nanotechnology. Two of the common tools are the focused ion beam-based lithography and secondary ion mass spectrometry (SIMS). Ion lithography can make structures in the sub 10 nm regime today. In the spatially resolved mode, SIMS can help obtain chemically specific, isotopically resolved elemental information, at a spatial resolution of 50 nm, which is impossible in the case of any other technique. As ion ejection is extremely surface-sensitive and surface damage can be reduced, nanometer thin films grown on surfaces can be analyzed with isotope specificity. This analysis can be done at trace levels and also quantitatively. These attributes make SIMS imaging extremely important.

In SIMS, one is concerned with the mass spectrometry of ionized particles resulting from the impact of primary particles on a surface. The surface can be liquid or solid and the primary particles may be electrons, ions, neutrals or photons. The secondary particles emitted from a surface are electrons, neutral or ionic atoms or molecules, and neutral or ionic clusters. In SIMS, one is concerned only with the ions. The neutrals ejected are post-ionized and subjected to mass analysis in some cases (sputtered neutral mass spectrometry, SNMS). The process of decoupling emission and ionization allows for quantitative elemental analysis. Most of the current applications are in dynamic SNMS and can be done by using lasers (see below for an explanation of the term "dynamic").

Secondary particle emission is referred to as 'sputtering'. Most of the particles emitted will be neutrals but a few ions are also emitted. Mass analysis of these ions is carried out in one of several ways, such as time-of-flight, quadrupole and magnetic sector-based methods.

SIMS can be dated back to 1910 when Sir J.J. Thompson observed the emission of positive ions from the surface of a discharge tube upon its impact with primary ions. In the present-day instruments, the principal components are: ion gun which gives ions or atoms of 0.5–50 keV, ion optics to transfer the ions to a mass spectrometer, followed by the detector. The mass spectrometer can be of the magnetic, quadrupole or time-of-flight type. A schematic of the instrument is shown in Fig. 2.25.

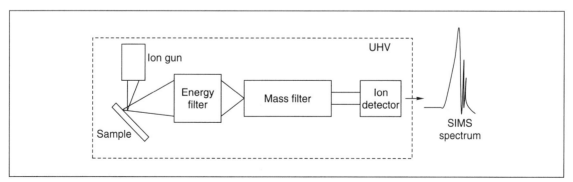

Fig. 2.25: *Schematic diagram of the main components of the SIMS experiment.*

In SIMS imaging, the mass spectral intensities collected from a surface are used to construct images. The spatial resolution of the method depends on the dimension of the sampled area. Thus it is necessary to have either a highly focused ion beam or a highly focused collection of ions. SIMS imaging is generally done in two ways. A focused ion beam is rastered across the sample and the spectrum at each point is collected. This method is called the 'microprobe mode' and its lateral resolution is of the order of sub-micron levels. The other mode is the microscope mode or the direct mode. Here the ions are collected from specific points on the sample simultaneously by electron lenses and are detected on a position-sensitive detector. The primary ion beam is not rastered.

The ion beams commonly used are Cs^+, O_2^+ or O^- or Ga^+. Although highly pointed beams of Ga^+ are possible, the secondary ion emission yield is small and therefore the first two are commonly used today. While Cs^+ enhances the negative ions, O_2^+ and O^- enhance the positive ions.

SIMS is operated in two modes. The first is the dynamic mode, wherein the primary ion flux is large and the rate at which the material is removed is high. This allows a fresh surface to get exposed to the primary beam continuously. In the case of static SIMS, only a small fraction (less than 0.1 per cent) of the surface is sampled. This results in minimal surface damage. High quality collection is necessary and therefore, the analyzer used in static SIMS is usually related to time-of-flight (TOF) based methods. As a result, static SIMS is also called TOF SIMS. In the dynamic SIMS, quadrupole and magnetic sector analyzers are used.

With the current innovations in instrumentation, imaging SIMS can acquire a lateral resolution of 50 nm. Its application areas include metallurgy, heterogeneous catalysis, biology, etc. Selective accumulation of ions of catalysts, incorporation of specific molecules in cell structures, and the separation of ions at grain boundaries, etc. are among the interesting problems that one can study. With sub-100 nm resolution, many of the cell organelles are accessible for high quality chemical imaging.

In the case of an ion or atom bombardment at a surface, the energy of the primary projectile is transferred through a billiard ball type process. This is followed by a collision cascade between the target atoms. Some of these collisions return to the surface, resulting in the emission of atoms and clusters, some of which are ionized in the course of ejection. The major concern in SIMS quantification is that the ionization coincides with sputtering. This means that calibration is a very important aspect in quantification. Particle bombardment changes the surface continually and the doses of bombardment have to be kept low to achieve meaningful surface analysis.

2.5.2 Other Mass Spectrometric Techniques of Interest to Nanoscience

Another method of carrying out SIMS is to use the fission fragments of ^{252}Cf to cause sputtering (plasma desorption mass spectrometry, PDMS). These fragments have MeV energies and when they collide at the rear of a thin foil coated with the analyte, molecular ions of high masses can get ejected. High-energy ions of $^{127}I^+$ from a tandem accelerator have also been used to achieve the same objective. High energy neutral atoms are also used for sputtering. This form of mass spectrometry, referred to as fast atom bombardment (FAB) mass spectrometry, is generally practised for organic analytes, with appropriate matrices. This results in soft ionization of the analyte species. 'Soft' implies ionization with lower internal energy for the ion formed, leading to lower fragmentation. Soft methods are especially important for the analyses of clusters as they are bound by lower energy and it is easy to fragment them during ejection to the gas phase. Today, the analysis of nanoparticles by mass spectrometry is becoming popular and the use of soft ionization methods is extremely important for achieving this goal. Laser desorption is another soft method used for desorption ionization. This is done by a technique called matrix-assisted laser desorption ionization mass spectrometry (MALDI MS), in which a low power ultraviolet laser (generally a 337 nm N_2 laser) is used to desorb the analyte. The analyte is mixed with a matrix so that the analyte concentration is small (typically in a 1:10 ratio). The mass analysis is generally done by the time-of-flight method as the laser is pulsed to help give a start time for data collection and the ions are collected at a high efficiency. An additional advantage of time-of-flight analysis is that there is no mass limit for the analysis in principle and the technique is well-suited for large biomolecules. The other soft ionization technique used is electrospray ionization mass spectrometry (ESI MS) wherein ions are produced from a solution, which is sprayed to form fine droplets in a high electric field. The solution sent through a fine needle is electrosprayed and the ions thus produced are generally multiply charged. This facilitates an analysis of large molecules in low mass range machines as their m/z value is small. Both these techniques are highly sensitive and femtomole analysis is possible with high resolution.

2.5.3 Focused Ion Beam (FIB)

The focused ion beam technique utilizes a liquid metal ion source and the interaction of high energy ion with the sample is used to investigate or modify the sample. Elastic and inelastic collisions take place when the ion beam strikes the sample. Elastic collisions lead to the sputtering of atoms and inelastic collisions yield secondary electrons and X-rays. Ions are also ejected. The ions and electrons ejected can be used for imaging the surface. The ion beam itself can be used for milling applications. In the presence of organometallic species, the ion beam can result in the deposition of materials and can be used for material repair applications. Ion beam milling can be used for sample preparation, especially for TEM.

Ions are particles with a higher momentum in comparison to the electrons used for a given acceleration energy, though they travel at a lower velocity. This leads to increased sputtering. The ion penetration depth is much smaller than that of electrons.

In SEM and FIB, the particles to be examined are scanned on the surface. The secondary particles (ions and electrons) are used for imaging in FIB. As the ions travel at much lower velocities, their Lorentz force is lower. (Lorentz force is the force exerted on a charged particle in an electromagnetic field. The particle will experience a force due to electric field qE, and due to the magnetic field, $qv \times B$. When combined, they give the Lorentz force: $F = q(E + v \times B)$, where E is the electric field, B is the magnetic field, q is the charge of the particle, v is its instantaneous velocity, and $\times$ is the cross-product.)

In addition to imaging, FIB is used for ion milling. Selective sputtering can be achieved by a method known as gas-assisted etching (GAE) in which one of the halogen gases is introduced to the work surface thus forming a volatile species. The decomposition of an already deposited organometallic molecule on the surface can lead to the deposition of certain atoms, such as Pt getting deposited from a Pt precursor molecule in the presence of the ion. The gas phase deposition process can be used to nucleate certain growth at the surface and the subsequent growth can be achieved by chemical vapor deposition (CVD). FIB can also be used for implantation.

In addition to single ion beam, an ion beam and an electron beam combination may be used, which will combine the capabilities of FIB and SEM. This can help achieve lithography, imaging and characterization together. Such an instrument has several applications in nanotechnology.

2.5.4 Photoelectron Spectroscopy

Photoelectron spectroscopy (PE) is concerned with a broad range of analytical techniques, all of which are based on Einstein's photoelectric equation, $h\nu = KE + \phi$, where $h\nu$ refers to the photon energy, ϕ refers to the work function and KE refers to the kinetic energy of the ejected electron. Depending on the type of photon used for photoemission, there are several kinds of photoelectron spectroscopic techniques. The principal PE techniques are X-ray photoelectron spectroscopy (XPS), which uses X-rays, and ultraviolet photoelectron spectroscopy (UPS) which uses soft X-rays and ultraviolet radiation for photoelectron emission. XPS is generally used for the analysis of solids and surfaces whereas UPS is also used for gaseous samples.

Photoelectron spectroscopy is related to a broader class of techniques referred to as electron spectroscopy. There are principally five analytical techniques in this branch, all of which deal with the kinetic energy of electrons ejected or reflected from materials. While photoelectron spectroscopy pertains to electron emission from materials, there are also electron spectroscopic techniques dealing with inelastically scattered electrons called electron energy loss spectroscopy (EELS). Kinetic energy analysis of Aüger electrons constitutes another electron spectroscopic technique, namely Aüger electron spectroscopy (AES). The last technique is called Bresstrahlung isochromat spectroscopy (BIS). While XPS, UPS and AES are used to investigate occupied energy states, BIS is used to study unoccupied states. EELS is used for the investigation of excitation between the occupied and unoccupied states and also for the study of molecular vibrations, especially when the molecule is adsorbed on a surface. Modern developments in XPS can be largely attributed to the work of Kai Siegbahn and his colleagues in Uppsala, Sweden. In recognition of his contributions he was awarded part of the Nobel prize in Physics for the year 1981. His father, Manne, has also been a Nobel laureate (1924).

There are several variations in each of the spectroscopic techniques mentioned. For example, in ultraviolet photoelectron spectroscopy depending upon the kind of photon source, different spectroscopic techniques such as HeI UPS, HeII UPS, laser-induced UPS, etc. exist. Each of these techniques can be applied for different kinds of samples. The instrumentation for these techniques can be different thus giving different types of information. Single colour spectroscopy can be replaced with multi-colour spectroscopy, analyzers of one kind can be replaced with others, energy analysis of the electron can be done near the threshold of ionization and wavelength can be varied over a window thus giving rise to a range of experimental techniques. Conventional photoelectron spectroscopy on stable gases or solids can be replaced with that of liquids, high temperature molecules, thin films, adsorbates, single crystals, etc., thereby offering wider sample coverage.

By studying the electron energy of a series of metals as a function of photon energy, Hertz proved the Einstein relation and determined the values of '*h*'. The early photoelectron spectrometers were made in the 1950s. In today's experiment as shown in Fig. 2.26, we have a photon source, which produces monochromatic radiation and is falling on a sample, which can be a solid, gas or liquid. The photoelectrons emitted are analyzed for their kinetic energies by an electron energy analyzer and are detected. The photoelectron spectrum is a plot of the intensity of the photoelectrons as a function of electron kinetic energy. The electron kinetic energy can be directly related to the ionization potential or the binding energy (in the case of a solid) by the expression, $E_{\text{kinetic energy}} = h\nu - \phi - E_\nu$, where ϕ is the work function as mentioned earlier, E_ν is the new quantity, which is the binding energy of the photoelectron. Binding energy refers to the energy of the level from which photoemission occurs, referenced to the Fermi energy, E_F, which is taken as zero in the case of solids. In a general discussion of photoelectron spectroscopy, the spectrometer and the sample are assumed to be in equilibrium so that their Fermi levels overlap. The quantity ϕ is the energy which is the difference between the Fermi energy and the vacuum level. The spectrometer may be used to study various aspects of photoemission such as the emission angle, spin of photoelectron, etc. In the simplest instrument, only the kinetic energy of the electron is investigated.

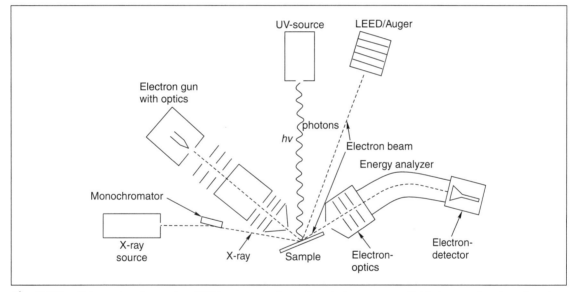

Fig. 2.26: *Schematic of a modern PES instrument. Several techniques are available in one chamber. A surface science apparatus of this kind has multiple techniques.*

In photoelectron spectroscopy, the electron kinetic energy that one normally encounters is of the order of 10–1000 electron volts. For electrons of this energy, the distance travelled before an inelastic encounter occurs is of the order of 20–30 Å. This distance corresponds to about ten atomic layers of the material, in most cases. Therefore, while analyzing kinetic energy of electrons, we observe that the electrons of maximum intensity which do not suffer energy loss due to inelastic collisions, are principally coming from the very top of the surface and therefore, the technique is extremely surface-sensitive. Thus photoelectron spectroscopy is inherently sensitive to nanometer thin samples. However, the irradiation area is generally much larger, of the order of several mm² in typical measurements. This can be reduced to facilitate microscopy. The distance just mentioned (the distance in which inelastic collisions do not occur) is called the inelastic mean free path and this quantity is very difficult to measure. In experimental situations, what one normally measures is the attenuation length wherein both elastic and inelastic contributions are included. This has been determined for a large number of materials and the universal attenuation length is shown in Fig. 2.7.

For the KEs normally encountered in photoelectron spectroscopy, the attenuation length is within 20 Å for most materials. Therefore, it can be seen that if a given surface is contaminated to a few tens of Å, it is difficult to obtain a photoemission signal from the underlying material. It may be noted that in poor vacuum conditions, such as 10^{-6} torr, several tens of overlayers of contaminants can be produced in just a few minutes of exposure of a clean surface into the system. Thus, in order to be surface-sensitive, it is necessary for the surface to be sufficiently clean and devoid of overlayers. Because of its extreme surface sensitivity, photoemission spectroscopy is referred to as a surface spectroscopy. Therefore, if one is interested in the surfaces of materials, photoelectron spectroscopy is done in ultra high vacuum (vacuum better than 10^{-9} torr) where it takes about a few tens of minutes to form a complete overlayer on the surface.

Photoelectron spectroscopy pertains to electronic structure. The ionization potential that one measures corresponds to the difference between the initial ground state from which photoelectron emission occurs and the final state produced upon photoemission. To a first approximation, one can assume that the only difference between the two is that there is one electron less in the latter. This gives the most commonly used approximation, called the Koopmans' approximation, stating that $IP(k) = -\varepsilon(k)$, where $\varepsilon(k)$ is the orbital energy of the kth orbital. This assumption disregards the orbital relaxation effects, correlation effects and relativity effects. Thus the observed ionization potential can be calculated if the other effects are also included in the calculation such that, $IP(k) = -\varepsilon(k) - \delta\varepsilon_{relaxation} + \delta\varepsilon_{correlation} + \delta\varepsilon_{relativistic}$.

Depending upon the kind of photon used for photoemission, the levels accessed are different. Generally, the photons used are He I, He II, Al K_α, Mg K_α, Na K_α, Si K_α, etc. The He I (21.22 eV) photon corresponds to the de-excitation of the $^1P_0$ to $^1S_0$ level of He. The same is the case with He II, which corresponds to the corresponding transition of He$^+$ (40.8 eV). The Al $K_{\alpha 1,2}$ corresponds to the de-excitation of 2p to 1s (1486.6 eV). The same is the case with Mg $K_{\alpha 1,2}$ (1253.6 eV). These X-ray sources do not penetrate matter too deep and are called soft X-rays. (Note: The $K_{\alpha 1,2}$ symbolism is explained in the section on X-ray diffraction.)

Most of the valence electrons are involved in bonding and it is possible to access these valence levels by using ultraviolet photons. The core levels, which are not participating in bonding, occur in the range of a few tens to 100 electron volts and can be accessed by soft X-rays. The deeper core levels of heavy elements can be accessed by hard X-rays. In photoelectron spectroscopy, one is generally concerned with energy levels of the order of 1000 eV and below.

The photoelectron spectrum is characterized by specific ionization energy and also by a width. The width has contributions from the inherent width of the state formed, the line width of the photon source used and the specific resolution of the analyzer. If all these are favourable, it is possible to measure the fine structures such as vibrational and rotational features accompanying photoionization in the case of molecules. In the case of materials and surfaces, vibrational features and other fine structures such as multiplet splitting and plasmon losses can also be measured.

Let us look at the spectrum of gaseous nitrogen as an example (Fig. 2.27). The spectrum corresponds to three different electronic states of ionized nitrogen. The potential energy curves of these ionic states are shown here with vibrational levels in them. Each band shows vibrational excitations. In fact the photoelectron spectrum is a reflection of the potential energy curves of the system. Using 21.22 eV photon, it is possible to access only these three levels. If we used sufficiently energetic photons, it would have been possible to look at other deeper levels. The photoelectron spectrum also gives an idea about the kind of energy levels involved in the ionization process. These levels actually correspond to orbitals, which participate in bonding. Ionization from these levels can give us information about the kind of bonding possible in the system. In this particular case, the first excitation corresponds to the removal of electron from a non-bonding orbital. The second corresponds to the removal of an electron from a bonding orbital. The third corresponds to the removal of an electron from a weakly anti-bonding orbital. Now it is quite clear why vibrational spacing of the first ionic state is quite similar to that of the ground state. There is no change in the bonding characteristics when an electron is removed from this level. For the second one, the vibrational spacing is smaller than that of the ground state. For the third one, the vibrational spacing is larger than that of the ground state. If an electron is removed from a non-bonding orbital, obviously the bonding characteristics

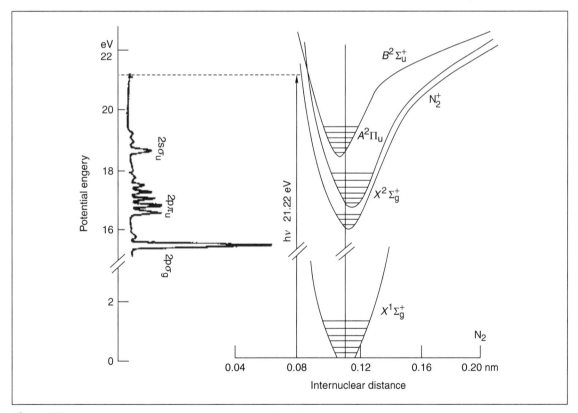

Fig. 2.27: *Photoelectron spectrum of nitrogen molecule showing that it is a reflection of the potential energy curves.*

of the molecule are not disturbed. Therefore, the internuclear distance remains unchanged and vibrational separation is largely the same. If one removes an electron from a bonding orbital, the state that is created has less bonding character, the internuclear distance can increase and therefore, the vibrational quantum may be reduced. If one removes an electron from an anti-bonding orbital, the state that is created has more bonding character and therefore the vibrational quantum is larger. It is also reflected in the equilibrium internuclear distance, which is unaffected in the first state, is shifted to the right in the second case, and is shifted to the left in the third case. Thus the photoelectron spectrum gives us a qualitative idea of the kind of bonding. In addition, the photoelectron spectrum is a mirror image of the electronic states. It is possible to reconstruct the photoelectron spectrum if one knows the states involved. It is also possible to reconstruct the states if we know the photoelectron spectrum. However, it has to be remembered that photoelectron spectroscopy is concerned with electronic transitions and therefore, the Franck-Condon rule is valid here. Thus the spectrum probes only the region very close to the equilibrium internuclear distance. The potential energy surfaces span a larger area and photoelectron spectra give very little information on the regions away from the minimum distance. This is because the molecule is always close to the minimum distance and the excitation is around it.

The earlier discussion was about ultraviolet photoelectron spectroscopy. In XPS, one is concerned about the core electrons. The important thing is that the core electron binding energies of atoms are very specific to them. In fact, the core electron binding energy of an atom can distinguish what that atom is. Due to this reason, the elemental characterization of a material is possible by photoelectron spectroscopy. In Fig. 2.28, we show the 1s photoelectron peak of the first row atoms.

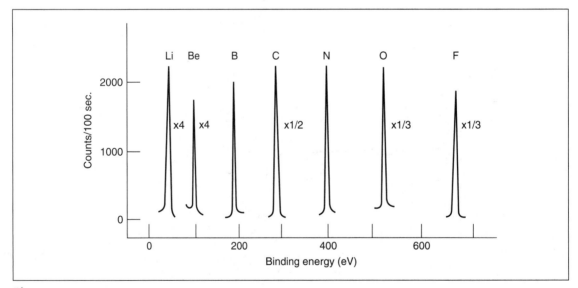

Fig. 2.28: *1s photoelectron feature of the first row atoms.*

From lithium to fluorine, it is seen that the 1s binding energy increases systematically. Besides, not only does the binding energy increase, but the intensity is also different. As can be seen, fluorine 1s is more intense than nitrogen 1s, and even if there is less amount of fluorine, it is possible to see it in the photoelectron spectrum. Not only does each level from a given atom have a specific energy but the level can also have different energies in different chemical situations. As in NMR spectroscopy, there are chemical shifts in X-ray photoelectron spectroscopy, which give us information about the kind of chemical environment. The photoelectron features of the core levels give quantitative information on the amount of that element present in the sample. Therefore, by using appropriate standards it is possible to obtain quantitative elemental information.

The photoelectrons emanating from the sample have the following three principal characteristics:

1. *The number distribution with kinetic energy*: A measurement of the number of photoelectrons with kinetic energy gives the energy distribution curve, which is normally referred to as the photoelectron spectrum. In normal photoelectron spectroscopy, the electron emission is performed at a fixed angle.
2. *The distribution of electron intensity as a function of the angle*: This measurement involves the study of intensity distribution of photoelectrons as a function of the electron take-off angle or photon propagation direction or with respect to both these parameters.
3. *The spin polarization or spin distribution of the photoelectron intensity*: This includes the preferential polarization of the electrons of the sample by an external magnetic field and studying the spin distribution of the ejected electron intensity. These measurements have not been done to a large extent, till recently.

In the photoelectron spectrum of solids, additional complications can result from the various inelastic processes that occur in the sample as the photoelectron leaves the material. This leads to a large background which may be overlapped with distinct peaks. These features are also informative and are often useful in providing a more detailed understanding of the electronic structure of the material under study.

X-ray photoelectron spectrum is normally measured with non-monochromatic X-ray sources; the principal limitation is the natural line width of the $K_{\alpha 1,2}$ radiations. The width of this radiation is about 0.7 eV for Mg $K_{\alpha 1,2}$ and about 0.8 eV for Al $K_{\alpha 1,2}$. If one goes up in the periodic table, the Na $K_{\alpha 1,2}$ lines has a width of about 0.4 eV. It is also possible to use neon, which has a natural line width of 0.2–0.3 eV and the K_α line occurs at 848.6 eV. The principal reason for the reduction in line width is the reduction in the $2p_{3/2}$–$2p_{1/2}$ spin orbit splitting and the increase in the lifetime of the 1s hole. However, neon K_α radiation is not generally used in X-ray photoelectron spectroscopy. In special cases, it is possible to use the $K_{\alpha 1,2}$ radiation of fluorine in highly ionic compounds and certain studies have been performed with this radiation.

It is possible to monochromatize the Al $K\alpha$ or Mg $K\alpha$ radiations, thereby reducing the half width considerably. It is possible to achieve a photoelectron peak as narrow as 0.3 or 0.4 eV by using monochromatized Al $K\alpha$ or Mg $K\alpha$ radiation. The consequent reduction in intensity as a result of monochromatization is overcome by the use of rotating anode X-ray sources as well as by special detection devices, which enhance sensitivity. In addition to the $K\alpha$ radiation, it is also possible to use Mξ radiation due to the $4p_{3/2}$ to $3d_{5/2}$ de-excitation process of elements from yttrium to molybdenum. These X-rays have energy in the range of 100–200 eV. Although these lines have substantially higher half widths of the order of 0.8 eV, the energy range is unique and therefore many photoelectron studies have been done with these. For example, Mξ radiation of yttrium corresponds to 132.3 eV (FWHM 0.5 eV), and zirconium to 151.4 eV (FWHM 0.8 eV). One of the problems is that several aspects have to be kept in mind in these anode designs. One is that the X-rays are soft and they have less ability to penetrate through matter. Therefore, thin polymeric windows have to be used. It is also necessary to use high excitation energies so that X-ray emission takes place from the interior of the material where the element has not undergone chemical changes. In contrast, if the emission takes place from the exterior of the material, there is a possibility of emission from oxide where the valence levels are broad and the natural line width of the radiation increases. It is also necessary to use continuous deposition of fresh anode material during the operation of the apparatus. Sometimes, there are a number of threshold effects when the photon energy is very close to the ionization energy. For these experiments, it is necessary to either utilize the Bremsstrahlung radiation of an X-ray spectrum or to go to synchrotron radiation wherein the wavelength can be tuned. The huge intensity of a synchrotron radiation, the polarization and the pulsed character of the light are all utilized for many of the measurements.

Ultraviolet photoelectron spectroscopy

Ultraviolet photons from gas discharge sources such as HeI (21.22 eV) and HeII (40.82 eV) have high intensity and an inherently small line width. It is also possible to carry out photoelectron spectroscopy with higher harmonics of several lasers. This facilitates spectroscopy with higher resolution, which is also achieved with other kinds of analyzers such as time-of-flight and magnetic bottle, which have a higher resolution. In general, the resolution that one can achieve in UPS is limited only by the analyzer. The best instrumental resolution, under 2 meV, has been reported with the common hemispherical analyzer (which is also used for XPS). UPS is an important tool used for understanding the changes in the electronic structure of materials when adsorption occurs. Changes in the work function and chemical properties of the material are captured very well by UPS.

The spectra from solids is normally referenced with respect to the Fermi energy. A binding energy of zero refers to the Fermi level. This kind of a reference scheme works well as the sample and the spectrometer are in electrical equilibrium when the measurement is done, that is to say that the sample is electrically connected to the spectrometer so that their Fermi surfaces overlap. It is important to note that this kind of a description is valid only in the case of a metallic sample connected to a metallic spectrometer. The energy level diagram of the sample and the spectrometer are shown in Fig. 2.29. The electron kinetic energy in the spectrometer is modified as the electron comes into the spectrometer. In the case shown in the figure, a deceleration is observed. This is because the work function of the spectrometer is higher than that of the sample. In general, the Kinetic energy in the spectrometer, $KE = h\nu - E_B(k) - \varphi_{spectrometer}$ where $h\nu$ is the energy of the excitation source, KE and $E_B(k)$ are the kinetic energy of the electron and the Fermi referenced binding energy of the kth level of the sample and $\varphi_{spectrometer}$ is the spectrometer work function. In general, the electron suffers a deceleration amounting to $\varphi_{spectrometer} - \varphi_{sample}$.

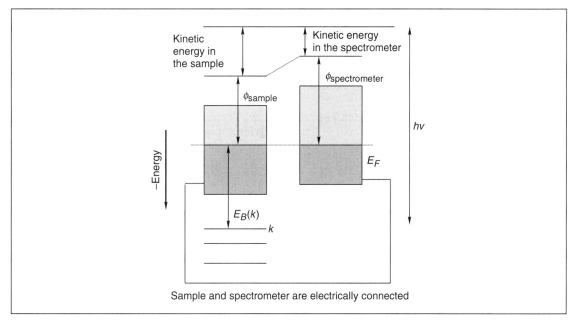

Fig. 2.29: *Schematic illustration of the sample and spectrometer energy levels.*

Aüger electron spectroscopy

The other kind of electron spectroscopy one uses is the Aüger electron spectroscopy. This refers to the energy analysis of the Aüger electron, thus named as the phenomenon was originally described by Aüger in 1925. The Aüger electron arises as a consequence of photoemission. When the core hole is produced, it is filled with a higher lying electron and the atom is now in an excited state. The excess energy is dissipated by the emission of another electron from a higher lying level. The electron kinetic energy is independent of the excitation energy of the photon used in creating the core hole, which is an important distinction from photoelectron spectroscopy where the kinetic energy of the electron increases with photon energy.

The Aüger electron ejected is labeled as KL_1L_2 in the case shown in Fig. 2.30. There can be several other Aüger electrons coming from the sample which can be labeled appropriately. All these Aüger electrons have characteristic energies, which can be used for identifying the element present in the sample or its chemical environment. The Aüger electron can be produced by other excitations also such as the electron and the ion. The electron excitation is convenient as the electron beam can be brought to a narrow size and the illumination of a microscopic area becomes possible. This allows for scanning Aüger microscopy, which analyzes the Aüger intensity of a peak as the electron beam is scanned to produce an image of the sample. The Aüger intensity decreases and X-ray fluorescence takes over at larger separations between the levels, and therefore Aüger is useful only for elements below $Z = 30$.

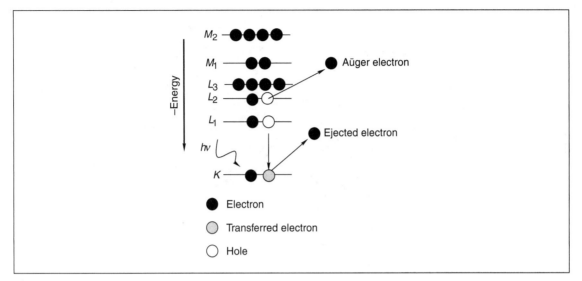

Fig. 2.30: *Schematic of the Aüger emission process.*

Surface enhancement

There are several ways to enhance the surface features in photoelectron spectroscopy. One of them is by varying the electron take-off angle (the angle between the sample plane and the electron direction). As the electron take-off angle varies, the depth of material sampled also varies as the electron attenuation length in a sample is a constant. In order to see truly surface signals, the spectrum has to be acquired at a very low take-off angle, of the order of a few degrees. The intensity enhancement of surface features is observable with reduced photon incident angles also. However, in order to observe the features, the angles have to be close to zero. Experimental limitations come into the picture when studies are done at low take-off angles.

Improved surface sensitivity can be achieved with lower photon energy as well. As photon energy is reduced, the kinetic energy of the electron is reduced, thereby reducing the electron attenuation length. Apart from the variation in the electron take-off angle, a change in photon energy also causes a variation in the ionization cross sections. The enhancement of these cross sections makes it possible to observe the chemical signatures accompanying the surface processes.

Complications

The photoelectron spectrum is complicated with several effects which accompany photoemission. They can be grossly categorized as multiplet splitting, multielectron excitation and other many body effects. Multiplet effects occur upon electron emission from a filled level of a material containing unpaired spins. There can be more than two distinct signatures in the photoelectron spectrum of one level due to the coupling of spin and orbital angular momenta. These will give several multiplet features which may not be properly observable due to reduced intensities or reduced separation. In addition to multiplet splitting, photoelectron spectrum can be complicated due to multielectron effects such as shake-up, shake-down and shake-off. In these phenomena, photoelectron ejection accompanies electron excitation involving other orbitals. In most such cases, electronic excitation occurs by reducing the electron energy (shake-up). De-excitation processes can also take place, thus increasing the electron energy (shake-down). Excitation can also lead to complete electron removal (shake-off). Apart from these discrete excitations, there can also be many electron effects accompanying photoemission. In addition, plasmons in metals can also be excited with photoelectrons. These excitations may occur during the creation of photoelectron or upon the travel of photoelectron through the material. These are called intrinsic and extrinsic excitations, respectively. It is difficult to classify whether an event is intrinsic or extrinsic through an experiment. Several other processes of importance accompany photoemission.

Photoelectron ejection can also accompany ligand to metal charge transfer transitions. This is observable in a number of cases, with the most well-known being the ligand to metal transitions in Cu^{2+} systems.

The photoelectron spectrum can also lead to vibrational excitations due to low energy of the vibrational transitions. These features are not clearly observed in the case of many materials. However, when the vibrational energies are large (molecules of lower atomic number), vibrational excitations accompanying photoelectron ejection can be resolved.

Labeling in photoelectron spectroscopy

Photoelectron spectra are labeled as per the electronic state produced by photoionzation. In the case of molecules, the molecular electronic state formed is used to assign the peak as seen in the spectrum of nitrogen. In the case of complex molecules, the orbital from which photoionization occurs is used to label the peaks (using Koopmans' approximation, which states that the ionization potential is the negative of the orbital energy from which ionization occurs). In the case of XPS, note that each electronic state with $l > 0$ with one unpaired electron is split into two doublet states with $j = l \pm 1/2$. Two conventions are followed in naming the states. The first uses nL_j symbolism, where n is the principal quantum number and $L = s, p, d, f, ...$ corresponding to $l = 0, 1, 2, 3, ...$ This would make $2p_{1/2}$ and $2p_{3/2}$ states upon photoemission from Ne. In the alternate method, these are labeled as $L2$ and $L3$, corresponding to $l = 1$ and $j = 1/2$ and $3/2$. In this, the shell notations, K, L, M, N, ... are used for $n = 1, 2, 3, 4, ...$ and subscript 1, 2, 3, 4 ... for lowest (l, j) to highest (l, j). The former is more commonly used.

Photoelectron microscopy

Photoelectrons can also be used for spatial information. There are several ways in which this can be done. In the first approach, a narrow probe beam (electrons or photons) is rastered over the sample. All the photoelectrons from the irradiated zone are collected as a function on the probe position. In the second approach, the whole sample is illuminated, but electrons from a smaller area are collected by a suitable aperture. The location of the sample from which data are collected is moved and an image is constructed. The third approach involves the illumination of the whole area and collection of the image with a position sensitive detector. The two kinds of images that are possible are spectromicroscopy and microspectroscopy. In spectromicroscopy, electrons of a specific energy range are collected. This implies energy resolved imaging used for compositional mapping. Microspectroscopy is a collection of spectral data as a function of the spatial location of the sample. The electronic structure of materials of smaller dimensions can be measured this way.

Focusing X-rays to narrow regions is difficult due to the difficulties involved in X-ray optics. It is possible to achieve photoelectron spectroscopy with 50 μm spatial resolution in this way. Imaging detectors can achieve a resolution of the order of 10 μm. However, with electrons one can get spatial resolution of the order of tens of nanometers. Scanning Aüger Microscopy (SAM) is carried out in this way. If sputtering is accompanied by SAM, three-dimensional imaging of the material composition is possible.

Another way of carrying out microscopy is through photoemission electron microscopy (PEEM). In normal photoelectron spectroscopy, the photon energy is much higher than the work function of the material. However, in the case of localized surface changes such as adsorption, only the local work function, near the area of adsorption is changed. If the sample is illuminated with a higher energy photon and the data are collected over a large area, only average effects will be seen. If, on the contrary, a low energy photon just enough to cause photoemission from the clean surface along with a higher resolution instrument is used, the region of higher work function such as an oxygen-adsorbed region will appear with a different contrast. Currently, this technique is being used to obtain sub-micron resolution in favourable cases.

2.5.5 Vibrational Spectroscopies

Vibrational spectra are characteristic of the material and are specific to the chemical bonds. Changes in the chemical characteristics of matter are reflected in the vibrational spectra. Vibrational energies are much smaller as compared to the chemical bond energies and even minute changes in the local atmosphere of a sample are reflected in the spectra. Spectroscopic information of this kind is commonly derived from infrared (IR) spectroscopy, Raman spectroscopy and electron energy loss spectroscopy (EELS). However, there are a few other less common techniques such as inelastic electron and neutron tunneling, helium scattering and sum frequency spectroscopy which may be used to obtain information on vibrations. All these are more involved techniques in terms of both money and effort. The first three techniques are more common and are used in various ways to analyse nanomaterials.

Vibrational spectroscopy is used to derive information on the vibrational excitation of molecules. Most of the time, the excitation is limited to the fundamental vibrational frequency. As the molecule is normally at the lowest vibrational level, namely $v = 0$, the harmonic oscillator approximation is a sufficient representation of the vibrational transitions. The transition frequency measured corresponds to the fundamental vibrational frequency, $v = 1/2\pi \sqrt{(k/\mu)}$, where k is the force constant of the bond and μ is the reduced mass of the system. There are several fundamental vibrational modes in the system depending on the symmetry. Several of these will be observed in the infrared and Raman spectroscopies. Altogether there are 3N-6 vibrational modes in a N atom containing non-linear molecule and 3N-5 modes in a linear molecule. The 6 and 5 correspond to the other degrees of freedom (translational + rotational) present in the molecule. The number of modes observed in Raman and IR depend on the symmetry of the system.

IR spectroscopy is possible only if the vibration changes the dipole moment of the molecule. The probability of a transition is proportional to the square of the transition dipole moment. $M_{vv'} = \int_{-\infty}^{+\infty} \psi(v) \mu \, \psi(v') d\tau$, where $\psi(v)$ and $\psi(v')$ are vibrational wavefunctions of states v and v' and μ is the dipole moment of the bond undergoing vibration. The value of the integral makes a vibration active or inactive in the IR. The symmetries of the molecule and the vibration decide this. The vibrational band intensity is a function of the intensity of the electric field of the radiation and its orientation with respect to the transition dipole. A study of the variation of infrared features as a function of the polarization of the light can be used to study the orientation of the dipole in a condensed system.

Electron energy loss spectroscopy is based on the inelastic collisions of a monochromatic beam of electrons and the study of the kinetic energy of the electrons. The energy loss of the sample corresponds to excitations in the sample. Electronic, vibrational and rotational excitations of the sample can be studied in EELS. Electronic excitations can be of the core levels or of the valence levels. While the former is referred to as inner shell EELS (ISEELS), the latter is referred to as EELS. The spectroscopy used in the study of vibrations is called high resolution electron energy loss spectroscopy (HREELS) and is generally used for the study of adsorbates. Due to experimental problems, it is not possible to observe all the excitations in one kind of spectrometer. Rotational spectra are not studied with EELS as electron energy analysis is not possible at the resolution required for rotational spectroscopy in most cases. The principal aspect of the spectrometer which limits resolution, is the electron energy analysis and the analyzer limits the instrumental performance. The analyzer performance is given in terms of the full width at half maximum (FWHM) which is generally of the order of a few meV in HREELS. A typical vibrational frequency occurs at 100 meV (806.5 cm^{-1}).

EELS involves three kinds of excitation mechanisms. The first is dipole excitation and the interaction between an electron and molecule is similar to light and matter. This leads to transitions similar to that observed in optical spectroscopy. This leads to specular scattering. The other kind of transition is impact-driven in which the electron behaves like a particle. The angular distribution is complex and non-allowed transitions can be excited in this process. The next kind is resonance excitation wherein the electron undergoes exchange with the sample. This results in isotropic scattering. Here again, the transitions which are not allowed can be excited. Thus a proper application of EELS can provide additional information which is not possible in the case of optical spectroscopy.

The infrared spectrum is measured by illuminating the sample with a polychromatic infrared light and by measuring the absorption of the sample. Two kinds of methodologies are normally used, one by a dispersive infrared spectrometer and another by using the Fourier transformation of the interferogram resulting from the interference of two light beams with different path lengths. The latter technique called FT-IR is the most common analytical method used these days. The Raman spectrum is measured by analyzing the scattered light coming from the sample. The illuminating light is generally in the visible region of the electromagnetic spectrum, which allows for focusing to a few hundreds of nanometers. If the light is focused on a sample using a SNOM aperture, it is possible to illuminate a single nano object and the Raman spectral measurement of this object becomes possible. The other illumination source used for vibrational spectroscopy, namely electrons, can be focused on the sample at smaller areas, as in the case of transmission electron microscopy. However, this beam is inherently of large energy and concomitantly of poor resolution, and therefore vibrational spectroscopy is impossible. At the same time, though poor in resolution, the features of inner shell excitation are useful in identifying the elemental constituents and nature of binding of materials at the nanometer regime. At lower excitation energies of the order of a few eV, the electron beam has a very narrow energy width, of the order of 1 meV or less, which allows the use of vibrational spectroscopy. This is normally performed on adsorbate molecules which are present on the surface at monolayer coverages. Because of this fact and also due to the fact that electrons are used, HREELS is done in ultra high vacuum. Due to the extreme surface sensitivity of the low energy electrons, the technique looks at the top monolayer only. In HREELS, one can observe low energy vibrational excitations, such as those representing the adsorbate-surface interactions. These are necessarily low energy vibrations as their force constants are small and the mass of the surface atom is large. It is not possible to see these low energy motions in infrared as far infrared spectroscopy is rather difficult at surfaces. These may be possible to observe in Raman, but regular Raman becomes difficult due to a poor scattering cross section.

Imaging materials with vibrational spectroscopies provides complete information about the sample. This is done in IR and Raman spectroscopies, and the instruments are called IR and Raman microscopes, respectively. Just as in XPS, in vibrational spectroscopy too, there is microspectroscopy and spectromicroscopy. The IR images of samples reveal the chemical details of the sample. Peak intensities, shifts, widths, etc. can be used for imaging. Figure 2.31 (Plate 3) shows a Raman image of a chemical vapor-deposited diamond. The spectrum of the sample is shown on the right side. Different regions of the sample give slightly different spectra with a distinguishable shift. These shifts correspond to the strain in the sample. The shift can be used to obtain a strain distribution in the sample. This is possible with a spatial resolution of the illumination source (of the order of $\lambda/2$). Sub-micron particles have been imaged by Raman spectroscopy using the vibrational band intensity. Using confocal techniques, it is possible to confine the scattering molecule within the interaction volume and to measure spectra from single molecules. This is possible only when the spectra can be enhanced, as in the case of surface-enhanced Raman spectroscopy.

2.5.6 Dynamic Light Scattering

Dynamic light scattering (DLS) is also called quasi-elastic light scattering (QELS) or photon correlation spectroscopy. This is one of the foremost techniques used to measure the radius of a particle in a medium. The motion of particles of micron or lower size is uncorrelated, i.e. they are random. As light scatters from such particles, there will be a shift in the phase of the scattered light which is random and as a result, when the scattered light rays from several particles are added together, constructive or destructive interference occurs. What we get is time-dependent fluctuation in the intensity of the scattered light. The scattering of light from particles undergoing Brownian motion also leads to a Doppler shift of the radiation, modifying the wavelength of the light. In a set-up, a laser light beam is sent through a sample containing particles. The sample has to be inhomogeneous in one of the several ways (such as due to the presence of particles, micelles, proteins, acoustic waves, etc.). The scattered light is received by a fast detector. If the intensity of the light is measured as a function of the scattered direction, we undertake what is called the static light scattering experiment. If the correlation of light intensity is measured as a function of time, we undertake a dynamic light scattering experiment.

The summary of the theory is that when the electric field of the light interacts with the molecules in the medium, an oscillating electric field is induced. The interaction leads to a shift in the frequency of the light and angular distribution of the scattered light, both of which are related to the size. If one assumes that the particles are in Brownian motion, one can apply the Stokes-Einstein equation and get the radii of the suspended particles; $a = k_b T/6\pi\eta D$, where k_b is be Boltzmann constant, D is the diffusion coefficient, η is the viscosity and T is the absolute temperature.

In a QELS measurement, the time-dependent fluctuations in the scattered light are measured. A quantitative measure of the fluctuation is the correlation function. A second order correlation function can be given as: $g^2(\tau) = <I(t)I(t+\tau)>/<I(t^2)>$, where $I(t)$ is the intensity at time t and $I(t+\tau)$ is the intensity at an incremental increase in time $t + \tau$. Brackets correspond to averaging over t. The correlation function can be analyzed to yield a decay rate Γ by using the equation, $g^2(\tau) = B + \beta \exp(-2\Gamma\tau)$, where B is the baseline at infinite decay and β is the amplitude at zero decay. The diffusion constant can be evaluated from $D = \Gamma/q^2$, where q is the magnitude of the scattering vector $(4\pi n_0/\lambda)\sin(\theta/2)$ where n_0 is the solvent index of refraction. Now the Stokes-Einstein equation can get the radii of the suspended particles.

In a typical experiment, only the wavelength and one scattering angle are used. In principle, the technique can distinguish the nature of particles, separated or aggregated, over a range of particle sizes. Typical measurements are done in the nanometer to one micrometer size regime.

2.6 X-Ray Diffraction

The genesis of XRD can be traced to the suggestion of Max von Laue in 1912 that a crystal can be considered as a three-dimensional diffraction grating. The suggestion was based on a Ph.D. thesis of Paul Ewald who considered a crystal as a three-dimensional array of oscillators separated at a distance. Experiments proved the suggestion of Laue. Methodologies of powder X-ray diffraction were developed independently in Germany and in the United States. Single crystal diffractometers were developed in the early 1950s. Although the method of X-ray diffraction is quantitative, in general, it is used for qualitative analysis. This form of analysis extends to all crystalline solids including ceramics, metals, insulators, organics, polymers, thin films, powers, etc. X-ray diffractometers can be used either for single crystals or for powders, with both using significantly different infrastructure. While single crystal diffractometers are used for the study of molecular structure, powder diffracrometers are used for the analysis of phases, though the latter can also be used to derive molecular information.

X-rays corresponds to electromagnetic radiation in the wavelength range of 1 Å. The wavelength range is below that of ultraviolet light and above that of gamma rays. This radiation is produced when charged particles are decelerated by metals, thus producing a continuum called Bremsstrauhlung radiation. X-rays are generally produced when electrons of several thousands of electron volts are decelerated or stopped by metals. This will produce a white radiation up to a threshold frequency corresponding to the kinetic energy of the particle. This threshold corresponds to a wavelength (in angstroms), $\lambda = 12399/V$ where V is the accelerating voltage of the electrons.

When particles such as electrons fall on matter with high energy, electrons can be ejected from various energy levels. Electron ejection from the core orbital is also accompanied by the emission of characteristic X-rays. In the case of electron from the 1s orbital an outer electron from the 2p or 3p orbital can fall down to occupy the vacant 1s orbital. This $2p \to 1s$ transitions leads to the emission of K_α radiation. A similar transition is possible from the 3p level resulting in K_β. The K_α is a doublet with $K_{\alpha 1}$ and $K_{\alpha 2}$ corresponding to electronic transition from the two possible spin states of the 2p electron ($2p_{3/2}$ and $2p_{1/2}$, respectively). In most of the diffraction experiments $K_{\alpha 1}$ and $K_{\alpha 2}$ are not separated and the statistically weighted average of the two wavelengths is taken. The wavelength of a given X-ray line depends on the atomic number.

The emission spectrum of a metal is shown in Fig. 2.32. Characteristic radiations are overlapped with the Bremsstruhlung. Various kinds of filter materials are used to avoid unwanted radiations. Monochromatization by diffraction can also be done to improve the optical purity of the radiation. In a typical X-ray tube used to generate X-rays, high energy electrons are accelerated to a target in an evacuated tube. Only a fraction of the incident electron energy is converted into X-rays. Most of it is converted to heat and efficient cooling of the anode is necessary to avoid it from melting. X-rays come out of the tube through a window made of small atomic number materials such as beryllium.

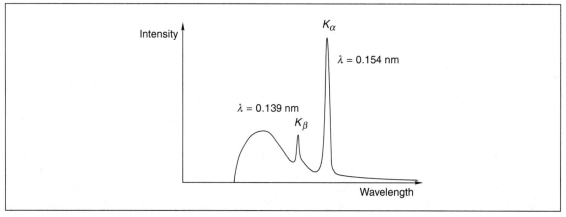

Fig. 2.32: X-ray emission spectrum from copper. Copper K_α corresponds to 1.54 Å.

Diffraction of light by crystals can be understood with the help of an optical grating consisting of several parallel lines drawn on a glass plate. As light is incident on the grating, each group will act as a line source and light will be radiated in all directions. Interference occurs between the waves and in a certain direction, constructive interference occurs. In Fig. 2.33, constructive interference is shown to occur in two directions, marked by lines. In the direction represented by the bottom arrow, the waves are in phase though each wave is shifted by one wavelength from the other. Between these two directions, in all the other directions interference occurs reducing the intensity. In the case of several line sources, as would be present in the case of a grating, interference occurs over several waves and no intensity can be seen between the directions shown. In the case of a grating, the condition of constructive interference is that the path length between the beams should be an integral multiple of the wavelength. This can be written as $n\lambda = d\sin\theta$, where d is the distance between the grooves and is θ the angle of observation. For the first order diffraction, $\lambda = d\sin\theta$. As the maximum value of $\sin\theta = 1$ and $\theta = 90°$, the first order diffraction will be observed at this angle. In general, the angle will be lower than 90° and for d less than λ only zero order direct beams will be observed. If d is much larger than λ, individual diffracted beams of different orders will be closed to each other and we get a diffraction continuum. In the case of visible light (4000 to 7000 Å wavelength), a grating spacing of 10000 to 20000 Å is used to observe diffraction.

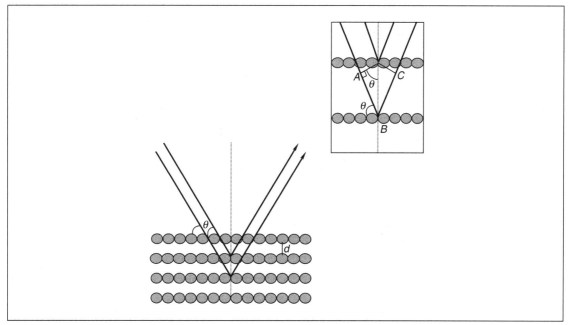

Fig. 2.33: *The conventional derivation of Bragg law. Between the directions shown, the path difference is a multiple of the wavelength and as a result, the intensity will be the maximum at the outward direction. The path length between the two lines, AB + BC = 2d sin θ (as shown in the inset).*

X-Ray Diffraction Three kinds of radiations are generally used for diffraction: X-rays, electrons and neutrons. Commonly, the characteristic X-ray used for diffraction is the copper $K\alpha$ radiation at 1.5418 Å wavelength. Two approaches are generally used for the analysis of X-ray diffraction data. These are the Laue equations and the Bragg's law. In the Laue equations, diffraction from a one-dimensional crystal may be treated in the same way as the diffraction by an optical grating. Upon projection, the grating is like an array of points similar to a crystal. The diffraction condition is again, $n\lambda = d\sin\theta$. In a crystal arrangement of atoms is periodic in all the three directions and three independent Laue equations can be written. The three equations have to be satisfied simultaneously for diffraction to occur.

In Bragg's law, a crystal is viewed as a plane containing several lattice points. The reflection of X-rays will take place from these planes with the angle of reflection being equal to the angle of incidence as shown in Fig. 2.33. The reflected beams are in phase when the path length between the beams is an integral multiple of the wavelength. The planes of light travelling after reflection will be in phase only when this condition is satisfied. This would mean that the distance, $ABC = n\lambda$ or $2d \sin\theta = n\lambda$ (shown in the inset). For all angles other than θ, destructive interference will occur leading to cancellation of the intensity. For crystals containing thousands of such planes, Bragg's law imposes severe restrictions on θ and the cancellation of intensities is usually complete. However, in cases where the number of diffracting planes is limited, the diffraction peak will broaden. In fact, this effect can be used to measure the particle size which is the basis of the Scherrer formula.

One should remember that the interaction of X-rays by atoms is a rather complex event involving electrons of the scattering centres. However, a simplistic picture of reflection is adequate to explain the observed phenomena.

The diffraction of X-rays is generally analyzed in terms of the experimental lattice of the crystal depicted in terms of the lattice vectors. There are relations between observed reflections and (hkl) values. The lowest d spacing observable is $d = \lambda/2$ since the maximum value of $\sin\theta = 1$. Although it appears that the number of lines observable is infinite, the possible number of planes is finite. The Miller indices used for calculating the d spacing can only have integral values. The largest d spacing corresponds to Miller indices such as (100), (010), etc. If the unit cell dimensions are known, the d spacing can be calculated by using the appropriate formula. There are relations such as $1/d^2 = h^2/a^2 + k^2/b^2 + l^2/c^2$ for orthogonal crystals ($\alpha = \beta = \gamma = 90°$). For tetragonal crystals ($a = b$), the equation is further simplified. For cubic crystals ($a = b = c$), the relation is $1/d^2 = (h^2 + k^2 + l^2)/a^2$. Several possible ($hkl$) combinations and their d spacings can be calculated. Depending upon the structure of the unit cell, several diffraction peaks may be absent in the diffraction patterns. Several diffractions are absent as a result of the symmetry such as non-primitive lattice type or certain elements of space symmetry. Consider the bcc lattice type. The (100) reflection has 0 intensity in the pattern. This is because at the Bragg angle for these planes, the body centre atoms diffract X-rays at 180° out of phase relative to the atoms in the corners. As the number of atoms in the corners is equal to those at the centre, the diffracted intensity gets completely cancelled. For a body-centred cell, all reflections for $(h + k + l)$ is odd are absent.

The X-ray diffraction experiment requires the following: a radiation, a sample and a detector for the reflected radiation. In each of these cases, there can be several variations. For example, the radiation can be of many kinds, a single monochromatic source or of variable frequency. The sample can be powder, single crystal, solid piece or a thin film. The detector can be of several kinds, ranging from a simple photographic plate to a sophisticated counter or an area detector. In a powder diffraction experiment, there are crystals arranged in all possible orientations in a finely powdered sample. The various lattice planes are also arranged in all possible orientations. For each crystal plane, there will be a number of orientations. The reflected X-rays may be collected on a photographic plate or by using a counter that is suitably connected to a recorder.

In the Debye-Scherrer method of diffraction, we use a monochromatic X-ray and a power sample with every possible set of lattice planes exposed to the radiation (Fig. 2.34). The diffracted radiation gives rise to a cone. The condition of diffraction is that the radiation is at an angle θ to the incident beam. The cone arises because there are several angular positions of the crystals. The cone is a result of several closely separated spots (inset). In the case of a finely ground sample, the spots will be replaced by a continuous line. Each (hkl) results in one core. The detector is moved in a circle to collect all the reflections corresponding to various (hkl).

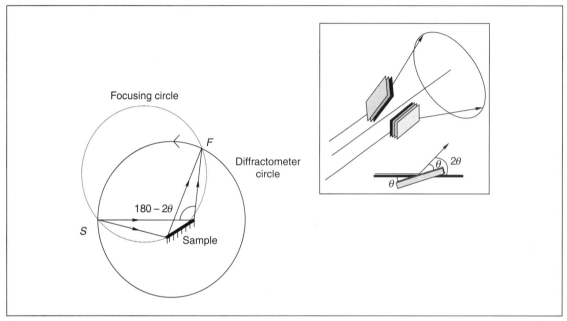

Fig. 2.34: *The Debye-Scherrer method of powder diffraction. S and F are source and detector, respectively. Two microcrystallites of different orientation with respect to the incident beam give diffracted rays that lie in a cone. The diffraction pattern is due to all the (hkl).*

In the modern diffraction method called diffractometry, a convergent beam strikes the sample and the intensity as a function of diffraction angle is measured. The position of the diffraction peak and the intensity at this point are the two factors used in the determination. Both these can be measured accurately and compared with standards in the literature. In fact, this is what one normally does in the phase identification work.

2.6.1 Intensities in X-ray Scattering

Intensities are important in X-ray analysis for determining unknown crystal structures and quantitative phase analysis. X-rays are electromagnetic waves and can interact with electrons, thus making them vibrate. The vibrating charge will emit electromagnetic radiation which is in phase (coherent) with the incident X-ray. Coherent scattering is similar to elastic collision and the wavelength of the X-ray is not changed. The intensity of the radiation scattered can be given by the Thomson equation, $I_p \alpha\ 1/2(1 + \cos^2 2\theta)$, where I_p is the scattered beam intensity at point p, and 2θ is the angle between the incident and the scattered beams. X-rays can also interact with matter and get inelastically scattered thus giving rise to Compton scattering. Loosely bound electrons are more involved with this and are responsible for the background in the diffraction measurements.

Scattering can be considered to occur from each electron. The total scattered intensity will be a sum of all the scattered intensities due to individual electrons. Thus the scattering factor or the form factor is proportional to the atomic number. The scattering factors of atoms of similar atomic numbers are close and therefore it becomes difficult to identify similar atoms. This becomes a problem in the single crystal diffraction of organic compounds containing C, N and O. This is also a problem in aluminosilicates due to the similarity of Si and Al. It is difficult to determine hydrogen in the presence of heavy elements. This is not a problem with neutrons as the scattering factors are not the functions of atomic numbers alone.

The structure factor or structure amplitude for each (hkl) can be defined and calculated from the measured intensities. This can be used to define a residual factor or R factor. The quality of the structure determination is understood in terms of the magnitude of the R factor, the lower the better. Generally it is in the range of .1 to .2. While solving the structure, one often determines the electron density maps. These maps are instructive for understanding bonding.

The most commonly used X-ray instrument is the powder diffractometer. It has a scintillation or Geiger counter. The detector spans a range of scattering angles. Generally it is a practice to mention 2θ, not θ, as the scattering angle. A 2θ range of 10 to 80 degrees is adequate for covering the most useful part of the pattern.

The d values can be readily calculated from the graph. The intensities are generally taken as peak heights unless an intensity analysis is performed where the peak area is taken. The peak of maximum height is taken as 100 and all the other peaks are scaled accordingly. A set of peaks and their heights is generally adequate for phase identification. In several cases, an accurate measurement of peak positions is needed. Preferred orientations exist for a number of materials and it is likely that only these peaks are manifested. In order to make all crystal planes observable (within the crystal system), it is important that the sample is finely ground. Several sample preparation methodologies are employed.

The convergent beam of X-ray is important for improved sensitivity and resolution. This is achieved by placing the source and detector at the circumference of a circle.

In focusing geometry, the configuration of the diffractometer is different. Here the source and the detector will form the circumference of a circle called the diffractometer circle and the surface of the sample must lie tangential to the focusing circle. While the detector has an angular speed of 2θ deg min^{-1}, the sample rotates in the same direction at an angular speed of θ deg min^{-1}.

2.6.2 Particle Size Effects

The normal diffraction line is of a finite width due to several factors such as the finite line width of the excitation source and the imperfections in the focusing geometry. The Bragg condition occurs when each plane in a crystal diffracts exactly one wavelength later than the previous plane. Constructive interference occurs due to this condition. When the incident ray at a larger angle, θ_1 than the diffraction angle, θ strikes the crystal plane, the phase lag is greater than the wavelength, λ to become $\lambda + \delta\lambda$. As the number of planes becomes $j + 1$, the cumulative phase lag, $\sum \delta\lambda$ could increase to become λ i.e., $j\,\delta\lambda = \lambda/2$. For the ray incident at the larger angle $\theta 1$, the diffracted rays from plane 1 and plane, $j + 1$ are 180° out of phase. As a result, there is no net intensity for the diffracted ray at this angle. Now let us understand that we have several planes in the crystallite and the rays diffracted from the set of planes, 1 through j are exactly cancelled by the planes $j + 1$ through $2j$, if there are $2j$ planes present in the crystallite. What this means is that the intensity of the diffracted beam will fall to zero at a finite angle, with a peak maximum as a result of this effect. One should note that there is also a phase difference, $\lambda - \delta\lambda$ which occurs for an angle $\theta 2$, smaller than θ. The width of the diffraction peak is therefore determined by the number of planes present in the crystallite. For large crystallite, j is large, $\delta\lambda$ is small and the width is negligible. The particle size effects seen as broadening of the diffracted lines is given by the Scherrer formula, $t = 0.9\,\lambda/(B\cos\theta)$, where t is the thickness of the crystallite in (angstroms) and θ is the Bragg angle. B is the line broadening, indicating the extra peak width of the sample in comparison to the standard, derived using the Warren formula, $B^2 = B_M^2 - B_s^2$, where M and S refer to specimen and standard. B's are measured in radians at half height. The sample and standard should have peaks close to each other.

Particle sizes up to 200 nm can be measured by using the Scherrer formula. In the range of 5–50 nm, the broadening is easy to determine. At larger particle sizes, the difference between the sample and the standard is small and at small particle sizes, the peak is difficult to distinguish from the background. For smaller particle sizes, low angle peaks are used for size determination as they are less broad as compared to large angle peaks.

In passing, it may be mentioned that the powder pattern may be shifted or broadened as a result of stresses present in the material. Due to uniform compressive stress, the d spacing may reduce and the peak may shift to larger angles. If the stress is non-uniform throughout the crystallite, broadening will occur. A composite of these effects is generally observed.

2.7 Associated Techniques

In addition to the above, various analytical techniques are also used for nano measurements. Broadly, any technique used for material characterization can be applied in this area too. Several of these techniques such as zeta potential will be discussed at appropriate places in the text where a discussion on these relates to the subject matter. Such studies imply that almost every tool is adaptable for nanomaterial investigations.

Review Questions

1. Why objects in the nanoscale cannot be seen by visible light? How do we see them?
2. What are the principal differences between electron and scanning probe microscopies?
3. Why is it not possible to image nano objects with infrared or X-rays? What are the current capabilities with these techniques? What are their specific advantages?

4. What are the characteristic properties of objects in the nanoscale? Which of those properties we use to examine them?
5. Every property possessed by bulk materials is also possessed by nano objects. So, how can one study nano objects uniquely?
6. How do we study properties of single nano objects?
7. What are such properties being investigated? Describe them with an example.
8. Propose an experiment to study the strength of a single chemical bond.
9. Single molecules are governed by laws of quantum mechanics. So their position will be uncertain if we examine them. Then how is it possible to observe and manipulate them?
10. How will nanotechnology work if positioned atoms and molecules do not stay at the specific location?
11. Are there properties which we cannot measure with the techniques described?
12. Are there other techniques, other than those described here, to study nano objects? Propose a new technique or modify a technique known to you for this study.

References

1. Goldstein, J.I., C.E. Lyman, D.E. Newburry, E. Lifshin, P. Echlin, L. Sawyer, D.C. Joy and J.R. Michael, (2003), *Scanning Electron Microscopy and X-ray Microanalysis*, Kluwer Academic/Plenum Publishers, New York.
2. Thomas, G., and Goringe, M.J., (1979), *Transmission Electron Microscopy of Materials*, John Wiley and Sons, New York.
3. Treacy, M.M.J., T.W. Ebbesen and J.M. Gibson, (1996), *Nature*, **381**, p. 678.
4. Binning, G., H. Rohrer, Ch. Gerber and E. Weibel, (1982), *Phys. Rev. Lett.*, **49**, p. 57.
5. Wildoer, W.G., L.C. Venema, A.G. Rinzler, R.E. Smalley and C. Dekker, (1998), *Nature*, **391**, p. 59.
6. Eigler, D.M., and E.K. Schweizer, (1990), *Nature*, **344**, p. 524.
7. Crommie, M.F., C.P. Lutz and E. Eigler, (1993), *Science*, **262**, p. 218.
8. Manoharan, H.C., C.P. Lutz and D. Eigler, (2000), *Nature*, **403**, p. 512.
9. Stroscio A., and W.J. Keiser, *Scanning Tunneling Microscopy*, Academic Press, 1993.

   SXM Techniques and Capabilities (From Ref. 9)
   (a) *Scanning Tunneling Microscope*, (1981), G. Binnig, and H. Rohrer, "Atomic Resolution images of Conducting Surfaces", G. Binning, H. Rohrer, Ch. Gerber and E. Weibel, *Phys. Rev. Lett.*, **49** (1982), pp. 57–61.

(b) *Scanning Near-field Optical Microscope* (1982), D.W. Pohl, "50 nm (Lateral resolution) Optical Images", D.W. Pohl, W. Denk and M. Lanz, *Appl. Phys. Lett.*, **44** (1984), p. 651; A. Harootunian, E. Betzig, A. Lewis and M. Isaacson, *Appl. Phys. Lett.*, **49** (1986), p. 674.

(c) *Scanning Capacitance Microscope* (1984), J.R. Matey, J. Blanc, "500 nm (Lateral Resolution) Images of Capacitance Variation", J.R. Matey and J. Blanc, *J. Appl. Phys.*, **57** (1984), pp. 1437–1444.

(d) *Scanning Thermal Microscope* (1985), C.C. Williams, H.K. Wickramasinghe, "50 nm (Lateral Resolution) Thermal Images", C.C. Williams and H.K. Wickramasinghe, *Appl. Phys. Lett.*, **49** (1985), pp. 1587–1589.

(e) *Atomic Force Microscope* (1986), G. Binning, C.F. Quate, Ch. Gerber, "Atomic Resolution on Conducting/Non-conducting Surfaces", G. Binning and C.F. Quate, *Phys. Rev. Lett.*, **56** (1986), pp. 930–933.

(f) *Scanning Attractive Force Microscope* (1987), Y. Martin, C.C. Williams, H.K. Wickramasinghe, "5 nm (Lateral Resolution) Non-contact Images of Surfaces", Y. Martin, C.C. Williams, H.K. Wickramasinghe, *J. Appl. Phys.*, **61** (1987), pp. 4723–4729.

(g) *Magnetic Force Microscopy* (1987), Y. Martin, H.K. Wickramasinghe, "100 nm (Lateral Resolution) Images of Magnetic Bits/Heads", Y. Martin and H.K. Wickramasinghe, *Appl. Phys. Lett.*, **50** (1987), pp. 1455–1457.

(h) *"Frictional" Force Microscope* (1987), C.M. Mate, G.M. McClelland, S. Chiang, "Atomic-scale Images of Lateral ("Frictional") Forces", C.M. Mate, G.M. McClelland, R. Erlandsson, and S. Chiang, *Phys. Rev. Lett.*, **59** (1987), pp. 1942–1945.

(i) *Electrostatic Force Microscope* (1987), Y. Martin, D.W. Abraham, H.K. Wickramasinghe, "Detection of Charge as Small as Single Electron", Y. Martin, D.W. Abraham, and H.K. Wickramasinghe, *Appl. Phys. Lett.*, **52** (1988), pp. 1103–1105.

(j) *Inelastic Tunneling Spectroscopy STM* (1987), D.P.E. Smith, D. Kirk, C.F. Quate, "Photon Spectra of Molecules in STM", D.P.E. Smith, G. Binning, and C.F. Quate, *Appl. Phys. Lett.*, **49** (1987), pp. 1641–1643.

(k) *Laser Driven STM* (1987), L. Arnold, W. Krieger, H. Walther, "Imaging by Non-linear Mixing of Optical Waves in STM", L. Arnold, W. Krieger, H. Walther, *Appl. Phys. Lett.*, **51** (1987), pp. 786–788.

(l) *Ballistic Electron Emission Microscope* (1988), W.J. Kaiser, "Probing of Schottky Barriers in nm Scale", W.J. Kaiser, *Phys. Rev. Lett.*, **60** (1988), pp. 1406–1409.

(m) *Inverse Photoemission Force Microscopy* (1988), H. Coombs, J.K. Gimzewski, B. Reihl, J.K. Sass, R.R. Schlittler, "Luminescence Spectra on nm Scale", B. Reihl, J.H. Coombs and J.K. Gimzewski, *Surface Science* **1988** (1989), pp. 211–212, 156–164.

(n) *Near Field Acoustic Microscope* (1989), K. Takata, T. Hasegawa, S. Hosaka, S. Hosoki, T. Komoda, "Low Frequency Acoustic Measurements on 10 nm Scale", K. Takata, T. Hasegawa, S. Hosaka, S. Hosoki and T. Komoda, *Appl. Phys. Lett.*, **55** (1989), pp. 1718–1720.

(o) *Scanning Noise Microscope* (1989), R. Moiler, A. Esslinger, B. Koslowski, "Tunneling Microscopy with Zero Tip-sample Bias", R. Möller, A. Esslinger, and B. Koslowski, *App. Phys. Lett.*, **55** (1989), pp. 2360–2362.

(p) *Scanning Spin–Precession Microscope* (1989), Y. Manassen, R. Hamers, J. Demuth, A. Castellano, "1 nm (Lateral Resolution) Images of Paramagnetic Spins", Y. Manassen, R.J. Hamers, J.E. Demuth and A.J. Castellano Jr., *Phys. Rev. Lett.*, **62** (1989), pp. 2531–2534.

(q) *Scanning Ion–Conductance Microscope* (1989), P. Hansma, B. Drake, O. Marti, S. Gould, C. Prater, "500 nm (Lateral Resolution) Images in Electrolyte", P.K. Hansma, B. Drake, O. Marti, S.A. Gould and C.B. Prater, *Science*, **243** (1989), pp. 641–643.

(r) *Scanning Electrochemical Microscope* (1989), O.E. Husser, D.H. Craston, A.J. Bare, O.E. Hüsser, D.H. Craston and A.J. Bard, *J. Vac. Sci. and Tech. B: Microelectronics and Nanometer Structures*, **6** (1989), pp. 1873–1876.

(s) *Absorption Microscope/Spectroscope* (1989), J. Weaver, H.K. Wickramasinghe, "1 nm (Lateral Resolution) Absorption Images/Spectroscopy", J.M.R. Weaver, L.M. Walpita, H.K. Wickramasinghe, *Nature*, **342** (1989), pp. 783–785.

(t) *Scanning Chemical Potential Microscope* (1990), C.C. Williams, H.K. Wickramasinghe, "Atomic Scale Images of Chemical Potential Variation", C.C. Williams and H.K. Wickramasinghe, *Nature*, **344** (1990), pp. 317–319.

(u) *Photovoltage STM* (1990), R.J. Hamers, K. Markert, "Photovoltage Images on nm Scale", R.J. Hamers and K. Markert, *Phys. Rev. Lett.*, **64** (1990), pp. 1051–1054.

(v) *Kelvin Probe Microscopy* (1991), M. Nonnenmacher, M.P. O'Boyle, H.K. Wickramasinghe, "Contact Potential Measurements on 10 nm Scale", M. Nonnenmacher, M.P. O'Boyle and H.K. Wickramasinghe, *Appl. Phys. Lett.*, **58** (1991), pp. 2921–2923.

10. Wong, S.S., A.T. Woolley, T.W. Odom, J.L. Huang, P. Kim, D.V. Vezenov and C.M. Lieber, (1998), *App. Phys. Lett.*, **73**, pp. 3465–3467.

11. Hong, S., J. Im, M. Lee and N. Cho, (2005), in *Handbook of Microscopy for Nanotechnology*, N. Yao and Z. L. Wang (eds), Kluwer Academic Publishers, New York.

12. Kam, Nadine Wong Shi, Theodore C. Jessop, Paul A. Wender and Hongjie Dai, (2004), *J. Am. Chem. Soc.*, **126(22)**, pp. 6850–6851.

13. Synge, E.H., *Phil. Mag.*, **6** (1928), p. 356, 11 (1931), p. 65, **13** (1932), p. 297.

14. Anshup, J. Sai Venkataraman, Chandramouli Subramaniam, R. Rajeev Kumar, Suma Priya, T.R. Santhosh Kumar, R.V. Omkumar, Annie John and T. Pradeep, (2005), *Langmuir*, **21**, pp. 11562–11567.

Additional Reading

1. Ibach, H., and D.L. Mills, (1982), *Electron Energy Loss Spectroscopy and Surface Vibrations*, Academic Press, New York.

2. Vickerman, J.C., (1997), *Surface Analysis: The Principal Techniques*, John Wiley, Chichester, Sussex.
3. Kolasinski, K.W., (2002), *Surface Science Foundations of Catalysis and Nanoscience*, John Wiley and Sons, Ltd. Chichester.
4. Hufner S., (1995), *Photoelectron Spectroscopy*, Springer-Verlag, Heidelberg.
5. Chen, C.J., (1993), *Scanning Introduction to Tunneling Microscopy*, Oxford University Press.
6. Fadley, C.S., (1978), in *Electron Spectroscopy: Theory, Techniques and Applications*, C.R. Brundle and A.D. Baker (eds), Volume 2, Academic Press, New York.
7. Smith, G.C., (1994), *Surface Analysis by Electron Spectroscopy*, Plenum Press, New York.
8. Kimura, K., S. Katsumata, Y. Achiba, T. Yamazaki and S. Iwata, (1981), *Handbook of HeI Photoelectron Spectra of Fundamental Organic Molecules*, Japan Scientific Societies Press, Tokyo.
9. Willard, H.H., L.L. Merritt, Jr., J.A. Dean and F.A. Settle, Jr., (1986), *Instrumental Methods of Analysis*, VIth Edition, CBS Publishers, New Delhi.
10. West, A.R., (1986), *Solid State Chemistry and its Applications*, John Wiley and Sons, New York.

PART THREE

Diversity in Nanosystems

Contents:

- Fullerenes
- Carbon Nanotubes
- Self-assembled Monolayers
- Gas Phase Clusters
- Semiconductor Quantum Dots
- Monolayer-protected Metal Nanoparticles
- Core-shell Nanoparticles
- Nanoshells

Chapter 3

FULLERENES

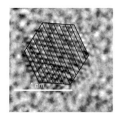

The science of fullerenes is rather old, but its understanding is important as it has, in many ways, heralded giant leaps in nanoscience. The discovery of these molecules in 1985 opened up a new area of science. Fullerenes are molecular forms of carbon, which are distinctly different from the extended carbon forms known for millennia. There are numerous molecular forms, all of which are spheroidal in structure. We sketch here the fascinating area of this new allotrope of carbon, which may be the only allotrope of any element discovered in the 20th Century, focusing primarily on C_{60}. The early history, synthesis and characterization, mass spectrometry, derivatization, orientational ordering, pressure effects, superconductivity, magnetism and photophysical properties of fullerenes and fullerene-based compounds are discussed in this chapter. The latest discoveries in this area are delineated at the end.

Learning Objectives

- What are fullerenes and what are their properties?
- How is the discovery of fullerenes related to the development of nanoscience and technology?
- What are the unusual properties of fullerenes?
- How can gas phase spectroscopy be used to study condensed phase properties?

3.1 Introduction

The role of serendipity in scientific discoveries is widely recognized. The story of buckministerfullerene (Ref. 1), C_{60} and of fullerenes, in general, is no exception. In their eagerness to understand the chemistry of interstellar molecules, scientists hit upon an unusual and extraordinary discovery in the history of chemistry. The molecule they discovered had sixty equivalent carbon atoms, which formed the pattern of a football that gave it the highest symmetry. Theoretical predictions (Ref. 2) of such structures were known since 1970. In chemistry, there is no other molecule formed by the same atom, which is as big as buckministerfullerene. For millennia, elemental carbon has been known to occur in two polymorphic forms, graphite and diamond. Graphite has two-dimensional layers of sp^2 hybridized carbon atoms interlinked by weak van der Waals forces. Since its interlayer interaction is weak, graphite is used as a lubricant. Diamond, however, is one of the hardest materials known to man. This property arises from the strong three-dimensional bonding in diamond in which each sp^3 hybridized carbon atom is bonded to four other similar atoms. Fullerenes constitute another allotrope of carbon, which is probably the first allotrope of any element discovered in recent times. Figure 3.1 shows the structures of graphite, diamond and C_{60}. There also exist other all-carbon molecules similar to C_{60} with cage structure, collectively called 'fullerenes' in honour of the famous American architect Buckminister Fuller whose geodesic domes are landmarks of 20th Century architecture. Other forms of carbon such as carbon rings (Ref. 3) are also receiving considerable attention.

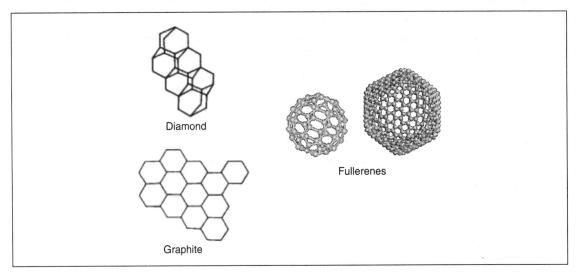

Fig. 3.1: *A schematic representation of the structures of graphite, diamond and fullerenes. While the two-dimensional sheets formed by hexagons are packed one over another in graphite, the diamond structure is three-dimensional. Only two fullerenes are shown. The smaller one is buckminsterfullerene, C_{60}. The bonds between the hexagons are more like double bonds showing the corannulene-type substructure. The double bonds are localized exocyclic to the pentagons giving [5]radialene character to the pentagons and cyclohexa-1,3,5-triene character to the hexagons.*

It is only natural for a molecule of such immense beauty to attract the entire scientific community (Refs 4, 5). The structure, spectroscopy, chemistry, materials science and applications of this molecule have been intensely investigated. Buckminsterfullerene thus became the subject matter of nine out of ten of the most cited papers in chemistry in 1991. In 1992, it scored a perfect ten by becoming the subject of ten out of ten papers. According to statistics (Ref. 1), one paper per week got published in this area from 1985 to 1990. After 1990, the figure has risen to one paper per day. In 1995, it was slightly less than two per day. The statistics of the recent years show that this subject continues to attract intense interest. According to webofscience (www.webofscience.com), the number of papers on the subject for the years, 2000, 2001, 2002, 2003, 2004 and 2005 are, 1255, 1350, 1254, 1334, 1313 and 1578, respectively.

Molecules with 70, 76, 82 and other numbers of carbon atoms were soon characterized. The developments in 1991 showed that these molecules are found only with tens and hundreds of atoms but also with thousands of atoms. These giant molecules of carbon occur as nanometer size tubes and balls which are called carbon nanotubes and onions, (Ref. 6), respectively. There is another chapter on this topic in this book itself. This chapter, gives an account of the chemistry, physics and materials science of this new form of carbon in a rather illustrative fashion. Numerous conferences have been held on the subject of fullerenes and even on specialized topics of fullerene chemistry and physics over the years. Several books (Refs 4, 5) have appeared on fullerenes and a comprehensive review of these is impossible here.

3.2 Discovery and Early Years

As noted earlier, the search for certain linear molecules of carbon normally found in the interstellar region called cyanopolyynes, was the starting point of this search (Ref. 7). Some of these molecules of the type $H-C\equiv C-C\equiv C-C\equiv N$ or HC_5N, have been synthesized in the laboratory. These molecules contain seven, nine, eleven and even up to 33 carbon atoms. This was the time when Prof. Smalley and his co-workers in Houston were working with a newly developed cluster source (Ref. 8), which used lasers for evaporation, supersonic molecular beam expansion for clustering, and photoionization mass spectrometry for detecting the products. Similar studies were also conducted by Dr. Kaldor and Dr. Cox in Exxon (Ref. 9). Their studies on graphite, which were preceded by those of Smalley and his colleagues, showed the presence of carbon clusters heavier than C_{33}, but interestingly no odd number cluster was seen. Soon after the beginning of these measurements, it was observed that species such as HC_7N and HC_9N are formed in the reaction of $C_n (n < 30)$ with H_2 and N_2. However, the major discovery was not the detection of cyanoplyynes, but of the unusually abundant species C_{60}, which dominated the mass spectrum under certain clustering conditions (Ref. 10). There were other heavier clusters too. It was found that these clusters were particularly unreactive as compared to the lower clusters. Reactivity and photofragmentation studies showed that the 60 atom cluster is extremely stable. The observed chemistry can be explained if one assumes that the graphitic sheets transform into a hollow chicken-wire cage similar to the domes of Buckminister Fuller (Ref. 11). Such a closed cage requires that (Ref. 12) $12 = 3\ n_3 + 2\ n_4 + 1\ n_5 + 0\ n_6 - 1\ n_7 - 2\ n_8 - ...$, where n_k represents the number of k-sided faces. For carbon, only values of k are 5 and 6, though 7 is also possible which has been detected in carbon nanotubes (Ref. 13). This means that there should be 12 pentagonal faces and the number of hexagonal faces is arbitrary. In C_{60}, there are 12 pentagonal faces and 20 hexagonal faces. Fullerenes, belonging to a class of closed cage molecules, have the general formula C_{20+2n6}. C_{60} has only one chemically distinct carbon atom. However, in C_{70}, there are five distinct carbon atoms. In larger fullerenes, there is the possibility of isomers and some fullerenes are also chiral. It is seen that fullerenes have many more isomers than was previously believed of (Ref. 14). However, because of the isolated pentagon rule (observed generally), in which pentagons are separated by hexagons, the number of isomers is limited.

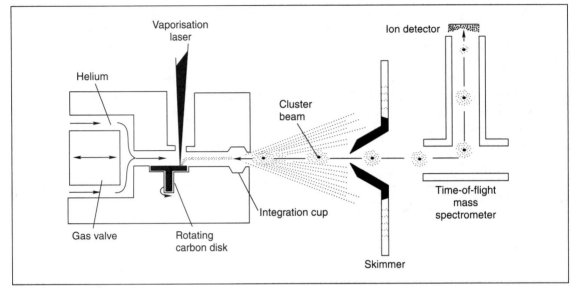

Fig. 3.2: *The experimental set-up used to discover C_{60}. The graphite disk is evaporated with a Nd:YAG laser and the evaporated carbon plasma is cooled by a stream of helium coming from a pulsed valve. The clusters of carbon are produced in the integration cup and are expanded into vacuum. The ions are detected by time of flight mass spectrometry.*

The experimental apparatus used in the discovery of C_{60} is shown in Fig. 3.2. This set-up is now commercially available for the study of a variety of clusters. A rotating disk of graphite is irradiated with a powerful laser to evaporate carbon. As the laser falls on the disk, a stream of helium gas is passed over the disk by releasing a valve. The gas carries the evaporated carbon species with it and during its passage to the nozzle, the species in the vapor undergoes clustering. The cluster beam emanating from the nozzle is selected by a skimmer. The clusters are then subjected to mass analysis by time of flight mass spectrometry. Under certain experimental conditions, the mass spectrum was very similar to that reported previously, showing a distribution of even-numbered species. But a variation in the experimental conditions, especially the introduction of the integration cup (see Fig. 3.2), increased the intensity of the sixty atom cluster to such a point that in some experiments, only C_{60} and C_{70} were seen (Fig. 3.3).

At the same time, experiments were also done using a Fourier transform ion-cyclotron resonance (FT–ICR) apparatus. In these set of experiments, mass selected cluster ions were subjected to reactions with a variety of gases. C_{60}^+ was extremely unreactive to gases such as O_2, NH_3 and NO. On the contrary, a cluster of other elements such as Si showed high reactivity. In fact, there was no evidence to show that Si_{60}^+ was special. A body of other experimental data was accumulated by the Houston group on the photophysics, photodetachment and optical spectrum of C_{60}. None of these studies, however, contradicted the proposed structure.

For nearly five years, C_{60} was truly a playground for chemical physicists. A large number of theoretical papers got published on the electronic structure (Ref. 15), reactivity (Ref. 16), magnetism (Ref. 17) and a

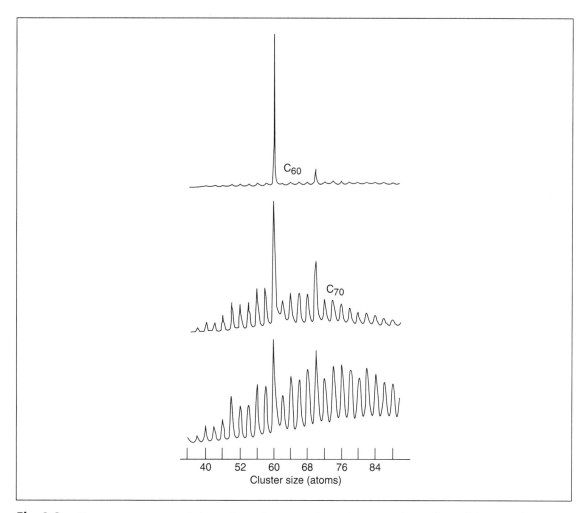

Fig. 3.3: *The mass spectrum of the carbon clusters under various experimental conditions. Under certain conditions, only C_{60} and C_{70} are seen (Adapted from, Ref. 1).*

number of other properties of C_{60}. The studies were simplified by the very high symmetry of the proposed structure. In fact, a close similarity between the observed peaks in the IR spectrum of an evaporated carbon soot and the theoretical frequencies (Ref. 18) made Kratschmer and his colleagues (Ref. 19) look for C_{60} in the soot. They were working on laboratory-produced carbon soot in order to understand the interstellar spectrum of carbonaceous materials. The soot they obtained after evaporating graphite resistively in an atmosphere of helium contained four bands in the infrared region. The absorption frequencies correlated well the proposed bands of C_{60}. The application of solvent extraction yielded significant quantities of fullerenes (Ref. 20) from the soot and a host of techniques were applied to the characterization of this newly made form of carbon.

3.3 Synthesis and Purification of Fullerenes

For a synthesis of fullerenes, all that is needed is a welding transformer, a chamber connected to a vacuum pump (even a single stage oil sealed rotary pump is adequate) and some graphite rods. The graphite electrodes are brought into close contact with each other and an arc is struck in an atmosphere of 100–200 Torr of helium or argon. To sustain the arc, a voltage of 20 V (ac or dc) may be necessary. For a graphite rod of 6 mm in diameter, about 50–200 A current may be consumed. Generally, spectroscopically pure graphite of high porosity is used in order to achieve a high evaporation rate. The soot generated is collected on water-cooled surfaces which could even be the inner walls of the vacuum chamber. After sustaining the arc for several minutes, the vacuum is broken and the soot is collected and soxhlet extracted for about 5–6 hours in toluene or benzene, resulting in a dark reddish-brown solution which is a mixture of fullerenes. Of the entire soot collected in this manner, 20–30 per cent is soluble. This soluble material is subjected to chromatographic separation (Ref. 21). Over the years, several simple methods avoiding time consuming chromatography have been discovered including a filtration technique over an activated charcoal–silica gel column (Ref. 22). About 80 per cent of the soluble material is C_{60}, which can be collected in one pass using toluene as the mobile phase. C_{70} can be separated by using toluene/o-dichlorobenzene mixture as the eluant. Repeated chromatography may be necessary to get pure C_{70}. C_{60} solution is violet in colour while that of C_{70} is reddish-brown. Higher fullerenes such as C_{76}, C_{78}, C_{82}, etc., require HPLC for purification (Ref. 23). The spectroscopic properties of several of these fullerenes are now known. Normally, the preparation of a gram of C_{60} from graphite requires about five hours of work. But it may take as much as 250 hours to make 1 mg in the case the higher fullerenes. C_{60} and C_{70} are now commercially available from several sources. C_{60} costs about $25 a gram but C_{70} is not yet affordable for synthetic chemists. Therefore, a majority of the studies reported are carried out on C_{60}. Fullerenes crystallized from saturated solutions retain solvent molecules, removing which may require long hours of vacuum drying. Crystal growth by vapor transport is an excellent method of growing millimeter-sized crystals devoid of solvent for sensitive measurements. For solid state spectroscopic measurements, it is better to use evaporated fullerene films in high or ultrahigh vacuum to avoid solvent contamination. Evaporation is also used as a method of purification as there are substantial differences in the onset of evaporation between C_{60} and C_{70}. Calixarenes (bowl-shaped macrocycles with hydrophobic cavities) have been used (Ref. 24) in the purification of fullerenes. See Fig. 3.4 for a schematic procedure of the synthesis and purification and fullerenes.

Arc evaporation is not a unique way of making C_{60}. Fullerenes have been found in flames (Refs 25, 26), upon chemical vapor deposition used to produce diamond (Ref. 27), in a 1.85 billion year old bolide impact crater (Ref. 28) as well from spacecrafts (Ref. 29). They have also been made from diamond (Ref. 30). However, no one has made them through chemical reactions though such a possibility has excited many organic chemists (Ref. 31). They have also been synthesized from camphor (Ref. 32). Mass spectrometry has shown that higher clusters of carbons can be formed through the laser evaporation of polymers (Ref. 33). Upon laser evaporation (Ref. 34), highly unsaturated carbonaceous ring systems produce C_{60}. There are several other exotic means of producing C_{60}. However, the total synthesis would indeed be a landmark in chemistry though approaches to this have been suggested (Ref. 35).

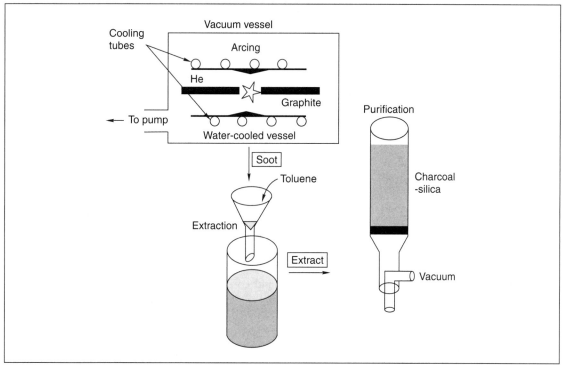

Fig. 3.4: *Schematic illustration of the processes involved in the synthesis and purification of fullerenes. Graphite rods are evaporated in an arc, under He atmosphere. The soot collected is extracted with toluene and subjected to chromatography.*

Synthesis and purification were followed by the characterization of fullerenes in term of a variety of spectroscopic techniques. The pivotal role played by mass spectrometry in characterizing fullerenes cannot be over-emphasized. Other techniques such as NMR showing a single line corresponding to the equivalence of the carbon atoms (Ref. 22), single crystal X-ray structure resolving the atomic positions (Ref. 36), characterization by UV\VIS (Refs 22, 37), IR and Raman (Ref. 38) spectroscopies, etc. soon followed. The predicted electronic structure was confirmed by HeI and HeII photoelectron spectroscopies (Ref. 39).

3.4 Mass Spectrometry and Ion/Molecule Reactions

Soon after a method of macroscopic synthesis of C_{60} was used, several ion/molecule reaction studies were carried out (Ref. 40). In the early days, fullerenes were expensive and a mass spectrometer was the ideal reaction vessel. C_{60}^- and $C_{60}H^+$ were produced by chemical ionization (CI) with methane (Ref. 41). It was shown that C_{60}^- was 15 times more abundant than $C_{60}H^+$, consistent with its high electron affinity. The methane CI spectrum showed protonated fullerenes and adducts of fullerenes with C_2H_5. Collision induced dissociation (CID) of C_{60}^+ in the keV range with helium showed the expected fragments of C_{58}^+, C_{56}^+, etc., just as in the earlier photofragmentation study. CID is a method of activating the ion by collisions with atoms or molecules. The fragmentation of C_{60}^{2+} yielded only C_n^{2+}, not C_n^+ in contrast to polycyclic aromatic hydrocarbons (PAHs), which undergo charge separation reactions. In addition to CID, surface-induced dissociation (Ref. 42) was also performed. This involves the collision of the ion with another surface for imparting energy transfer, which is higher in efficiency than CID. Collisions on both silicon and graphite surfaces showed that C_{60} does not dissociate appreciably at collision energies in the range of 200–300 eV. A number of studies have confirmed this unusual stability of C_{60}, which has been attributed to the 'resilience' of the molecule. Similar properties have been predicted for its hydrides also (Ref. 43). C_{60} collision also leads to delayed ionization and thermionic emission (Ref. 44). Surface collision experiments have been reported on higher fullerenes and metallofullerenes as well (Ref. 45). Mass spectrometry has also been used to study the thermodynamic properties of fullerenes (Ref. 46).

In addition to fragmentation, endohedral complex formation was also observed when high energy collisions were performed (Ref. 47). Eight keV collision of helium with C_{60}^+ produced a number of C_{n+4}^+ mass peaks. When the collision gas was changed to ^3He, the peaks shifted by one mass unit showing that the peak was due to the addition of helium atoms to C_{60}. Other scientists conducted experiments in different types of hybrid tandem mass spectrometers and confirmed the results. Ion kinetic energy and CID measurements showed unambiguously that $C_{60}@He^+$ (the @ symbolism implies that the species prior to @ is within the cage of the fullerene after the symbol) is an endohedral complex. Similar measurements were repeated with C_{60}^{2+} and C_{60}^{3+}. C_{70}^+ and C_{84}^+ were also shown to form endohedrals. $C_{60}@Ne^+$ and $C_{60}@Ar^+$ are hard to observe in conventional spectrometers due to large energy losses, but they have been seen in specially designed instruments (see also Section 3.6 on endohedral complexes).

In addition to the work on pure C_{60}, mass spectrometry has been extensively used to characterize the derivatives of C_{60} formed by reactions (Ref. 48). The identification of Birch reduction products of C_{60} was done with electron impact (EI) mass spectrometry. The products of C_{70} were also studied with the help of EI. Reaction products with fluorine showed mass peaks at $C_{60}F_{36}^+$ and $C_{70}F_{40}^+$. These products fragmented with the elimination of F, CF_3 and C_2F_5. Methylated C_{60} showed products with 1 to 24 methyl groups. C_{60} and C_{70} were found to add to aromatic molecules such as benzene, toluene, xylene, anisole and bromobenzene (Ref. 49).

3.5 Chemistry of Fullerenes in the Condensed Phase

Originally it was thought that C_{60} is an aromatic molecule because it has about 12,500 possible resonance structures. However, it should be remembered that in systems where pentagons are near hexagons, the system avoids double bonds in pentagons. The presence of double bonds in pentagons reduces the bond distances, thereby increasing the strain. In the case of C_{60}, there is only one structure which avoids double bonds in pentagons. This means that the delocalization of electrons is poor and C_{60} is poorly aromatic (Ref. 50). This 'poorly aromatic' classification immediately suggests a certain type of chemistry. C_{60} can be visualized in terms of corannulene subunits with two distinct chemical bonds. The 60 6–6 bonds (between hexagons) with a bond length of ca. 1.38 Å, have more double bond character than the 60 6–5 bonds (between pentagons and hexagons) of bond length ca. 1.45 Å, which are more like single bonds. This means that the pentagonal rings are extremely strained and the insertion of a double bond in the 6–5 ring can cause instability to the tune of 8.5 kcal/mol (Ref. 51). However, a description of its chemistry in terms of sub-structures such as radialene and paracyclene (see Fig. 3.1) does not reflect the chemical reality fully since these sub-structures are planar while fullerenes are spherical. Fullerenes are strained and continuous, and describing their chemistry in terms of a strained weakly aromatic molecule may be more appropriate (Ref. 52).

Since fullerenes have only carbon atoms, they cannot be used to achieve substitution reactions. However, such reactions can be achieved on their derivatives. The cage consists of sp^2 hybridized carbon atoms which have $-I$ inductive effect. Therefore, fullerenes strongly attract electrons and react readily with nucleophiles. These reactions are similar to those of poorly conjugated alkenes. A major problem concerns the number of additional products created by each reagent. A given product can also have a large number of structural isomers. In addition, they may have very little solubility in organic solvents. Many of them do not have high stability due to the strain caused by the addition of other products, and therefore, they may revert to the parent fullerenes under mass spectrometric examination. Only a few of them crystallize easily to facilitate single crystal examination. Besides, the study has to be performed with a very small quantity of material, often measured in milligrams, though the situation has improved substantially in recent years. Sometimes, the purification procedures leave only tiny quantities of pure compound at the end. Many a time the purification of isomers becomes very difficult which makes a complete understanding of the chemistry extremely time-consuming, and often impossible. Therefore, a large number of reactions have still not been fully studied. However, there are certain investigations which stand out and have resulted in the creation of unique products. Because of its comparatively easy availability and its unique symmetry, C_{60} has been subjected to more detailed examinations.

Early electrochemical studies have suggested that fullerene C_{60}, undergoes six reversible reductions (Ref. 53) corresponding to the complete filling of the t_{1u} LUMO. The early C_{60} chemistry revolved around this high electron affinity of the cage. Adducts of C_{60} with radicals, nucleophiles, carbenes and dienophiles have been reported. In organometallic chemistry functionalization leading to η^2 complexes of transition metals (Refs 36, 54) was reported early. The alkylation of C_{60} leading to methanofullerenes (Ref. 55) is being intensely pursued by the groups of Wudl, Rubin and Diederich. A detailed account of this can be found in a book by Hirsch (Ref. 4). These reactions have been shown to result in fullerene polymers (Ref. 56), fullerene dendrimers (Ref. 57) (see Fig. 3.5), fullerene-based HIV protease inhibitors (Ref. 58), fullerene nucleotide conjugates (Ref. 59) and a number of other potentially useful materials. The electrochemical and photophysical characteristics of fullerene adducts can be technologically interesting. Although the chemistry of C_{60} has been intensely investigated, that of higher fullerenes is beginning to attract attention. The product distribution is more complex because of the presence of a large number of non-equivalent bonds.

The fullerene cage itself could include other elements such as boron (Ref. 60) and nitrogen (Ref. 61). The evaporation of boron nitride doped graphite disk produced boron doped fullerenes (Ref. 60) such as $C_{60-n}B_n^+$ with $n = 0$ to 6. Photodissociation showed that these species are resistant to fragmentation, but reactions with ammonia showed the acidic behavior of boron atoms. Species such as $C_{60-n}B_n(NH_3)_n^+$

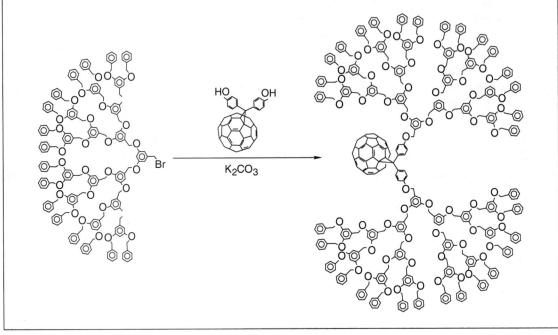

Fig. 3.5: *Schematic showing the synthesis of dendritic methanofullerene, using the established dentritic growth procedures. Reprinted with permission from Wooley, et al. (Ref. 51). Copyright (1993) American Chemical Society.*

were observed showing that boron has been incorporated into the cage. Substituted fullerenes such as $C_{59}NH$ are now prepared by synthetic chemistry (Ref. 62).

Photoionization time of flight mass spectrometry has been used to show that clusters of C_{60} are indeed formed during vaporization (Ref. 63). Intensity anomalies in the mass spectrum have been detected corresponding to $(C_{60})_n^+$ where n = 13, 19, 23, 35, 39, 43, 46 and 55. The numbers suggest that the closed shell clusters n = 13 and 55 are probably icosahedra. The magic numbers are similar to the clusters of Xe and Ar.

Fullerenes undergo coalescence reactions in the gas phase. In laser evaporation of fullerene films, it was found that the mass spectra show peaks at very high masses above m/z 720 with enhanced intensities in the range of $(C_{60})_n^+$ (Fig. 3.7) (Ref. 64). This proposes the fusion of C_{60} cages to form larger cage structures. The identities of the products were also studied by surface-induced dissociation which showed that the clusters are hard to dissociate and no parent fullerene is formed. This excludes the interpretation that the peaks could be due to difullerene-like structures. The reaction is supposed to be taking place in the excited dense plasma. Similar coalescence reactions have been reported by others also. Another type of reaction which has not been observed before has been the addition of C_{60} and C_2 in the gas phase (Ref. 65). In this study, a C_{60} film was bombarded by high energy ions. The mass spectrum of the produced ions shows peaks at C_{60}, C_{60+2}, C_{60+4}, etc. all the way up to C_{108}, with the mass limit of the instrument as shown in Fig. 3.6. This suggests that the C_n fragments add back to C_{60} in the plasma region immediately above the condensed phase where ion/molecule reactions occur. In fact such addition reactions are known to occur without any activation barrier (Ref. 66).

3.6 Endohedral Chemistry of Fullerenes

The cavity in the 7 Å diameter buckministerfullerene molecule has intrigued chemists ever since the discovery of fullerenes. Later, Smalley and his colleagues (Ref. 67) showed that laser evaporation of graphite

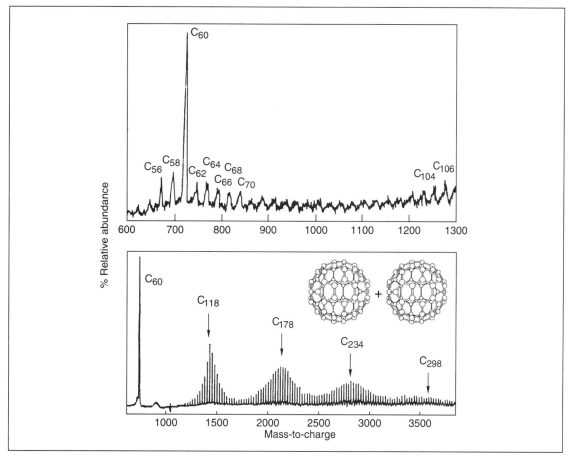

Fig. 3.6: *(Bottom) Mass spectrum of a laser evaporated C_{60} film showing coalescence of fullerenes. Mass peaks are seen at $(C_{60})_n$ (Ref. 64). (Top) Collision of high energy ions on C_{60} results in the addition of C_2s to C_{60}. The mass spectrum here shows the addition of a number of such species (Ref. 65). Combined figure originally published in, T. Pradeep, Current Science, 72 (1997) 124.*

impregnated with La_2O_3 can give $La@C_{82}$ and the product can be extracted in toluene. The molecule was shown to be a radical exhibiting EPR hyperfine structure (Ref. 68). Mass spectral evidence for other endohedrals was soon available. High energy collision experiments involving fullerene ions and noble gases were shown to result in noble gas encapsulated fullerenes (Ref. 47). For a substantial period of time, the characteristic mass spectral fragmentation pattern was the only experimental evidence for the existence of endohedral fullerenes. The fact that He could be put into the cage prompted Saunders to probe the chemistry of endohedrals through ^{3}He NMR (Ref. 69). $Rg@C_{60}$ (Rg = rare gas) molecules have been synthesized by a high temperature–high pressure method (Ref. 70). Studies showed the high diamagnetic shielding of the inner C_{60} surface indicating the existence of a high degree of aromaticity (Ref. 71). The ^{3}He endohedral chemistry has been extended to the study of derivatives of C_{60} (Refs 71, 72) and isomers of higher fullerenes (Ref. 15). Results show that there are at least eight C_{78} isomers and nine C_{84} isomers, which is much higher than the number originally suggested. $La@C_{82}$, $La@C_{80}$, $La@C_{76}$ and a number of Y, Sc and Mn endohedrals have been prepared and purified. The electrochemical, electronic and magnetic properties of these materials have been investigated (Refs 73, 74). $Gd@C_{82}$ is paramagnetic down to 3 K (Ref. 75). There are theoretical predictions about the various electronic properties of endohedrals but the experimental confirmation of these predictions is awaited. Another development in the endohedral chemistry is the preparation of $Li@C_{60}$ by low energy ion-beam collision of Li^+ on a C_{60} film (Ref. 76).

3.7 Orientational Ordering

As regards the several properties of fullerenes, a widely investigated aspect is their orientational ordering. Generally molecules of a high point group symmetry crystallize with a certain degree of orientational disorder. Some lower fullerenes are highly spherical and are held together by weak van der Waals forces, and their molecular orientations need not be ordered as in the case of an ordered crystal. This high asymmetry is proved by the fact that large thermal parameters are required to fit the observed diffraction patterns. This asymmetry is the reason why C_{60} had to be fixed with a molecular handle by complexation to refine the atomic positions as was done by Hawkins (Ref. 36).

In the pure form, C_{60} rotates freely in the lattice because of its high symmetry and weak intermolecular interactions. This may lead to interesting properties such as the unusual flow of liquids over C_{60} film (Ref. 77). This rotation can be frozen at low temperatures, resulting in an orientational ordering. Upon reaction or by co-crystallization, which inhibits the free rotation of C_{60} molecules, the atomic positions have been studied. At room temperature, C_{60} gives a sharp NMR signal at 143 ppm and the signal broadens at low temperatures as a result of chemical shift anisotropy (Ref. 78). This is due to the dynamic nature of the disorder, and the orientational correlation times are of the order of 16 picoseconds. The orientationally disordered face centred cubic (fcc) phase undergoes a transition to a simple cubic (sc) phase at 248 K (Ref. 79). In Fig. 3.7, a differential scanning calorimetric trace of the transition is shown. In the sc phase, the rotation exists only along the preferred axis. NMR studies have shown that below the ordering temperature, the molecules jump between preferred orientations over a barrier of 3000 K (Ref. 78). Phase transitions can be modeled which show that simple Lennard–Jones (L–J) potentials reproduce experimental behavior. Charge-transfer effects are, however, important (Ref. 80). Neutron diffraction studies indeed strongly support this suggestion in which it is found that the electron rich bonds of one molecule are located close to the electron poor pentagon rings of the adjacent molecule (Ref. 81).

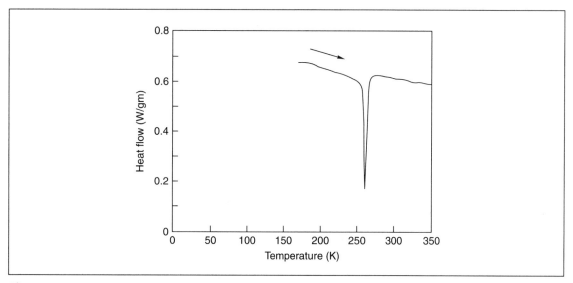

Fig. 3.7: *Results of a different scanning calorimetric measurement of a powder sample of C_{60}. The arrow indicates that the data were taken upon warming. Reprinted with permission from Heiney, et al. (Ref. 79). Copyright (1991) by the American Physical Society.*

Orientational ordering results can be seen in IR and Raman spectroscopies also (Ref. 82). A glassy phase has been observed around 80 K when molecular rotations are completely frozen (Ref. 83). It may be mentioned that many of the derivatives of C_{60} are also spherical and phase transitions of this type may be observed in them. We have (Ref. 84) observed such a transition in $C_{60}Br_{24}$.

C_{70} undergoes two-phase transitions at 337 and 276 K, and these transitions have been studied by IR and Raman spectroscopies (Ref. 85). Diffraction studies suggest a monoclinic structure for the lowest temperature phase and a rhombohedral or hexagonal structure for the in-between phase (Ref. 86). Complete orientational freezing occurs only at 130 K. Upon phase transition, the intramolecular phonons undergo hardening, a general feature observed for both C_{60} and C_{70}. Resistivity measurements have proved that there are two phase transitions in C_{60} and three in C_{70} (Ref. 87).

3.8 Pressure Effects

Both C_{60} and C_{70} are soft solids and their compressibilities are comparable to the c-axis compressibility of graphite. Upon the application of pressure, the orientational ordering temperature increases at a rate of

10 K/kbar (Ref. 88). Around 20 Gpa, C_{60} undergoes a transition to a lower symmetry structure (Ref. 89). Above 22 Gpa, the material undergoes amorphization and the phase shows evidence of sp^3 hybridized carbon. In C_{70}, amorphization take place around 12Gpa and only sp^2 hybridized carbon is seen (Ref. 99). This suggests that while polymerization via Diels-Alder addition might be taking place in C_{60}, no such reaction occurs in C_{70}. Polymerization upon light irradiation has been observed in C_{60} (Ref. 91). Under extremely high pressures, it is possible to convert C_{60} to diamond at low temperatures (Ref. 92). Such a conversion is seen in C_{70} also.

At fairly low pressures below the amorphization pressure, the photoluminescence band originally at 1.6 eV undergoes a redshift (Ref. 93). At a pressure of 3.2 Gpa, the band merges with the background corresponding to the collapse of the photoluminescent gap. This arises due to the broadening of the HOMO and LUMO which is a result of shortening of the inter-molecular distance and a concomitant increase in the inter-ball hopping integral. This study has repercussions for the low temperature electrical properties found in doped fullerenes.

3.9 Conductivity and Superconductivity in Doped Fullerenes

Doping with potassium, corresponding to the stoichiometry K_3C_{60}, produces a superconductor with a transition temperature of 19.8 K (Ref. 94, 95) (Fig. 3.8). The K_xC_{60} phase was earlier found to be metallic

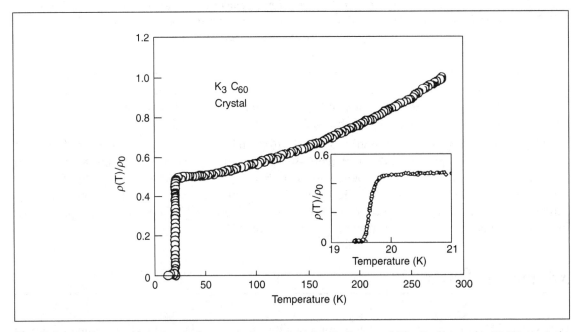

Fig. 3.8: *Normalized DC electrical resistivity $\rho(T)$ of a K_3C_{60} single crystal. The T_c observed is 19.8K. ρ_0 is the resistivity at T = 280 K. Reprinted with permission from Xiang, et al. (Ref. 95). Copyright (1992) AAAS.*

(Ref. 96). It is interesting to note that doping graphite with potassium also results in superconductivity, but at extremely low temperatures. However, the doping of Na and Li does not result in superconductivity, but other alkali metals do produce superconductors with a high transition temperature (T_c). The mixture of metals also shows the same phenomenon. Since the t_{1u} LUMO is triply degenerate, six electrons can be pumped into C_{60}. Six more can be placed in the t_{1g} LUMO. These electrons can be easily transferred from metal atoms placed in the tetrahedral and octahedral voids of fcc C_{60}. The highest T_c organic superconductor, Rb_2CsC_{60}, has a T_c of 31 K (Ref. 97). A_6B_{60}(bcc) and A_4C_{60} (bct) phases are non-superconducting (Ref. 98). Electron spectroscopy shows that the continuous filling of the LUMO takes place upon doping but on prolonged exposure, the LUMO band is shifted below E_f, making it an insulator (Ref. 99). Upon exposure to transition-metals, the d states grow near E_f and incomplete metal to C_{60} charge–transfer is observed (Ref. 100). Studies show that T_c is a function of inter-ball separation. Thus, intramolecular phonons are responsible for the superconducting ground states in these materials. Superconductivity in A_3C_{60} has been treated as a special case of the BCS theory with intermediate electron-phonon coupling (Ref. 101). Ca_3C_{60} (Ref. 102), Ba_6C_{60} (Ref. 103) and Sm_3C_{60} (Ref. 104) are superconducting. Superconductivity in higher fullerenes has not been reported probably due to lower symmetry.

3.10 Ferromagnetism in C_{60}.TDAE

Another important property which arises as a result of charge–transfer is ferromagnetism. Organic ferromagnetism is a rare phenomenon and is proved conclusively only in a few systems. So far there is only one donor that is known to produce ferromagnetism as a result of complexation with C_{60}, which is TDAE, tetrakisdimethylaminoethylene (Ref. 105). The T_c observed is 16.1 K, which is the highest T_c found for an organic material. The material belongs to the monoclinic system with short C_{60}–C_{60} contact along the c axis (Ref. 106). The quasi one-dimensional behavior has been suggested by susceptibility measurements also (Ref. 107). Single electron transfer has been confirmed by EPR (Refs 107, 108) and Raman studies (Ref. 109). The magnetic and electrical transport properties of C_{60}.TDAE have been reported on single crystals also (Ref. 110). A number of organic donors have been investigated for potential organic magnetism. Although all of them show pronounced magnetic correlation, no ferromagnetism has been reported (Ref. 111). In C_{60}.TDAE spontaneous magnetization has been verified by the zero-field muon relaxation technique as well (Ref. 112).

3.11 Optical Properties

C_{60} trapped in molecular sieves gives intense red light (Ref. 113). The light-transmitting properties of C_{60} embedded silica films have been investigated (Ref. 114). C_{60} is found to be an interesting material for non-linear optics, especially for third harmonic generation (Ref. 115). It would also be interesting to study C_{60} derivatives in this regard. The third harmonic generation optical susceptibility of C_{70} film is interpreted in terms of the totally symmetric ground state along with two excited states: a one-photon state and a two-photon state (Ref. 116). Photoconduction and photoinduced electron transfer, optical switching properties and second harmonic generation are also thoroughly investigated. Several conferences have devoted exclusive sessions on the subject of photophysics of fullerenes (Ref. 117).

3.12 Some Unusual Properties

In this section, are briefly summarized some of the fascinating discoveries in this area.

C_{60} shows anomalous solubility behavior. Normal solutes show a linear temperature dependence of solubility. In most cases, the solubility increases upon warming. In inorganic systems, unusual phenomena are seen due to solute-solvent interactions. Studies of solubility of C_{60}, studied in a number of solvents, indicate that there is a solubility maximum at near room temperature (Ref. 118) (280 K), dissolution is endothermic below room temperature and exothermic above the latter. The results have been interpreted as occurring due to a phase change in solid C_{60} modified by wetting due to the solvent. The solubility maximum in organic materials is unprecedented.

Ordinarily materials have a critical point above which the distinction between liquid and solid disappears. The liquid is not seen below the triple point where solid and vapor co-exist. C_{60} is one substance in which no liquid phase exists according to theoretical investigations of the phase diagram (Ref. 119), though the question is has still not been resolved (Ref. 120). If at all such a liquid phase does exist, it will be only at high temperatures above 1800 K where the molecules may not be stable.

The functionalization of C_{60} makes it soluble in water and the interaction of organofullerenes with DNA, proteins, and living cells, has been investigated. The interesting biological activity of organofullerenes is due to their photochemistry, radical quenching, and hydrophobicity to form one- to three-dimensional supramolecular complexes (Ref. 121). The cytotoxicity of water-soluble fullerene species is sensitive to surface derivatization. The lethal dose of fullerene changed over seven orders of magnitude with relatively minor alterations in fullerene structure. Under ambient conditions in water, fullerenes can generate superoxide anions and it has been suggested that these oxygen radicals are responsible for membrane damage and subsequent cell death (Ref. 122).

Several new forms of fullerenes have been made. The simplest dimer of C_{60}, namely C_{120}, linked by a cyclobutane ring alone, has been synthesized by the solid-state mechanochemical reaction of C_{60} with potassium cyanide (Ref. 123). X-ray structural analysis shows that the C_4 ring connecting the cages is square rather than rectangular. The dimer dissociates cleanly into two C_{60} molecules on heating, but in the gas phase, during mass-spectrometric measurements, it undergoes a successive loss of C_2 units, shrinking to even-numbered fullerenes such as C_{118} and C_{116} in a sequence similar to that of other fullerenes. The smallest possible fullerene is C_{20}, which consists solely of pentagons. But the extreme curvature and reactivity of this structure make it unstable. While a few isomeric structures, namely bowl and ring, have been identified, no cage structure is known so far. The cage-structured fullerene C_{20} can be produced from its perhydrogenated form (dodecahedrane $C_{20}H_{20}$) by replacing the hydrogen atoms with bromine atoms, followed by gas phase debromination (Ref. 124). The gas phase C_{20} clusters have been characterized by using mass-selective anion photoelectron spectroscopy.

Most of the metallofullerenes prepared so far have been based on C_{82}, and have incorporated most of the lanthanide elements, but there has been some debate about the endohedral nature of these compounds. Various observations such as scanning tunneling microscopy, extended X-ray absorption fine structure, transmission electron micrscopy and electron spin resonance, have strongly suggested that the metal atoms are indeed inside the fullerene cages. There are also theoretical suggestions that endohedrals do exist. However, no structural model has confirmed the endohedral nature of metallofullerenes. The endohedral nature of the metal atom has been proved through the use of synchrotron X-ray powder diffraction of $Y@C_{82}$, (Ref. 125).

Several advances have been made in the synthesis of endohedral fullerenes. One of the most recent ones is the synthesis of C_{60} with molecular hydrogen inside. This has been done by a four-step reaction completely closing an open cage fullerene (Ref. 126).

Notable quantities of C_{60} have been synthesized in 12 steps from commercially available starting materials by using rational chemical methods (Ref. 127). A molecular polycyclic aromatic precursor bearing chlorine substituents at key positions forms C_{60} when subjected to flash vacuum pyrolysis at 1100°C. No other fullerenes are formed as by-products.

Review Questions

1. What made the discovery of fullerenes possible?
2. What are the principal properties of C_{60}?
3. What is the condition for a geometrical structure to be closed-caged?
4. Due to sphericity, are there specific properties possible with C_{60}?
5. Is it possible to understand the chemistry of C_{60} starting from that of ethylene? If so, what are the differences between the two?
6. Why C_{60} chemistry is investigated in the gas phase?
7. Are the molecules similar to C_{60}, formed by other elements?
8. What are the principal challenges in understanding the chemistry of fullerenes?
9. Are there other closed cage objects formed by carbon? Can one propose a few chemically consistent structures?
10. Why only cages? How about rings and boxes? Are such structures possible and what would be their properties?

References

1. For a historical review of this area see, Aldersey-Williams, (1995), Hugh, (1995), *The Most Beautiful Molecule, The Discovery of the Buckyball*, John Wiley and Sons, Inc., New York.
2. Osawa, E., *Kagaku (Kyoto)*, **25** (1970), p. 854; D.A. Bochvar and E.G. Gal'pern, *Proc. Acad. Sci. USSR*, **209** (1973), p. 239.
3. Hunter, J., J. Fye and M.F. Jarrold, *J. Phys. Chem.*, **97** (1993), p. 3460; Gert von Helden, P.R. Kemper, N.G. Gotts and M.T. Bowers, *Science*, **259** (1993), p. 1300; Gert von Helden, Ming-Teh Hsu, P.R. Kemper and M.T. Bowers, *J. Chem. Phys.*, **95** (1991), p. 3835.
4. Jena, P., B.K. Pao and S.N. Khanna, (1987), *Physics and Chemistry of Small Clusters*, Plenum Press, New York, H.W. Kroto, J.E. Fischer and D.E. Cox, (eds) *The Fullerenes*, Pergamon, Oxford, (1993): W.E. Billups and M.A. Ciufolini, (eds) *Buckministerfullerene*, VCH, New York, (1993); K.M. Kadish and R.S. Ruoff, (eds) *Fullerenes: Recent Advances in the Chemistry and Physics of Fullerenes and Related Materials*, Electrochemical Society, Pennington, New Jersey, (1994); J. Baggot, *Perfect Summetry: The Accidental Discovery of Buckministerfullerene*, Oxford University Press, New York, (1994); J. Cioslowski, *Electronic Structure Calculations on Fullerenes and their Derivatives*, Oxford University Press, New York, (1995); M.S. Dresselhaus, G. Dresselhaus and P.C. Eklund, *Science of Fullerenes and Carbon Nanotubes*, Academic, New York, (1996); M.C. Flemings, J.B. Watchman Jr., E.N. Kaufman and J. Giordmaine, (eds) *Annual Review of Materials Science*, **24**, 1994.
5. A. Hirsch, (1994), *The Chemistry of Fullerenes*, Stuttgart, Thieme.
6. S. Iijima, *Nature*, **354** (1991), p. 56; P.M. Ajayan and T.W. Ebbesen, *Nature*, **358** (1992), p. 220.
7. For an account of the early researches in this area, please refer to H.W. Kroto, *Science*, **242** (1988), p. 1139; R.F. Curl and R.E. Smalley, *Science*, **242** (1988), p. 1017.
8. Brucat, P.J., L.S. Zheng, C.L. Pettiette, S. Yang and R.E. Smalley, *J. Chem. Phys.*, **84** (1986), p. 3078; S.C. O'Brien, Y. Liu, Q. Zhang, J.R. Heath, F.K. Tittel, R.F. Curl and R.E. Smalley, *J. Chem. Phys.*, **84** (1986), p. 4074.
9. Rohlfing, E.A., D.M. Cox and A.J. Kaldor, *J. Chem. Phys.*, **81** (1984), p. 3322.
10. Kroto, H.W., J.R. Heath, S.C. O'Brien, R.F. Curl and R.E. Smalley, *Nature*, **318** (1985), p. 162.
11. Marks, R.W., (1960), *The Dymaxion World of Buckminister Fuller*, Reinhold, New York.
12. Jones, D.E.H., *New Sci.*, **3** Nov. (1966). p. 245; D.E.H. Jones, *The Inventions of Daedalus*, Freeman, Oxford, (1982), pp. 118–119.
13. Mackay, A.L. and H. Terrons, *Nature*, **352** (1991), p. 762; T. Lenosky, X. Gonze, M. Tester and V. Elser, *Nature*, **355** (1992), p. 333.
14. Saunders, M., H.A. Jimenéz-Vazques, R.J. Cross, W.E. Billups, C. Gessenberg, A. Gonzalez, W. Luo, R.C. Haddon, F. Diederich and A. Herrmann, *J. Am. Chem. Soc.*, **117** (1995), p. 9305.
15. Haddon, R.C., L.E. Brus and K. Raghavachari, *Chem. Phys. Lett.*, **125** (1986), p. 459; R.L. Disch and J.M. Schulman, **125** (1986), p. 465.

16. Zhang, Q.L., S.C. O'Brien, J.R. Heath, Y. Liu, R.F. Curl, H.W. Kroto and R.E. Smalley, *J. Phys. Chem.*, **90** (1986), p. 525; S.W. McElvany, H.H. Nelson, A.P. Baronowski, C.H. Watson and J.R. Eyler, *Chem. Phys. Lett.*, **134** (1987), p. 214; F.D. Weiss, J.L. Elkind, S.C. O'Brien, R.F. Curl and R.E. Smalley, *J. Am. Chem. Soc.*, **110** (1988), p. 4464.

17. Elser, V. and R.C. Haddon, *Phys. Rev.*, A **36** (1987), p. 4579.

18. Santon, R.E. and M.D. Newton, *J. Phys. Chem.*, **92** (1988), p. 2141; Z.C. Wu, D.A. Jelski and T.F. George, *Chem. Phys. Lett.*, **137** (1987), p. 291; A.D.J. Haymet, *J. Am. Chem. Soc.*, **108** (1986), p. 319; D.E. Weeks and W.G. Harter, *J. Chem. Phys.*, **90** (1989), p. 4727.

19. Krätschmer, W., K. Fostiropoulos and D.R. Huffmän, *Chem. Phys. Lett.*, **170** (1990), p. 167.

20. Krätschmer, W., L.D. Lamb, K. Fostiropoulos and D.R. Huffmän, *Nature*, **347** (1990), p. 354.

21. Taylor, R., J.P. Hare, A. Abdul-Sada and H.W. Kroto, *J. Chem. Soc., Chem. Commun.*, **20** (1990), p. 1423.

22. Issacs, L., A. Wehrsig and F. Diederisch, *Hew. Chem. Acta.*, **76** (1993), p. 1231; W.A. Scrivens, P.V. Bedworth and J.M. Tour, *J. Am. Chem. Soc.*, **114** (1992), p. 7917; A. Govindaraj and C.N.R. Rao, *Fullerene Sci. Technol.*, **1** (1993), p. 557.

23. Diederisch, F. and R.L. Whetten, *Acc. Chem. Res.*, **25** (1992), p. 119; F. Diederich, R. Ettl, Y. Rubin, R.L. Whetten, R.D. Beck, M.M. Alvarez, S. Anz, D. Sensharma, F. Wudl, K.C. Khemani and A. Koch, *Science*, **252** (1991), p. 548; K. Kikuchi, N. Nakahara, T. Wakabayashi, M. Honda, H. Matsumiya, T. Moriwaki, S. Suzuki, H. Shiromaru, K. Saito, K. Yamauchi, I. Ikemoto and Y. Achiba, *Chem. Phys. Lett.*, **188** (1992), p. 177.

24. Atwood, J.L., G.A. Koutsantonis and C.L. Raston, *Nature*, **368** (1994), p. 229.

25. Howard, J.B., J.T. Mckinnon, Y. Makarovsky, A. Lafleur and M.E. Johnson, *Nature*, **352** (1991), p. 139.

26. Pradeep, T. and C.N.R. Rao, *Curr. Sci.*, **61** (1991), p. 432.

27. Chow, L., H. Wang, S. Kleckly, T.K. Daly and P.R. Buseck, *Appl. Phys. Lett.*, **66** (1995), p. 430.

28. Becker, L., J.L. Bada, R.E. Winans, J.E. Hunt, T.E. Bunch and B.M. French, *Science*, **265** (1994), p. 642.

29. Radicati di Brozolo, F., T.E. Bunch, R.H. Fleming and J. Macklin, *Nature*, **369** (1994), p. 37.

30. Yosida, Y., *Appl. Phys. Lett.*, **67** (1995), p. 1627.

31. Scott, L.T., M.S. Bratcher and S. Hagen, *J. Am. Chem. Soc.*, **118** (1996), p. 8743.

32. Mukhopadhyay, K., K.M. Krishna and M. Sharon, *Phys. Rev. Lett.*, **72** (1994), p. 3182.

33. Brenna, J.T., W.R. Creasy, *J. Chem. Phys.*, **92** (1990), p. 2269; E.E.B. Campbell, G. Ulmer, B. Hasselberger, H.G. Busmann and I.V. Hertel, *J. Chem. Phys.*, **93** (1990), p. 6900.

34. McElvany, S.W., M.M. Ross, N.S. Goroff and F. Diederich, *Science*, **259** (1993), p. 1594.

35. Mehta, G., S.R. Shah and K. Ravikumar, *J. Chem. Soc., Chem. Commun.*, **12** (1993), p. 1006; G. Mehta, G.V. Sharma, R. Kumar, M.A. Krishna, T.V. Vedavyasa and E.D. Jemmis, *J. Chem. Soc., Perkin Trans.*, **1** (1995), p. 2529.

36. Hawkins, J.M., A. Meyer, T.A. Lewis, S. Loren and F.J. Hollander, *Science*, **252** (1991), p. 312.
37. Aijie, H., M.M. Alvarez, S.J. Anz, R.D. Beck, F. Diederich, K. Fostiropoulos, D.R. Huffmän, W. Krätschmer, Y. Rubin, K.E. Schriver, K. Sensharma and R.L. Whetten, *J. Phys. Chem.*, **94** (1990), p. 8630; J.P. Hare, H.W. Kroto and R. Taylor, *Chem. Phys. Lett.*, **117** (1991), p. 394.
38. Bethune, D.S., G. Meiger, W.C. Tang and H.J. Rosen, *Chem. Phys. Lett.*, **174** (1990), p. 219.
39. Lichtenberger, D.L., K.W. Nebeshy, C.D. Ray, D.R. Huffmän and L.D. Lamb, *Chem. Phys. Lett.*, **176** (1991), p. 203.
40. For a review on the mass spectrometry of fullerenes, see S.W. McElvany and M.M. Ross, *J. Am. Soc. Mass Spectrom.*, **3** (1992), p. 267.
41. McElvany, S.W. and J.H. Callahan, *J. Phys. Chem.*, **951** (1991), p. 6186.
42. Campbell, E.E.B., G. Ulmer, B. Hasselberger, H.G. Busmann and I.V. Hertel, *J. Chem. Phys.*, **93** (1990), p. 6900; R.D. Beck, P. St. John, M.M. Alvarez, F. Diederich and R.L. Whetten, *J. Phys. Chem.*, **95** (1991), p. 8402.
43. Rathna, A. and J. Chandrasekhar, *J. Mol. Struct.*, **327** (1994), p. 255.
44. Campbell, E.E.B., G. Ulmer and I.V. Hertel, *Phys. Rev. Lett.*, **67** (1991), p. 1986.
45. Yeretzian, C., R.D. Beck and R.L. Whetten, *Int. J. Mass Spectrom. Proc.*, **135** (1994), p. 79; R.D. Beck, P. Weis, G. Brauchle and J. Rockenberger, *Rev. Sci. Instr.*, **66** (1995), p. 4188.
46. Baba, M.S., T.S.L. Narasimhan, R. Balasubramanian, N. Sivaraman and C.K. Mathews, *J. Phys. Chem.*, **98** (1994), p. 1333.
47. Weiske, T., D.K. Bohme, J. Hrusak, W. Kratschmer and H. Schwarz, *Angew. Chem. Int. Ed. Engl.*, **30** (1991), p. 884; R.C. Mowrey, M.M. Ross and J.H. Callahan, *J. Phys. Chem.*, **96** (1992), p. 4755; S.W. McElvany, *J. Phys. Chem.*, **96** (1992), p. 4935.
48. Haufler, R.E., J. Conceicao, L.P.F. Chibante, Y. Chai, N.E. Byrne, S. Flanagan, M.M. Haley, S.C. O'Brien, C. Pan, Z. Xiao, W.E. Billups, M.A. Ciufolini, R.H. Hauge, J.L. Margrave, L.J. Wilson, R.F. Curl and R.E. Smalley, *J. Phys. Chem.*, **94** (1990), p. 8634; J.M. Wood, B. Kahr, S.H. Hoke, L. Dejarme, R.G. Cooks and D. Ben-Amotz, *J. Am. Chem. Soc.*, **113** (1991), p. 5907; S.H. Hoke, J. Molstad, B. Kahr and R.G. Cooks, *Int. J. Mass Spectrom. Ion Process.*, **138** (1994), p. 209.
49. Bohme, D.K., *Chem. Rev.*, **92** (1992), p. 1487; S. Petrie, G. Javahery, J. Wang and D.K. Bohme, *J. Am. Chem. Soc.*, **114** (1992), p. 6268; S. Petrie, G. Javahery, H. Wincel and D.K. Bohme, *J. Am. Chem. Soc.*, **115** (1993), p. 6290.
50. Taylor, R. and D.R.M. Walton, *Nature*, **363** (1993), p. 685.
51. Matsuzawa, N., D. Dixon and T. Fukunaga, *J. Phys. Chem.*, **96** (1992), p. 7594.
52. Haddon, R.C., *Science*, **261** (1993), p. 1545.
53. Dubois, D. and K.M. Kadish, *J. Am. Chem. Soc.*, **113** (1991), p. 7773; Q. Xie, F. Arias and L. Echegoyen, *J. Am. Chem. Soc.*, **115** (1993), p. 9818; C. Jehoulet, A.J. Bard and F. Wudl, *J. Am. Chem. Soc.*, **113** (1991), p. 5456.

54. Balch, A.L., V.J. Catalano, J.W. Lee and M.M. Olmstead, *J. Am. Chem. Soc.*, **114** (1992), p. 5455; R.S. Koefod, M.F. Hudgens and J.R. Shapley, *J. Am. Chem. Soc.*, **113** (1991), p. 8957; A.L. Balch, J.W. Lee, B.C. Noll and M.M. Olmstead, *J. Am. Chem. Soc.*, **114** (1992), p. 10984; P.J. Fagan, J.C. Calabrese and B. Malone, *J. Am. Chem. Soc.*, **113** (1991), p. 9408; P.J. Fagan, J.C. Calabrese and B. Malone, *Acc. Chem. Res.*, **25** (1992), p. 134.
55. Suzuki, T., Q. Li, K.C. Khemani, F. Wudl and O. Almarsson, *Science*, **254** (1991), p. 1186.
56. Shi, S., K.C. Khemani, Q. Li and F. Wudl, *J. Am. Chem. Soc.*, **114** (1992), p. 10656.
57. Wooley, K.L., C.J. Hawker, J.M.J. Frechet, F. Wudl, G. Sardanov, S. Shi, C. Li and M. Rao, *J. Am. Chem. Soc.*, **115** (1993), p. 9836.
58. Friedman, S.H., D.L. Decamp, R.P. Sijbesma, G. Srdanov, F. Wudl and G.L. Kenyon, *J. Am. Chem. Soc.*, **115** (1993), p. 6506; R.F. Schinazi, R. Sijbesma, G. Srdanov, C.L. Hill and F. Wudl, *Antimicrob. Agents Chemother*, **37** (1993), p. 1707.
59. Toniolo C., *et al.*, *J. Med. Chem.*, **37** (1994), p. 4558.
60. Guo, T., D. Jin and R.E. Smalley, *J. Phys. Chem.*, **95** (1991), p. 4948.
61. Pradeep, T., V. Vijayakrishnan, A.K. Santra and C.N.R. Rao, *J. Phys. Chem.*, **95** (1991), p. 10564.
62. Keshavarz-K, M., R. Gonzalez, R.G. Hicks, G. Srdanov, V.I. Svdanov, T.G. Collins, J.C. Hmmelen, C. Bellavia-Lund, J. Pavlovich, F. Wudl and K. Holczer, *Nature*, **383** (1996), p. 147.
63. Martin, T.P., U. Naher, H. Schaber and U. Zimmermann, *Phys. Rev. Lett.*, **70** (1993), p. 3079.
64. Yeretzian, C., K. Hansen, F. Diedrich and R.L. Whetten, *Nature*, **359** (1992), p. 44.
65. Pradeep, T. and R.G. Cooks, *Int. J. Mass Spectrom. Ion Process*, **135** (1994), p. 243.
66. Yi, J.Y. and J. Bernholc, *Phys. Rev.*, **B48** (1993), p. 5724.
67. Chai, Y., T. Guo, C. Jin, R.E. Haufler, L.P.F. Chibante, J. Fure, L. Wang, J.M. Alford and R.E. Smalley, *J. Phys. Chem.*, **95** (1991), p. 7564.
68. Johnson, R.D., M.S. de Vries, J. Salem, D.S. Bethune and C.S. Yannoni, *Nature*, **355** (1992), p. 239; T. Suzuki, Y. Maruyoma, T. Kato, K. Kikuchi and Y. Achiba, *J. Am. Chem. Soc.*, **115** (1993), p. 11006.
69. Saunders, M., H.A. Jimenéz-Vazques, R.J. Cross, S. Mroczkowski, M.L. Gross, D.E. Giblin, R.J. Poreda, *J. Am. Chem. Soc.*, **116** (1994), p. 2193; M. Saunders, R.J. Cross, A. Hugo, H.A. Jimenéz-Vazques, R. Shimshi and A. Khong, *Science*, **271** (1996), p. 22.
70. See for a review of rare gas encapsulated fullerenes, Saunders, M., H.A. Jimenéz-Vazques, R.J. Cross and R.J. Poreda, *Science*, **259** (1993), p. 1428; M. Saunders, H.A. Jimenéz-Vazques, R.J. Cross, S. Mroczkowski, D.I. Freedberg and F.A.L. Anet, *Nature*, **367** (1994), p. 256.
71. Saunders, M., H.A. Jimenéz-Vazques, R.J. Cross, W.E. Billups, C. Gesenberg and D. McCord, *Tetrahedron Lett.*, **35** (1994), p. 3869; A.B. Smith III, R.M. Strongin, L. Brard, W.J. Romanow, M. Saunders, H.A. Jimenéz-Vazques and R.J. Cross, *J. Am. Chem. Soc.*, **116** (1994), p. 10831.
72. Saunders, M., H.A. Jimenéz-Vazques, B.W. Bangerter and R.J. Cross, *J. Am. Chem. Soc.*, **116** (1994), p. 3621.

73. Suzuki, T., S. Nagase, K. Kobayashi, Y. Maruyama, T. Kato, K. Kikuchi, Y. Nakao and Y. Achiba, *Angew. Chem. Int. Ed. Engl.*, **34** (1995), p. 1094.
74. Kikuchi, K., S. Suzuki, Y. Nakao, N. Nakahara, T. Wakabayashi, H. Shiromaru, K. Saito, I. Ikemoto and Y. Achiba, *Chem. Phys. Lett.*, **216** (1993), p. 67.
75. Funasaka, H., K. Sakurai, Y. Oda, K. Yamamoto and T. Takahashi, *Chem. Phys. Lett.*, **232** (1995), p. 273.
76. Tellgmann, R., N. Krawez, S.H. Lin, I.V. Hertel and E.E.B. Campbell, *Nature*, **382** (1996), p. 520.
77. Campbell, S.E., G. Leungo, V.I. Srdanov, F. Wudl and J.N. Israelachivili, *Nature*, **382** (1996), p. 520.
78. Yannoni, C.S., R.D. Johnson, G. Meijer, D.S. Bethune and J.R. Salem, *J. Phys. Chem.*, **95** (1991), p. 9.
79. Heiney, P.A., J.E. Fischer, A.R. McGhie, W.J. Romanow, A.M. Denenstein, J.P. McCauley Jr. and A.B. Smith III, *Phys. Rev. Lett.*, **66** (1991), p. 2911.
80. Chakrabarti, A., S. Yashonath and C.N.R. Rao, *Chem. Phys. Lett.*, **215** (1993), p. 591; M. Sprik, A. Cheng and M.L. Klein, *J. Phys. Chem.*, **96** (1992), p. 2027; J.P. Lu, X.P. Li and R.M. Martin, *Phys. Rev. Lett.*, **68** (1992), p. 1551.
81. David, W.I.F., R.M. Ibberson, J.C. Matthewman, K. Prassides, T.J.S. Dennis, J.P. Hare, H.W. Kroto, R. Taylor and D.R.M. Walton, *Nature*, **353** (1991), p. 147.
82. Chase, B., N. Herron and E. Holler, *J. Phys. Chem.*, **96** (1992), 4262; P.H.M. van Loosdrecht, P.J.M. van Bentum and G. Meijer, *Phys. Rev. Lett.*, **68** (1992), p. 1176.
83. Akers, K., K. Fu, P. Zhang and M. Moscovits, *Science*, **259** (1992), p. 1152; F. Gugenberger, R. Heid, C. Meingast, P. Adelmann, M. Braun, H. Wudl, M. Haluska and H. Kuzmany, *Phys. Rev. Lett.*, **69** (1992), p. 3774.
84. Resmi, M.R., L. George, S. Singh, U. Shankar and T. Pradeep, *J. Mol. Struct.*, **435** (1997), p. 11.
85. Varma, V., R. Seshadri, A. Govindaraj, A.K. Sood and C.N.R. Rao, *Chem. Phys. Lett.*, **203** (1993), p. 545; N. Chandrabhas, K. Jayaraman, D.V.S. Muthu, A.K. Sood, R. Seshadri and C.N.R. Rao, *Phys. Rev.*, **B47** (1993), p. 10963.
86. Christides, C., I.M. Thomas, T.J.S. Dennis and K. Prassides, *Europhys. Lett.*, **22** (1993), p. 611; M.A. Verheijin, H. Meeks, G. Meijir, P. Bennema, J.L. de Boer, S. van Smaalen, G. van Tendeloo, S. Amelinckx, S. Muto and J. van Landuyt, *Chem. Phys.*, **166** (1992), p. 287.
87. Ramasesha, S.K., A.K. Singh, R. Seshadri, A.K. Sood and C.N.R. Rao, *Chem. Phys. Lett.*, **220** (1994), p. 203.
88. Samara, G.A., J.E. Schriber, B. Morosin, L.V. Hansen, D. Loy and A.P. Sylwester, *Phys. Rev. Lett.*, **67** (1991), p. 3136.
89. Duclos, S.J., K. Brister, R.C. Haddon, A. Kortan and F.A. Thiel, *Nature*, **351** (1991), p. 380.
90. Chandrabhas, N., A.K. Sood, D.V.S. Muthu, C.S. Sundar, A. Bharati, Y. Hariharan and C.N.R. Rao, *Phys. Rev. Lett.*, **73** (1994), p. 3411; C.N.R. Rao, A. Govindaraj, H.N. Aiyer and R. Seshadri, *J. Phys. Chem.*, **99** (1995), p. 16814.

91. Zhou, P., Z.H. Dong, A.M. Rao and P.C. Eklund, *Science*, **259** (1993), p. 955.
92. Nunez-Regueiro, M., P. Monceau, A. Rassat, P. Bernier and A. Zahab, *Nature*, **354** (1994), p. 289.
93. Sood, A.K., N. Chandrabhas, D.V.S. Muthu, A. Jayaraman, N. Kumar, H.R. Krishnamurthy, T. Pradeep and C.N.R. Rao, *Solid State Commun.*, **81** (1992), p. 319.
94. Hebard, A.F., M.J. Rosseinsky, R.C. Haddon, D.W. Murphy, S.H. Glarum, T.T.M. Palstra, A.P. Ramirez and A.R. Kortan, *Nature*, **350** (1991), p. 600.
95. Xiang, X.D., J.G. Hou, G. Briceno, W.A. Vareka, R. Mostovoy, A. Zettl, V.H. Crespi and M.L. Cohen, *Science*, **256** (1992), p. 1190.
96. Haddon, R.C., A.F. Hebbard, M.J. Rosseinsky, D.W. Murphy, S.J. Duclos, K.B. Lyons, B. Miller, J.M. Rosamilia, R.M. Flemming, A.R. Kortan, S.H. Glarum, A.V. Makhija, A.J. Muller, R.H. Eick, S.M. Zahurak, R. Tycko, G. Dabbagh and F.A. Thiel, *Nature*, **350** (1991), p. 320.
97. Tanigaki, K., T.W. Ebbesen, S. Saito, J. Mizuki, J.S. Tsai, Y. Kubo and S. Kuroshima, *Nature*, **352** (1991), p. 222; S.P. Kelty, Chia-Chun Chen and C.M. Leiber, *Nature*, **352** (1991), p. 223.
98. Chalet, O., G. Ozlanyi, L. Forro, P.W. Stephens, M. Tegze, G. Faigel and A. Janossy, *Phys. Rev. Lett.*, **72** (1994), p. 2721; S. Pekker, A. Janossy, L. Mihaly, O. Chauvet, M. Carrord and L. Forro, *Science*, **265** (1994), p. 1077.
99. Benning, P.J., J.L. Martins, J.H. Weaver, L.P.F. Chibante and R.E. Smalley, *Science*, **252** (1991), p. 1417.
100. Santra, A.K., R. Seshadri, A. Govindaraj, V. Vijayakrishnan and C.N.R. Rao, *Solid State Commun.*, **85** (1993), p. 77; J.H. Weaver, *J. Phys. Chem. Solids*, **53** (1992), p. 1433.
101. Varma, C.M., J. Zaanen and K. Raghavachari, *Science*, **254** (1993), p. 989.
102. Kortan, A.R., N. Kopylov, S. Glarum, E.M. Gyorgy, A.P. Ramirez, R.M. Fleming, F.A. Thiel and R.C. Haddon, *Nature*, **355** (1992), p. 529.
103. Kortan, A.R., N. Kopylov, S. Glarum, E.M. Gyorgy, A.P. Ramirez, R.M. Fleming, F.A. Thiel and R.C. Haddon, *Nature*, **360** (1992), p. 566.
104. Chen, X.H. and G. Roth, *Phys. Rev.*, **B52** (1995), p. 15534.
105. Allemand, P.M., K.C. Hemani, A. Koch, F. Wudl, K. Holczer, S. Donovan, G. Gruner and J.D. Thompson, *Science*, **253** (1991), p. 301.
106. Stephens, P.W., D. Cox, J.W. Lauher, L. Mihaly, J.B. Wiley, P.M. Allemand, A. Hirsch, K. Holczer, Q. Li, J.D. Thompson and F. Wudl, *Nature*, **355** (1992), p. 331.
107. Seshadri, R., A. Rastogi, S.V. Bhat, S. Ramasesha and C.N.R. Rao, *Solid State Commun.*, **85** (1993), p. 971.
108. Tanaka, K., A.A. Zhakidov, K. Yoshizawa, K. Okahara, T. Yamabe, K. Yakushi, K. Kikuchi, S. Suzuki, I. Ikemoto and Y. Achiba, *Phys. Lett.*, **A64** (1992), p. 221.
109. Muthu, D.V.S., M.N. Shashikala, A.K. Sood, R. Seshadri and C.N.R. Rao, *Chem. Phys. Lett.*, **217** (1994), p. 146.
110. Suzuki, A., T. Suzuki and Y. Maruyama, *Solid State Commun.*, **96** (1995), p. 253.

111. 1Klos, H., I. Rystau, W. Schutz, B. Gotschy, A. Skiebe and A. Hirsch, *Chem. Phys. Lett.*, **224** (1994), p. 333.
112. Asai, Y., M. Tokumoto, K. Tanaka, T. Sato and T. Yamabe, *Phys. Rev.*, **B53** (1996), p. 4176.
113. Gu, G., W. Ding, G. Cheng, W. Zang, H. Zen and Y. Du, *Appl. Phys. Lett.*, **67** (1995), p. 326.
114. Maggini, M., G. Scorrano, M. Prato, G. Brusatin, P. Innocenzi, P. Guglielmi, A. Renier, R. Signorini, M. Meneghetti and R. Bozio, *Advan. Mater.*, **7** (1995), p. 404.
115. Kajzar, F., C. Taliani, R. Zamboni, S. Rossini and R. Danieli, (1993), *Proc. Of the SPIE-The Int. Soc. Optical Eng.*, San Diego, U.S.A. 2025, 352 and other papers in the Conference Proceedings.
116. Kajzar, F., C. Taliani, R. Danieli, S. Rossini and R. Zamboni, *Phys. Rev. Lett.*, **73** (1994), p. 1617.
117. See for, e.g., *Proc. of the SPIE-The Int. Soc. Optical Eng.*, (1995) San Diego, U.S.A.
118. Ruoff, R.S., R. Malhotra, D.L. Heustis, D.S. Tse and D.C. Lorentz, *Nature*, **362** (1993), p. 140.
119. Cheng, A., M.L. Klein and C. Caccamo, *Phys. Rev. Lett.*, **71** (1993), p. 1200.
120. Hagen, M.H.J., E.J. Meijer, G.C.A.M. Mooij, D. Frenkel and H.N.W. Lekkerkerker, *Nature*, **365** (1993), p. 425.
121. Nakamura, E., H. Isobe, *Acc. Chem. Res.*, **36** (2003), pp. 807–815.
122. Sayes, C.M., J.D. Fortner, W. Guo, D. Lyon, A.M. Boyd, K.D. Ausman, Y.J. Tao, B. Sitharaman, L.J. Wilson, J.B. Hughes, J.L. West, V.L. Colvin, *Nano Lett.*, **4** (2004), pp. 1881–1887.
123. Wang, G.W., K. Komatsu, Y. Murata, M. Shiro, *Nature*, **387** (1997), pp. 583–586.
124. Prinzbach, H., A. Weller, P. Landenberger, F. Wahl, J. Worth, L.T. Scott, M. Gelmont, D. Olevano, B. von Issendorff, *Nature*, **407** (2000), pp. 60–63.
125. Takata, M., B. Umeda, E. Nishibori, M. Sakata, Y. Saito, M. Ohno, H. Shinohara, *Nature*, **377** (1995), pp. 46–49.
126. Komatsu, K., M. Murata, Y. Murata, *Science*, **307** (2005), pp. 238–240.
127. Scott, L.T., M.M. Boorum, B.J. McMahon, S. Hagen, J. Mack, J. Blank, H. Wegner, A. de Meijere, *Science*, **295** (2002), pp. 1500–1503.

Additional Reading

Recent Books

1. Kadish, K.M. and R.S. Ruoff (eds) (2000), *Fullerenes: Chemistry, Physics, and Technology*, Wiley VCH.
2. Hirsch, A., M. Brettreich and F. Wudl, (2005), *Fullerenes: Chemistry and Reactions*, Wiley VCH.
3. Kroto, H.W., (2002), *The Fullerenes: New Horizons for the Chemistry, Physics and Astrophysics of Carbon*, Cambridge University Press.
4. Guldi, D.M. and N. Martin, (2002), *Fullerenes: From Synthesis to Optoelectronic Properties*, Springer.

5. R. Taylor (ed.) (1995), *The Chemistry of Fullerenes*, World Scientific.
6. Osawa, E., (2002), *Perspectives of Fullerene Nanotechnology*, Springer.
7. Shinar, J., Z.V. Vardeny, Z.H Kafafi (eds) (1999), *Optical and Electronic Properties of Fullerenes and Fullerene-Based Materials*.
8. See also ref. 4.

This work was originally published in, Current Science, **72** (1997) 124.

Chapter 4

CARBON NANOTUBES

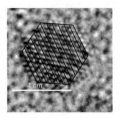

Carbon nanotubes are among the most extensively researched materials today. Research in this area is throwing up numerous surprises. This is the most versatile material, with the properties ranging from optical absorption and emission on one hand to the mechanical properties of bulk materials such as Young's modulus, on the other. The various aspects of science such as chemistry, physics, biology and material science are creating numerous possibilities for application. An overview of carbon nanotubes and their applications is presented here.

Learning Objectives

- What are carbon nanotubes?
- How do you make them?
- What are their properties?
- Can you fill the nanotube?
- What do you use carbon nanotubes for?
- Do carbon nanotubes have industrially relevant applications?

4.1 Introduction

Carbon is responsible for creating the most diverse variety of compounds. It has more allotropes than any other element. The most recent additions to this list are fullerenes and *nanotubes*. The sp^2 hybridized state of carbon makes two-dimensional structures and the most studied of them is its allotrope, graphite. The other well-known allotrope, diamond has sp^3 hybridized atoms. The two-dimensional sheets made of sp^2 hybridized carbon can curl, just like a piece of paper, and make cylinders. By using hexagons alone, carbon cannot yield closed three-dimensional structures. The inclusion of pentagons results in a closed-cage structure; at least six pentagons are needed on each sides of the cylinder, thereby making a closed pipe. This is called a carbon nanotube as the diameter of such a tube is typically in the nanometer range. The tube can be closed or open and the length can be several hundred times the width. The aspect ratio typically encountered is of the order of 100. The longest nanotubes can have lengths of the order of micrometers.

A single sheet of graphite is called graphene. A carbon nanotube is produced by curling a graphene sheet. Just like a sheet of paper, planar carbon sheets can also curl in a number of ways. This makes the carbon sheet helical around the tube axis. If we fold the graphene symmetrically as shown in Fig. 4.1, the hexagons in the resulting tube will be neatly arranged side to side as shown by the arrow. Imagine that the graphene is folded differently, at an angle. This results in a tube in which the hexagons form a coil around the tube axis. One can see that there are infinite ways of folding the graphene sheet, thereby resulting in tubes of different helicities. All these are different kinds of tubes. Since the extent of *helicity* varies, numerous tube structures are possible, which results in both variety and diversity in the properties of the tubes. The electronic structure of the tube also varies as the helicity changes.

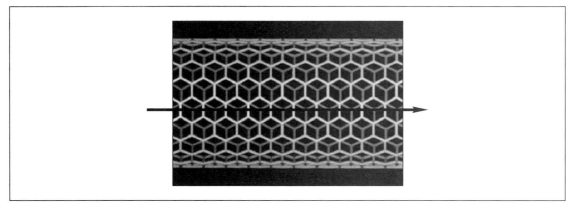

Fig. 4.1: *A part of the nanotube. The tube is highly symmetrical and is made from a graphene sheet.*

The structure of a cylindrical tube is best described in terms of a tubule diameter d and a chiral angle θ as shown in Fig. 4.2. The *chiral vector* $\mathbf{C} = n\mathbf{a}_1 + m\mathbf{a}_2$ along with the two parameters d and θ define the tube. The unit vectors $\mathbf{a}_1$ and $\mathbf{a}_2$ define the graphene sheet. In a planar sheet of graphene (a single sheet of graphite), carbon atoms are arranged in a hexagonal structure, with each atom being connected to three neighbours. In Fig. 4.2, each vertex corresponds to a carbon atom. The vector $\mathbf{C}$ connects two crystallographically equivalent points. The angle θ is with respect to the zigzag axis, and it is 30° for the armchair tube. If we roll over from one end of the tube to the other end, we obtain a cylinder. The rolling can be done in several ways. The bond angles of the hexagons are not distorted while making the cylinder. The properties of the tube get modified depending on the chiral angle θ and the diameter d.

The tubes are characterized by the (n, m) notation, with the tube constructed in Fig. 4.2 being (4, 2). Here the vector $\mathbf{C} = 4\mathbf{a}_1 + 2\mathbf{a}_2$. It is made by making four translations along the zigzag direction and two translations at 120° from the zigzag axis, as shown in Fig. 4.2. There are numerous ways in which the tubes can be rolled. While the $(n, 0)$ tubes are called 'zigzag tubes' where θ is zero, the (n, n) tubes are called 'armchair tubes' where θ is 30°. These two types of tubes have high symmetry and a plane of symmetry perpendicular to the tube axis. Any other tube (n, m) is a chiral tube, which can be either left-handed or right-handed. The tubes will be optically active to circularly polarized light, circulating along the tube axis.

The two important tube parameters, d and θ can be found from n and m.

$d = C/\pi = \sqrt{3}r_{C-C}(m^2 + mn + n^2)^{1/2}/\pi$ and $\theta = \tan^{-1}[\sqrt{3}m/(m + 2n)]$, where r_{C-C} is the C-C distance of the graphene layer (1.421 Å) and C is the length of the chiral vector. Due to the symmetry of the graphene layer, several tubes, although having different (n, m) notations are indeed the same. A tube of $(0, n)$ is the same as $(n, 0)$. The tube diameter will increase with an increase in n and m.

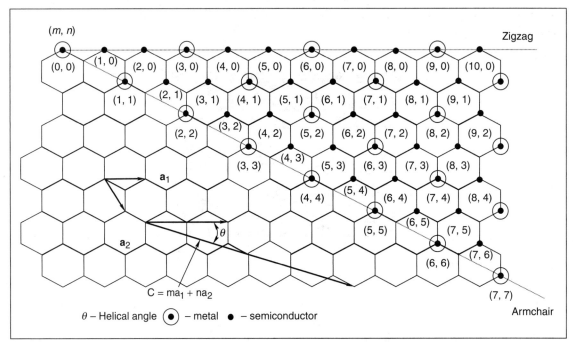

Fig. 4.2: *Illustration of the notations used in understanding carbon nanotubes and the indexing scheme used in carbon nanotubes.*

Hitherto the discussion was confined to single nanotubes, which are called single-walled nanotubes. However, the experimentally observed tubes are also multiwalled, i.e. several tubes are stacked one within the other. In the nanotube assemblies of this kind, there is no three-dimensional order between the graphite layers as in the case of bulk graphite. This is due to the rotational freedom existing between the tubes which is called turbostratic constraint. This lack of three-dimensional order within a multiwalled nanotube (MWNT) has been found in atomically resolved STM measurements. It is not possible to fit any arbitrary tube in a given tube due to lack of space. For a tube to fit into another, there must be a gap of at least 3.44 Å between the layers. We can fit a (10, 0) tube in a (19, 0) tube, but not in a (18, 0) tube. This is because in order to insert a 7.94 Å diameter tube, the larger diameter tube has to have a diameter of 14.82 Å or larger [(7.94) + 2(3.44) Å]. The diameters of (19, 0) and (18, 0) tubes are 15.09 Å and 14.29 Å, respectively.

4.2 Synthesis and Purification

Carbon nanotubes were first noticed in the graphitic soot deposited on the negatively charged electrode used in the arc–discharge synthesis of fullerenes. In the Kratschmer-Huffman procedure (Ref. 1), the graphite rods are evaporated in a dynamic atmosphere of helium (helium is leaked in while the vacuum system is pumped). Typically a pressure of 130 torr of helium is used and the arc is run at 30 V dc with current being maintained at ~180 A. The carbon deposited on the cathode has a soft inner core and a hard outer cover. The core containing MWNTs is extracted and suspended in suitable solvents. The tubes are seen as empty cylinders lying perpendicular to the electron beam along with amorphous carbon material. The interlayer gap is 0.34 nm, close to the spacing found in graphite. The very first images taken by Iijima (Ref. 2) are shown in Fig. 4.3 along with the imaging geometry. The tube's inner diameter, interlayer spacing, length as also chiral angle θ can be determined from the TEM images. While the first three are straightforward from a high resolution image, the determination of the chiral angle necessitates measurement of the interference pattern of the parallel planes and is usually not done during the routine TEM examination of nanotubes.

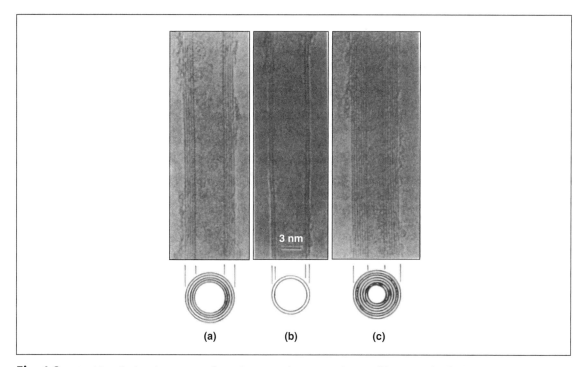

Fig. 4.3: *Multiwalled carbon nanotubes of various diameters observed by Iijima (Ref. 2). Cross-sectional view of the tube is also shown. Copyright Nature.*

Nanotubes are mostly found with closed ends on either side, though open tubes are also seen. Thus these are three-dimensional closed-cage objects, and may be considered as elongated fullerenes. In order to make a closed-cage structure, there must be at least 12 pentagons according to the Euler's theorem, considering only pentagons and hexagons. The hexagons make the elongated body of the tube and the ends contain both hexagons and pentagons, with a minimum of six pentagons on each face. However, the tube body and the ends can have defects. While pentagons result in positive curvature, heptagonal defects result in negative curvature. Both these types of defects have been observed. The former makes a larger tube smaller while the latter can remove this curvature. Various kinds of end tube morphologies have been found, which are illustrated in Fig. 4.4.

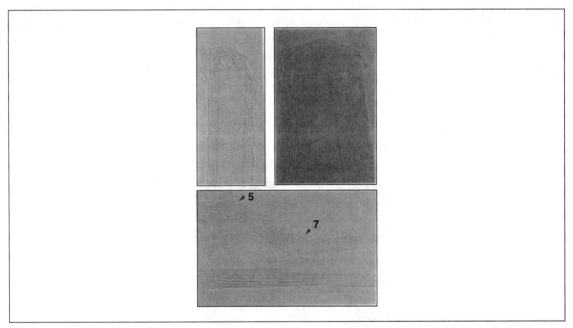

Fig. 4.4: *Transmission electron microscopic images of multiwalled tubes showing the various tip morphologies (Ref. 3). Defects incorporating pentagons (marked 5) and heptagons (marked 7) are shown. While a pentagon gives positive curvature, a heptagon gives negative curvature. Reprinted from Ajayan and Ebessen (Ref. 3). Used with permission from IOP publishing.*

Various modifications to the arc–discharge process are reported in the literature for the synthesis of nanotubes. In the process a smaller diameter (typically 3 mm) anode evaporates on the face of a larger diameter (6 mm) cathode in a direct current arc–discharge apparatus. The bowl that grows on the cathode contains multiwalled tubes. The bowl can be broken and ground, and the nanotubes may be suspended in a suitable solvent and deposited on the TEM grids for examination.

The incorporation of transition metals in catalytic amounts results in the formation of single-walled nanotubes. The catalytic metal is added into the anode. The most common metals used are iron and nickel, but it is better to use a mixture of transition metals. Several bimetallic systems such as Co–Ni, Co–Pt and

Ni–Y have been tried for this purpose. Web-like deposits are found around the cathode or the cooler regions of the reaction vessel. These materials contain significant quantities of single-walled nanotubes. These are seen in the form of ropes containing 5–100 individual SWNTs with amorphous carbon and nanoparticles of the metal/metal–carbon compounds. Optimized synthesis utilizes an Ni–Y catalyst in the atomic ratio of 4:1 and several grams of SWNT containing material can be prepared.

Laser evaporation is another way in which a large yield of SWNTs can be produced (Ref. 4). It is possible of synthesize SWMTs by heating a mixture of graphite with Fe and Ni catalysts at a temperature of 1200°C and irradiating the material with laser. The yield of the nanotubes is about 50–70 per cent of the product. Nanotubes thus synthesized are found to form ropes in which individual tubes organize into a hexagonal assembly. This clearly shows the homogeneity of the tubes synthesized.

Chemical vapour deposition is another useful way in which the synthesis of SWNTs and MWNTs can be achieved (Ref. 5). Here, an organometallic precursor is mixed with a carbon containing feed gas, it is pyrolyzed in a quartz tube and the nanotubes are collected from the cooler end of the reaction vessel. The feed gas may contain several species and is often mixed with an inert gas. Nanotubes are also grown on solid catalytic substrates such as SiO_2, quartz, alumina, etc., which contain transition metal precursors. Such approaches are important for making supported MWNT assemblies for specific applications. By feeding suitable precursor species, it is possible to incorporate other atoms such as nitrogen into the nanotube structure, by substitution. It is also possible to change the morphology of the tubes by changing the precursors.

Both MWNTs and SWNTs are formed with significant quantities of carbonaceous material. One way of separating the tubes from the carbon mass is to heat-treat the product. Although all the carbon forms react with oxygen, they do so at different rates. All the amorphous carbon materials can be burnt off by heating the soot at 750°C for half an hour (Ref. 6). At the end of this process, only less than 1 per cent of the original material is left, but the product thus obtained is essentially a mixture of nanotubes. The existence of a large number of defects in amorphous carbon make it react at a higher rate, in comparison to nanotubes. Acid-based cleaning procedures can also be used.

In a number of applications, it is important to have aligned nanotubes oriented perpendicular to the surface. One of the approaches that has received significant attention in recent times is the synthesis of aligned nanotube bundles on substrates (Ref. 7). Here a two-furnace approach is used along with metallocenes and organic precursors. Compact aligned nanotube bundles can be obtained by introducing acetylene during the sublimation of ferrocene. Such assemblies grown on substrates, especially in a patterned fashion, have important applications such as field emission displays, see Fig. 4.6 (Plate 4).

4.3 Filling of Nanotubes

The nanotubes obtained directly from the synthetic processes are closed on both the ends. The ends can be opened by suitable chemistry. One of the methods used is acid treatment which oxidizes the ends and leaves behind the oxide containing functionalities. The common functional groups are —COOH and —OH. These may be removed by heating the tubes at 600°C in flowing Ar. Other methods such as treating with liquid bromine followed by heat treatment, are also used. There are several ways to fill the open tubes with materials. In one, the nanotubes are soaked in a concentrated solution of the desired metal salt, dried and fired in a reducing atmosphere at high temperatures to form metals in the nano form (Ref. 8). This has been done with metals such as Au and Ag. Filling can also be done from the melt of the filling material, if the surface tension is less than 100–200 nN/m. This leads to long crystals of the filled material, filling uniformly inside the tube. A number of different materials are found to be stuck into the nanotube cavity.

There are also other ways of filling nanotubes (Ref. 9). The simplest one is to use the arc evaporation process with graphite anodes filled with appropriate metals. This produces metals or metal carbides inside the tubes. Pyrolysis of organic molecules over metals can also produce MWNTs with metals or metal carbides. In most of such pyrolytic efforts, the objective is to make pure nanotubes by using catalytic processes.

Filled SWNTs have also been made. The general protocol for these is the same as that for MWNTs, with the opening followed by filling with suitable precursor species and firing at the appropriate temperature in a suitable atmosphere. Most of the time, such filling leads to coverage of the filled material both inside as well as over the surface of the nanotube. A selective purification method is thus required to remove the filled material from the outer surface of the nanotube.

Nanotubes may be used as templates to fill materials. In such strategies, the tubes are fired after filling so as to burn off the carbon and obtain nanorods or tubes of the required materials (Ref. 10).

4.4 Mechanism of Growth

The process of nanotube growth has still not been fully understood. The presence of MWNTs and SWNTs in uncatalyzed and catalyzed conditions, respectively, indicate that two different growth mechanisms may be operative. In an open-end mechanism, in which atoms are continuously added to the growing end, the dangling bond energy is stabilized by interaction between the adjacent layers. The bond may be breaking and forming at the periphery of an open-ended tube. In the case of SWNTs, catalysts are important and it appears that catalyst atoms decorate the growing end, which absorb and incorporate the incoming carbon atoms into the nanotube structure. The most recent suggestion (Ref. 11) pertaining to this mechanism is that carbon fibres grow on nickel nanocrystals through reaction-induced reshaping of the particles. The nucleation and growth of the graphene layers occur along with the dynamic formation and restructuring of mono-atomic step edges at the nickel surface. The surface diffusion of carbon and nickel atoms takes place during the growth of the nanotubes.

4.5 Electronic Structure

Nanotubes can have distinctly different electronic properties depending on the chirality. Early calculations predicted that they can be semiconducting or metallic depending on the type of structure. While armchair tubes are always metallic, others can be semiconducting or metallic. Obviously as the diameter increases, tubes resemble graphite, which can be metallic. The curling of the graphite layers and a decrease in the number of layers cause changes in the electronic structure of the metallic tubes, as compared to those of graphite. The presence of defects on the body of the tube can alter the electronic structure and can make regions of specific electronic properties, such as metallic and semiconducting. This would help one make nanoscopic device structures within one tube itself. During one such attempt, a single tube was seen to possess a *y*-junction. Each arm of this tube can have different electrical transport properties, thereby making a transistor possible within one tube (Ref. 12).

Early theoretical studies predicted drastic change in the properties with change in the tube indices. In the figure below (Fig. 4.5) the electronic density of states of (12, 8) and (10, 10) tubes are presented. The density of states shows a distinct gap in the (10, 0) tube, while no gap exists in the (12, 8) tube. Gapless conduction is thus possible in this tube, making it metallic. The (10, 10) tube is semi-conducting and in general, the band gap varies depending on the tube indices.

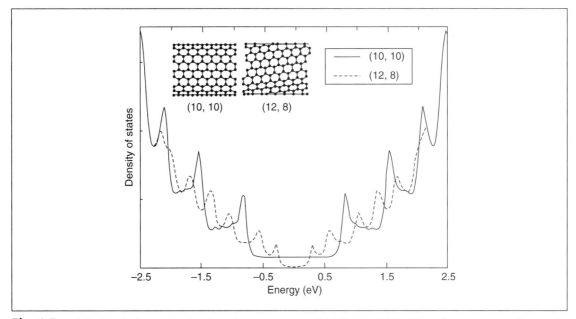

Fig. 4.5: *Calculated electronic density of states for two SWNTs (Ref. 13). The tubes corresponding to the indices (10, 10) and (12, 8) are shown. Used with permission from the author.*

The electronic structure of the tubes has to be probed with tools in order to ensure that nothing else is sampled. This is indeed a difficult task as probes such as photons are typically larger in dimension than the nanotubes. Electron energy loss spectroscopy is thus superior to photon based techniques as the incident electron beam is comparable to the dimensions of the nanotube. In addition to the graphite-like plasmon features at 7 and 25 eV, nanotubes also show features in the range of 10–16 eV. This is attributed to the low dimensionality of the tubes and also to the dimensional cross-over from one to three dimensions.

The 1s resonance in the inner shell electron energy loss spectrum shows features at 285 and 291 eV, due to 1s $\rightarrow \pi^*$ and 1s $\rightarrow \sigma^*$ resonances. These are similar to graphite in MWNTs but are distinctly different in SWNTs (Ref. 14).

Raman scattering is found to be highly diameter-dependent (Ref. 15). There are specific vibrations in the 100–1600 cm^{-1} region which are strongly diameter dependent. The features of specific tubes can be enhanced at specific excitation energies in resonance Raman experiments. This is a manifestation of one dimensionally confined electronic structure. Raman spectroscopy can be used to determine many properties of nanotubes such as tube diameter due to its extreme sensitivity to various tube parameters.

In a synthesis both metallic and semiconducting SWNTs are formed simultaneously. It has been shown recently that chemical processes can be used to separate metallic SWNTs from the others (Ref. 16).

4.6 Transport Properties

Scanning tunneling spectroscopy has shown that the band gaps of the nanotubes vary from 0.2 to 1.2 eV (Ref. 17). The gap varies along the tube body and reaches a minimum value at the tube ends. This is due to the presence of localized defects at the ends due to the extra states. The measurements on SWNTs show the helicity and size-dependent changes in the electronic structure.

The transport properties of MWNTs and NWNTs have been measured (Ref. 18). However, the principal problem in these measurements relates to the need for making proper contacts. Due to large contact resistances, it is not possible to obtain meaningful information without four probe measurements. The conductive behaviour of MWNTs was consistent with the weak two-dimensional localization of the carriers. The inelastic scattering of carriers from lattice defects is more significant than carrier–carrier or carrier–phonon scattering. In SWNTs, conduction occurs through discrete electronic states that are coherent between the electrical contacts (hundreds of nanometers). This means that nanotubes can be treated as quantum wires, at least at very low temperatures.

4.7 Mechanical Properties

The strength of the carbon–carbon bond is among the highest and as a result, any structure based on aligned carbon–carbon bonds will have the ultimate strength. Nanotubes are therefore the ultimate high-strength carbon fibres. The measurement of Young's modulus gave a value of 1.8 TPa (Ref. 19). The theoretical prediction is in the range of 1–5 TPa, which may be compared to the in-plane graphite value of 1 TPa. It is difficult to carry out measurements on individual nanotubes. The problem with MWNTs is that the individual cylinder can slide away thus giving a lower estimate for Young's modulus. It is also possible for individual SWNTs to slip from a bundle, thereby again reducing the experimentally measured Young's modulus. Measurements based on vibration spectroscopy, AFM and transmission electron microscopy can be used in determining estimates, and all of them come up with nearly the same numbers (see Chapter 2).

One of the important properties of nanotubes is their ability to withstand extreme strain in tension (up to 40 per cent). The tubes can recover from severe structural distortions. The resilience of a graphite sheet is manifested in this property, which is due to the ability of carbon atoms to rehybridize. Any distortion of a tube will change the bonding of the nearby carbon atoms and in order to come back to the planar structure, the atoms have to reverse to sp^2 hybridization. If the tube is subjected to elastic stretching beyond a limit, some bonds are broken. The defect is then redistributed along the tube surface.

4.8 Physical Properties

Nanotubes have a high strength-to-weight ratio (density of 1.8 g/cm^3 for MWNTs and 0.8 g/cm^3 for SWNTs). This is indeed useful for lightweight applications. This value is about 100 times that of steel and over twice that of conventional carbon fibres. Nanotubes are highly resistant to chemical attack. It is difficult to oxidize them and the onset of oxidation in nanotubes is 100°C higher than that of carbon fibres. As a result, temperature is not a limitation in practical applications of nanotubes.

The surface area of nanotubes is of the order of 10–20 m^2/g, which is higher than that of graphite but lower than that of mesoporous carbon used as catalytic supports where the value is of the order of 1000 m^2/g.

Nanotubes are expected to have a high thermal conductivity and the value increases with decrease in diameter (Ref. 20). The thermal conductivity of single nanotubes were shown to be comparable to diamond and in-plane graphite (Ref. 21).

4.9 Applications

The use of nanotubes as electrical conductors is an exciting possibility. A nanotube-based single molecule field effect transistor has already been built (Ref. 22). The performance of this device is comparable to that of semiconductor-based devices, but the integration of this into circuits will require a lot of effort. One of the problems associated with such devices is the need to make contacts and adopt newer kinds of approaches. It has been seen that it is possible to fabricate-nanotube-based connectors. Such interconnectors between structures patterned on substrates have also been made (Ref. 23).

It is possible to construct a heterojunction by having a junction between nanotubes of different helicities. This approach facilitates the creation of a device with one molecule. An approach for making devices of this kind has been developed in Bangalore. The pyrolysis of a mixture of metallocenes with thiophene yields excellent quality junction nanotubes (Ref. 24). Various metallocenes such as ferrocene, cobaltocene and nickelocene have also been tried. The interesting aspect is that the three arms of the Y junction can have different helicities thus yielding a molecular transistor. Although plenty of 'y' type junctions are synthesized, the transistor action was demonstrated only recently (Ref. 25).

One of the areas of immediate commercial application of nanotubes is CNT-based field emission displays. Here CNTs act as electron emitters at lower turn on voltages and high emissivity. The turn on voltage on a practical device developed by Samsung is 1 V/micrometer and the brightness measured is of the order of 1800 candela/m^2 at 4 V/micrometer (Ref. 26). With the addressing of individual pixels, the method will soon be ready for commercial exploitation. An image of a running video on a 5-inch diagonal display is shown in Fig. 4.6 (Plate 4). The emission is due to patterned SWNTs. One of the things needed in the process of commercial application is the capability to make aligned CNTs on substrates, which is possible with CVD-based methods. Field emission is stable in air, and electron emission is stable for several hours. Therefore, there is no practical limitation in CNTs for this application.

Nanotube tips can be used as nanoprobes. The possibility using AFM and STM tips has also been demonstrated. The functionalization of tips can be used in chemical force microscopy wherein a chemical functionality interacts with an appropriate one on the substrate. Such studies help one deduce information such as the strength of a chemical bond. Being flexible, the probes are not susceptible to frequent crashes, unlike in the case of normal STM tips. The tubes can also penetrate into crevices, which facilitates sub-surface imaging.

Research is also being carried out to assess the ability of carbon nanotubes to store hydrogen (Ref. 27). The storage occurs both in between and inside the nanotube bundles. Research has shown that the amount of hydrogen stored is comparable to that in the best storage materials such as metal hydrides. The application of this kind of storage will be important for fuel cell applications for automobiles where the storage of hydrogen is one of the critical factors. However, in order to make this feasible, it is important to store hydrogen to the extent of 5 per cent of the nanotube mass. Achieving this quantity appears to be a problem, though the intake and release are feasible.

A flow sensor using nanotubes has also been demonstrated (Ref. 28). In this device, liquids flow through aligned single-walled nanotubes supported on a substrate. The flow generates an electrical potential of the order of a few millivolts. The potential is sensitive both to the flow velocity as well as the dipole moment of the analyte. Gas flow sensors have also been developed along similar lines. Nanotube-based gas sensors have also been developed which utilize the narrow channels of the tubes that are of a comparable dimension to molecules (Ref. 29). Nanotube-based filters have also been demonstrated (Ref. 30). Here a liquid containing a mixture of molecules such as petroleum is separated into the components by filtration. Such an approach makes it possible to filter out bacteria, viruses and chemicals from water. The most important aspect in the development of such a filter is the fabrication of a mechanically stable filter with aligned carbon nanotubes.

4.10 Nanotubes of Other Materials

In principle, any planar structure should be able to curl and make a tubular structure. Certain clays such as christolite and imogolite are found in tubular form. The structural characteristics of CNTs such as helicity and rotational disorder, are found in these clays too. The first nanotube structure found with inorganics has been reported with WS_2 and MoS_2. These structures consist of alternating layers of W/Mo and S. They have an excellent lubricating property, and they roll on the substrate. Nanotubes have been made with BN and BCN as well as with $B_xC_yN_z$.

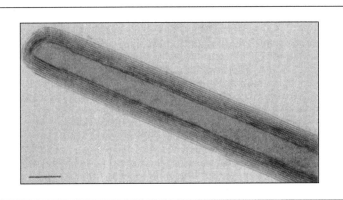

Fig. 4.7: *Electron micrograph of part of a WS_2 based nanotube (Ref. 30). The tube is assumed to be hollow. The contrast within the tube is attributed to the outer wall, perpendicular to the tube. The scale bar is 10 nm. Copyright Nature. Used with permission from the author.*

A variety of polyhedral and tubular structures of WS_2 have been obtained by heating a thin tungsten film in H_2S (Ref. 31). The tubes observed are hollow and are closed at ends. An image of a tube is shown in Fig. 4.7. The curling of a graphite-like sheet of WS_2 leads to the creation of defects. Such defects can be nucleated by high temperature treatment. Structures other than tubes, such as onions, have also been made in this way. These are, in general, called inorganic fullerenes. The properties of inorganic fullerenes and inorganic nanotubes have been thoroughly researched. Analogous to carbon nanotubes, several properties of these systems have also been studied.

Review Questions

1. How would one classify carbon nanotubes? What are the various kinds of carbon nanotubes?
2. How would one get the (n,m) indices from the diameter?
3. What are the diameter-dependent properties of nanotubes?
4. Why is it not possible to inset an arbitrary tube into a given tube?
5. What are the other materials which can form nanotubes?
6. Are there specific properties of nanotubes which will be different in multiwalled and single-walled nanotubes?
7. What are the unique properties of nanotubes and how would one study those?

8. Can one extend the knowledge of the chemistry of fullerenes into carbon nanotubes? What are such properties?
9. How can be the properties of 'tubes' of cylinders used in the case of nanotubes?
10. From the everyday examples of macroscopic objects similar to tubes, such as iron pipes, suggest a few properties of nanotubes which could be investigated.

References

1. Kratschmer, W., L.D. Lamb, K. Fostiropoulos and D.R. Huffman, *Nature*, **347**, **354** (1990).
2. Iijima, S., *Nature*, **354** (1991), p. 56.
3. Ajayan, P.M. and T.W. Ebessen, *Rep. Prog. Phys.*, **60** (1997), p. 1025.
4. Guo, T., P. Nikolaev, A. Thess, D.T. Colbert and R.E. Smalley, *Chem. Phys. Lett.*, **243** (1995), p. 49.
5. Amelinckx, S., X.B. Zhang, D. Bernaerts, X.F. Zhang, V. Ivanov and J.B. Nagy, *Science*, **265** (1994), p. 635.
6. Tsang, S.C., P.J.F. Haris and M.L.H. Green, *Nature*, **362** (1993), p. 520.
7. Rao, C.N.R., R. Sen, B.C. Satishkumar and A. Govindaraj, *Chem. Commun.*, (1998), p. 1525.
8. Chen, Y.K., A. Chu, J. Cook, M.L.H. Green, P.J.F. Haris, R. Heesom, M. Humphries, J. Sloan, S.C. Tsang and J.C.F. Turner, *J. Mat. Chem.*, **7** (1997), p. 545.
9. Serpahin, S., D. Zhou, J. Jiao, J.C. Withers and R. Roufty, *Nature*, **362** (1993), p. 503.
10. Gundiah, G., G.V. Madhav, A. Govindaraj and C.N.R. Rao, *J. Mater. Chem.*, **12** (2002), p. 1606.
11. Helveg, S., C. Lopes–Cartes, J. Sehested, P.L. Hansen, B.S. Clausen, J.R. Rostrup–Nielsen, F. Abid-Pedersen and J.K. Norskov, *Nature*, **427** (2004), p. 426.
12. Satishkumar, B.C., P.J. Thomas, A. Govindaraj and C.N.R. Rao, *Appl. Phys. Lett.*, **77** (2000), p. 2530.
13. Ajayan, P.M. in *Nanostructured Materials and Nanotechnology*, (2002), Hari Singh Nalwa, (ed.) Academic Press, San Diego.
14. Bursill, L.A., P.A. Stadelmann, J.L. Peng and S. Prawer, *Phys. Rev. B.*, **49** (1994), p. 2882.
15. Hiura, H., T.W. Ebessen, K. Tanigaki and H. Takahashi, *Chem. Phys. Lett.*, **202** (1993), p. 509.
16. Maeda, Y., S. Kimnra, M. Kanda, Y. Harashima, et al., *J. Am. Chem. Soc.*, **127** (2005), p. 10287.
17. Dresselhaus, M.S., G. Dresselhaus and P.C. Eklund, (1996), *Science of Fullerenes and Carbon Nanotubes*, Academic Press, New York.
18. Langer, L., V. Bayot, E. Grivei, J.P. Issi, J.P. Heremans, C.H. Olk, L. Stockman, C. Van Haesendonck and Y. Bruynseraede, *Phys. Rev. Lett.*, **76** (1996), p. 479.
19. Treacy, M.M.J., T.W. Ebessen and J.M. Gibson, *Nature*, **381** (1996), p. 678.

20. Fuji, M., X. Zhang, H.Q. Zie, H. Ago, K. Takashashi, T. Iknta, H. Abe and T. Shimizn, *Phys. Rev. Lett.*, **95** (2005) p. 65502.
21. Hone, J., M. Whitney, C. Piskoti and A. Zettel, *Phys. Rev.*, **59** (1999), p. R2514.
22. Tans, S.J., A.R.M. Verschueren and C. Dekker, *Nature*, **393** (1998), p. 49.
23. Homma, Y., Y. Kobayashi, T. Ogino and T. Yamashita, *Appl. Phys. Lett.*, **81** (2002), p. 2261.
24. Satishkumar, B.C., P.J. Thomas, A. Govindaraj and C.N.R. Rao, *Appl. Phys. Lett.*, **77** (2000), p. 2530.
25. Bandaru, P.R., C. Dario, S. Jin, and A.M. Rao, *Nature Materials*, **4** (2005) p. 663.
26. Chung, D.S., S.H. Park, H.W. Lee, J.H. Choi, S.N. Cha, J.W. Kim, J.E. Jang, K.W. Min, S.H. Cho, M.J. Yoon, J.S. Lee, C.K. Lee, J.H. Yoo, J.M. Kim, J.E. Jung, Y.W. Jin, Y.J. Park and J.B. You, *Appl. Phys. Lett.*, **80** (2002), p. 4045.
27. Dillon, A.C., K.M. Jones, T.A. Bakkedahl, C.H. Kiang, D.S. Bethune and M.J. Heben, *Nature*, **386** (1997), p. 377.
28. Ghosh, S., A.K. Sood, and N. Kumar, *Science*, **299** (2003), p. 1042.
29. Modi, A., N. Koratkar, E. Lass, B. Wei, and P.M. Ajayan *Nature*, **424** (2003) p.171.
30. Srivastava, A., O.N. Srivastava, S. Talapatra, R. Vajtai and P.M. Ajayan, *Nature Materials*, **3** (2004), p. 610.
31. Tenne, R., L. Margulis, M. Genut and G. Hodes, *Nature*, **360** (1992), p. 444.

Additional Reading

1. Rao, C.N.R. and A. Govindaraj, *Acc. Chem. Res.*, **35**, (2002), pp. 998–1007. (Carbon nanotubes)
2. Rao, C.N.R. and Minakshi Nath, *Dalton Trans.* (2003), pp. 1–24. (Inorganic nanotubes)
3. Dai, L. (ed.) (2006) *Carbon Nanotechnology: Recent Developments in Chemistry, Physics, Materials Science and Device Applications*, Elsevier.
4. Rao, C.N.R. and A. Govindaraj, (2005) *Nanotubes and Nanowires*, Royal Society of Chemistry.

Chapter 5

SELF-ASSEMBLED MONOLAYERS

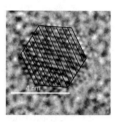

Self-assembled monolayers are important nanostructured systems which are two-dimensional nano assemblies. The structure of this assembly is such that it facilitates precise control of molecules. Various spectroscopic, scattering and imaging techniques have been used to understand the structure of self-assembled monolayers in detail. These assemblies have been used in a number of applications, mostly in the area of sensors. A protoptypical molecular nanomachine has been by built by using SAMs. The diversity of SAMs allows almost anything to be grown on them through appropriate chemistry.

Learning Objectives

- What are the various kinds of monolayers?
- What are self-assembled monolayers? What are their properties?
- What are their applications?
- How can one use them for nanotechnology?

5.1 Introduction

The bottom-up approach of manufacturing nano devices has been demonstrated very well. The most celebrated example is the iron corrals and the molecular abacus made by IBM researchers (Ref. 1). However, the use of such an approach to make devices, i.e. placing atoms one at a time to form a functional structure, cannot have a high throughput. The alternate approaches for making functional nanostructures must involve self-assembly. In this approach, once the process begins, structures are formed without external intervention. The structure organizes itself, on the basis of external conditions. The information required to form the structure is contained in the molecules themselves. The structure is organized by utilizing weak interactions such as hydrogen bonding, and van der Waals interactions, and there are numerous interactions of this kind in a structure which makes it stable. These are the interactions which make and sustain life, and there are numerous examples of such interactions in the world around us.

Monolayers are single-molecule thin layers prepared on surfaces. They can be assumed to be molecularly thin sheets of infinite dimension, just like ultra thin foils. They are among the simplest chemical systems on which nanotechnological approaches can be practised. This feasibility of using monolayers arises from the simplicity of their design and molecular structure and from the user's ability to manipulate them at ease. The possibility of bringing about patterns of nanometer spatial resolution allows one to incorporate multiple functions within a small area.

The concept of monolayers was first introduced by Irving Langmuir in 1917, during his study of amphiphiles in water (Ref. 2). While spreading the amphiphiles on water, he found that the film formed had the thickness of one molecule. Later Katherine Blodgett was able to transfer the monolayer onto a solid support (Ref. 3). The spontaneous formation of a monolayer was first reported by Zisman, et al. in 1946 (Ref. 4). They observed the spontaneous monolayer formation of alkyl amines on a platinum surface. The field observed a tremendous growth when in 1983 Nuzzo and Allara found that ordered monolayer of thiols can be prepared on a gold surface by the adsorption of di-*n*-alkyl disulfides from dilute solutions (Ref. 5).

The name self-assembled monolayers (SAMs) indicates that their formation does not require the application of external pressure. The study of SAMs generates both fundamental as well as technological interest. Nature uses the same process of self-assembly to produce complex architectures. One such example is the formation of the cell membrane from lipid molecules through self-assembly. SAMs are ideal systems which can answer fundamental questions related to interfacial properties like friction, adhesion and wetting. They have been used to alter the wetting behavior of the condenser plates in steam engines. This is because the drop-wise condensation of steam enhances the efficiency of the engine as compared to the film-like condensation. In the latter case, the film acts as an insulator between the metal plate and steam. The two approaches used to make a monolayer of a molecule on a metal surface are discussed as follows.

1. The Langmuir—Blodgett Technique   The Langmuir film is prepared by spreading amphiphilic molecules on a liquid surface. Considerable order can be achieved in these films by applying pressure. The film is then transferred to a solid substrate. The various steps involved during the preparation of Langmuir–Blodgett (L–B) films are shown in Fig. 5.1.

2. Self-assembly   As mentioned earlier, the formation of a self-assembled monolayer does not require the help of an external driving force. Such a monolayer is formed when the metal (or any other substrate) surface is exposed to a solution containing the surfactant (Fig. 5.2).

5.2 Monolayers on Gold

Alkanethiolate monolayers (Ref. 6) grown on coinage metals (Au, Ag, Pt)—the molecular sheet is made of thiolate species (with long alkyl chains) and the substrate is one of the above metals—are structurally simple and easy to construct, which is why a larger number of studies have been conducted on them. It is important to mention that there are several other kinds of monolayers and a review of these is available in (Ref. 7).

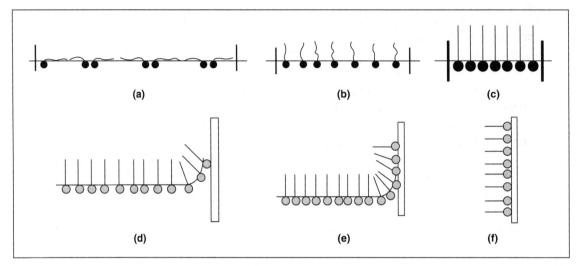

Fig. 5.1: *Langmuir–Blodgett methodology. (a) Surfactant in water. The black dot represents the head and the line represents the tail. The molecules are disordered. (b) Partially compressed monolayer by the pushing of a barrier across the film surface. (c) Ordered monolayer by the application of pressure. (d) Immersion of the substrate in an ordered film. (e) Transfer of the monolayer onto the substrate. (f) Densely packed monolayer on the substrate.*

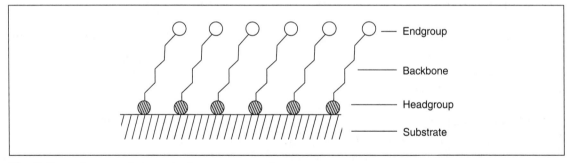

Fig. 5.2: *Schematic showing the basic structure of 2D SAMs.*

Gold has been the preferred substrate for a number of reasons. It is easy to make a thin gold film by thermal evaporation, and almost all kinds of supports can be used for film growth in the case of gold. For gold, there is no stable oxide at room temperature, though Au_2O_3 can be made by ozone exposure to Au films at room temperature. None of the commonly present gases of the atmosphere undergoes chemisorption on gold surfaces. Repeated solvent washes are adequate to make a gold surface atomically clean in most cases. The removal of carbonaceous deposits on gold can also be done by oxidizing them with a 'piranha' solution. This is a 1:3 mixture of 30 per cent H_2O_2 and concentrated H_2SO_4 at 100°C. The solution is, however, highly reactive and should be handled cautiously. It is known to be an explosive mixture when kept in closed containers. Another way of cleaning a gold surface is by repeated cycling between

−0.3 and +1.5 V versus Ag/AgCl in a voltammetric cell. During the positive potential sweep, the surface is oxidized and in the negative sweep, the oxide is reductively removed. Cyclic voltammetric measurements carried out in this way facilitate the calculation of the effective surface area from the peak area.

5.2.1 Preparation

SAMs are prepared by dipping substrates such as evaporated gold films into a millimolar solution of the surfactant. The gold films of 500–2000 Å thickness are made on substrates. The solution is normally made in hexane for long-chain surfactants and in ethanol for short chain surfactants. The process of assembly of monolayers on the surface involves two stages. During the first stage, the surfactants are rapidly pinned on the surface, followed by a slow reorganization step, during the second stage, extending over several hours (Ref. 8). The exact kinetics of both these steps depend on parameters such as the concentration of the solution, length of the alkyl chain, etc. It is a standard practice to leave the substrate in the solution, face up ('face' refers to the gold coating), for periods ranging from 12–24 hours for complete self-assembly. Before being used, the monolayer surfaces are washed and blown dry with nitrogen. The kinetics of initial pinning can be monitored by a quartz crystal microbalance whereas the extent of organization can be studied through infrared spectroscopy. A monolayer of a mixture of alkane thiols can be prepared starting with a mixed solution of the concerned thiols. A mixed thiol monolayer can also be formed through a ligand place exchange reaction in which one surfactant molecule replaces another that is already present on the surface. Monolayers can be formed by microcontact printing wherein the thiol solution is used as ink on a stamp that is prepared from a polymer. This approach is especially useful in preparing patterned surfaces (see Section 5.5). Other methods that have been used to make SAMs include vapour phase deposition on molecules, LB methodology and potential-assisted deposition.

5.2.2 Structure

Two types of sites are available for the thiol chemisorption on the Au(111) surface, the on-top site and the hollow site. *Ab-Initio* calculations have shown that the charge of sulphur at the hollow site is −0.7e and that at the on-top site is −0.4e. Hence the hollow site is the energetically favorable one from the point of view of charge-transfer. Thiolate can migrate between the two adjacent hollow sites. This can happen either through the on-top site or through the bridge site. In both cases, the excited state will be polar. The excited state during such migration will then be stabilized by polar solvents. This is confirmed by the fact that ordered SAMs are formed in ethanol. Figure 5.3 shows the overlayer structure of the monolayer with sulphur atoms occupying alternative hollow sites above the Au(111) layer, giving a hexagonal ($\sqrt{3} \times \sqrt{3}$) $R30°$ unit cell (the symbolism refers to the crystallographic structure of an overlayer).

In this assembly, alkyl chains are in close contact with each other and the chains are fully stretched to form a zig-zag assembly. As a result of this close contact and due to the large distance between the sulphur atoms, the chains tilt. The tilt angle is 34° in the case of Au(111), but it is 5° in Au(100). The chains have rotational freedom at room temperature, which means there is no three-dimensional order in the arrangement of different chains. The adjacent chains are rotationally disordered and this persists at a low temperature. In fact, the freezing of the rotational disorder can be seen in low temperature infrared spectroscopy (Ref. 9). The presence of very bulky groups at the tail end can affect this order, as molecules may not get attached to all the available surface sites. The structure of alkane thiol monolayers on Au(111) is shown schematically in Fig. 5.3.

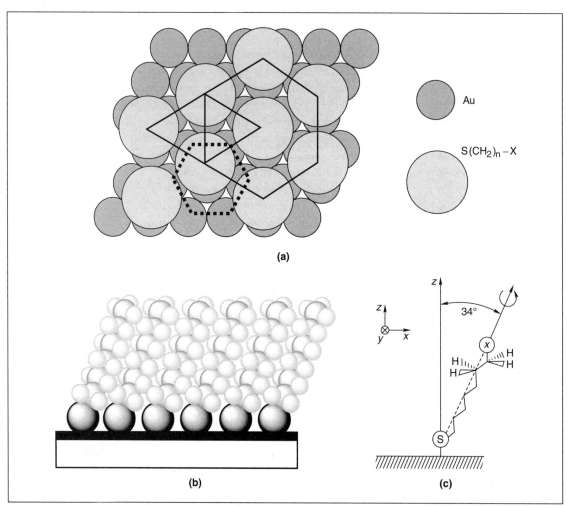

Fig. 5.3: *Schematic structure of a monolayer of alkanethiol on Au(111). (a) Layer of hexagonally arranged gold atoms (dotted hexagon). Each corner of this dotted hexagon corresponds to the center of the gold atom. On top of this surface, the alkanethiol molecules chemisorb. The structure of the sulphur atoms is also hexagonal (solid hexagon), as indicated. These atoms sit on three-fold sites created by the gold atoms. (b) Assembly of thiolate chains. This will form a two-dimensional sheet on the surface. (c) Extended chain. It shows a zig-zag assembly. The tilt angle is shown. It is also shown that the chains have rotational freedom. X is a functional group.*

As mentioned above, this structure is present for several hundreds of angstrom square area. The underlying surface itself is not ordered over an extended area and the typical grain size (within this region, the surface structure is uniform) is of the above dimension. This limits the growth of an ordered SAM over the entire area of the substrate. At the edges of the grain, the monolayer structure is not ordered due to defects in the structure of the gold. There are ways to increase the area of the ordered region such as high temperature annealing.

SAMs can be formed on other surfaces such as Cu, Ag, Pt, GaAs, etc. The coverage of molecules on these surfaces is in the order Cu>Ag>Pt>Au>GaAs, which is a direct consequence of the adsorbate structure. On Ag(111), the overlayer structure is the same as that of gold, but the tilt angle is only 12° and not 34°. This reduced tilt angle reduces the contact distance between the chains (4.1 Å and not 4.95 Å). On a GaAs(100) surface, a large tilt angle of 57° is reported (Ref. 10).

The chain length affects the order of the alkyl chain assembly. In the case of chains of longer thiols, the van der Waals interaction between the chains is large to enable all the chains to stand up. This makes the carbon chain assembly all-trans and reduces the number of defects. 'All-trans' means that the carbon atoms on either end of a C–C bond are trans to each other. This makes the alkyl chain appear like a zig-zag ladder-like structure. However, as the chain length decreases, the van der Waals interaction becomes weak and defects occur. These defects imply the incorporation of gauche conformations. The extent of order in the alkyl chain assembly manifested depends on the technique used for its investigation. The best tool to see the order is surface infrared spectroscopy, which shows a red shift in the methylene C–H stretches as a function of order. Peaks at 2918 and 2846 cm^{-1} are characteristic of ordered alkyl chain assembly. Decreased order is observed when the chain length is less than 11, and when the length is less than 6, the assembly is assumed to be disordered. When the groups at the chain ends have a smaller cross-sectional area than that of the alkyl chain (20 Å), the hydrocarbon chain assembly is the same as that of the alkyl thiol. This is the case with all monolayers with chain ends such as –OH, –NH$_2$, –CONH$_2$, –CO$_2$H, –CO$_2$CH$_3$ in addition to simple –CH$_3$. However, in the case of larger end groups, the chains cannot pack as efficiently as in the case of simple thiols.

The structure of the monolayer varies with the structure of the gold below it. Au(111) is thermodynamically the most stable surface due to the largest surface density of gold atoms. Therefore, in the case of evaporated or annealed gold, the surface is principally Au(111). A film of this kind can be grown on mica wherein the surface will be atomically flat. An atomic force microscopic image of octadecanethiolate monolayer grown on this Au(111) film is presented in Fig. 5.4. The image scale is 3.02 × 3.02 nm. The nearest neighbor is located at a distance of 0.52 × 0.03 nm (a) while the next nearest neighbor is located at a distance of 0.90 ± 0.04 nm (b) (Ref. 11).

Two binding modes of alkanethiol on a hollow site have been observed—one with the Au-S-C bond angle of 180° called the sp mode, and the other with Au-S-C bond angle 104° (sp^3 mode) (see Fig. 5.5). The energy difference between the two modes is 2.5 kcal mol^{-1}. Thus the thiolate can change from one mode to the other without much difficulty.

The overlayer structure of thiols on Ag(111) shows different types of lattice as shown in Fig. 5.6. The overlayer is of ($\sqrt{7} \times \sqrt{7}$) $R10.9°$ structure. The tilt angle in the case of thiols on Ag(111) surface is small as compared to gold. The observed value is about 5°. In some cases, the values were even close to zero indicating that the chains are almost perpendicular to the surface. Also, the S-S distance in the case of SAMs on Ag(111) is close to 4.41 Å. Thus the van der Waals forces will be stronger in the case of silver.

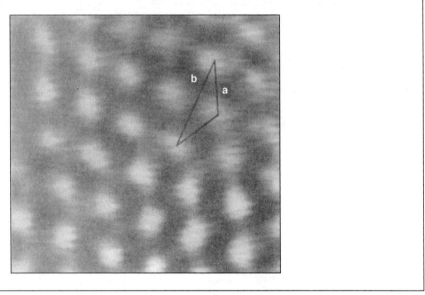

Fig. 5.4: *AFM image of octadecanethiolate on Au(111) (Ref. 11). Reprinted with permission from Alves, et al. (Ref. 11). Copyright (1992) American Chemical Society.*

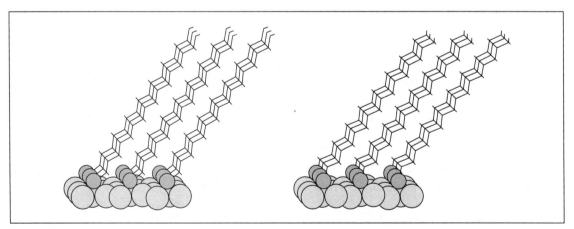

Fig. 5.5: *Nine molecule section of full coverage $C_{16}H_{33}SH$ monolayer on Au(111) based on molecular mechanics energy minimization calculation showing the tilted chains. Reprinted with permission from Ulman (Ref. 7). Copyright (1996) American Chemical Society.*

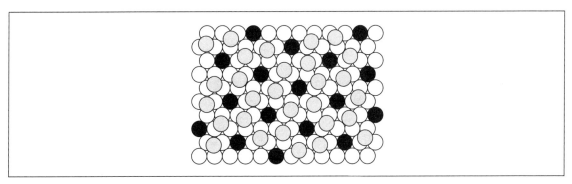

Fig. 5.6: *Overlayer structure of thiol on Ag(111). The open circles represent the silver atoms. The grey and black circles represent the sulphur at the hollow site and on-top site, respectively. Reprinted with permission from Ulman (Ref. 7). Copyright (1996) American Chemical Society.*

Another interesting result was observed in the case of the adsorption of fatty acids on metal oxide surfaces. The carboxylate group adsorbed symmetrically on the surface of silver oxide whereas in the case of aluminium oxide, the fatty acid molecules adsorbed asymmetrically with the tilt angle close to zero (see Fig. 5.7).

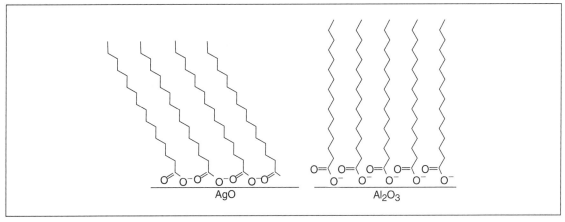

Fig. 5.7: *Orientation of alkyl chains on AgO and Al_2O_3 surfaces. In the case of AgO, the chains are tilted whereas in Al_2O_3, the chain is perpendicular to the surface. Reprinted with permission from Ulman (Ref. 7). Copyright (1996) American Chemical Society.*

The mechanism of thiol adsorption onto a gold surface is still not clear. One possibility is to treat it as an oxidative addition of R-S-H to the metal surface followed by the reductive elimination of the H_2 molecule from the metal surface. In an experiment, we have detected the evolution of hydrogen upon thiol chemisorption on gold surfaces.

$$\text{R-S-H} + \text{Au}_n \rightarrow \text{R-S-Au}^+ + \text{Au}_{n-1} + 1/2\text{H}_2$$

The energy of adsorption calculated from the RS-H (87 kcal mol^{-1}), H-H (104 kcal mol^{-1}) and the Au-SR bond energy (40 kcal mol^{-1}) taking the homolytic Au-SR bond strength is –5 kcal mol^{-1}. This value is in close agreement with the adsorption energy that Schlenoff calculated by using electrochemical data. This suggests that the value of 40 kcal mol^{-1} for the Au-S bond is probably correct.

Piezoelectric oscillators such as quartz crystal microbalance (QCM) can detect mass changes of the order of nanograms. This has been used to study the kinetics of monolayer adsorption. QCM is also used to detect the mass changes during the binding of monolayers with other molecules (for the basic principles of QCM, see Chapter 15).

Reflection-absorption infrared spectroscopy (RAIRS) can be used to derive useful information regarding the orientation of the alkyl chains on a metal surface. The spectrum is accumulated in the grazing reflection mode. In this mode, the incoming plane polarized laser light strikes the metal surface at a large angle of incidence. Only transition dipoles having a component perpendicular to the metal surface will be IR-active. If the chains are perpendicular to the metal surface, then $v_a(CH_2)$ and $v_s(CH_2)$ vibrations would be parallel to the plane of the surface and will be absent in RAIRS. In experiments, we see methylene vibrations indicating that the chains are tilted by a certain angle with respect to the normal surface.

The thickness of the monolayer is estimated by using ellipsometry. In this technique, a plane polarized laser light is used to probe the thickness. The plane polarized light reflected by a metal surface suffers both a phase change (Δ) as well as an amplitude change (Ψ). By comparing the covered and uncovered metal surfaces, one can find out the thickness as well as the refractive index of the monolayer. Porter, *et al.* (Ref. 11) studied a series of alkanethiols with n = 1, 3, 5, 7, 9, 11 and 21 to check the dependence of n (chain length) on the thickness of the monolayer. They found that for chains with $n \geq 7$, the measured thickness was less than the expected value. This indicates that the short chain members behave like a liquid with considerable disorder. Hence the values are less when compared to a chain in the all-trans configuration.

Angle dependent reflection of the *P* polarized light is used in surface plasmon resonance to study the thickness of the monolayer. At a certain angle of incidence called 'plasmon angle', laser light selectively excites the surface plasmon oscillations of the metal. This results in the absorption of light and a decrease in reflectance. The plasmon angle is very sensitive to the changes in the refractive index near the metal surface. This helps one detect the thickness of the monolayer on a metal surface.

It is essential to have an idea about the physisorption and chemisorption energetics for an understanding of the growth of the monolayer on a surface. The desorption of various thiols on Au(111) has been investigated by using specular helium scattering in combination with temperature-programmed desorption. In this technique, a helium beam reflected from the Au(111) plane is fed into a mass spectrometer tuned to m/z 4. A clean gold surface is able to reflect up to 30 per cent of the impinging helium beam. The adsorption of molecules onto the gold surface causes an increase in the diffuse scattering of the helium beam, consequently decreasing the specular intensity. When the molecules start desorbing, a greater part of the gold surface will be exposed to the helium beam, which increases the specular reflection. The intensity of the specular signal will be maximum at the point of maximum desorption. The enthalpy of desorption can be calculated by using the Redhead equation from the value of temperature of maximum desorption (T_{des}) and heating rate, β, obtained from the temperature ramp experiment. From this, $E_{des} = R_g T_{des}[\ln(\nu T_{des}/\beta) - 3.64]$. R_g is the gas constant. ν is the pre-exponential factor in the Arrhenius expression. The typical value is 1×10^{13} s^{-1}. Two peaks were observed in the temperature-programmed desorption (TPD) of thiols on Au(111). The first peak occurring at a lower temperature in Fig. 5.8 is due to physisorption, while the second corresponds to chemisorption. The value for chemisorption was the same for most of the alkanethiols except for compounds wherein the steric effect can affect the bonding between the adsorbate and the substrate. Also the dialkyl sulphides do not show the chemisorption peak. This indicates that the dialkyl sulphides only physisorb on the surface. As the chain length increases, the van der Waals interaction increases and the contribution due to physisorption increases. For chains with $n = 14$, the two peak areas are comparable. For higher chain lengths, the van der Waals contribution becomes higher than that due to chemisorption. Since a long chain molecule is held on a surface by physisorption for a sufficiently long time, it has a greater chance of crossing the chemisorption energy barrier and thus of getting chemisorbed.

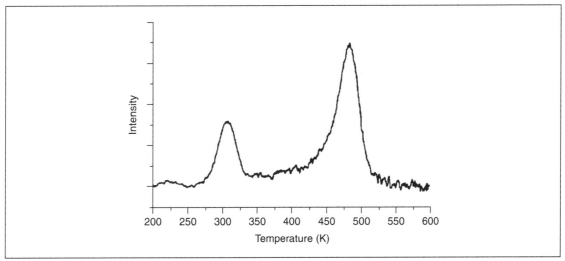

Fig. 5.8: *TPD spectrum derived from helium atom reflectivity signal as a function of temperature for hexanethiol on Au(111). Low temperature peak, which changes with the chain length, corresponds to physisorption and the high temperature peak, which is unaffected by the chain length variation, corresponds to the chemisorbed state. Reprinted with permission from Schreiber, et al. (Ref. 12). Copyright (1998) American Physical Society.*

In STM, the tunneling of the electron between the tip and the surface is used to image the surface. In AFM, the image is due to the force acting between the tip and the substrate. Both these techniques give atomic resolution. STM has been used to study the overlayer structure of adsorbates on metal surfaces (see Fig. 5.9).

In XPS, X-ray is used to knock out the core electrons. Deep core electrons do not participate in bonding. Their energies are characteristic of the atoms from which they originate. Hence XPS has been used to find the elemental composition of the monolayer. At lower take-off angles, the ejection of photoelectrons from the atoms at the top layer will be much more as compared to those from the deeper layers. Hence angle-dependent XPS can be used to find the composition of the monolayer.

The capacitive charging current in an inert electrolyte can be used to measure the thickness of the monolayer by the equation $C_{ML} = \varepsilon_0 \varepsilon_r / d$, where d is the thickness of the monolayer and C_{ML} is the capacitance of the electrode covered with a monolayer. ε_0 is the permittivity of the free space and ε_r is the dielectric constant of the separating material. By adding a redox couple, one can observe the Faradaic current. This gives the number of defects in the monolayer.

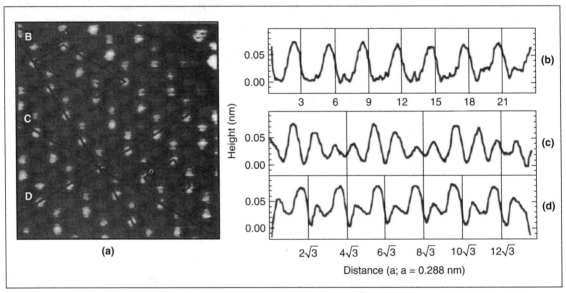

Fig. 5.9: STM image of octanethiol on Au(111). Fig. 5.9(b) is the plot of the cross section labeled B in Fig. 5.9(a) running along the Au nearest neighbour direction. C and D are cross-sectional plots of Fig. 5.9(a) along the Au next-nearest-neighbour directions. Reprinted with permission from Poirier (Ref. 13). Copyright (1994) American Chemical Society.

5.3 Growth Process

5.3.1 Growth from the Solution Phase

Various steps in the growth of monolayers are shown in Fig. 5.10.

The growth of the monolayer can be explained by using the Langmuir growth law. The rate of growth is proportional to the number of available sites given by the equation, $d\theta/dt = k(1 - \theta)$, where θ is the

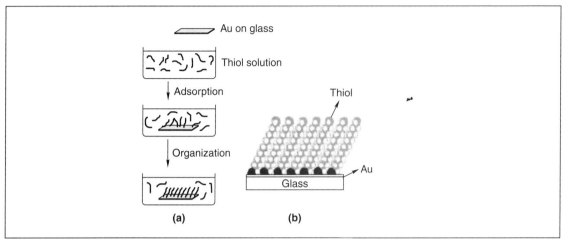

Fig. 5.10: (a) Various steps during the formation of self-assembly. Gold coated glass slide is dipped in an ethanolic solution of thiol. The initial chemisorption process is very fast. This is followed by a slow step during which organization happens which takes several hours. (b) Structure of organized monolayer using a space filling model.

fraction of sites occupied and k is the rate constant. Sum frequency generation (SFG) studies show three distinct steps during the growth of the monolayer. The first step corresponds to the chemisorption of the head group onto the metal surface and takes place very fast. Then the alkane chains start ordering into all-trans configuration, which is slower than the first process. During this straightening of the alkyl chain, the signal due to the d_- mode (antisymmetric CH_2 vibration) decreases in intensity. The gradual reorientation of the terminal methyl group during the final step is indicated by the slow evolution of r^+ mode (symmetric CH_3). The evolution of different modes is shown in Fig. 5.11.

5.3.2 Growth from the Gas Phase

The growth of monolayer from the gas phase in UHV allows one to study the process by using various *in-situ* measurements. The study by low energy electron diffraction (LEED) shows that the first phase, occurring immediately after dosing with the adsorbent molecule, is the stripped phase. On continued deposition, the structure changes to the standing phase with C (4×2) lattice.

The growth of mercaptohexanol monolayer on a gold surface has been investigated by using STM (see Fig. 5.12). Exposing the surface to low concentration of mercaptohexanol gives strips as shown in Fig. 5.12 (pointing finger, Fig. 5.12(b)). In these strips, the sulphur atoms will be sitting in the next-nearest-neighbour three-fold hollow sites and the alkane chain will be parallel to the substrate. On increasing the dosing, this stripe starts growing and covers the whole surface (Fig. 5.12(c) and Fig. 5.12(d)). Towards the end of saturation, a new feature called 'islands' starts appearing (pointing finger, Fig. 5.12(e)). From this point onwards, the growth starts taking place in the perpendicular direction and the alkane chain will be

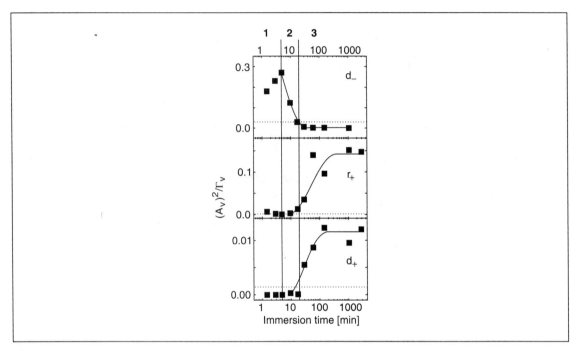

Fig. 5.11: *Intensities of the various modes upon the a adsorption of docosanethiol on Au(111). The three regions indicate the three steps during the growth process. Intensity of the features is plotted as a function of immersion time. Reprinted with permission from Schreiber F. (Ref. 14). Copyright (2000), with permission from Elsevier.*

perpendicular to the substrate in these islands. The Au vacancies will appear as deep pits after the island formation (Fig. 5.12(f)).

Even though the stripped phase with the alkane chain parallel to the surface, and the standing up phase with the alkane chain perpendicular to the surface, are the two important phases during the growth of the monolayer, metastable phases exist between the two extremes. This has been shown in Fig. 5.13.

5.3.3 Stability and Surface Dynamics

The thermal stability of SAMs depends on: (1) the strength of surface binding, and (2) the strength of lateral interaction. For alkanethiols on gold the thermal stability increases as a function of the chain length. While butanethiol monolayers desorb starting from a temperature of 75°C, octadecane monolayers desorb at temperatures ranging from 170–230°C (Ref. 15). Increasing the lateral $\pi - \pi$ interaction increases the thermal stability. Electrochemically the stability range of –0.1 to +0.1 V is very high, providing a large electrochemical window for most applications. This implies that a number of electrochemical processes can be conducted without monolayer desorption. The exposure of shorter chain thiols in solution can

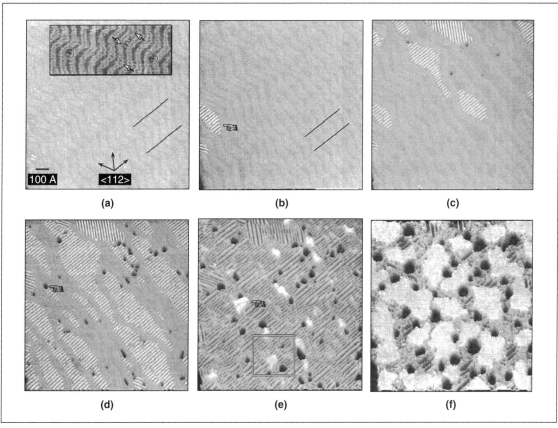

Fig. 5.12: *Constant current STM topographs showing the growth of mercaptohexanol monolayer from gas phase on Au (111) surface. (a) Clean Au (111) surface. (b) Stripped phase islands. (c) Striped phase growth. (d) Stripped phase growth showing Au vacancies. (e) Growth of standing up phase at the cost of the stripped phase. (f) Standing up phase growth at saturation limit. Reprinted with permission from Schreiber F. (Ref. 14). Copyright (2000) with permission from Elsevier.*

lead to the removal of defects on the SAM surface. This occurs by the exchange of monolayers at the grain boundaries, called 'ligand place exchange' and the diffusion of the monolayers, which occurs over a time window of several hours. The exposure of reactive gases such as ozone can affect the stability of the monolayer as the thiolate group can be oxidized (Ref. 16).

5.4 Phase Transitions

As a result of the van der Waals interaction, alkane thiol monolayers form crystalline phases on the metal surface. But as the temperature of the system increases, the orientational disorder increases. This weakens

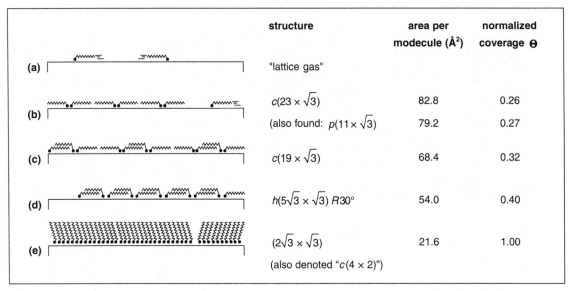

Fig. 5.13: *Various phases that can exist on a metal surface. The corresponding lattice structure, area per molecule and surface coverage have also been given. Reprinted with permission from Schreiber F. (Ref. 14). Copyright (2000) with permission from Elsevier.*

the van der Waals interactions, which give rise to liquid-like phases. Reflection absorption IR spectroscopy (RAIRS) has been used to study the phase transitions in planar SAMS. The position of the symmetric and anti-symmetric stretching mode of the methylene mode is indicative of the crystallinity of the alkane chains. The phase transitions in HS-$(CH_2)_{21}CH_3$/Au have been studied in detail by Bensebaa, *et al.* (Ref. 17). When the temperature was increased, the peak corresponding to d_+ showed broadening as well as a decrease in intensity. The same trend was shown by the d_- mode also. The melting is complete at 411 K. The position of the d_+ and d_- bands at this point exactly matches with that of the liquid alkanethiol confirming the liquid-like isotropic phase at this temperature.

5.5 Patterning Monolayers

SAMs are ideal templates for surface modifications. With appropriate modifications, patterned surfaces can be produced. Several methods are used for the patterning of monolayers. These belong to two kinds of routes: (1) decomposition, and (2) composition.

5.5.1 Decomposition

In this case, an already formed monolayer is given a desired pattern by decomposing a part of it. Patterning of the surface with a desired functionality in a specific area is important for molecular recognition. Among the several applications, one of particular importance is biorecognition wherein a biologically important material is anchored to the surface. This can be used for DNA recognition, as for instance, the binding of a DNA strand on the surface can be used to recognize the complementary strand in the solution. A change in the property of the monolayer or the substrate can be used to identify the presence of the analyte. In one of the approaches used, a mixed monolayer with mercaptopropionic acid (MPA) and mercaptohexanoic acid (MHA) was prepared first. The resultant monolayer formed segregated regions. The electrochemical potential for reductive desorption for MPA is lower than that of MHA and as a result, MPA can be selectively desorbed from the surface. By exposing a thiolated DNA, it can be bound to the region vacated by MPA (Ref. 18).

Lithography is another means of achieving patterned monolayers. The traditional lithography pathway is to expose the surface to be patterned to ultraviolet (UV) light at specific locations. On a SAM surface, UV exposure in the presence of oxygen leads to an oxidized SAM and this region can be removed by washing, thus exposing the bare metal surface. The exposed areas can be deposited with a new monolayer or can be etched away by an etchant. This process has been demonstrated on SAMs with patterns as small as 100 nm (Ref. 19). However, the problem with such methods is that the wavelength of light and patterns cannot be drawn beyond the resolution of the light used. If one wants to go to lower limits, wavelength of light has to be reduced, but in the X-ray regime, it is difficult to obtain reliable optics for manipulating light. This problem also affects the semiconductor industry.

Another lithographic method involves the use of particle beams. In this method, neutral atoms, ions or electrons are used to remove part of the surface. Patterns as small as 100 nm have been drawn with neutral Cs and Ar beams (Ref. 20).

In another method called 'nanoshaving', the adsorbate is removed physically (Ref. 21). This is achieved by scanning an atomic force microscope (AFM) tip at a load that is higher than the displacement threshold. The features can be as small as 20 nm. By conducting the shaving in a solution of another thiol, the second thiol can be deposited in regions vacated by the first, and a pattern can be created. The process is called 'nanografting'. A similar methodology can be used with scanning tunneling microscopy (STM) where the tip potential is kept high to cause thiol desorption.

5.5.2 Composition

In this case, a monolayer is made in a controlled fashion. Two kinds of approaches are used in composition. In the first called 'microcontact printing' (Ref. 22), a mask is generated on a polymer from the master. The mask is then inked with a thiol. Upon contact with the surface, the mask transfers the ink to specific areas of the surface. The exposed metal surface can be deposited with another thiol or the exposed material can be etched away by an etchant. Unlike in the case of beam lithography methods, microcontact printing has been used to make patterns on non-planar surfaces. The principal problem in this case, however, appears to be one of finding ways to reduce the dimension of the patterns.

The second approach is called 'dip-pen lithography' (Ref. 23). Here an AFM tip coated with a surfactant (ink) is drawn over a wet surface. The water meniscus touching the tip travels with it and solubilizes some of the molecules. In this approach, the ink can be made of a variety of materials including nanoparticles. The approach need not necessarily make a monolayer.

If the substrate can be controlled in such a way that deposition occurs only in the selected areas, patterns can be drawn. In one such approach, the potential was controlled on an array of electrodes. If this is done on selected areas of an already coated electrode, a part of the monolayers can be desorbed and another monolayer can be coated on such locations. It is also possible to control the potential on specific areas of an uncoated electrode to accelerate or decelerate deposition. Both these approaches have been used to make patterned monolayers.

5.6 Mixed Monolayers

One can obtain a mixed monolayer by mixing two monolayer forming species in an appropriate ratio. If the chemical constitutions of the two monolayer-forming entities are similar in terms of the alkyl chain length and the functionality, the mixed monolayer is similar to a two-dimensional alloy (two-dimensional in the sense that the film is planar, while the thickness is molecular). However, chemical differences between the species can lead to the segregation of one entity. If the two chemical constituents are separated, it is possible to anchor two different kinds of materials at these locations by using selective molecular chemistry.

5.7 SAMS and Applications

5.7.1 Sensors

The interest generated by the study of SAMs has shifted from fundamental studies to technology. The potential applications of SAMs include molecular recognition and wetting control. The chemical properties of the monolayers can be used for sensing applications. There are two important elements in a sensor, of which the first is the chemically selective recognition layer, while the second is the signal transducer which provides a signal that can be monitored. A number of approaches have been used to make selective recognition layers, which depends on the species to be detected. The fact that several of these sensing elements can be located on a given area provides the capability to sense several species simultaneously (see Chapter 12).

A variety of metal ion sensors can be made by functionalizing the metal surface with a ligand with high specificity to a particular ion. Sensors selective to Cu^{2+} were made by functionalizing the gold surface with 2,2'-thiobisethyl acetoacetate. The selective detection of perchloroethylene in the presence of other molecules such as trichloroethylene, tetrachloromethane, chloroform, and toluene was achieved by using modified resorcin(4)arene as the monolayer (see Fig. 5.14). The incorporation of perchloroethlelene into the receptor site results in mass changes in terms of nanograms, which are detected with the help of a quartz microbalance oscillator.

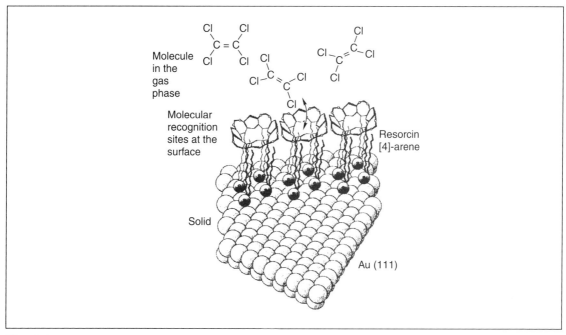

Fig. 5.14: *Molecular recognition by a monolayer of resorcinol on Au (111) surface. The receptor site showed high selectivity to perchloroethylene when dosed with a mixture of halocarbons. Reprinted with permission from Schierbaum, et al. (Ref. 24). Copyright (1994) AAAS.*

Among the most common examples of SAM-based sensors are enzyme biosensors. As enzymes are catalytic in action, they are also called catalytic biosensors. Here an enzyme acts as the recognition element. The signal is transduced by detecting either the molecule consumed or that generated. An example of this kind of sensor is the glucose biosensor which uses glucose oxidase (GOD), to oxidize glucose to gluconolactone. In this process, the enzyme is reduced and a mediator used in the process gets the enzyme back to the original state. In nature, O_2 is the oxidizer and H_2O_2 is produced in the process. The mediators used typically in experiments are ferrocene and ferricyanide, and their change is monitored electrochemically. The reduced form of the mediator gets oxidized at the electrochemical surface and the current generated is proportional to the amount of glucose oxidized. The immobilization of the enzyme is more controlled when it takes place on a SAM surface.

Enzyme immobilization on the monolayer surface can be achieved by several means. One method is to covalently modify the SAM by selected reactions. An approach used for this is described below. Here a mercaptopropionic acid monolayer is made on Au. The monolayer is activated by reacting with 1-ethyl-3(3-dimethylaminopropyl)carbodiimide hydrochloride (EDC) and N-hydroxysuccinamide (NHS). This process makes a succinamide ester on the monolayer surface, helping amine groups of the enzyme to bind to the surface, thereby forming an amide linkage. There are other approaches such as the use of electrostatic interactions. The approach mentioned here is illustrated in Fig. 5.15. An excellent review is available on the preparation of sensors on monolayers (Ref. 25).

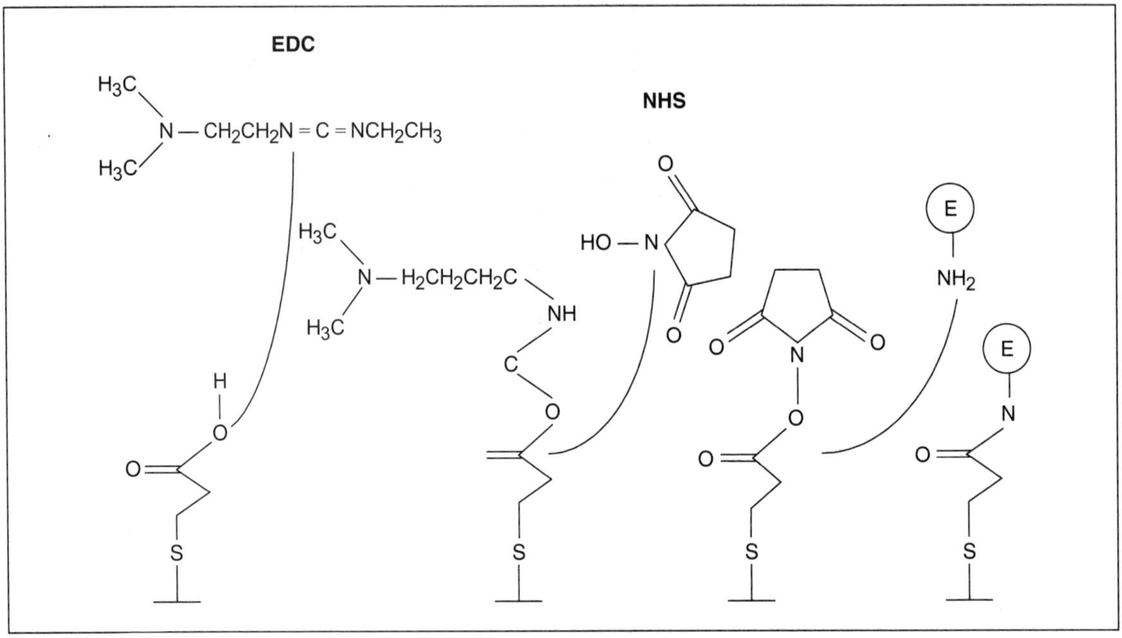

[Figure: chemical scheme showing EDC and NHS coupling steps for attaching an enzyme E to a thiolated surface]

Fig. 5.15: *An approach used to achieve enzyme immobilization. E is an enzyme.*

5.7.2 Affinity Biosensors

In these sensors, a biorecognition molecule is the sensor element, which is specific to the analyte molecule. Depending on the sensor molecule, which can be in the form of antibodies, proteins, DNAs, etc. various analyte species can be detected. The recognition event needs to be sensed. The most common approaches for sensing are surface plasmon resonance and quartz crystal microbalance (QCM). In the former, the change in the surface plasmon resonance of a thin film of gold as a result of the biorecognition event is used to sense the event. In the QCM method, the change in the oscillation frequency of a quartz crystal as a result of the mass accumulation on a gold film deposited on its surface, is used to sense the biorecognition event.

The biorecognition molecule must be immobilized on the surface. One protocol used for this purpose is to thiolate the biomolecule. In the case of a DNA the 5' end can be thiolated by a mercaptohexyl unit and a solution of the DNA upon exposure to the gold surface forms a monolayer. However, the single-stranded DNA (ssDNA) lies on the gold surface and as a result, the hydridization efficiency is reduced to 10 per cent. By exposing the thiolated DNA surface to mercaptohexane (MCH), the locations wherein the DNA is not pinned, are occupied by the monolayer, which makes the DNA stand up. This facilitates specific binding and the hybridization efficiency is enhanced to 100 per cent (Ref. 26).

The immobilization of biomolecules can be used for various applications. By immobilizing an antibody specific to a bacterial strain, the sensor can detect that specific bacterium. This has been demonstrated (Ref. 27) for *salmonella paratyphi* with specificity to other serogroups of *salmonella* and the detection limit found was 1.7×10^2 cells/mL. Various kinds of species can be detected on the basis of the immobilized biomolecule.

An affinity biosensor can behave like a nanomachine. This has been demonstrated with an ion channel biosensor. Here a gramicidin ion channel has been used as a sensor and the transduction is achieved by measuring the conductivity (Ref. 28). Gramicidin is an example of a channel found in cell walls. It is an unusual peptide, having alternating D and L amino acids. In lipid bilayer membranes, as in the case of a cell wall, gramicidin dimerizes and folds as a right handed β helix. The dimer just spans the bilayer. In the mechanism of ion transport through membranes, it has been found that the ion transport rate depends on gramicidin because the gramicidin channel functions when the proteins reversibly dimerize on the membrane, thereby opening up a channel. This is illustrated in Fig. 5.16.

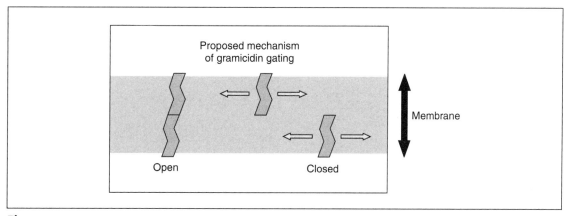

Fig. 5.16: *Mechanism of opening and closing of ion channels by gramicidin. Gramicidin (shaded) moves on the membrane and locking of two units opens the ion channel.*

In the sensor, the gramicidin (IG) is thiolated and immobilized on the surface. It is separated with thiols which act as spacers (ST). A lipid bilayer is anchored onto the SAM through tethered lipids (TL), which penetrate into the lower half of the bilayer. The upper half of the bilayer has mobile gramicidines (MG) with pendent biotin groups. Membrane spanning lipids (MSL) are also anchored to biotin groups. Antibody fragments (Fab) are linked to MG and MSL units through streptavidin (S) using biotin. In the open state, the MG units are mobile and as a result, MG and IG get linked, thereby opening up the ion channel. This leads to a large increase in conductivity. When analyte molecules (A) are present, MG cannot diffuse freely, as these molecules are locked in position. Depending on the antibody fragments used, this sensor can work for different analytes. Its use has been demonstrated in the case of hormones, bacteria and certain sequences of DNA (Ref. 29). The working principle of the sensor is illustrated in Fig. 5.17.

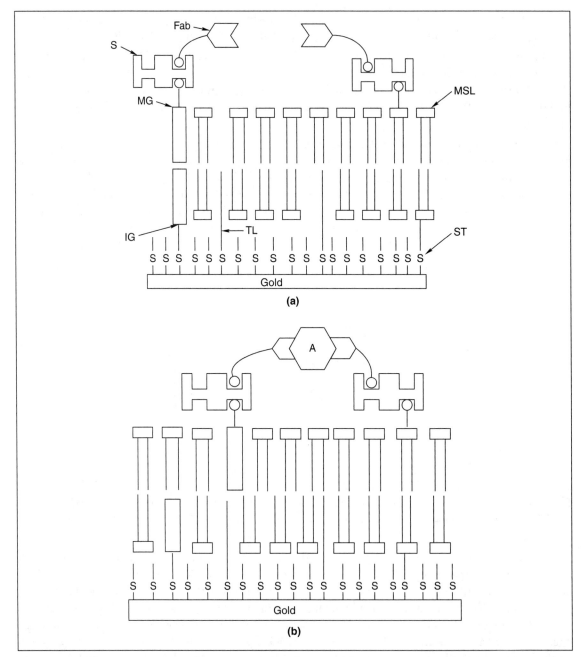

Fig. 5.17: *Schematic illustration of the ion channel based bio sensor (Ref. 28). When MG and IG coincide, a large increase in ionic conductivity is observed as gramicidin ion channel opens up (a). When the analyte blocks the antibody fragments, the movement of gramacidins become impossible, reducing the ionic conductivity (b). Adapted from Ref. 28. Copyright Nature.*

5.7.3 Chemical Sensors

Depending on the kind of recognition element, a sensor may be biochemical or chemical. In the chemical category, the recognition element is non-biological in origin and is synthesized in the laboratory. There are several chemical sensors which use SAMs and utilize a variety of properties. Electrochemical, capacitance, plasmon resonance and mass are the most common properties utilized for transduction of the recognition event.

One of the most common sensing events is the quantitative detection of transition metal ions, especially in presence of other ions. The detection of Cu(II) in presence of Fe(III) is an example. The four-coordination preference of Cu^{2+} can be provided by attaching a tetradentate ligand and the other end of the molecule can be anchored onto the gold surface. The sensor sites can be kept separated by preparing a mixed SAM. The presence of other ions does not influence the detection as they do not get attached to the electrode surface. A variety of ligands and metal ions can be used to implement this approach.

Molecular imprinting is another approach that has been developed for sensing. Here the shape of a molecule is imprinted on the monolayer surface and when the location is occupied by the analyte molecule, it is recognized as a sensing event. In one of the approaches demonstrated, a mould for the analyte barbituric acid, in the form of thiobarbituric acid, is made to bind the surface. The shape is imprinted on the surface by co-adsorbing another monolayer forming thiols. When the analyte molecule is exposed to the surface, it can bind to the cavities imprinted on the surface. This binding event changes the capacitance of the surface. Several molecules such as cholesterol, barburates, quinines, etc. have been detected in this way (Ref. 30).

5.7.4 pH Sensing is Ion Sensing

SAMs have been used for pH sensing by several investigators. The general approach is to have two electroactive species on the monolayer surface. While one is pH-sensitive, the other is insensitive and acts as a reference. This is seen in the case of quinine and ferrocene, both of which are immobilized on the monolayer surface. While in the first, the oxidation and reduction shift linearly with pH, the second does not show any response to pH. It is also possible to make the pH-independent electrochemistry of ferrocene pH-dependent by having a bifunctional molecule. In one of the approaches used, a ferrocene carboxylic acid is linked to the surface through a –S linkage. The oxidized state of ferrocene is stabilized by the deprotonation of the carboxylic acid group, which makes the redox chemistry of ferrocene pH-dependent.

5.7.5 Corrosion Prevention

When used as coatings, SAMs offer corrosion resistance. In this regard, several monolayers have been made on surfaces such as Au, Cu, Fe, etc. The chain length of the molecule has to be long enough to offer effective protection. The monolayer coating is inadequate in offering complete protection against ion penetration. The permeability of ions through the assembly poses a serious problem, thereby necessitating an additional coating on the monolayer. An approach that has been tried to resolve this problem is to make a siloxane polymer of the SAM.

5.7.6 Other Areas

Self-assembled monolayers on a gold surface constitute an ideal system for practising fundamental studies related to the electron transfer process. The electrode surface used for this purpose is modified with electoactive monolayers through self-assembly. The attachment of thiols with the active terminal group allows further derivatization through classical organic reactions. In one attempt, cycloaddition was also used to derivatize a surface modified with thiol containing azide at the termini (Fig. 5.18).

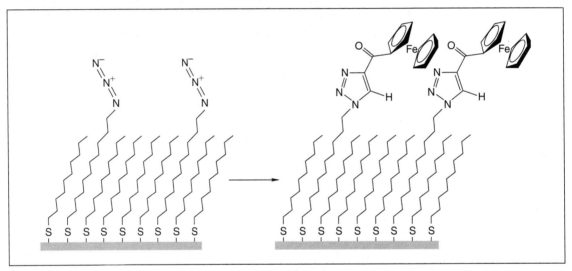

Fig. 5.18: *Electrode surface before and after cycloaddition. Reprinted with permission from Collman, et al. (Ref 31). Copyright (2004) American Chemical Society.*

Chemical force microscopy (CFM) combined with chiral discrimination by a molecule can be used to distinguish different chiral forms of the same molecule. In this technique, an AFM tip is functionalized with a chiral molecule. This chiral probe is then used to discriminate between the two chiral forms of the same molecule on a surface. The gold-coated AFM tip is functionalized with a chiral probe by using acylated phenyl glycine modified with alkanethiol. The changes in the friction or adhesion forces are used to distinguish between the two enantiomers of mandelic acid.

5.7.7 Wetting Control

The wetting properties of a surface can be modified, to a great extent, by coating it with a monolayer of molecules. This is one of the important applications of SAMs. A low coverage surface is made intentionally by the self-assembly of alkanethiol with a bulky end group on an Au(111) surface. The end group is then hydrolyzed to form a carboxy terminated monolayer with low surface coverage. The carboxylate groups generated by the hydrolysis can be attracted to the gold surface by applying an electric field. Thus the surface can now be made hydrophilic or hydrophobic by changing the electrochemical potential. The working of the sensor is schematically shown in Fig. 5.19 (Plate 4).

5.7.8 Molecular Electronics

SAMs are also used to make electrical contacts. By using a self-assembled monolayer of dithiol on gold, one can make a surface with pendent thiol groups and can also attach a gold nanoparticle. This attachment with a covalent linkage showed four orders of magnitude higher current than when the nanoparticle was physisorbed on the SAM surface at the same tip bias voltage. This showed that monolayer-based electrical contacts are feasible. In various studies, different kinds of interactions such as van der Waals, hydrogen bonding and covalent interactions have been made between monolayers attached to metallic surfaces such as gold and mercury. These studies have shown that the electron transfer rates increased in the order, van der Waals > hydrogen bonding < covalent.

In all these experiments, it is necessary to make a contact. This is done through monolayers. The experimental protocol is illustrated in Fig. 5.20. In the experiment, a nanoparticle solution is exposed to a monolayer making nanoparticles sit on the monolayer. The current flowing between the tip and the surface is measured at a bias voltage.

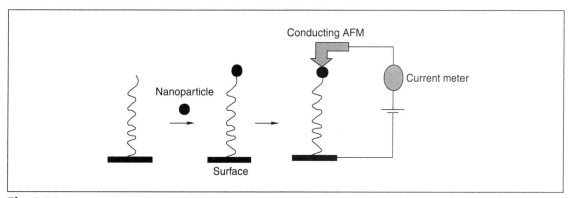

Fig. 5.20: *Experimental approach for making electrical contact to a nanoparticle.*

5.7.9 Templates

SAMS are excellent templates on which nanofabrication can be done. A given structure can be produced on the surface by using a number of methodologies. Various nanoscopic objects can be used for producing a structure. Several of these objects are detailed below.

Nanoparticles Monolayer-protected nanoparticles (see Chapter 8) are new kinds of materials. In this class of systems, monolayers are grown on the surface of nanoparticles of metals, semiconductors and insulators. The monolayer protection allows the nanoparticle to preserve its size and shape. Through a judicious choice of the functionality on the nanoparticle surface, one can bind the material on to an SAM in a well-defined fashion. The nanoparticle surface can be coated with various molecules having additional properties. The chemistry discussed on the planar monolayer surface can be eminently done on the nanoparticle surface as well. One of the important aspects in the use of nanoparticles is that with monolayer functionalization, nanoparticles can be incorporated in any matrix such as polymers. It is also possible to attach nanoparticles on solid surfaces through a monolayer anchor. This allows one to produce device structures with nanoparticles.

Nanotubes By functionalizing carbon nanotubes, one can attach them to SAMs through specific chemistry. This facilitates the arrangement of aligned nanotubes on monolayer surfaces. In this way, nanotubes can be made available on stable supports. They can be used as devices for applications such as field emission.

Crystal Growth As mentioned earlier, SAMs can nucleate crystal growth and such low temperature chemical routes are important for making ultra-thin layers of materials. By creating a functional molecular surface at specific locations, it is possible to grow another material at these locations. Several examples of these are known. The formation of tetragonal zirconia on SAMs has been reported (Ref. 33). In Fig. 5.21 a SEM image of ZrO_2 crystals grown on a SAM is shown. The crystals are highly faceted, and signify one of the very early examples of low temperature growth of ordered materials. The SAMs act as templates on which initial nucleation occurs. There are also examples of this kind of assisted growth on metal nanoparticles, wherein a ZrO_2 shell is grown on silver nanoparticles by solution chemistry (Ref. 34).

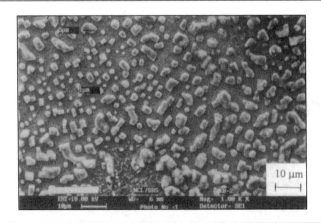

Fig. 5.21: *Scanning electron micrograph of tetragonal ZrO_2 grown on self-assembled monolayers. From Bandyopadhyay, et al. (Ref. 33). Used with permission from the author.*

By patterning an SAM with mercaptohexanoic acid (MHA) and the balance with mercaptohexanol (MCH) it has been shown that calcite crystals are grown selectively at locations of MHA instead of MCH (Ref. 35). Several other materials have also been grown in this way.

Polymers One can build polymers on an organized structure. The variety in polymeric structures is diverse so that almost anything can be grown on a SAM surface through an appropriate choice of precursors.

Complex Molecules One can also think of placing biological molecules at specific locations just as in materials chemistry. The most common biomaterials are proteins and placing them on metal surfaces is important. However, proteins placed directly on surfaces get denatured, which is why placing them on SAMs is a feasible alternative. The problem, however, is that SAMs have a non-specific affinity for proteins, which needs to be controlled. Ethylene glycol units on the surface prevent non-specific protein adsorption and SAMs with 4 to 7 poly(ethyleneglycol) units are used for this purpose. Specific thiolated monolayers can be grown on surfaces and the protein can have a biorecognition function. Such an approach can be used to adhere cells onto monolayers. This is mediated by proteins of the extracellular matrix (ECM) such as fibronectin and collagen. These proteins can be anchored to surfaces by using patterned monolayers. All the unpatterned region of the SAM is covered with poly(ethylenegycol) monolayer so that no protein adsorption occurs. Cells can then be attached to the protein modified sites. The size of the islands affects the cell growth. This kind of capability, along with the spatial control that is possible in monolayer growth, will make it possible to have lab on a chip.

Layer-by-layer-Structure SAMs involve growing layers. If such monolayers can be grown one over the other, step by step, microscopic thickness can be generated. For achieving this, it is important to have the link the monolayers through covalent chemistry. In one such approach, a benzenedimethane thiol (BDMT) monolayer was linked to another thiol through the covalent chemistry of the pendent thiol group of the BDMT monolayer (Ref. 36).

Review Questions

1. Why are these monolayers referred to as, 'self-assembled'? Are there 'force-assembled' monolayers?
2. What are the various aspects which determine the stability of a monolayer?
3. Why self assembled monolayers are ideal systems for probing fundamental phenomena? State a few such phenomena not described here.
4. Why SAMs are difficult to study?
5. Propose a few other monolayer systems other than alkane thiols on gold.
6. How do we study the kinetics of self organization by experimental means?
7. Monolayers are crystalline assemblies. How does one study the changes in the crystalline order, especially as a function of temperature?

8. Suggest a few uses of monolayers not described in this chapter.
9. Are there day to day examples of monolayers?
10. Suggest a method to study the strength of single chemical bonds using monolayers.

References

1. Atomic and Nano Technology, IBM Research, http://www.research.ibm.com/atomic.
2. Langmuir, I., *J. Am. Chem. Soc.*, **39** (1917), p. 1848.
3. Blodgett, K., *J. Am. Chem. Soc.*, **57** (1935), p. 1007.
4. Bigelow, W.C., D.L. Pickett and W.A. Zisman, *J. Colloid Interface Sci.*, **1** (1946), p. 513.
5. Nuzzo, R.G. and D.L. Allara, *J. Am. Chem. Soc.*, **105** (1983), p. 4481.
6. Ulaman, A., (1991), *An Introduction to Ultra-thin Organic Films from Langumir-Blodgtt to Self-Assembly*, Academic Press, London.
7. Ulman, A., *Chem. Rev.*, **96** (1996), p. 1533.
8. Chidsey, C.E.D., G.Y. Liu, P. Rowntree and G. Scoles, *J. Chem. Phys.*, **91** (1989), p. 4421.
9. Dubois, L.H. and R.G. Nuzzo, *Annu. Rev.*, *Phys. Chem.*, **43** (1992), p. 437.
10. Sheen, C.W., J.X. Shi, J. Martensson, A.N. Parikh and D.L. Allara, *J. Am. Chem. Soc.*, **114** (1992), p. 1514.
11. Alves, C.A., E.L. Smith and M.D. Porter, *J. Am. Chem. Soc.*, **114** (1992), p. 1222.
12. Schreiber, F., A. Eberhardt, T.Y.B. Leung, P. Schwartz, S.M. Wetterer, D.J. Laurich, L. Berman, P. Fenter, P. Eisenberger and G. Scoles, *Phys. Rev.*, **B 57** (1998), 12476.
13. Poirier, G.E., M.J. Tarlov, *Langmuir*, **10** (1994), 2853.
14. Schreiber, F. *Proc. Sur. Sci.*, **65** (2000), 151.
15. Sandhyarani, N. and T. Pradeep, *Vacuum*, **49** (1998), p. 279.
16. Sandhyarani, N. and T. Pradeep, *Chem. Phys. Lett.*, **338** (2001), pp. 33–36.
17. Bensebaa, F., T.H. Ellis, A. Badia and R.B. Lennox, *J. Vac. Sci. Technol. A*, **13** (1995), p. 1331.
18. Satjapipat, M., R. Sanedrin, and F.M. Zohu, *Langmuir*, **17** (2001), p. 7637.
19. Behm, J.M., K.R. Lykke, M.J. Pellin and J.C. Hemminger, *Langmuir*, **12** (1996), p. 2121.
20. Younkin, R., K.K. Berggren, K.S. Johnson, M. Prentiss, D.C. Ralph and G.M. Whitesides, *Appl. Phys. Lett.*, **71** (1997), p. 1261.
21. Liu, G.Y., S. Xu, and l. Qian, *Acc. Chem. Res.*, **33** (2000), p. 457.
22. Xia, Y. and G.M. Whitesides, Angew. *Chem. Int. Ed.*, **37** (1998), p. 551.
23. Piner, R.D., J. Zhu, F. Zu, S. Hong and C.A. Mirkin, *Science*, **283** (1999), p. 661.

24. Schierbaum, R.D., T.E. Weiss, E.U. Thoden van Velzen, J.F.J. Engbersen, D.N. Reinhoudt and W. Göpel, *Science*, **265** (1994), p. 1413.
25. Shipway, N., E. Katz and I. Willner, *Chemphyschem*, **1** (2000), p. 18.
26. Levicky, R., T.M Herne, M.J. Tarlov and S.K. Satija, *J. Am. Chem. Soc.*, **120** (1998), 9787.
27. Fung, Y.S. and Y.Y. Wong, *Anal. Chem.*, **73** (2001), p. 5302.
28. Cornell, B.A., V.L.B. Braach–Maksvytis, L.G. King, P.D.J. Osman, B. Raguse, L. Wieczorek and R.J. Pace, *Nature*, **387** (1997), p. 580.
29. Lucas, S.W. and M.M. Harding, *Anal. Biochem.*, **282** (2000), p. 70.
30. Mirsky, V.M., T. Hirsch, S.A. Piletsky and O.S. Wolfbeis, *Angew. Chem.-Int. Ed.*, **38**, (1999), p. 1108.
31. Collman, J.P., N.K. Devaraj and E.D. Chidsey, *Langmuir*, **20** (2004), p. 1051.
32. Lahann, J., S. Mitragotri, T. Tran, H. Kaido, J. Sundaran, S. Hoffer, G.A. Somorjai and R. Langer, *Science*, **299** (2003), p. 371.
33. Bandyopadhyay, K., S.R. Sainkar and K. Vijayamohanan, *J. Am. Cer. Soc.*, **82** (1999), p. 222.
34. Eswaranand, V. and T. Pradeep, *J. Mat. Chem.*, **12** (2002), p. 2421.
35. Aizenberg, J., A.J. Black and G.M. Whisides, *Nature*, **398** (1999), p. 495.
36. Murty, K.V.G.K., M. Venkataramanan and T. Pradeep, *Langmuir*, **14** (1998), p. 5446.

Additional Reading

1. Pradeep, T. and N. Sandhyarani, *Pure and Appl. Chem.*, **74** (2002), pp. 1593–1607.
2. Sandhyarani, N. and T. Pradeep, *Int. Rev. Phys. Chem.*, **22** (2003), pp. 221–262.
3. J.J. Gooding (2004) in *Encyclopaedia of Nanoscience and Nanotechnology*, **1**, pp. 17–49, H.S. Nalwa, (ed.) American Scientific Publishers.

Chapter 6

GAS PHASE CLUSTERS

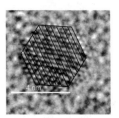

Clusters belong to a new class of systems and are studied to understand several aspects related to the science of nanomaterials. Nanoparticles are nucleated at the cluster stage and it is important to understand these systems in the naked state using mass spectrometry in order to know the origin of molecular and electronic structure in bulk systems. Almost everything forms clusters and the diversity in cluster systems is vast. The methods of preparation of clusters and their diverse variety are discussed in this chapter. While exploring clusters, scientists have discovered new molecules such as C_{60}. The chemistry of clusters in the gas phase still constitutes an active area of investigation.

Learning Objectives

- Why do you study clusters?
- How do you make them and study them?
- What are the commonly found cluster types?
- What are the basic techniques of gas phase cluster spectroscopy?
- How can one understand the properties of clusters?

6.1 Introduction

Clusters belong to a new category of materials; in size they fall between bulk materials and their atomic or molecular constituents. Sometimes they are considered to constitute a new form of matter, as their properties are fundamentally different from those of discrete molecules and bulk solids. They are systems of bound atoms or molecules, existing as an intermediate form of matter, with properties that lie between those of atoms (or molecules) and bulk materials. Depending on the kind of constituent units, they are called either atomic or molecular clusters. The term, 'molecular clusters' also implies clusters which behave like super molecules. Clusters include species existing only in the gas phase or in the condensed phase or in both. Clusters identified first in the gas phase have been synthesized later in the condensed phase and vice versa. They can have a net charge (ionic clusters) or no charge (neutral clusters) at all. The finite aggregates of atoms or molecules constituting clusters are bound by forces which may be metallic, covalent, ionic, hydrogen bonded or van der Waals in character and can contain up to a few thousand atoms (called the *nuclearity* of the cluster). As a result, they have regular shapes (such as icosahedra). However, they also exist in a spherical shape. Usually, one can distinguish between different types of clusters by the nature of the forces between the atoms, or by the principles of spatial organization within the clusters. Clusters containing a well-defined number of transition metal atoms have unique chemical, electronic and magnetic properties, which vary dramatically with the number of constituent atoms, the type of element and the charge on the cluster. Clusters differ from bulk materials in terms of the presence of a magic number of atoms or molecules they contain. Magic numbers signify electronic and structural stability. Although this chapter is written from the perspective of gas phase spectroscopy, the reader may note that clusters are also made in the condensed phase. Examples of such clusters include molecules containing metal clusters within, with various kinds of protecting ligands, which are being investigated in several labs including ours. These studies are even more exciting now as such clusters containing a few atoms of noble metals, such as Au_{25} for example, can be made in the laboratory as a bulk powder, stored and investigated over long periods. In essence, they are similar to many laboratory chemicals.

As shown in Fig. 6.1 clusters can be depicted as a state between isolated atoms or bulk solid. What is also implied is that it is possible to get them from either side, from atoms or molecules or from both.

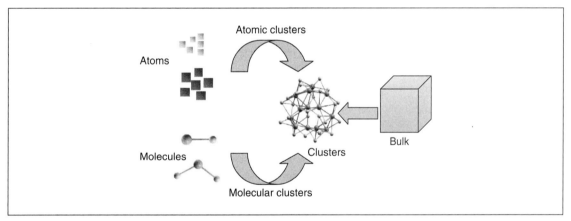

Fig. 6.1: *Schematic representation of cluster placed in between atom, molecule and bulk material. From left to right the dimension of the constituent matter increases. Clusters can be produced from atomic or molecular constituents or from the bulk material. The variously colored balls are all different kinds of atoms. In relation to the dimensions of atoms and the cluster shown, the material should be a thousand times larger than the one depicted.*

From a general perspective, there are two broad reasons why we investigate clusters. There are several associated reasons as well. 1. As clusters bridge the gap between molecules and materials, the evolution in properties of molecules as they become materials can be understood by investigating clusters. 2. Properties such as chemical reactivity and catalysis depend strongly on specific geometry, electronic structure, etc., and clusters help us to understand such fundamental phenomena. Such understanding can make large economic impact, for example, in terms of making the most appropriate, greener, cleaner and economical catalyst.

One can distinguish the properties of atoms molecules of clusters as listed in Table 6.1.

Table 6.1: *Difference between atom, molecule, clusters and bulk materials*

Property	Atom/Molecule	Cluster	Bulk material
Size of constituting entities	Few Angstroms (Å)	Angstrom (Å) to nanometer (nm)	Microns to higher
Number of constituents (n)	1 for atom, many for molecule	2 to several thousands	Infinite
Electronic structure	Confined (quantized)	Confined (quantized)	Continuous
Geometric structure	Well-defined and predictable	Well-defined and predictable	Crystal structure decides
Example	$Na/NaCl/C_6H_6$	C_{60}, $(NaCl)_n$	Bulk gold, silver

6.2 History of Cluster Science

The importance of clusters was first proposed by the Irish-born chemist Robert Boyle in his book, *The Sceptical Chymist* published in 1661 (Ref. 1). In it Boyle was critical of Aristotle's four element theory of matter and proposed that it exists in the form of 'corpuscles'. He thought about it, "minute masses or clusters that were not easily dissipable into such particles that composed them." 'Clusters' for him were not collections of atoms or molecules as both were unknown then.

During the last several decades, cluster science has grown to become a field of interdisciplinary study. Improvement in experimental techniques such as mass spectrometry and advancement in computational power (and methods) have increased the interest in cluster science. After the discovery of buckminsterfullerene, C_{60}, the field of cluster science witnessed an enormous growth. The use of clusters goes back to several centuries. Examples of application include photography (AgBr clusters in films) and glass works (for staining); see Chapter 1 for more details. There are several eastern societies which used small particles in medicines. Rayleigh recognized that colors of stained glasses are due to the scattering of light by small metal particles, clusters of atoms embedded in the glass.

6.3 Cluster Formation

In any material, there are atoms on the surface. But the number of surface atoms is very small in comparison to the bulk. If one takes a one cm^3 metal block, there are about 2×10^{23} atoms in it, assuming a radius of

1 Å for the atom. On the surface of this cube, there are only about 1.5×10^{16} atoms. The fraction of surface atoms is one in a million and therefore they do not make measurable influence in the properties investigated. However, in the case of a cluster, this fraction can be of the order of one and that makes significant difference in the properties. Assuming a spherical cluster, the fraction of surface atoms is (number of atoms on the surface/total number of atoms), $F = 4/n^{1/3}$, where n is the total number of atoms. One can see that, $F = 0.3$ for $n = 1,000$, $F = 0.2$ for $n = 10,000$ and $F = 0.04$ for $n = 1,000,000$. The surface atoms are unsatisfied in their valencies and they are extremely reactive due to this reason. Therefore, many of the clusters cannot be kept in the free state. They have to be made *in-situ*, in experimental apparatus where the properties are investigated. Therefore, almost all the studies are conducted in vacuum or in presence of rare gases.

Gas phase clusters are generated in cluster sources. There are many kinds of cluster sources. Some of them are listed below.

- Laser vaporization-flow condensation source
- Pulsed arc cluster ion source
- Laser ablation cluster source
- Supersonic (free jet) nozzle source
- Knudsen cell (effusive source)
- Ion sputtering source
- Magnetron sputtering source
- Gas aggregation/Smoke source
- Liquid metal ion source

Some of the more common methods are described below. Additional reading material listed at the end of the chapter may be consulted for more details.

6.3.1 Laser Vaporization

The laser vaporization source is a pulsed cluster source which is used to produce small- and medium-sized clusters. The resultant cluster may be formed from any element or compound. This method typically combines laser ablation and supersonic jet expansion. In the laser vaporization source, vapor is generated by pulsed laser ablation of a rod of the starting material. An intense pulsed UV laser is used here (typically third or fourth harmonic of $Nd:YAG$). Each 10 ns pulse vaporizes 10^{14}–10^{15} atoms per mm^2 of the target. Since the use of lasers for cluster generation also leads to ionization, this source also generates neutral, cationic and anionic clusters which can be investigated directly by mass spectrometry, without post-ionization. In fact what is produced by laser evaporation is a plasma. Pulsing helps in (1) to get an intense light capable of evaporation of materials directly breaking their bonds in the lattice, (2) produces a pulse of clusters suitable for time of flight analysis and (3) pulsed laser firing and subsequent expansion of the evaporated plasma into vacuum is generally done in presence of a carrier gas, which is also pulsed, reducing the pumping requirements.

6.3.2 Pulsed Arc Cluster Ion Source

Pulsed arc cluster ion sources (PACIS) are related to laser vaporization sources. Instead of the laser here the cluster precursor is vaporized by an intense electrical discharge. Cluster beams generated in this way are significantly more intense in comparison to laser vaporization. Nearly 10 per cent of the clusters formed by using this technique are charged and again, post ionization is not necessary for mass analysis.

6.3.3 Supersonic (Free Jet) Nozzle Sources

Supersonic nozzle sources are of two main types, unseeded and seeded. In the first type, clusters of inert gases, molecules and low boiling metals (e.g. Hg) are formed. In the other type, the metal is vaporized (with a vapor pressure of 10–100 mbar) in an oven and the vapor is mixed with (seeded) an inert carrier gas at a pressure of several atmospheres (10^5–10^6 pa) at a temperature of 77–1500 K. The metal/carrier gas mixture is then expanded through a nozzle (with diameter of 0.03–0.1 mm) into high vacuum (10^{-1}–10^{-3} Pa), which creates a supersonic beam. Nozzles with rectangular opening have been used to generate two-dimensional cluster beams (normally they have a disk-like cross section), which are necessary for certain studies. The cluster growth stops at least immediately after the nozzle exit, when the vapor density reduces drastically. Such sources produce intense continuous cluster beams of narrow energy spread, suitable for high resolution spectroscopy. Seeding produces large clusters while in the absence of a carrier gas, smaller clusters (> 10 atoms) are formed. Intensity of the beam and the smaller energy spread are the significant aspects of this kind of sources.

6.3.4 Gas-Aggregation or Smoke Sources

The source utilizes the property of aggregation of atoms in inert media. The vapors generated by one of the several means are introduced into a cold inert gas at a high pressure of the order of 1 torr. The species, originally at a high temperature, are thermalized. The gas phase is supersaturated with the species and they aggregate. These sources produce continuous beams of clusters of low-to-medium boiling (< 2000 K) metals. By controlling the kinetics of quenching and aggregation, various cluster sizes can be produced.

6.3.5 Knudsen Cell

The Knudsen cell produces a continuous, low flux beam of clusters. The velocity of the species is low (subsonic). In the cell having a small aperture, a volatile solid or liquid is heated; the cell itself is held in a vacuum vessel. In design, this is similar to a smoke source. At the low vapor pressures produced, their mean free path is greater than the collision diameter of the aperture, as a result of which there are very few collisions before particles leave the cell. The energy resolution of the cluster beam formed in the effusive sources is poor. The angular spread is also larger. In these sources, as the aperture is small, the solid-gas mixture is nearly at equilibrium. The cluster intensity (I) falls exponentially with an increase in the cluster nuclearity (N) according to the equation, $I(N) = ae^{-b/N}$, where a and b are parameters. The intensities in a smaller window of masses are related to the stability of the clusters. For example, in antimony, Sb_4 dominates than Sb_3 or Sb_5.

6.3.6 Liquid Metal Ion Source

These sources are primarily used to produce clusters of multiple charges, with low-melting metals. A needle held above the melting point of the metal to be studied. The tip of it is wetted with the metal and it is held at a potential. Very high electric fields at the tip of the needle (due to smaller dimension) cause the emission of a spray of tiny droplets from it. Similar sources are used as ion sources (Chapter 2). Hot, multiply charged droplets undergo evaporative cooling and fission to smaller sizes. Fission occurs as Coulomb repulsion between the charges become larger than the binding energy of the drop itself.

In addition to the above sources sputter sources are used in which a high energy ion beam is used to sputter atoms, ions and clusters from a surface. A schematic illustration of various cluster sources is given in Fig. 6.2.

6.4 Cluster Growth

Cluster growth occurs in two stages.

Nucleation   The nucleation can be homogeneous or heterogeneous. Heterogeneous implies the nucleation occurs on foreign objects, dust particles, etc. In it, collision between like or unlike atoms occurs such that the thermal energy is lower than the binding energy of the species formed. Dimer formation occurs when the third body involved in the collision removes the excess internal energy as kinetic energy.

$$A + A + A\ (KE_1) \rightarrow A_2 + A\ (KE_2 > KE_1), \text{like molecules}$$
$$A + A + B\ (KE_1) \rightarrow A_2 + B\ (KE_2 > KE_1), \text{un-like molecules}$$

The dimer acts as a seed for further condensation and additional growth occurs.

Growth   Initial growth occurs by the aggregation of atoms or molecules one at a time. Coalescence of clusters results in the formation of larger clusters.

$$A_N + A \rightarrow A_{N+1}$$
$$A_N + A_M \rightarrow A_{N+M}$$

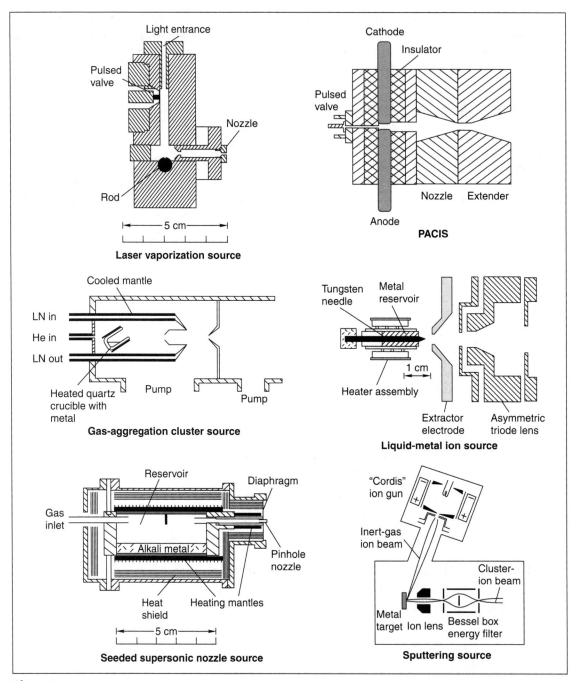

Fig. 6.2: *Various clusters sources. Reprinted with permission from, W.A. de Heer. (1993) Rev. Mod. Phys., 65, 611. Copyright (1993) by the American Physical Society.*

6.5 Detection and Analysis of Gas Phase Clusters

The formed clusters can exist as neutrals or ions (both positively and negatively charged). Various mass spectrometers are used to detect the ionic clusters. The clusters which exist in solid or in liquid state can be analyzed by several spectroscopic, microscopic or diffraction techniques. Here we will discuss the mass spectrometric studies on clusters, because we are focusing only on gas phase clusters.

Mass spectrometers are unique devices used to study the exact constitution of the clusters which exist in the gas phase. From the mass we can easily calculate the empirical formula of the cluster. Wien filter, time of flight (TOF), quadrupole mass filter (QMF), and ion cyclotron resonance (ICR) are the normal kinds of mass spectrometric techniques used to study clusters (although magnetic sector instruments can also be used). These techniques are briefly discussed below.

6.5.1 Wien Filter

This is a low resolution ($\Delta m/m \sim 10^{-2}$), low mass range (less than m/z 1500) instrument. Here mass separation is accomplished using crossed homogeneous electric (E) and magnetic (B) fields, perpendicular to the ionized clusters beam, which travels along the axis of the filter (Fig. 6.3(a)). The net force acting on a charged cluster with mass M, charge Q, and velocity v vanishes if $E = Bv$. These cluster ions are accelerated by a voltage V, so that they have energy QV (where Q is the charge). In the filter, clusters with $M/Q = 2V/(E/B)^2$ are undeflected, while others with different M/Z undergo deflection and are lost (Ref. 2). The undeflected cluster ions are selected and detected, after a slit. As can be seen, the mass resolution of a Wien filter depends on the velocity spread of cluster ions, the strength of the fields, and the slit width. In order to obtain high resolution, narrow slits and strong fields are required as also large acceleration potentials, which reduces the initial velocity spread.

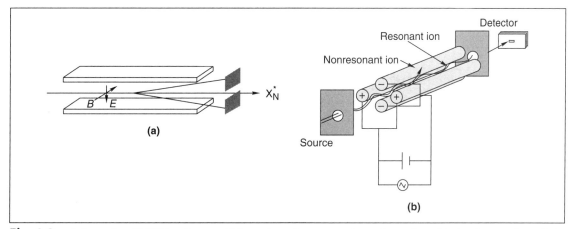

Fig. 6.3: *Schematic of (a) Wien filter and (b) quadrupole mass analyzer. Ions of one kind only pass through for a given condition.*

6.5.2 Quadrupole Mass Filter (QMF)

The QMF is the most widely used type of mass spectrometers today because of its ease of use, compactness and low cost. The principle of QMF is based on the achievement of a stable trajectory for specific ions in a hyperbolic electrostatic field. For easiness of manufacture, cylindrical rods are often used (See Fig. 6.3(b)). An idealized QMF consists of four parallel hyperbolic rods, such filters are also available. To a pair of diagonally opposite rods, a potential consisting of a *dc* voltage and a superimposed *rf* voltage is applied. To the other pair of rods, a superimposed *dc* of opposite polarity and an *rf* voltage of 180° phase shift is applied. The potential ϕ_o applied to opposite pairs of rods is given by:

$$\pm \phi_o = U + V \cos \omega t$$

where U is a *dc* voltage and $V \cos \omega t$, the time-dependent *rf* voltage in which V is the *rf* amplitude and ω, the *rf* frequency. At given values of U, V and ω, only certain ions can have stable trajectories so as to pass through and reach the detector. The various m/z values, capable of passing through the mass filter, are decided by the ratio, U/V. All the other ions will have unstable trajectories (i.e., they will have large amplitudes in *x*- or *y*-direction) and will be lost. The ion trajectory is decided by the Mathieu equations, from which Mathieu parameters a_u, and q_u can de determined.

$$a_u = a_x = -a_y = 4zU/m\omega^2 r_o^2 \qquad q_u = q_x = -q_y = 2zV/m\omega^2 r_o^2$$

where m/z is the mass-to-charge ratio of the ion and r_o, half the distance between two opposite rods. Mass scanning on a QMF means changing U and V at a constant ratio, $a/q = 2U/V$, while keeping the *rf* frequency, ω fixed. The resolving power of a QMF depends on the number of cycles experienced by an ion within the *rf* field (during its flight), which, in turn, depends on the ion velocity. Thus, the resolution will increase with an increase in mass, as ions of higher mass move with lower velocity. However, ion transmission will decrease due to the longer time the higher mass ions spend in the quadrupole. Resolution will decrease when ion is accelerated to a higher potential. The advantages of a QMF are its good transmission, high scan speed, and wide acceptance angle to facilitate high sensitivity. Due to these, it is coupled to several instruments such as gas chromatographs.

6.5.3 Time of Flight (TOF) Mass Filter

TOF has emerged as an efficient mass analyzer. There is no mass limit in this instrument. The TOF of an ion is related to its mass. A set of ions differing in mass, if accelerated through a given extraction voltage, will have varying flight times (Fig. 6.4(a)). Through the use of fast electronic circuits and by the incorporation of a reflection electric region (called reflectron) high mass resolution is possible. TOF can be used to study the dissociation of metastable clusters. A reflectron can also be used to investigate dissociation in the field-free region so that slower processes exhibited by the ion may be observed. This kind of dissociation in the field-free region is called, 'Post Source Dissociation' (PSD). Here ions are accelerated in an appropriate electric field of the order of kilovolts, and then the ions enter into the field-free region. These ions get fragmented in the reflectron region. The parent species with greater kinetic energy have a longer path length than that of daughter ions (Fig. 6.4(b)). By subjecting the ions to different potentials, different fragments may be observed. Finally all the daughter peaks in various regions have to be combined to get the fragmentation pattern of particular clusters. Recently, many clusters and their fragments have been studied by the TOF method. This method of fragmentation is applied for the sequencing of DNA, proteins and larger peptides. Nowadays, with the advancement in electronics, the TOF has an arrangement called the Nielsen-Bradbury gate (Ref. 3). It can act as a gate which allows only a particular mass into the detector region. This can be used to study the particular clusters in the presence of many other clusters from the same source. In Fig. 6.4(b) we present a pictorial representation of the fragmentation which occurred in the reflectron region. The molecular ion peak MH^+ gets reflected and focused correctly towards the detector in the reflector region.

Possible fragmentation channels are,

$$MH^+ \rightarrow AH^+ + B$$
$$MH^+ \rightarrow A + BH^+$$

The fragments BH^+ and AH^+ are poorly focused. This happens while we apply the same potential to both the accelerating plate as well as the reflectron (what is called the mirror ratio of 1). If we change the potential in the reflectron, leaving the accelerating voltage as such, we can focus the daughter ions, instead of the parent ions, towards the detector. So by collecting various daughter ions this way, we can easily study the fragments formed from the parent ion. Combining all the daughter ion spectra we can get a complete picture of the fragmentation of the parent ion.

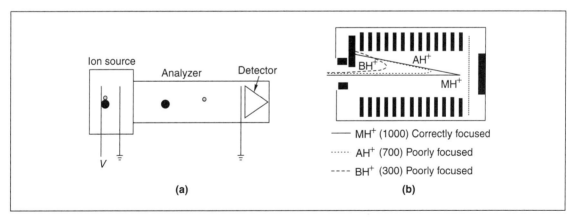

Fig. 6.4: (a) Linear TOF mass analyzer and (b) PSD mode analysis of fragments formed in the reflectron region of TOF. Masses of the ions analyzed and their trajectories are shown.

6.5.4 Ion Cyclotron Resonance

Ion cyclotron resonance (ICR) is a unique technique wherein we can, in principle, perform all the mass spectrometric studies in a single cell. They include, mass analysis, ion selection, ion interaction and product ion mass analysis. All these are possible by a sequence of pulses which are used to initiate the various events. Ions are trapped in the cell by a combination of a static quadrupole electric field and a highly uniform magnetic field for strength, B. Due to the magnetic field, ions move at the cyclotron frequency. The equations of interest are,

$F = zv \times B$ (× refers to a cross product)

$\omega_c = zB/2\pi m$,

$m/z = B/2\pi \omega_c$

where F is the Lorentz Force felt by the ion when entering the magnetic field, v is the incident velocity of the ion, ω_c is the induced cyclotron frequency, m is the mass of the ion and z is its charge. A measurement of cyclotron frequency can be used to determine the mass-to-charge ratio (m/z) of the ions. In the ICR instrument, the cyclotron resonance frequencies are measured by exciting the ion cloud with an electrical pulse, applied on the excitation plates (Fig. 6.5). A sensitive parallel plate capacitor then picks up the electric signal from trapped ions. This signal is accumulated and then Fourier transformed. An analysis of the frequency components and their amplitudes help determine the masses and relative abundances of the ions. Unlike the other techniques, the ions are detected non-destructively which facilitate a repetitive mass analysis of the same collection of charged entities. Methods by which ions can be trapped and collided with molecules for fragmentation or reaction are also used. By combining various pulse sequences, it is possible to trap specific ions and study their chemistry. Due to the superior resolution and possibility of combining with cluster sources, and extreme sensitivity, FT-ICR is the ideal technique for cluster mass spectroscopy. However, it is highly expensive.

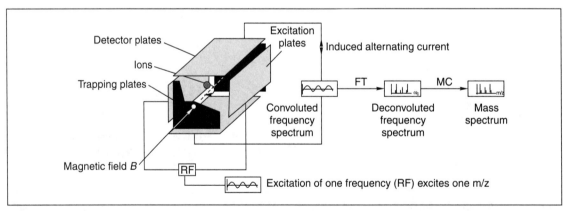

Fig. 6.5: *A schematic of FT-ICR-MS showing the ion trapping, detection and signal generation.*

6.6 Types of Clusters

Most of the elements in the periodic table form clusters. Alkali metal, coinage metal and rare gas clusters are more thoroughly investigated. Clusters can be classified according to both the type atoms of which they are made and the nature of the bonding in these clusters. We can classify the clusters by their composition; for example, if the clusters are formed from metallic elements they are called metallic clusters. A summary of the various cluster types and properties are given in Table 6.2.

Table 6.2: *Various cluster types and their properties*

Type	Examples	Nature of binding	Binding energy/mole
Ionic clusters	$(NaCl)_n$, $(CsI)_n$	Ionic bonds (Strong binding)	~ 50–100 kcal
Covalent clusters	C_{60}, Si_n	Covalent bonding (Strong binding)	~ 20–100 kcal
Metal clusters	Au_n, Na_n, Ag_n, …	Metallic bond (Moderate to strong binding)	~ 10–50 kcal
Molecular clusters	$(H_2O)_n$	Molecular interactions, hydrogen bonding, van der Waals, etc.	< 10 kcal
van der Waals clusters	Ar_n, Xe_n, …	Polarization effects (Weak binding)	< 5 kcal

These are explained in greater detail below.

6.6.1 Metal Clusters

Metal clusters are formed from alkali metals, alkaline earth metals and transition metals. Metal clusters may be formed from single metallic element or from more than one metal, giving rise to intermetallic or nanoalloy clusters. Some of the metal clusters are discussed below.

Neutral sodium clusters are produced in a gas aggregation source. Metallic sodium is heated in an oven to a temperature of about 400°C. The hot sodium vapor (partial pressure ~ 0.1 mbar) expands into a low vacuum He-atmosphere (several mbar, $T \sim 77$ K) where it condenses due to super-saturation. Clusters are formed and they are directed into a differentially pumped section followed by an interaction region, with additional differential pumping. The cluster velocity is related to the source conditions and ranges from 200 to 400 m/s. A typical mass spectrum of Na_n is shown in Fig. 6.6(a) (Ref. 4). As can be seen, clusters up to about 150 atoms are seen.

Various kinds of metal clusters such as silver, aluminum, copper and nickel are known. Recently such clusters are made from molecular precursors. High aggregation of silver clusters from silver trifluoroacetate can be achieved in matrix assisted laser desorption ionization (MALDI) conditions using 2-(4-hydroxyphenylazo) benzoic acid (HABA) as the matrix. MALDI-TOF mass spectrum of silver cluster is shown in Fig. 6.6(b) (Ref. 5). The clusters formed can also be nanoalloys.

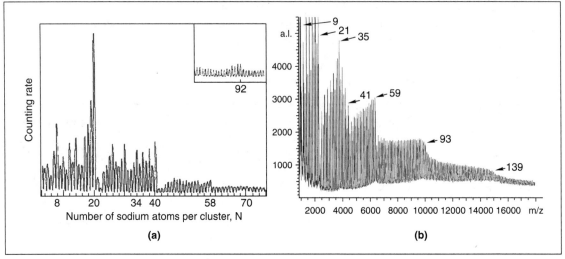

Fig. 6.6: *(a) Mass spectrum of sodium clusters, N = 4–75. Inset shows N = 75–100 region. (b) MALDI spectrum of silver clusters produced from silver trifluoroacetate. Reprinted with permission from, (a) W.D. Knight, K. Clemenger, W.A. de Heer, W.A. Saunders, M.Y. Chou, and M.L. Cohen. (1984) Phys. Rev. Lett. 52, 2141. (b) S. Kéki, L. Szilágyi, J. Török, G. Deák, and M. Zsuga. (2003) J. Phys. Chem. B, 107, 4818. Copyright (1984 and 2003) by the American Physical Society and American Chemical Society, respectively.*

6.6.2 Semiconductor Clusters

Semiconductor clusters are generated from elements which are semiconductors in nature such as silicon, carbon and germanium. The discovery of the fullerene, C_{60} a carbon cluster, stimulated researchers to explore the possibility of a number of semiconductor clusters. Carbon has the tendency to form a greater variety of clusters as compared to other elements. The bonding in these clusters is covalent in nature. The earlier carbon clusters were produced by using an electric discharge between graphite electrodes. The generated carbon clusters were detected by mass spectrometers (Ref. 6). C_{60} was discovered in such experiments in an FT-ICR, with laser desorption ionization. Fullerenes, discovered in the gas phase were later made in the condensed phase. This subject is discussed in greater detail in Chapter 3.

Next to carbon, silicon clusters have been studied widely. The first reported silicon clusters were generated by laser flash evaporation, quenched in a carrier gas and then cooled by supersonic expansion. Photofragmentation studies on mass selected silicon clusters were conducted. The reactivity of mass selected silicon clusters has been studied widely by using ion trap mass analyzers (Ref. 7).

Apart from carbon and silicon, other semiconductor elements such as germanium also form clusters. Both silicon and germanium also form nanoparticles, which are interesting today in view of their luminescence which can be tuned across a large window of wavelength. These are, however, investigated in the condensed phase.

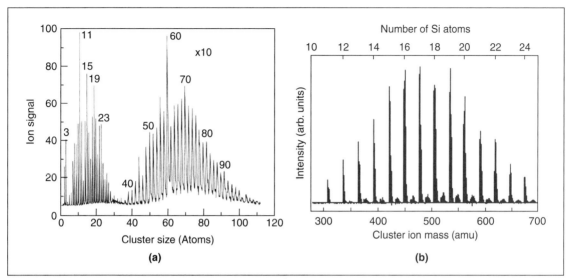

Fig. 6.7: (a) Photoionization (PI)-TOF mass spectrum of carbon clusters. (b) Mass spectrum of silicon clusters. Reused with permission from, (a) E.A. Rohlfing, D.M. Cox, A. Kaldor. (1984) J. Chem. Phys., 81, 3322. Copyright 1984, American Institute of Physics. (b) S. Maruyama., M. Kohno, and S. Inoue. (1999) Thermal Science & Engineering, 7, 69.

Apart from the bare metal clusters, metal oxides (e.g. MoO_3, WO_3, V_2O_5, FeO, LiO, MgO, PuO), metal chalcogenides (e.g. MoS_2, WS_2, TeS, FeS, ZnS, MoTe, Nb_2S_2, VS_4) and metal halides (e.g. $NiCl_2$, NaCl), are also known to produce clusters. The structural changeover from cyclic structure to the cage structure of MoO_3 clusters in gas phase has been reported recently (Ref. 8). A recent report shows the formation of magic number closed-cage clusters, from inorganic materials such as MoS_2 and WS_2 (Ref. 9). These negatively charged clusters, with the formula $Mo_{13}S_{25}$, are likely to be inorganic fullerenes with a cavity inside and may be formed by the curling of nanoflakes of MoS_2.

6.6.3 Metcars

These are closed-cage clusters made of metals and carbon. Various such clusters, such as Mo-C, Ti-C, Hf-C, V-C, Cr-C, etc. are known. Metcars were discovered by Castleman, *et al.* (Ref. 10) by laser vaporization of titanium metal in the presence of methane gas. The first cluster discovered had a stoichiometry, Ti_8C_{12}. This discovery has led a number of researchers to investigate analogues clusters. These clusters are called as 'metallocarbohedrenes' or met-cars. Photodissociation mass spectra of met-car ions show fragment ions with the loss of metal atoms. The chemical reactivity of met-car ions was found to be very high towards polar molecules like H_2O, NH_3, CH_3OH, etc. Several studies have been done on metcars, and the field has been reviewed recently (Ref. 11). Although the material is yet not synthesized in the laboratory as a bulk powder, the structure of the molecule is fairly well understood. This has a closed-cage as shown in Fig. 6.9 (Ref. 12).

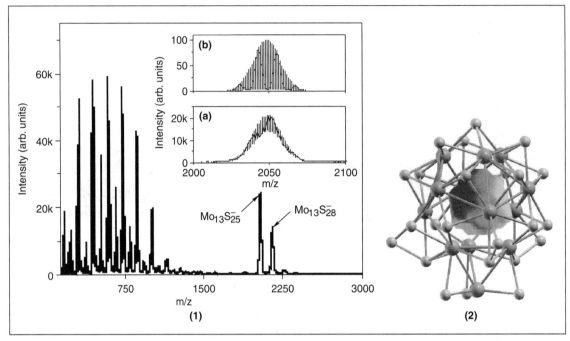

Fig. 6.8: *(1) Laser desorption ionization (LDI) mass spectrum of MoS_2 in the negative mode showing magic closed cage clusters. Inset: Experimental spectrum (a) shows the expected isotope distribution for $Mo_{13}S_{25}^-$ (b). (2) Atomic structure of the $Mo_{13}S_{25}$ cluster. A cloud in the center clearly showing the void space enclosed inside the cage-like structure of the $Mo_{13}S_{25}$ cluster. From the author's work, published in Ref. 9, Copyright (2005) American Chemical Society.*

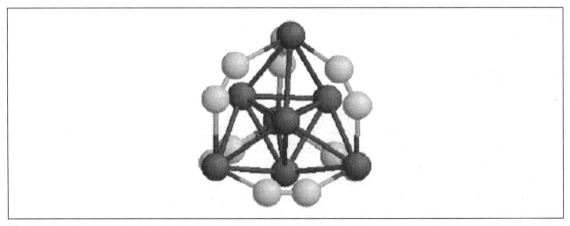

Fig. 6.9: *Optimized tetrahedral structure of the Ti_8C_{12} metcar. Titanium atoms are shown as dark and carbon atoms as light spheres. From Joswig, et al. (Ref. 12). Reproduced by permission of the PCCP Owner Societies.*

6.6.4 Rare Gas or Noble Gas Clusters and Magic Numbers

Rare gas clusters are the earliest clusters detected in molecular beam experiments. Detailed studies were carried out on these clusters as they were easy to make and their physical properties facilitated easy investigation. These rare gas elements have fully filled electronic configuration, as a result of which they are inert. Fig. 6.10 shows the mass spectra of positively charged Ar, Kr and Xe clusters (Ref. 13).

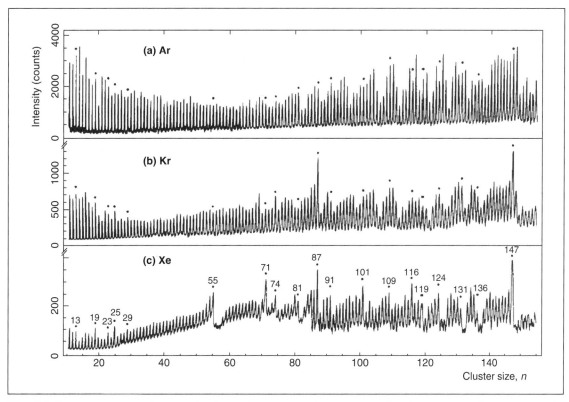

Fig. 6.10: *Mass spectra of positively charged Ar, Kr, Xe clusters. Reused with permission from, W. Miehle, O. Kandler, T. Leisner, and O. Echt. (1989) J. Chem. Phys., 91, 5940. Copyright 1989, American Institute of Physics.*

In the rare gas clusters spectrum, we see that a few cluster peaks have higher intensity than that of the nearer clusters. The nuclearities corresponding to those intense peaks are termed as magic numbers. The magic numbers in the mass spectra may arise due to the size-dependent binding energy of rare gas cations and the fragmentation processes that occur after ionization. The electronic structure of neutral rare gas clusters is different from that of the charged clusters. When we analyze the mass spectrum, the clusters with numbers N = 13, 19, 25, 55, 71, 87 and 147 are found to have high intense peaks. These magic-numbered clusters can be understood in terms of cluster structures consisting of polyhedral shells of atoms around a central atom. For a geometric cluster composed of K icosahedral shells, the magic number of atoms is given by,

$$N(K) = 1 + \sum_{K=1}^{K} (10K^2 + 2)$$

Or, upon expansion,

$$N(K) = 1/3(10K^3 + 15K^2 + 11K + 3)$$

This equation explains the intense peaks at $N = 13$ ($K = 1$), $N = 55$ ($K = 2$) and $N = 147$ ($K = 3$), found in the mass spectrum (Fig. 6.10). The outer shells of these icosohedral clusters are known as Mackay icosohedra. These magic numbers are consistent with calculations on neutral rare gas clusters, using model interatomic potentials such as the Lennard-Jones potential, which predict a growth sequence based on maximizing the number of nearest neighbour contacts, so as to maximize the total cluster geometry.

6.6.5 Ionic Clusters

The term 'ionic clusters' signifies those clusters derived from ionic solids having large differences in electronegativity, such as NaCl, CsCl, etc. Ionic clusters may exist with positive or negative charge. Ionic clusters can be generated by methods like heating or laser vaporization of ionic compounds in a stream of cold inert gas. The studies on ionic clusters have made it possible to determine the size at which ionic clusters begin to acquire the properties of solids. Figure 6.11 depicts an example of ionic clusters from CsCl and CsI detected by mass spectrometry (Ref. 14). The size distribution is completely different in this case. Clusters with larger intensities correspond to the formation of structures similar to bulk solids.

6.7 Properties of Clusters

All properties vary with size, as all of them are dependent on energy. The energy of the system is affected by the fractional surface atoms and that makes properties change. It was proposed that the variation of physical and chemical properties can be predicted on the basis of cluster size equations (CSEs) (Ref. 15). These are of the form, $\chi(n) = \chi(\infty) + An^{-\beta}$ where χ is a property, which is a function of the number of atoms, n and A and β are constants ($0 \leq \beta \geq 1$). $\chi(\infty)$ corresponds to the bulk value. A variation of the properties can be given as in Fig. 6.12. In the cluster size regime, properties vary discontinuously while in the larger size regime, there is a smooth variation.

The properties of clusters explain the transition from single atoms to the solid state. This transition can be carefully examined with clusters. For example, one can ask the question when does a cluster of a metal indeed become a metal. One can systematically increase the cluster size and find out when specific features emerge in certain spectroscopic techniques. It is important to remember that such studies can also be done in the condensed phase with techniques such as STM.

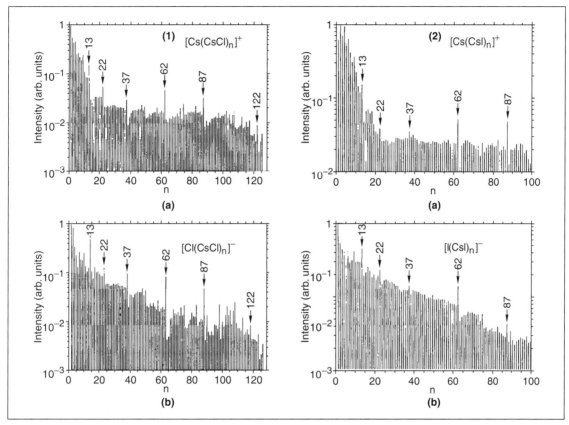

Fig. 6.11: *Mass spectra of (1) CsCl (2) CsI in (a) positive and (b) negative ion modes. Reprinted with permission from, Y.J. Twu, C.W.S. Conover, Y.A. Yang, L.A. Bloomfield. (1990) Phys. Rev. B., 42, 5306. Copyright (1990) by the American Physical Society.*

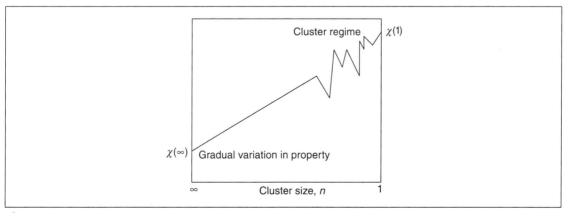

Fig. 6.12: *Variation in properties as predicted by cluster size equations.*

There are numerous properties which make clusters interesting. We will list a few examples. Mercury clusters show very interesting properties with respect to the size of clusters. They show a transition from van der Waals to metallic clusters (Ref. 16). We note that such changes are expected in a number of clusters but only a few are investigated in a large size range.

6.7.1 Mercury Clusters

The clusters are generated by a molecular cluster beam source. A typical mass spectrum of the clusters is shown in Fig. 6.13(a) (Ref. 16). A monochromatized radiation (typically from a synchrotron) is used to photoionize the neutral cluster beam. The light coming out from the undulator (one of the insertion devices in a synchrotron, used for enhanced light intensity) provides more than 10^{13} photons/sec/m rad. The photoionization efficiency (PIE) curve of each mass selected cluster ion is monitored by the variation

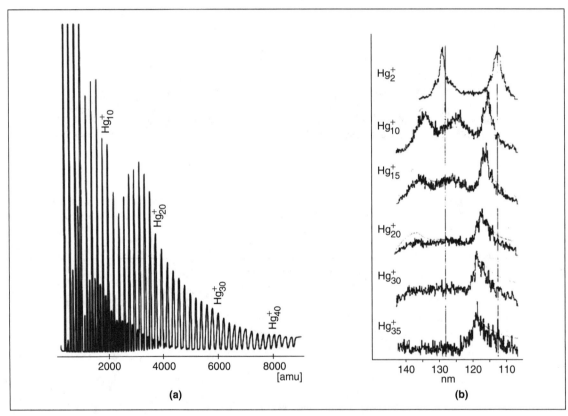

Fig. 6.13: (a) Typical mass spectrum of mercury, obtained by electron impact ionization. (b) The recorded photoionization efficiency curves obtained for some mercury clusters. Reprinted with permission from, C. Bréchignac, M. Broyer, Ph. Cahuzac, G. Delacretaz, P. Labastie, J.P. Wolf, L. Wöste. (1988) Phys. Rev. Lett. 60, 275. Copyright (1988) by the American Physical Society.

of the photon energy (Fig. 6.13(b)). The PIE curve of the atom is recorded between the ionization corresponding to the ejection of one s electron or one d electron. For small clusters with Hg_n ($n \leq 12$), the two autoionization lines are well resolved and appear to shift with respect to the corresponding atomic transition. For cluster size $n = 13$, the $1/n$ dependence is no longer observed in PIE. The gradually increasing shift for both lines illustrates the deviation from van der Waals bonding in larger cluster size (in which isolated atomic features are expected). In the size range $13 \leq n \leq 20$ the lines broaden significantly and their shifts show a deviation from the linear behavior. The spin-orbit splitting of the 5d levels increases with an increase in the cluster size. In the larger cluster size regime, $n > 20$, the spectral line shape is markedly asymmetric and is indicative of a transition from molecular to bulk metal-like properties.

6.7.2 Optical Properties

Optical properties of isolated clusters in the gas phase are rarely investigated. One can obtain information on the optical properties from photodetachment spectroscopy (technique by which electron removal of negatively charged species is investigated, see Chapter 2). From this one learns about the separation of energy levels of the neutral cluster and thereby get information on the optical absorption properties. It is also possible to do spectroscopy of isolated clusters (such as fluorescence) to obtain information on optical transitions. However, most of the optical studies are done in the condensed phase. For metals, the optical absorption gives beautiful colors, which has been the subject of investigation for a long time. In a metal cluster, in the metallic regime, there are free electrons which distribute throughout the cluster. As a result they are susceptible for external electric field. When a cluster is irradiated by light, which has a wavelength much larger than the cluster size, the electrical field is uniform as far as the cluster is concerned. The field induces collective oscillations of the electrons in the cluster. This aspect is discussed in greater detail in Chapter 9.

6.7.2 Ionization Potential

This is one of the well-studied properties of clusters. It was with this the shell structure and stability of clusters were confirmed. Ionization potential and electron affinity have been used to find the origin of metallicity in clusters (Ref. 17). At a critical size regime, the electron affinity of clusters become similar to the bulk. This has been investigated in metals such as copper and mercury. In the case of copper, metallicity was shown to appear in clusters as small as several hundreds of atoms. Origin of metallicity in mercury clusters has been investigated by photodetachment spectroscopy. By studying Hg_n^- ($n = 3$–250) clusters (Ref. 18), it has been shown that the s-d excitation band gap decreases with increase in size and by extrapolation, it was suggested that the band gap goes to zero at $n = 400 \pm 30$, when the cluster will behave like a metal. Such studies have been done on other transition metals too.

6.8 Bonding in Clusters

Here we discuss briefly two theoretical models, which are used to describe the structure and bonding of clusters. The structure varies between metallic clusters and noble gas clusters even though both have the same number of atoms. What is the reason for this kind of structural change? One can take for example the clusters formed from sodium and argon. Computations have been done to predict the structural and binding properties.

The comparison of these two types of clusters is interesting because their formation and binding are completely different. Delocalized electrons exist in a metal cluster. The situation is different in the case of noble gas clusters. All the noble gas atoms have a closed valence electron shell, as a result of which the valence electrons are localized near the ions, and the binding in these clusters arises due to van der Waals forces, which act between atoms in the cluster. With advances in computational capabilities many of the larger clusters are amenable for all-electron calculations. From such studies, total electronic structure and bonding of several clusters are now understood. The bonding in these systems is distinctly different from the traditional molecular systems. Often a specific valence state cannot be defined. For examples in the case of a 25 atom gold cluster stabilized by ligands, the valence state of the atoms involved in binding with ligand is not what we encounter in gold compounds. Similarly the metal-metal bonds are distinctly different from that of bulk gold.

Basically, two kinds of theoretical models are applicable for metallic clusters, of the type Na_n in order to predict their various properties. Properties like ionization potential and electron affinity vary with respect to the size of clusters. The two models are: 1) the Jellium model and 2) the liquid drop model. The Jellium model was originally developed to explain the structures and stability of atomic nuclei. This model was used to describe the electronic structure of the atom. The applicability of the Jellium model over a wide range from an atom to cluster makes it a major unifying concept. It is a quantum mechanical model with the quantization of electron energy levels arising due to the boundary conditions imposed by the potential. In this model, the metal cluster is considered as a uniform, positively charged sphere filled with electron gas. The Schrödinger equation is solved for an electron constrained to move within the cluster sphere under the influence of an attractive mean field potential due to the nuclei or ionic cores. This is in contrast to the classical Liquid Drop model, wherein there is no electronic structure. The solutions to the Schrodinger equation are the energy levels, $\psi_{n,l,m}(r, \theta, \phi) = R_{n,l}(r) Y_{lm}(\theta, \phi)$ and the energies (symbols have the same meaning as in the case of hydrogen atom). Therefore, the Jellium model gives rise to an electronic shell structure for clusters consisting of up to several thousands of atoms. The Jellium potential may be empirical, or alternatively *ab initio* effective potentials may be used. The potential may be modified for situations where spherical symmetry is not observed. Many of the predictions of this model have been verified with experiments, especially with alkali metal clusters.

A simpler model, the Liquid Drop model (LDM) has also been developed for metal clusters. This is an electrostatic model, in which the metal cluster is represented as a uniform conducting sphere. Variations of properties with size can be predicted by developing scaling laws using this model. According to the LDM, the IP should decrease as the cluster size gets larger (i.e. it requires less energy to remove an electron from a larger cluster than from a smaller one).

Review Questions

1. Are there specific properties we understood with the aid of clusters?
2. Why clusters are less stable in comparison to bulk materials?
3. Why do molecules form clusters?
4. Why clusters are generally investigated in vacuum?
5. Propose a method to make nanoparticles in bulk starting from gas phase clusters.
6. What are the important methods to make clusters in the gas phase?
7. Propose a study to understand a property of a bulk material using clusters not discussed in this volume.
8. Propose a method to make gas phase clusters, not discussed here.
9. Are there specific properties, other than those described here to study clusters?
10. List another way of classifying the various clusters, other than that described here.
11. What are the problems in making ternary clusters in the gas phase?

References

1. Boyle, R., (1661), *The Sceptical Chymist: or Chymico-Physical Doubts and Paradoxes*, London.
2. de Heer., W.A., (1993), *Rev. Mod. Phys.*, **65**, p. 611.
3. Bradbury, N.E., R.Nielsen, (1936), *A. Phys. Rev.*, **49**, p. 388.
4. Knight, W.D., K. Clemenger, W.A. de Heer, W.A. Saunders, M.Y. Chou and M.L. Cohen, (1984), *Phys. Rev. Lett.*, **52**, p. 2141.
5. Kéki, S., L. Szilágyi, J. Török, G. Deák and M. Zsuga, (2003), *J. Phys. Chem. B*, **107**, p. 4818.
6. Rohlfing, E.A., D.M. Cox, A. Kaldor, (1984), *J. Chem. Phys.*, **81**, p. 3322.
7. Maruyama, S., M. Kohno and S. Inoue, (1999), *Thermal Science & Engineering*, **7**, p. 69.
8. Singh, D.M.D.J., T. Pradeep, (2004), *Chem. Phys. Lett.*, **395**, p. 351.
9. Singh, D.M.D.J., T. Pradeep, J. Bhattacharjee, U.V. Waghmare, (2005), *J. Phys. Chem. A*, **109**, p. 7339.
10. Guo, B.C., K.P. Kerns, A.W. Castleman, Jr., (1992), *Science*, **255**, p. 1411.
11. Rohmer, M.M., M. Bénard, J.M. Poblet, (2000), *Chem. Rev.*, **100**, p. 495.
12. Joswig, J.O., M. Springborg and G. Seifert, (2001), *Phys. Chem. Chem. Phys.*, **3**, p. 5130.
13. Miehle, W., O. Kandler, T. Leisner and O. Echt, (1989), *J. Chem. Phys.*, **91**, p. 5940.
14. Twu, Y.J., C.W.S. Conover, Y.A. Yang, L.A. Bloomfield, (1990), *Phys. Rev. B.*, **42**, p. 5306.

15. Jortner, J., *Z. Phys. D* (1992), **24**, p. 247.
16. Bréchignac, C., M. Broyer, Ph. Cahuzac, G. Delacretaz, P. Labastie, J.P. Wolf, L. Wöste, (1988), *Phys. Rev. Lett.*, **60**, p. 275.
17. Cheshnovsky, O., K.J. Taylor, J. Conceicao, R.E. Smalley, (1990), *Phys. Rev. Lett.*, **64**, p. 1785.
18. Busani, R., M. Folkers, O. Cheshnovsky, (1998), *Phys. Rev. Lett.*, **81**, p. 3836.

Additional Reading

1. Roy L. Johnston, *Atomic and Molecular Clusters*, Taylor and Francis, London (2002).
2. Boris M. Smirnov, *Clusters and Small Particles in Gases and Plasmas*, Springer-Verlag, New York (2000).
3. Paul–Gerhard Reinhard, Eric Suraud, *Introduction to Cluster Dynamics*, Wiley–VCH Verlag GmbH & Co. KGaA, Weinheim (2004).
4. Faraci, G. and P. Selvam in *Encyclopedia of Nanoscience and Nanotechnology*, Edited by H.S. Nalwa, American Scientific Publishers (2004).

Chapter 7

SEMICONDUCTOR QUANTUM DOTS

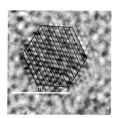

Quantum dots are the very first extensively researched nanoparticle systems. We learned many of the properties of electronically confined systems through such investigations. Their optical, photophysical, photochemical, biological and catalytic properties have opened up numerous application possibilities. Several of these have been realized such as the dye sensitized solar cells which utilize the electronic properties of these materials. The applications in biology using these materials are some of the most extensively researched areas in nanobiology, covered separately in this book. In this chapter we look at the reasons for the unique properties of nanocrystals, their synthesis, experimental investigations and applications.

Learning Objectives

- What are quantum dots?
- What is quantum confinement?
- How can one make and study quantum dots?
- What properties of quantum dots have been discovered?
- What are their applications?

7.1 Introduction

Semiconductor quantum dots are signify a class of materials in which quantum confinement effects are investigated in greater detail. They are also referred to as 'semiconductor nanocrystals'. In fact these constitute a sub-class of a broad family of nanoparticles, which include semiconductor, metal, insulator, organic, etc. particles. 'Quantum dots' is a term referred only to semiconductor particles, while 'nanocrystal' can be any inorganic entity in which there is a crystalline arrangement of constituent atoms/ions. In a macroscopic semiconductor crystal, the energy levels form bands. The valence band is filled and the conduction band is completely empty at 0 K. The bands are separated with a specific energy gap, E_g. When an electron gets excited due to thermal excitations, an electron–hole pair is created. The electron in the conduction band and the hole in the valence band can be bound when they approach each other at a finite distance. This bound pair is called an 'exciton', which is delocalized throughout the crystal. The Bohr radius of the exciton can be given as: $a = h^2\varepsilon/4\pi^2 e^2[1/m_e + 1/m_h]$, where ε is the dielectric constant of the material, m_e and m_h are the effective masses of electron and hole, respectively, and e is the elementary charge. Quantum size effects are manifested when the length of the nanocrystal made, d, is comparable to the exciton radius, a. This is ~ 56 Å for CdSe. Note that the unit cell dimensions of the semiconductor are much smaller than the characteristic length. The de Broglie wavelength in materials, $\lambda = h/mv$, is in the range of nanometers and strong confinement effects are manifested only when the particle dimension approaches this value. At this dimension, most materials have structures similar to those of their bulk counterparts, at least in the core of the particle.

The electronic structure of materials is strongly related to the nature of the material. In a three-dimensional object of large size, the electronic structure is not restricted by the dimension of the material. The wavelength of electrons is much smaller than the typical length of the material. When the electronic motion is confined in one dimension, and it is free in the other two dimensions, it results in the creation of 'quantum wells' or 'quantum films'. The quantum well notation implies that the electrons feel a potential well as they are trapped in the film. The quantum film notation is self-explanatory. Here the density of states shows a step-like behaviour. In the case of a one-dimensional system, i.e. when the electrons are free to move only in one direction, we get a situation wherein the density of states shows a Lorenzian line shape. Such a situation can be seen in carbon nanotubes. If the electrons are confined to a point, we get a zero-dimensional system, wherein the electrons are not free to move at all. Here we get states which are molecular in nature. The situation is schematically depicted in Fig. 7.1. What is shown in Fig. 7.1 is that while the density of states is smoothly varying in bulk materials, it shows discontinuities in confined systems. This will lead to steps in two-dimensional confinement, singularities in one-dimensional confinement and discrete lines in zero-dimensional confinement.

Some of the properties which change drastically as a function of size are the optical properties, including both the absorption and emission of light, which is evident from Fig. 7.1. Nanocrystals have discrete orbitals. The energy of the first level will be shifted from the position of the bulk value by $h^2/8m_e a^2$ where a is the diameter of the particle. Remember that the particle in a box model predicts the energies of the levels as, $n^2h^2/8m_e a^2$ ($n = 1, 2, ...$). The simplest model of a quantum dot would be a particle in a sphere model, assuming that the nanocrystal is a sphere. This does not make a difference in the energy level description mentioned above. The energy gap increases with a decrease in a. As a consequence of this, the CdSe nanocrystals emit light anywhere from 4500 to 6500 Å, so that any color from blue to red is achievable, depending on the size of the particle.

It may be worthwhile to describe the issue of confinement again. These particles are called quantum *dots* as their electrons are *confined* to a point in space. They have no freedom in any dimension and electrons are said to be localized at a point, implying that a change in all directions changes the properties (in reality, a dot is a three-dimensional object comprising several hundreds or thousands of atoms, with finite shape). Compare this tiny unit of matter with semiconductor structures which are grown by thin film evaporation methods, which facilitate the creation of ultra thin films wherein the thickness is comparable to *a*, the diameter mentioned above. Such a material is called a 'quantum well', implying that the electron is confined within a two-dimensional area, which is said to be a 2D confinement. It is important to recall that the well itself is made by evaporation and the material of interest is confined inside another material of larger dielectric constant. Confinement along one direction results in a quantum wire, leaving the electron free to move only along one direction. This is known as a 1D confinement. In the quantum dot, there is no freedom along any direction.

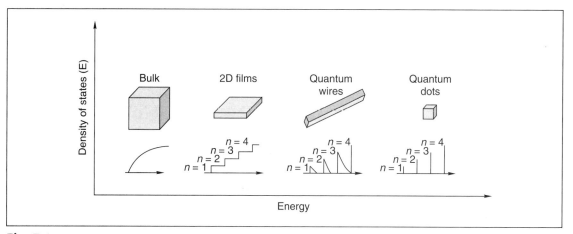

Fig. 7.1: *Quantization of the electronic density of states as a result of variation in the dimensionality of materials.*

An ideal quantum dot is realizable only when the electronic states within the dot face a discontinuity at the edge of the material. Due to this, the electron within the dot feels an insurmountable barrier at the edge. When a material is truncated at the surface, the surface atoms have unsatisfied valencies. In order to reduce the surface energy, the surface reconstructs, which leads to energy levels in the forbidden gap of the semiconductor. The electrical and optical properties of the material are degraded by these traps. In an ideal semiconductor nanocrystal, the surface atoms are bonded to other materials in such a way as to remove the defects. This is what is done when a dot is covered with a material of larger band gap. In an ideal quantum dot, when there are no defect sites for charges to get trapped, the quantum yield of luminescence will be very high, nearly one. The emission will also be sharp. Better light emission occurs as the electrons and holes are confined spatially. This is achieved by chemically protecting the surface with proper protecting molecules called 'capping agents'. Since the surface of the nanocrystal can be modified by using various capping molecules, these materials can be adapted to suitable media, including biology. They are therefore ideal probes in a biological environment.

An important aspect of nanocrystals (NCs) is that they link molecules and bulk materials. The properties change continuously as a function of size in the regime of NCs. Thus if the diameter is changed from 11.5 nm to 1.2 nm, the band gap of CdSe NCs can be changed from 1.8 eV to 3 eV. This corresponds to light emission from red to blue. The size regime of nanocrystals is difficult to pinpoint and depends on the material. In the case of most materials, it lies in the range of 100 to 1,00,000 atoms. The lowest size regime arises because at this point the structure changes to that of molecules (and consequently the properties). In the larger size regime, the system is nearly bulk-like as its energy level spacing is comparable to that of thermal energy.

7.2 Synthesis of Quantum Dots (QDs)

The kind of method used for the synthesis of quantum dots depends on the properties. Ideally, nanocrystals should have the following properties:

1. Monodispersity
2. Possibility of further chemical derivatization
3. High degree of crystallinity and specificity (avoiding polymorphic phases)
4. Chemical integrity
5. Lack of defects

There are several methods in literature for the chemical synthesis of nanocrystals. All of them can be grouped under certain broad classes. It may be noted that there are also other approaches such as 'biological synthesis', which has been receiving some attention recently (Ref. 1).

7.2.1 General Strategies

In the synthetic protocol, two kinds of general approaches are used, the top-down approach and the bottom-up approach. In the top-down method, the bulk material is brought into a smaller dimension by various tools. Numerous methods are included in this category such as the various tools of patterning used to make structures in semiconductor electronics. These include chemical etching, optical lithography, use of particle beams (such as electron, ion and atom), etc. Apart from these, various methods are used to create ultra thin films of the material under investigation. These include thin film evaporation, molecular beam epitaxy, etc. These tools will not be discussed in the context of nanomaterials as it is difficult to obtain large quantities of materials by using them. However, there are other methods of creating powders starting from bulk materials. These techniques utilize various methods of milling and grinding, of which ball milling is the most commonly used one.

In the bottom-up approach, one can use gas phase or liquid state approaches. In the gas phase approach, the material to be synthesized is mixed in the atomic state in the gas phase itself. For this, the atoms of the materials are produced *in-situ* in an evaporation apparatus. This may also involve reacting atoms of one element with a gas phase species (such as oxygen). The prepared material in the gas phase is condensed to get the bulk material. At the stage of condensation, stabilizers may be added. Monodisperse metal particles are made in this way. Bulk powders of several oxides can be conveniently made by this route. This methodology may also be referred to as top-down as the synthesis starts from the bulk materials, which are subsequently evaporated.

The principal synthetic strategy used to make nanoparticles in a solution can be classified as 'arrested precipitation'. Here at some stage of the growth of the particles, the surface is stabilized and further growth is arrested. This is commonly done by surfactants which bind to the surface of the growing nanocrystal. This synthetic approach, which is used for various kinds of materials, is similar to that used for metal nanoparticles. Various kinds of surfactants, both ionic and covalent, have been used for this purpose. Arresting of the growth can be achieved in cavities such as those of micelles, zeolites, membranes, etc. Arresting can also be brought about by sudden variation in temperature and pressure conditions leading to quenching.

All the nanoparticles formed are stabilized in the solution by the presence of an electrical double layer when the protecting agents present are ionic in nature. This is typically seen in the case of metal nanoparticles such as gold when the reduction is done with sodium citrate. The surface of the nanoparticle is protected with anions and cations. In the specific case mentioned, the anions are citrate and chloride, while the cations are sodium and protons (as gold is completely reduced). The electrical potential created by the double layer is large so that Coulomb repulsion prevents aggregation. On closer contact between the particles, the interaction potential rises sharply due to charge repulsion. However, there is van der Waals attraction at larger distances and the net result is that there is a weak potential minimum at a moderate distance. This stabilizes the nanoparticle dispersion. These dispersions cannot be concentrated beyond a point as the particles agglomerate. However, in dilute solutions, many colloidal solutions are stable for extended periods, such as in the case of gold. If they are precipitated, metallic mirrors are obtained. In the case of semiconductor materials, the bulk material is formed. However, it is possible to change the surface cover to a covalent one and achieve subsequent purification. In the case of covalently bound ligands, there is no net charge and the particle behaves like a molecule. These materials can be precipitated, re-dispersed and stored for extended periods.

Some of the techniques used for synthesis are discussed in more detail in the following sections.

7.2.2 Synthesis in Confined Media

In this approach, nanoparticles are synthesized in a space that is already available. The chemical reaction occurs inside a reactor, which is prepared by one of the several ways. Among the various confined media, reverse micelles, Langmuir–Blodgett films, zeolites, porous membranes, clays, etc. are worth mentioning. They all have spaces of the order of nanometers, in which ions can be put. A proper stoicheometric mixture can be provided and the reaction conditions can be altered to produce the required nanocrystal. The chemistry in the solution phase can be conditioned to allow the usage of proper molecules for surface passivation.

For example, one can conduct the reaction in reverse micelles (water in oil). Here oil is the majority phase while water is the minority phase, and nanoscopic containers are formed by micelles. One can add metal ions and organometallic reagents into the nanoreactors, leading to the following reaction:

$$CdCl_2 + Se(SiMe_3)_2 \rightarrow CdSe + 2Me_3SiCl$$

The medium can have capping agents as a result of which the surface prepared is passivated. The oil/water ratio and the temperature conditions can be adjusted to obtain suitable particle dimensions. The materials can be separated from the solvent system and further annealed by heating in a higher boiling solvent to improve the crystallinity.

In confined cages such as those of zeolites, the maximum dimension of the particle formed is fixed. In a typical approach, the exchangeable ions such as Na^+ in the zeolite are exchanged with Cd^{2+} by washing the zeolite with a Cd^{2+} solution. The sample is then exposed to H_2S. Depending on the extent of Cd^{2+} present in the sample, various nanoparticle sizes are obtained. The same method can be used in membranes such as Nafion® which have empty spaces of the order of nanometers.

7.2.3 Molecular Precursors

In the precursor route, nanorystal seeds are prepared in a medium which can control the growth of the particles by co-ordinating with it. The solvent universally chosen for this approach is trioctylphosphine oxide (TOPO), which has high thermal stability and can co-ordinate with inorganic surfaces. The co-ordinating TOPO can be exchanged with other ligands after isolation. The nanocrystals prepared this way can undergo Ostwald ripening (growth of larger particles at the expense of smaller ones) as a result of which monodisperse particles can be formed. A number of materials have been prepared by using the TOPO route. In a standard method, the metal ion precursors are added into hot TOPO while stirring continuously (an injection of the ion precursors is made into a hot solution, as seen in Fig. 7.2). The difficulty of using pyrophoric compounds at elevated temperatures constitutes a risk and to avoid this, a single compound delivering both the constituent elements in the semiconductor has been developed. Cadmium dithiocarbomates (e.g. $Cd(S_2CNEt_2)_2$) can produce CdS nanocrystals. A variation of chemicals can get other semiconductors. The synthetic protocol can be modified by changing the solvent (using a mixture of surfactants instead of TOPO), synthesis parameters such as volume and the number of injections, etc. These modifications have resulted in the formation of various morphologies, apart from the common spherical particles. It may be noted that even the spherical particles are faceted.

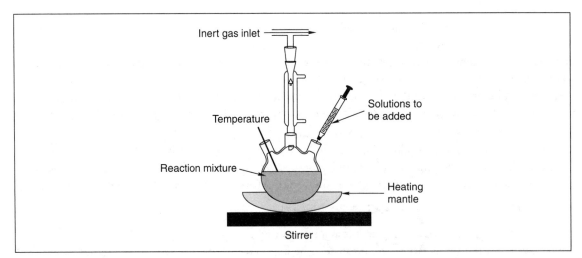

Fig. 7.2: *Typical chemical synthesis approach for making nanoparticles, especially QDs. These are made at higher temperatures, produced by a heating mantle. The process is carried out in an inert atmosphere. Solutions may be added simultaneously.*

7.2.4 Chemical Synthesis Using Clusters

Nanocrystals are large molecules. Therefore, it is natural to think of building them starting from atoms. This atomic building approach has many advantages, one of which is the removal of all possible defects while the other is the possibility to have truly monodisperse materials. A larger cluster can be constructed from a smaller one. Using a starting cluster compound, [(NMe$_4$)$_4$Cd$_{10}$S$_4$(SC$_6$H$_5$)$_{16}$] the cluster, Cd$_{32}$S$_{14}$(SC$_6$H$_5$)$_{24}$.4DMF has been prepared (Ref. 2). This is a cluster with a well-defined structure similar to the structure of bulk CdS. The absorption spectrum of the cluster is blue shifted in comparison to the bulk CdS (358 nm). More complex structures have been built by using this approach.

It is important to consider the crystalline state of the nanocrystal while deciding its properties. The effective masses of electrons and holes depend on the crystal structure of the material, as a result of which all the optical, photophysical and photocatalytic properties will be affected by the structure. In the case of CdS and ZnS, there are two distinct crystalline forms, cubic (zincblende) and hexagonal (wurtzite) structures. CdS exists in the bulk in the wurtzite form while ZnS is found in the cubic form. When synthesized, CdS mostly adopts the cubic form, which is metastable. However, hexagonal CdS can be synthesized in the nano form. By controlling the surface functionalization, ZnS has been made in different forms. When protected with compounds such as 1-hexanethiol, 1-decanethiol, benzoic acid, etc., ZnS nanocrystals exist in the cubic form, while in the unprotected case, they exist in the hexagonal form. When the surface of the nanocrystal is not bound to the interacting molecules, one gets the metastable hexagonal phase. Thus it appears that surface stabilization leads to the formation of the cubic phase. Obviously, surface energy plays a significant role, as the reduction in surface energy by surface stabilization leads to the formation of the stable phase. It is important to note that the energy difference between the cubic and hexagonal phases is small (3.2 kcal mol^{-1}), which facilitates transformation.

Current wet chemical methodologies can give highly monodisperse particles. High monodispersity implies a distribution of less than 5 per cent in the particle dimensions. The typical capping agent is TOPO or trioctylphosphine selenide (TOPSe), in the case of CdS and CdSe particles, respectively. These groups bind to the surface Cd atoms and the coverage changes, depending on the size of the particle. While almost all the surface Cd atoms are protected in a smaller crystallite, only half of them are covered in the case of a flat surface. This is because as the size decreases, the surface area increases and molecules get adequate space to arrange themselves.

7.2.5 Modification of the Surface of Nanocrystals

The surface modification of nanocrystals is important for the following reasons:

1. It removes surface states, thus making near band gap emission possible. Due to the removal of defects, the emission becomes narrow.
2. It adds chemical versatility to the system allowing it to become part of a larger structure. This is important in the use of such materials in polymers, inorganic matrices, etc. It is also important in making the system biocompatible.

3. Surface passivation and suitable functionalization make the system chemically inert and thermally stable. In fact, various attributes can be added to the nanocrystal by appropriate modification. This includes chemical and biological compatibility, hydrophobicity, hydrophylicity, etc.

The most common approach used is that of functionalizing the surface of the nanocrystal with various molecules, which can be thiols, amines, alkyl or silyl groups, etc. In all these cases, the chemical functionality of the molecule interacts with the metal atom at the surface of the nanocrystal. The other approach involves the functionalization of the molecule further. For example, aminopropyltriethoxysilane coupling agent can be used to functionalize the surface of the nanocrystal. The surface will have ethoxy groups which can be hydrolyzed. By using a silica forming precursor, one can form a silica shell around the nanocrystal. The coating of CdS with a thin layer of HgS and further with CdS-yielding CdS/HgS/CdS structures, has also been reported. This structure would be called a 'quantum dot–quantum well' structure in the sense that a quantum dot is covered with a two-dimensional structure with a large dielectric barrier. This construction becomes possible as cubic CdS has a lattice constant similar to that of cubic HgS.

One of the important advantages of such a shell structure is the retardation of charge recombination in semiconductors upon photoexcitation. The electron that is created in the large band gap material upon photoexcitation can be stabilized in the lower lying conduction band of the other material. The direction of electron flow can be controlled in this fashion and a rectifying action has been observed (Ref. 3).

7.2.6 InP Nanoparticles

As an example of the preparation of semiconductor nanocrystals, we will discuss the protocol used for InP quantum dots. This is called a group III–V semiconductor as In belongs to group III and phosphorus to group V. The indium precursor used is indium oxalate, indium chloride or indium fluoride. Trimethylsilylphosphine is used as the phosphorus precursor. A mixture of TOPO and trioctyl phosphine (TOP) is used as the colloidal stabilizer. The precursor species are mixed in an atomic ratio of 1:1 to make an InP precursor in the presence of the colloidal stabilizer. TOP is used in a ratio that is ten times larger than that of TOPO. The mixed solution which forms the transparent precursor is heated at 250–300°C for three days to obtain the colloidal solution. Depending on the size range desired and the extent of crystallinity, the dispersion can be heated to varying time/temperature. The decomposition temperature of the precursor is >200°C. The particle size can be modified by controlling the rate of decomposition. The material synthesized has a monolayer cover of TOPO/TOP. This makes the material disperse in a hexane:butabol mixture (0.9:0.1 by volume) containing 1 per cent TOPO. Precipitation of the nanocrystals can be achieved by adding methanol to the dispersion. This process is repeated. Dispersion and precipitation are carried out to remove unreacted materials or other products. TOPO is added as repeated washing can remove it from the surface of the nanocrystal and thereby affect colloidal stability. The protective cover can be replaced with thiols, amines, fatty acids, sulfonic acids, polymers, etc. It is also possible to bind the surface with proteins. Almost any functionality can be added by suitable chemistry. The material thus prepared can be precipitated, purified and redispersed as in the case of metal nanoparticles. Depending upon the functionality they will be dispersable in appropriate solvents; polar or non-polar. The polarity of the solution can be changed gradually. This approach helps precipitate particles gradually. The narrower the distribution of the parent material, the narrower are the precipitated particles as well.

The preparation method can get *n*-doped, *p*-doped and undoped particles. N-doping is achieved by the addition of S, while *p*–doping is done by adding Zn. These are done by mixing suitable precursors to the InP precursor-forming solution. Such InP:S and InP:Zn nanoparticles are also stable.

7.3 Electronic Structure of Nanocrystals

The electronic structure of complex systems can be understood on the basis of simpler systems. For example, graphite can be understood starting from an assembly of benzene fragments, or diamond can be thought of as an assembly of tetrahedrally connected sp^3 carbon atoms. Benzene has discrete π and $\pi\star$ states. As the number of rings increases, the highest occupied molecular orbital moves up in energy while the lowest unoccupied orbital moves down in energy. As a result, the energy of the transition decreases. This can be seen from ethylene to polyenes and from benzene to pentacene. The energy difference between the π and $\pi\star$ levels decreases and eventually it becomes smaller than thermal energy. At this stage, the levels are considered to be merged to form bands. In the case of an infinite ring containing solid, as in the case of a single sheet of graphite, the gap reduces to zero and the top of the valence band (the HOMO) touches the bottom of the conduction band (LUMO). This makes it a semimetal. Other similar situations may be envisaged in the kind of infinite chains such as polyenes and polyynes. When the gap cannot reduce to zero, we obtain situations categorized as semiconductors or insulators. The reduction in the energy gap with an increase in the number of electrons is illustrated in Fig. 7.3.

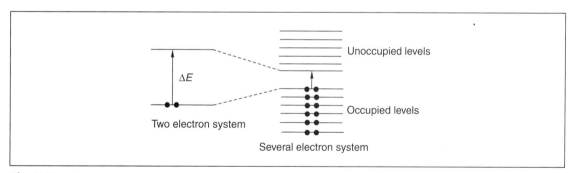

Fig. 7.3: *The change in the electronic energy levels of the system when the number of structural units increases, as in the case of a change from one double bond to many double bonds. The energy gap between the levels, corresponding to the first excitation energy, ΔE decreases. When the number of double bonds increases significantly, the energy levels merge and the gap becomes comparable to that of thermal energy.*

This discussion obviously implies that a material will behave differently when the size regime is smaller than that required for bulk properties. Although this discussion was in the context of electronic properties, every other property will change as most of them are electronic in origin. This variation makes the question, 'At what size do properties become bulk-like', valid. This depends on the material of choice. In the case of elements such as carbon, with the smaller size regime containing a few tens to hundreds of carbon atoms, we have molecules called fullerenes. Their electronic structure is similar to that of the molecules which we are familiar with. Even in the regime of hundreds of atoms, their bonding and several other properties bear close similarity with those of bulk materials such as graphite. This is true of many elements. In the case of silicon, smaller clusters are entirely different and a bulk-like structure is exhibited only in the larger size regime of 10^3 atoms. Although the lattice constants may be similar to bulk, the electronic properties may not reach the bulk limit. In a few cases, this bulk-to-molecular changeover has been investigated and has been discussed in Chapter 6 on gas phase clusters.

The above discussion suggests that the bulk band gap increases with a decrease in the size of the material. Electronic transitions across the band gap result in the spectroscopic properties of the system. The variation in absorption edge can be seen as a function of the particle size. This aspect has been investigated in detail. In addition to the states which are created outside the band gap, in a real nanocrystal, there are also states within the band gap, which are created as a result of defects. It is possible to calculate the diameter of the nanoparticle from the position of the absorption edge. The nanocrystal can be considered as a three-dimensional box and the energy gap can be related to the diameter of the particle or the width of the well.

While dealing with nanocrystals, one has to distinguish between three different kinds of size regimes. Various effects in these size regimes have been the subject of investigation, but we will not discuss this in great detail. The size regimes depend on the nanocrystal radius, r and the bulk exciton radius, a. In the strong confinement regime, $r \ll a$, the Coulombic interaction between the electron and the hole is much smaller than the confinement energies, and therefore, electron and hole can be considered as separate particles. A weak confinement regime, $r \gg a$, occurs when the electron and hole motions are strongly correlated, as electron–hole interactions are significant in comparison to the confinement energies. In the intermediate regime, $r \sim a$, the electronic structure depends strongly on both quantum confinement and Coulomb interaction.

The simplest model used to represent the energy states of a nanocrystal is a spherical quantum well, with an infinite potential barrier. Although the model is simple, if we include the Coulombic interaction between the electron and the hole, analytical solutions for the Schrödinger equation will not be possible. Disregarding the e–h interaction is possible in the strong confinement regime, as confinement energies scale with d^{-2} (as energy goes as n^2/d^2) while Coulomb interaction scales with d^{-1}. This results in states with distinct n, l, m quantum numbers referring to various symmetry, orbital angular momentum, its projection, respectively (similar to electrons in orbitals of an atom). The wave functions are represented as products of several terms. The energies of the states can be given as:

$$E_{l,n}^{e,h} = h^2 n^2 / 8\pi^2 m_{e,h} d^2$$

where n is a quantum number. The exact nature of the wave function and the quantum number are not introduced here. The wave functions correspond to the S, P, D, ..., etc. states depending on the orbital angular momentum, l. There is one more quantum number, m which decides the degeneracy of the states. A pictorial representation of the energy states is given in Fig. 7.4. The energies are measured from the bottom of the conduction (valence) band for electrons (holes). The energy increases as one goes higher in the quantum number. Since the electron mass is much smaller than that of the hole ($m_h/m_e \sim 6$ in CdSe), the electron levels are separated more widely than the hole levels.

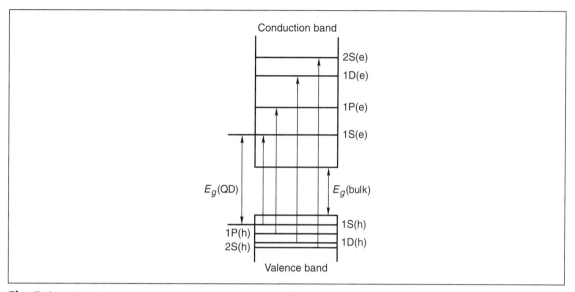

Fig. 7.4: *The electronic states of a nanocrystal. The allowed optical transitions are marked.*

Electronic transitions are possible between various energy levels. However, the wave functions corresponding to different n and/or l are orthogonal and therefore it is not possible to observe all these transitions. Optical transitions between states of the same symmetry can, however, be observed. The intensity of the transition will be related to the degeneracy of the states in question. The transitions observed are far more complex than those described by the spherical quantum well model. The description given here is inadequate to describe the hole states. Spin—orbit and Coulomb e-h interactions have to be considered to improve the energy level picture.

One of the important aspects to be considered while interpreting experimental spectra is the size range of particles prepared in a typical synthesis. Spectroscopic size selection can be done by using techniques such as fluorescence line narrowing (FLN), spectral hole burning and photoluminescence excitation (PLE). In these techniques, a narrow energy window is used for excitation (first two) or detection (last). This makes the technique sensitive only to a specific particle size. Size selection with the red region of the spectrum is preferable as it helps select the particles of the largest size in the ensemble.

7.4 How Do We Study Quantum Dots?

A quantum dot material prepared by one of the methods described before will be a powder. One needs to characterize its physical, structural, electronic and other properties to qualify it as a nanomaterial.

Several of the tools outlined in the earlier chapters may be used in this regard. We shall illustrate the most essential aspects of the characterization of quantum dots through the use of some of the tools. Several other tools may also be used for additional information. These methods may be used for any nanomaterial, irrespective of their nature.

7.4.1 Absorption and Emission Spectroscopy

Absorption and emission spectroscopy are performed to understand the quantum confinement of the system. The spectra are measured in absorption and fluorescence spectrometers for absorption and emission, respectively. For these measurements, one typically uses a solution in an appropriate solvent which does not have characteristic absorption or emission in the region of interest. Solution phase experiments are preferred, though it is possible to measure the spectra in other forms such as thin films, powders, etc.

The absorption and emission spectra of InP nanoparticles of 32 Å mean diameter are shown in Fig. 7.5. The absorption spectrum shows a characteristic peak at 590 nm, due to the excitonic absorption (formation of electron–hole pair). The bulk material has an onset of absorption at 918 nm (1.35 eV). The exciton radius of InP is about 10 nm and the particles presented here show strong confinement, which means that the absorption spectrum shifts considerably as a function of size. The spectrum shows a characteristic higher energy transition, above the first excitonic absorption. This indicates the presence of smaller particles, which show lower wavelength absorption. The extent of shift in the absorption spectrum can be used to calculate the particle dimension as shown earlier.

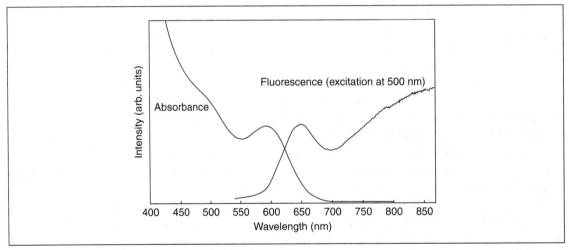

Fig. 7.5: *Absorption and emission spectra of InP nanoparticles. Reused with permission from Mićić, et al. (Ref. 4). Copyright 1996, American Institute of Physics.*

The photo-luminescence spectrum of the same sample shows two bands when excited at 500 nm. Two emissions with maxima at 655 nm and above 850 nm are seen. The first one is due to the band gap emission and the other is attributed to radiative surface states produced by phosphorus vacancies.

The optical properties of the quantum dots are heavily affected by the surface of the particles. In the case of such small clusters, the reactivity of the surface atoms will make a thin oxide layer on the surface. This quenches the light emission. In the case of bulk semiconductors, the surface is cleaned by an acid etching. A similar process is undertaken for quantum dots too, by using a mild acid solution (e.g. 5 per cent HF, 10 per cent H_2O in methanol). This creates a nascent surface which shows an intense band edge emission. However, the emission deteriorates over a period of time. The surface can be well protected with ligands which bind strongly such as thiols. In such cases, the material can be stored in air and the solution shows stable emission even after a month.

Both the absorption and emission spectra shift as a function of size. This is a characteristic feature evident in all the nanocrystal samples. Emission has been studied in a range of nanocrystal materials such as ZnO, CdS, CdSe, etc. In the case of CdS, emission occurs in the red region (>600 nm) of the electromagnetic spectrum. The spectrum is attributed to sulphur vacancies. The excitonic emission occurs when more charge carriers are created as in the case of a laser excitation. The emission occurs due to detrapping of electrons. Traps function as charge reservoirs and contribute to an increase in the emission time scale.

Emission is referred to as 'global emission' when the excitation energy is much higher than the absorption maximum of the sample. Note that particles of several diameters are present in the sample and by choosing an energy that is higher than the absorption maximum, a greater percentage of the samples can be excited. Both the photoabsorption and luminescence show enormous size dependence. The linewidths in luminescence can be reduced if the range of particles excited can be reduced. This can be achieved by reducing the size distribution in the sample. In a given size distribution, if the excitation energy is reduced, the range of particles excited reduces and the line width narrows. This technique is called 'fluorescence line narrowing' (FLN).

The PL shows a long lifetime of 28–73 ns at 298 K for 3 nm InP particles. The lifetime increases to 173–590 ns at 13 K. It appears that the emission occurs from a spin-forbidden state. As a result of the larger electron–hole exchange interaction in the excited state, relative to bulk, the excited state (excitonic in nature) splits into a triplet and a singlet. The triplet is lower in energy. However, excitation occurs to the higher lying singlet due to selection rules and relaxation occurs to the triplet from where it emits.

7.4.2 Life Time and Dynamics of the Excited States

The excited states have to decay eventually. The excited state dynamics (subject dealing with stability and rate of decay of the excited states) have been investigated in detail. Radiative and non-radiative processes occur. There can be several ways in which the excited energy can decay and non-radiative processes dominate in the case of nanocrystals whose surfaces are not passivated. Surface defects reduce the quantum efficiency of radiative decay. The defects occur in the form of unsatisfied valencies which provide an easy sink for the charge carriers. In the case of well-passivated nanocrystals, the quantum efficiency can be close to unity, which means that they are almost like dyes. The excited states can lead to charge separation and the charge could reside with an adsorbate species. This is the most critical aspect which decides the photocatalysis of semiconductor nanoparticles.

7.4.3 X-ray Diffraction

X-ray diffraction is the principal method used to identify the phases present in a solid state material. In the case of a nanomaterial, one is making not a new phase, but smaller dimension crystallites of an already known phase. As the dimension of the crystal reduces, the diffraction peaks broaden and in a very small crystallite, there may not be enough planes to diffract. The problem is discussed in Chapter 2. This is shown in the case of InP particles in Fig. 7.6. The extent of broadening can be used to find the diameter of the particles. Therefore, X-ray diffraction as a function of particle dimension is generally carried out. The size of the particles can be found by using the Scherer formula. This may be compared with the data from other techniques such as transmission electron microscopy, scanning electron microscopy, scanning probe microscopy, neutron scattering, dynamic light scattering, etc. The temperature and pressure dependence of powder diffraction is also investigated, which is used to identify phase transitions.

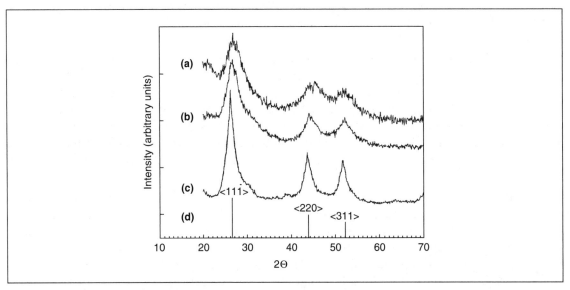

Fig. 7.6: *X-ray diffraction patterns of colloidal InP quantum dots as a function of particle size, (a) 2.5 nm, (b) 3.5 nm and (c) 4.5 nm. The data are compared with the data of bulk InP of zincblende structure (d). While the peak positions and intensities are the same as that of the bulk, the peaks broaden with a decrease in particle size. Reprinted with permission from Mićić, et al. (Ref. 5). Copyright (1995) American Chemical Society.*

7.4.4 Transmission Electron Microscopy (TEM)

TEM is the most important characterization tool for a nanomaterial, as nothing can be more convincing than seeing the object. In addition to observing the shape of the object, TEM can reveal the microscopic structure and atomic composition by using energy dispersive X-ray analysis. In addition, electron energy loss spectroscopy and energy-filtered imaging can provide additional information on the atomic constitution of the materials. These analyses can be done from an area of the order of 1 nm in diameter.

In TEM, it is important to study a large cross section of the sample, not merely one particle. This large area image, as shown in Fig. 7.7, gives the particle size distribution. Generally, a histogram of particle size distribution is plotted, which is more representative of the material synthesized. At higher magnifications, the lattice structure of individual nanoparticles is resolved.

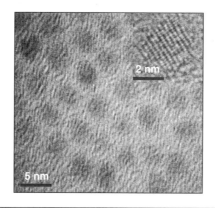

Fig. 7.7: *A collection of CdSe nanoparticles synthesized by chemical route. A magnified image of a single particle is shown as the inset. This gives a lattice-resolved image of the particle. Individual lattice points are observable. Data from the author's laboratory.*

7.4.5 Ancillary Techniques

Several other techniques are routinely used for the characterization of such materials. They include thermogravimetry, differential scanning calorimetry, X-ray photoelectron spectroscopy, Raman spectroscopy, infrared spectroscopy, etc. Thermogravimetry measures the thermal loss/gain of the material when it is subjected to heating at a constant rate. This gives the kinetics of thermal events such as decomposition, reaction, etc. This is also useful in estimating the extent of surface coverage in nanomaterials as the cover is lost in most cases before other thermal processes. Differential scanning calorimetry can be used for evaluating phase changes in the material. Due to the nano dimension, the phase changes occur at a lower temperature, which has been attributed to an increase in the surface energy of the system, which makes the phase unstable. Therefore, phase transitions such as melting occur early. X-ray photoelectron spectroscopy is useful since it provides direct information on the electronic structure. In most of the cases, it is useful in understanding the valence state of the elements present in the sample. Raman and infrared spectroscpies are useful in finding the vibrations in the sample, which are characteristic of the molecular and crystal structures. Thus a combination of techniques is used to characterize the system completely. There are several other more refined techniques such as small angle X-ray scattering (SAXS), small angle neutron scattering (SANS), and dynamic light scattering (DLS), which are useful in understanding the particle size distribution as well as shape in solutions. X-ray extended absorption fine structure (EXAFS) is used to study the co-ordination and local structure of materials, especially when they are amorphous in nature and where X-ray diffraction is not useful.

7.5 Correlation of Properties with Size

The absorption band edge shifts to the blue as a result of a decrease in the size of the particle. This effect shown in CdS particles is depicted in Fig. 7.8. The effective band gap increases, which explains this shift. The bulk bad gap is 2.42 eV (512 nm) and for all the particles, the measured band gap is higher than this. For several other materials, the band gap has been determined. In all the cases, systematic shifts have been observed. Models have been developed to correlate this shift with the diameter of the particles.

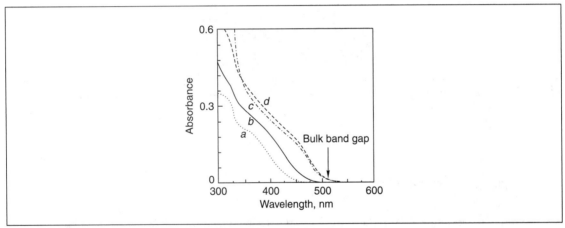

Fig. 7.8: *Shift in the absorption spectra of CdS measured in solution, as a function of the particle dimension. The curves, a, b, c and d correspond to particle dimensions, <25, 30, 37 and >42 Å, respectively. Reprinted with permission from Kamat, et al. (Ref. 6). Copyright (1987) American Chemical Society.*

The simplest of these approaches is to consider the particle in a box model. For one electron confined in a three-dimensional box, the energy levels will be:

$$E_n = h^2/8m_e L^2 (n_x^2 + n_y^2 + n_z^2)$$

where

$L = Lx = Ly = Lz$, the length of the box in three dimensions,

n_x, n_y, n_z = the quantum numbers,

m_e = electron mass.

For a spherical semiconductor quantum dot, the energy expression can be derived by using the confined Wannier exciton Hamiltonian and we get:

$$E_n = E_g + n^2 h^2 / 8 m^\star R^2 - e^2 / 4\pi \varepsilon_0 \varepsilon R$$

where;

E_g = band gap of the bulk semiconductor,
n = quantum number (1, 2, 3 ...),
h = Planck's constant,
$m^\star = (m_e \times m_h)/(m_e + m_h)$,
m_e = effective mass of electron, m_h = effective mass of hole,
R = radius of the quantum dots,
ε = dielectric constant of the semiconductor.

For the electronic transition from the valence band to the conduction band ($n = 1$), we can write (note that we have only one electron in the system):

$$E_g(R) = E_g + h^2/8R^2 (1/m_e + 1/m_h)$$

This neglects the third term.

The increase in the band gap can be given as:

$$\Delta E_g(R) = E_g(R) - E_g = h^2/8R^2 (1/m_e + 1/m_h)$$

As shown here, there will be a $1/R^2$ dependence of the shift in the band gap.

Just like the absorption spectral shift, corresponding shifts are observed in emission too. In the case of CdS, the red emission seen is a result of the sulphur vacancies. The peak shifts to blue as a result of a decrease in the particle size. The peak intensity increases and a blue shift occurs as the temperature is decreased. The emission occurs as the electrons are de-excited from the traps. This results in delayed emission.

7.6 Uses

Semiconductor nanocrystals find applications in a number of areas. A summary of the various applications is presented in Fig. 7.9. As can be seen, there are applications in almost every area. Light can be absorbed by using nanoparticles, especially those with band gaps in the visible region. The absorbed light makes an electronic excitation possible in the material, leaving the electron free in the conduction band. This free electron can move throughout the material and can be collected at an electrode. The electron may be put back into the material by a separate event, which creates a category of solar cells. By using efficient dye molecules which have LUMO levels placed near the conduction band of the semiconductor, it is possible to inject the charge from the excited dye molecule into the semiconductor. This has been an active area for some time and the device is referred to as 'dye sensitized solar cell' (Ref. 7).

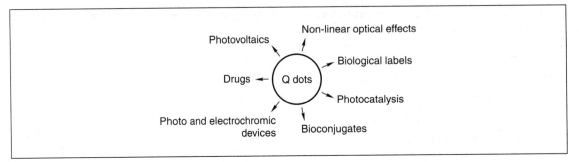

Fig. 7.9: *Diverse applications of quantum dots.*

7.6.1 Chemical Properties

Electron injection into the conduction band can lead to numerous possibilities. The surface of the particles will be covered with functional groups such as hydroxyls in the case of oxides. The hydroxyl groups take up the holes and form hydroxyl radicals, which can result in radical mediated oxidation. Pollutants in water and soil can be efficiently degraded in this way. Photocatalysis can also lead to reduction, using the electron in the conduction band which generates hydroxyl ions or oxide species. All these can make both oxidative and reductive processes feasible. Organic synthesis has utilized the power of semiconductor particles. To quote an example, aromatic ketones and olefins have been converted into alcohols and corresponding saturated compounds. The available electron can be used to fix gas phase CO_2 into organic compounds. In an attempt, CdS nanocrystals have been used to fix CO_2 into benzophenone, acetophenone and benzyl halides producing various compounds (Ref. 8). Various other means of photocatalysis such as the reduction of nitrogen to ammonia, nitrogen oxides to nitrogen, decomposition of pollutants, etc. have also been attempted. In almost all these cases, TiO_2 is used as the photocatalyst. The applications in this field are expanding enormously.

7.6.2 Single Electron Devices

Quantum dots can be used in single electron devices. In order to understand the elementary details of such a system, we consider the example of an electroneutral dot, so-called because the number of electrons that it has is the same as the number of holes. In order to put an electron inside, one has to apply a weak force. Tunneling is the most common means of putting a charge into a device of this kind as the device most often has an insulating barrier around it. The charge that the dot now possesses, $Q = -e$, produces an electric field, ε which repels any incoming negative charge. The fundamental charge is only 1.6×10^{-19} C, but the field it produces on the surface of a dot can be large, of the order of 140 kV on a 10 nm diameter particle. A more important parameter in discussing single electron phenomenon has been recognized as the charging energy, $E_c = e^2/C$. At smaller dimensions, the electronic energy states in the material are quantized. The energy required for the addition (of charge) is a sum of the charging energy and the electron kinetic energy in the material. In the larger size regime, of the order of 100 nm, the addition energy is dominated by the charging energy and the energy required for addition is of the order of μeV. Thus thermal energy is enough to do the job. Therefore, single electron effects will be manifested only at lower temperatures when thermal energy is small. However, in the lower size regime of the order of 10 nm, addition energy will be of the order of meV and will be dominated by electron kinetic energy or will become comparable to the charging energy.

Single electron transition and corresponding effects are manifested dramatically in current-voltage measurements of nanocrystals. Imagine that a device structure is constructed in such a way that a single nanoparticle is trapped between an electron source and an electrode. As the electrode is separated at a large distance such that appreciable tunneling does not occur, a potential U is applied. If one measures the charge of the particle, Q as a function of the 'external charge', $Q_e = CU$, we get a step-like function. This is called the 'Coulomb staircase'. Q is a step function of U, with the distance between the neighboring steps $= e$. $\Delta Q = e$ or $\Delta U = e/C$. However, when thermal energy becomes comparable to the charging energy, the thermal fluctuations smear out the staircase.

Such a staircase implies that electrons can be transferred one at a time. However, the device by itself cannot be used for properties such as rectification or memory. These properties can be attained only by the use of more complicated devices such as single electron transistors, single electron traps, etc. The fundamental aspect of such device structures is the capability to charge or discharge nanosized regions selectively.

Review Questions

1. Can a quantum dot be made with a metallic element?
2. How is quantum confinement manifested in various measurements?
3. How would one make and stabilize a quantum dot?
4. What are the different types of quantum dots investigated?
5. What makes quantum dot luminescence attractive?
6. How do we correlate absorption spectra with size of the quantum dot?
7. What are the unique chemical properties of quantum dots? Give specific examples and illustrate how these are possible.
8. Why functionalize these particles?
9. How do make biocompatible quantum dots?

References

1. Sastry, M., A. Ahmad, M.I. Khan, *et al.*, *Current Science*, **85** (2003), p. 162.
2. Herron, N., J.C. Kalabrase, W.E. Farneth and Y. Wang, *Science*, **259** (1993), p. 1426.
3. Liu, D. and P.V. Kamat, *J. Electroanal. Chem. Interfacial Electrochem.*, **347** (1993), p. 451.
4. Mićić, O.I., J.R. Sprague, Z. Lu and A.J. Nozik, *Appl. Phys. Lett.*, **68** (1996), p. 3150.
5. Mićić, O.I., J.R. Sprague, C.J. Curtis, K.M. Jones, J.L. Machol and A.J. Nozik, *J. Phys. Chem.*, **99** (1995), p. 7754.
6. Kamat, P.V., N.M. Dimitrijevic and R.W. Fessenden, *J. Phys. Chem.*, **91** (1987), p. 396.
7. O'Regan, B. and M. Gratzel, *Nature*, **353** (1991), pp. 737–739.
8. Kanemoto, M., H. Ankyu, Y. Wada and S. Yanagida, *Chem. Lett.*, **2113** (1992); H. Fugiwara, H. Hosokawa, K. Murakoshi, Y. Wada, S. Yanagida, T. Okada and H. Kobayashi, *J. Phys. Chem. B.* **101** (1997), p. 8270.

Additional Reading

1. Brus, L., *J. Phys. Chem.*, **90** (1986), p. 2555.
2. Henglein, A., *Chem. Rev.*, **89** (1989), p. 1861.
3. Alivisatos, A.P., *J. Phys. Chem.*, **100** (1996), p. 13226.
4. Nirmal, M. and L. Brus, *Acc. Chem. Res.*, **32** (1999), p. 407.
5. Klabunde, Kenneth J., (ed.) (2001), *Nanoscale Materials in Chemistry*, Wiley, New York.
6. Nalwa, Hari Singh (ed.) (2001), *Nanostructured Materials and Nanotechnology*, (ed.) (2002), Academic Press, New York. See articles of P.V. Kamat, K. Murakoshi, Y. Wada and S. Yanagida; O.I. Mićić and A.J. Nozik as well as V.L. Klimov in the book dealing with the subject.
7. Several articles in Nalwa, Hari Singh (ed.) (2004), *Encyclopedia of Nanoscience and Nanotechnology*, Academic Press, New York.
8. Liz-Marzan, Luis M. and P.V. Kamat, (2003), *Nanoscale Materials*, Kluwer Academic Publishers, Boston.

Chapter 8

Monolayer-Protected Metal Nanoparticles

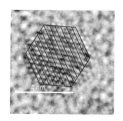

The synthesis of redispersible nanomaterials has expanded the scope of colloidal chemistry. They can be treated just as molecules, which may be stored as powders, dispersed in solvents, reacted with suitable molecules, etc. In effect, they act as reagents. Such materials can be characterized by spectroscopic and microscopic techniques. Any property can be incorporated in such materials by functionalizing them with suitable protecting molecules. This creates a lot of possibilities for the applications of such systems. The assembly of such systems into ordered structures is interesting. Such ordered arrays are expected to show unusual properties that are entirely different from those of individual nanocrystals.

Learning Objectives

- What are the differences between clusters with and without monolayers?
- How does functionality make a difference?
- What are the applications of clusters?
- What are metal cluster superlattices?

8.1 Introduction

In 1857 Faraday made colloidal gold by reducing the aqueous solution of $AuCl_4^-$ with phosphorus in CS_2 (Ref. 1). Since then several methods have become available for the synthesis of colloidal gold particles. The most popular one is the citrate reduction method of Turkevitch (Ref. 2). Here a solution of the gold or silver salt (typically 1 mM) is boiled with a higher concentration (typically 1 M) of sodium citrate for a few minutes. This results in the formation of metal colloids of 10–50 nm diameter, and the size can be varied by altering synthetic parameters. This colloidal solution is stable for several months (the gold colloid is much more stable than that of silver). Here the stability is due to an electrical double layer surrounding the metal surface. This layer is dynamic and the colloid is stable as long as the conditions are not altered greatly. For example, if we precipitate the colloid, the material cannot be re-dispersed. Precipitation leads to aggregation as the stabilizing ionic layer is easily disturbed. A stable colloidal particle which can be precipitated, dried and re-dispersed, or in effect one that has similar characteristics as a molecule, generated interest for a long time. The 3D monolayers or monolayer-protected metal clusters

(MPCs) belong to this category of materials. They are molecular materials, wherein the constituting monolayer-protected clusters or nanoparticles behave like molecules. 3D is to distinguish these from the corresponding 2D monolayers, which are grown on planar surfaces.

8.2 Method of Preparation

By combining this two-phase method of Faraday and the technique of phase transfer catalysis, Brust, et al., prepared dodecanethiol-protected gold nanoparticles with a core size in the range 1–3 nm (Ref. 3). $AuCl_4^-$ was transferred to toluene using tetra octyl ammonium bromide as the phase transfer agent. The phase-transferred Au^{3+} is then reduced in the presence of the surfactant, octadecanethiol using $NaBH_4$ as the reducing agent (see Fig. 8.1).

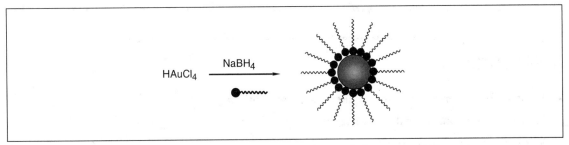

Fig. 8.1: *Schematic showing the Brust method of preparing monolayer-protected clusters.*

The overall reaction is:

$$AuCl_4^-(aq) + N(C_8H_{17})_4^+ (toluene) \rightarrow N(C_8H_{17})_4^+ AuCl_4^- (toluene)$$

$$mAuCl_4^-(toluene) + nC_{12}H_{25}SH(toluene) + 3me^- \rightarrow (Au_m)(C_{12}H_{25}S)_n(toluene) + 4mCl^{-1}(aq)$$

The material formed is a dark brown powder with a waxy texture. This may be referred to as Au@DDT, the @ symbolism means that DDT covers Au (DDT is dodecane thiol, $C_{12}H_{25}SH$). The material is redispersible in common organic solvents and can be purified by gel filtration chromatography with sepharose 6B/toluene. The size of the gold core can be controlled by varying the metal ion to ligand ratio. A large thiol/gold ratio leads to the formation of smaller nanoparticles. Various ligands other than thiols have been used to prepare 3D SAMs.

8.3 Characterization

Various tools have been used to characterize SAMs. These include UV/vis spectroscopy, transmission electron microscopy, X-ray diffraction, mass spectrometry, infra-red spectroscopy, X-ray photoelectron spectroscopy, nuclear magnetic resonance spectroscopy and differential scanning calorimetry.

UV-vis absorption spectroscopy has been used to characterize the core of the material. Noble metal nanoparticles show surface plasmon resonance due to the coherent oscillation of the conduction band electrons excited by electromagnetic radiation (Fig. 8.2). This happens when the core size is comparable to the mean free path of the electron. The frequency of the plasmon absorption band (ω_p) is related to the free electron density by the equation $\omega_p^2 = \pi N e^2 / m$, where N is the free electron density, e is the electron charge and m is its effective mass. Mie explained this phenomenon by solving Maxwell's equations for the absorption and scattering of the electromagnetic radiation by spherical particles (Ref. 4). More details on this can be found in the chapter on Core Shell Nanoparticles (Chapter 9).

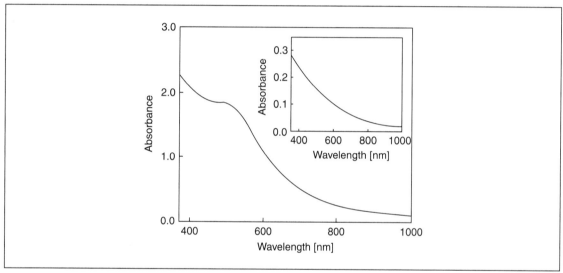

Fig. 8.2: *UV-vis spectrum of Au@hexanethiol showing the presence of surface plasmon resonance at 520 nm. Inset shows the spectrum for sub-nanosized metal clusters for which the plasmon is very weak. From the author's work.*

The plasmon absorption band is characteristic of the size and shape of nanoclusters. Anisotropic particles such as metal nanorods show the presence of two plasmon features, namely 'transverse', due to the coherent oscillation along the short axis, and 'longitudinal' due to the oscillation along the long axis. The intensity of the longitudinal plasmon band is large and its position varies with the length of the rod.

Techniques like TEM, XRD, AFM, STM, etc. have been used to derive information on the structure of the metal core. High resolution transmission electron microscopy at lattice resolution on size-selected gold nanocrystallites showed FCC packing of atoms in the gold core with a mixture of particle shapes with predominantly truncated octahedral, cuboctahedral and icosahedral structures. The octahedral core will have eight (111) planes and the six corners will be truncated by (100) planes. The high radius of curvature of the core suggests that the monolayer chain density decreases as we move away from the metal core. Thus the terminal methyl groups have enough orientational freedom. This is again supported by the fact that the hydrodynamic radius of the metal nanoparticles is lower than that calculated by assuming the monolayer to be a straight chain. TEM gives the mean diameter (d) of the particles which can be used to calculate the mean number of atoms (N) in the core using the relationship, $N_{Au} = 4\pi(d/2)^3/V_{Au}$.

X-Ray photoelectron spectroscopy has been used to find out the oxidation state of gold in the material. The Au $4f_{5/2}$ (87.5 eV) and Au $4f_{7/2}$ (83.8 eV) values are characteristic of Au^0 in the material. This shows that largely gold is present in the Au^0 state in the core. The absence of the Au(I) band at 84.9 eV (for Au $4f_{7/2}$) ruled out the possibility of gold sulphide character of the Au-S bond.

IR spectroscopy is a powerful tool used to study the structure of the adsorbate on the metal surface. An increase in the intensity of the methylene stretching vibrations with an increase in the chain length indicates that the structural integrity of the alkanethiol was maintained during the cluster formation. The position of the d_+ and d_- methylene bands indicates if the alkane chain is crystalline or not. Thiols with carbon atoms of less than eight show a liquid-like structure. In the case of longer thiols, the values 2850 cm^{-1} and 2920 cm^{-1} (for d_+ and d_-) are close to the values of crystalline alkanes. The high crystallinity indicates the all-trans arrangement of the methylene chains. The merging of the r_+ and r_- bands indicates that the terminal methyl group has enough rotational freedom. The disappearance of S-H stretching frequency at 2650 cm^{-1} indicates that the thiols attach to the metal surface as thiolates.

^{13}C NMR spectra of alkane thiol stabilized clusters for three different alkyl chain lengths showed that the peak narrows down as the distance of the carbon from the gold surface increases. For example, among C_6, C_{12} and C_{16} capped clusters, the line width of the methyl resonance at 14 ppm was minimum for hexadecane thiol capped gold. ^{13}C NMR spectrum of octane thiol capped gold nanoparticles showed peaks only for carbon beyond C_γ. The resonances corresponding to the carbons close to the gold surfaces, namely C_α, C_β and C_γ, did not show up in the spectra. This is because they were flattened into the base line. The factors which can affect the distance dependent broadening are dependence of spin lattice relaxation rate on magnetic field anisotropy, chemical shift anisotropy and residual heteronuclear dipolar interactions. Among these factors, the major contribution was from residual heteronuclear dipolar interactions.

MPCs display current due to double layer charging of the metal core. This is due to the extremely small sub-atto Farad capacitance of MPCs. The double layer charging occurs as a series of one electron, approximately evenly spaced, current peaks as shown in Fig. 8.3. This results from the single electron changes in the charge state of the core. For the differential pulse voltammetry of a dilute solution of MPCs, peaks will be observed for the addition or subtraction of each electron. The space between the successive peaks in the differential pulse voltammetry is e/C_{dlu}, where e is the electron charge and C_{dlu} is the double layer capacitance.

One can charge the core of the MPCs electrochemically in a solvent in the presence of a supporting electrolyte. The solvent is then removed and the supporting electrolyte is washed away. Such positively or negatively charged MPCs can act as oxidizing as well as reducing agents. These charged MPCs are reasonably stable and can be handled like usual compounds. Murray, et al., (Ref. 6) studied ligand exchange reactions on electrochemically charged MPCs. They found that the rate as well as the extent of exchange increased with an increase in the positive charge on the core. Mixing MPCs with different core charges results in the transfer of electrons between the MPCs resulting in a solution with a potential determined by the stoichiometry of the mixture and Nernst equation.

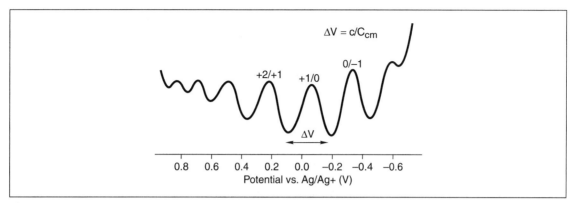

Fig. 8.3: *Differential pulse voltammetry of 0.1 mM C_6MPCs in CH_2Cl_2 measured at a 1.6 mm diameter Pt working electrode using 50 mM Bu_4NClO_4 as the supporting electrolyte. Reprinted with permission from Song, et al. (Ref. 5). Copyright (2002) American Chemical Society.*

The DSC of the alkanethiol-capped gold nanoparticles with alkane thiols with chain lengths varying from C_{12} to C_{20} are shown in Fig. 8.4. The broad endotherm is caused by the order–disorder transition associated with the hydrocarbon chain. The transition temperature as well as the transition enthalpy

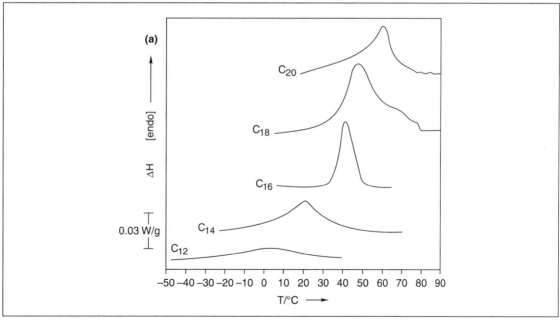

Fig. 8.4: *Differential scanning calorimetry of alkanethiol capped gold clusters showing the peak corresponding to the alkyl chain melting. The melting temperature as well as the transition enthalpy increase with an increase in the chain length of the thiol. Reprinted with permission from Templeton, et al. (Ref. 6). Copyright (2000) American Chemical Society.*

increase with the chain length. Alkyl chains below 12 carbons do not show the melting of the hydrocarbon chain. Apart from this, another melting is observed above 100°C. This is due to the small fraction of the superlattice (see later) present in the sample.

The enthalpy of the superlattice melting can be very high due to the high cohesive energy of such systems. As one can see from the cartoon structure of the superlattice in Fig. 8.5, the ordered portion of the alkyl chain can be either at the end or in between the two clusters.

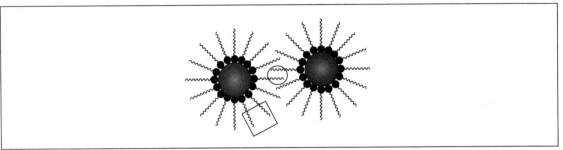

Fig. 8.5: *Cartoon of thiol protected gold clusters showing the possible regions of crystallinity. Circle shows the region of crystallinity because of the interdigitation of alkyl chains and square shows the possibility of interstitial folding.*

8.4 Functionalized Metal Nanoparticles

The synthesis of functionalized nanomaterials has been receiving considerable attention during the past few years. Functionalization can be either through a ligand exchange reaction or using modified thiol as the capping agent in the Brust method. Various photoactive molecules have been attached to the surface of gold using the thiol end group and such structures are shown in Fig. 8.6. These include derivatives of porphyrenes (Ref. 7), fullerenes (Ref. 8), pyrenes (Ref. 9), stilbenes (Ref. 10), fluorenes (Ref. 11) resorcinarenes (Ref. 12), azobenzenes (Ref. 13), etc.

Photoswitchable gold nanoparticles with a double shell structure in which the inner shell is made of spiropyran, have been used to control the binding and release of the outer shell of amino acids. Under dark conditions, a majority of the spiropyran exist in the close ring form. This non-polar form does not have color. When irradiated with light, spiropyran (SP) changes to the highly polar-colored merocyanin (MC) form. This open ring merocyanin can form a complex with amino acids which helps in forming a further layer of amino acid around the gold nanoparticles (Fig. 8.7). Such systems can be possible candidates for light-mediated binding and release of amino acid derivatives.

Another interesting property was observed in pyrene methyl amine-capped gold nanoparticles. The fluorescence of pyrene was found to increase when attached to the metal surface. This was entirely different from the normal trend wherein heavy metals quench the fluorescence. The binding of nitrogen onto the

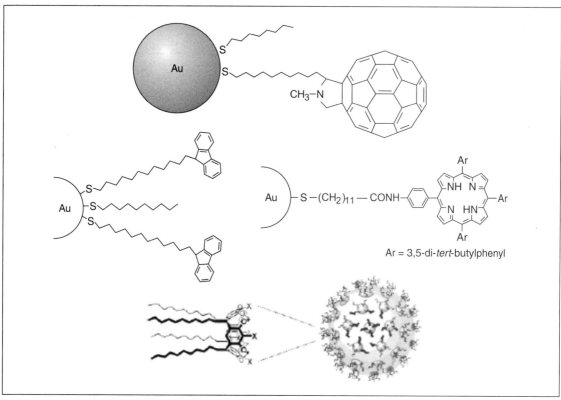

Fig. 8.6: *Various functionalized nanoparticles, with their chemical functionalities.*

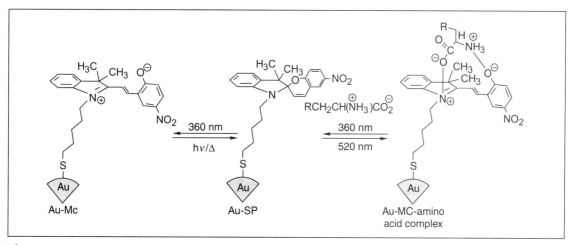

Fig. 8.7: *Reversible binding of amino acids with spiropyran capped gold nanoparticles. Reprinted with permission from Ipe, et al. (Ref. 14). Copyright (2003) American Chemical Society.*

gold suppresses the electron transfer between the nitrogen and the pyrene ring as shown by the arrows in Fig. 8.8. This results in enhanced fluorescence.

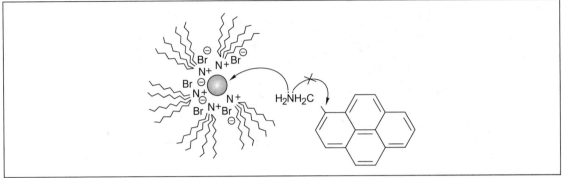

Fig. 8.8: *Gold nanoparticles-assisted enhancement of fluorescence in pyrene methyl amine. Due to the attachment of nitrogen onto the nanoparticles, conjugation between lone pair on nitrogen and the pyrene ring is blocked. This is indicated by an arrow with a cross mark. Reprinted with permission from Thomas and Kamat (Ref. 15). Copyright (2000) American Chemical Society.*

8.5 Applications

Gold nanoparticles have very high extinction coefficients, and the color change during the transition of nanoparticles from the dispersed to the aggregate state is quite evident. Hence the coupling of nanoparticles with ionophores gives better sensors. The sensing of a particular ion by the ionophores will be seen as a color change. Crown ethers in combination with gold nanoparticles have been used as K^+ ion sensor. K^+ has the ability to form 1:2 adducts with crown ethers attached to adjacent nanoparticles as shown in Fig. 8.9. This results in aggregation and the color changes from dark brown to purple (Ref. 16).

Bidentate ligands attached to gold nanoparticles provide it specificity to a particular ion due to its capability to form chelate complexes. The chelation leads to aggregation and color change, which can be used to find the concentration of the ion. The scheme in Fig. 8.10(a) shows the Li^+ ion assisted aggregation of gold nanoparticles. The corresponding TEM picture is shown in Fig. 8.10(c).

Toxic heavy metal ions such as lead, cadmium and mercury can be detected by using gold nanoparticles. In the presence of metal ions, gold nanoparticles aggregate through chelation of the ions with the carboxyl group on adjacent nanoparticles as shown in Fig. 8.11. This ion templated chelation process changes the color of the solution. It also changes the Rayleigh scattering response from the medium. This chelation can be reversed through the addition of a strong metal ion chelator such as EDTA.

Temperature-sensitive polymers have been grafted onto the surface of nanoparticles to make molecular thermometers. A commonly used functional polymer is poly-N-isopropyl acryl amide. This polymer has a lower critical solution temperature (LCST) of 35°C. Below 35°C, hydrogen bonding between the water

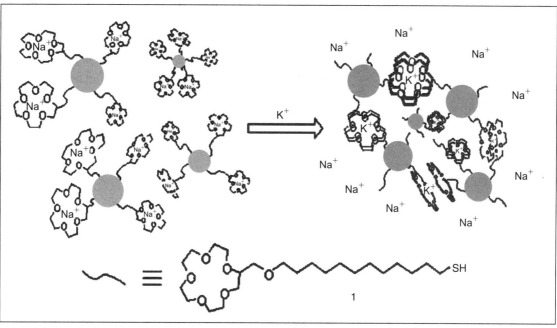

Fig. 8.9: *Schematic showing the sensing of potassium ion by crown ether functionalized gold nanoparticles. The addition of K⁺ leads to the formation of adducts which brings about the aggregation of nanoparticles resulting in a color change. Reprinted with permission from Lin, et al. (Ref. 16). Copyright (2002) American Chemical Society.*

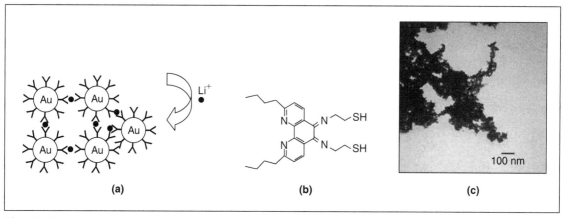

Fig. 8.10: *Selective detection of lithium ion by gold nanoparticles. The structure of the ligand used is shown in Fig. 8.10(b). Fig. 8.10(c) shows the TEM picture of the aggregate. Reprinted with permission from Obare, et al. (Ref. 17). Copyright (2002) American Chemical Society.*

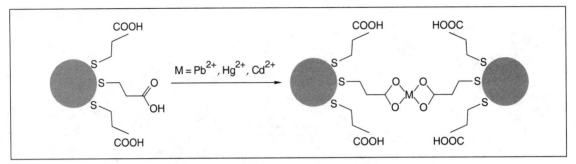

Fig. 8.11: *Ion-assisted chelation for the detection of heavy metals, Picture taken from the Table of Contents entry link of Ref. 18. Copyright (2001) American Chemical Society.*

and the polar groups makes the polymer soluble in water. Above 35°C the hydrophobic interaction predominates and throws out all the water molecules, thus acting as a molecular thermometer. The coil-to-globule chain transition taking place as a result of the changes in the surrounding temperature is schematically shown in Fig. 8.12.

8.6 Superlattices

Superlattice is a periodic, synthetic multi-layer, wherein a unit cell, consisting of successive layers that are chemically different from their adjacent neighbors, is repeated. These materials are characterized by their double periodicity in the structure, periodicity of atoms in the angstrom level, and periodicity of nanocrystals in the nanometer level.

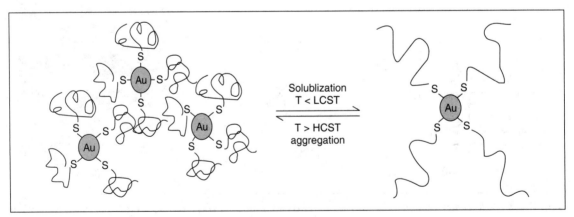

Fig. 8.12: *Temperature sensor based on gel-coated nanoparticles. HCST refers to higher critical solution temperature. Reprinted with permission from Zhu, et al. (Ref. 19). Copyright (2004) American Chemical Society.*

In order to allow the superlattice formation to occur, the van der Waals interaction between the particles should be sufficiently strong to create a secondary minimum M_2 in the potential energy–distance curve as shown in Fig. 8.13. The secondary minimum is very shallow for smaller particles. Hence during the growth of the particles, the system crosses the small energy barrier P and is taken to stable minimum M_1 which results in irreversible aggregation.

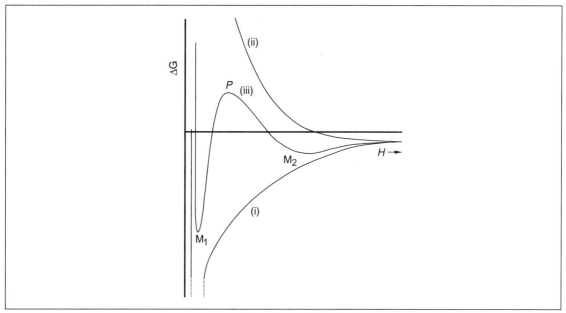

Fig. 8.13: *Potential energy–distance curve showing the resultant interaction energy (iii) due to contribution from attractive (i) as well as repulsive forces (ii). M_2 refers to the shallow minimum which can lead to particles associated reversibly and M_1 refers to the stable minimum leading to irreversible aggregation. H is the interparticle distance and ΔG denotes the change in Gibbs free energy. From Ref. 20.*

Monodispersity is another important factor that controls the formation of superlattice. Even though the Brust method gives redispersible materials, the particle distribution is very large. Hence size selective separation is required to make monodisperse particles. The method developed by Murray, *et al.*, 1993, (Ref. 21), facilitates the synthesis of monodisperse particles. Here the metal–organic precursors are injected into a hot solution containing co-ordinating ligands like trioctyl phosphine. Extensive research has taken place on CdSe superlattices that have been prepared by using this method (Ref. 22). Figure 8.14 (Plate 5) shows the superlattice prepared from CdSe nanoparticles.

When a drop of the solution of monolayer-protected cluster in a volatile organic solvent is placed on a plane surface and allowed to evaporate under controlled conditions, the particle starts coming closer, the individual nanocrystals start nucleating and the crystal starts growing. This is followed by an annealing process which results in the formation of superlattices.

In hydrophobically-modified SAMS, hard sphere repulsion and van der Waals forces assist the formation of superlattices. The self-organization of monolayers leads to an assembly of the chains on the crystal planes of the nanocrystals. Such assemblies of adjacent clusters can interact, forming a well-organized superlattice structure as shown in Fig. 8.15. The structure of such assemblies and the phase behavior of such systems has been the subject of intense investigation in our research group (Ref. 24). The superlattice structure undergoes a first order phase transition and can reverse into a liquid phase. Such a liquid phase exists for a narrow range of temperatures. Beyond a particular temperature, the alkyl chains possess orientational freedom in the superlattice structure.

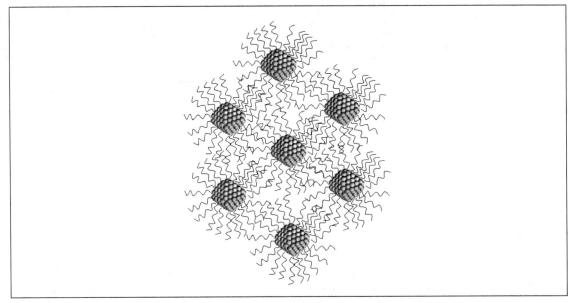

Fig. 8.15: *Pictorial representation of a superlattice.*

The formation of a superlattice is easy if the surface of the nanoparticle is modified with thiols with the terminal group capable of forming hydrogen bonds. Kimura has observed the hydrogen bond-mediated formation of superlattices in mercaptosuccinic acid modified gold nanoparticles (Ref. 26). The superlattice formation in this case is believed to be assisted by the water clusters trapped in the tetrahedral and octahedral cavities that are created by the hexagonal close packing of the constituent nanocrystals. The optical micrograph of the crystals thus obtained is shown in Fig. 8.16.

Digestive ripening is another way of making 'superlattice' (Ref. 27). In this method, the already prepared nanoparticles are refluxed in a solvent with ligands capable of acting as digestive ripening agents. This includes thiols, amines, alcohols, alkanes and silanes. Among the ripening agents, thiols are found to be better than the others. In this process, larger nanocrystals grow at the cost of smaller ones. Thus the growth takes place in size, and not in number. The TEM picture of the particles during each step is shown in Fig. 8.17.

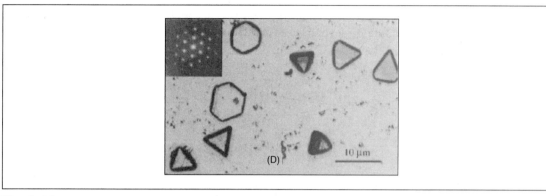

Fig. 8.16: *Picture showing the optical microscope images of the superlattices. Inset shows the low angle diffraction form one superlattice. Reprinted with permission from Wang, et al. (Ref. 26). Copyright (2004) American Chemical Society.*

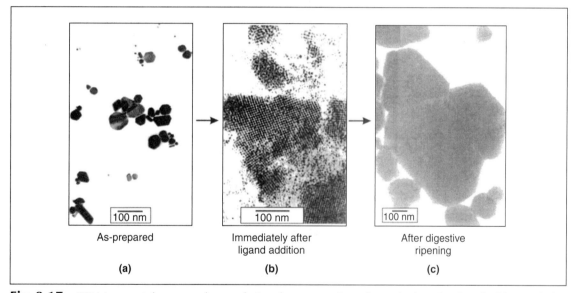

Fig. 8.17: *TEM images taken at each step during the preparation of superlattice. Reprinted with permission from Prasad, et al. (Ref. 27). Copyright (2003) American Chemical Society.*

Superlattices result from collective interactions, which is different from individual nanocrystals. One of the interesting phenomena observed in a superlattice is the metal–insulator transition. Heath and coworkers observed this phenomenon while compressing a film of silver nanocrystals using a Langmuir trough (Ref. 22). At a large interparticle distance, the nanocrystals are electrically isolated and the Coulomb band gap is large. When the film is compressed, the interparticle distance (D) decreases. At large values of pressure, D is comparable to the diameter ($2R$) of the particles. The coupling between the particles is very high at this point and the Coulomb gap disappears as shown in Fig. 8.18.

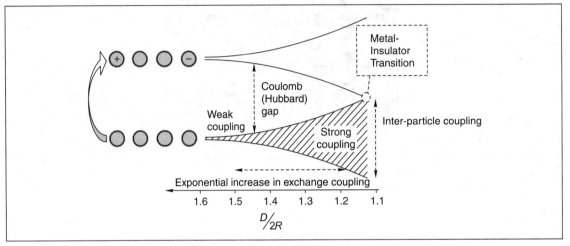

Fig. 8.18: *Schematic showing the metal insulator transition in superlattice. Reprinted with permission from the Annual Review of Material Science, Volume 30 © 2000 Annual Reviews. www.annualreviews.org©.*

Review Questions

1. Why nanoparticles need a protective layer of molecules?
2. What are the principal differences between a planar monolayer and a monolayer on a metal nanoparticle?
3. What are the principal properties of metal nanoparticles?
4. How would one characterize the monolayer and how can one characterize the core?
5. How do we know that a nanoparticle is indeed metallic?
6. Why most of the investigations are on gold particles?
7. What are the applications of gold particles?
8. What are the unique features of the Brust method?
9. What makes a metal cluster superlattice?
10. When does a metal cluster become a quantum dot? Are there examples?

References

1. Faraday, M., *Phil. Trans.*, **147** (1857), p. 145.
2. Turkevitch, J., P.C. Stevenson and J. Hiller, *Discuss Faraday Soc.*, **11** (1951), p. 55.
3. Brust, M., M. Walker, D. Bethell, D.J. Schffrin and R.J. Whyman, *J. Chem. Soc., Chem. Commun.*, **801**, (1994).
4. Mie, G., *Ann. Phys.*, **25** (1908), p. 377.
5. Song, Y. and R.W. Murray, *J. Am. Chem. Soc.*, **124** (2002), p. 7096.
6. Templeton, A.C., W.P. Wuelfing and R.W. Murray, *Acc. Chem. Res.*, **33** (2000), p. 27.
7. Imahori, H., M. Arimura, T. Hanada, Y. Nishimura, I. Yamazaki, Y. Sakata and S. Fukuzumi, *J. Am. Chem. Soc.*, **123** (2001), p. 335.
8. Fujihara, H. and H. Nakai, *Langmuir*, **17** (2001), p. 6393.
9. Wang, T., D. Zhang, W. Xu, J. Yang, R. Han and D. Zhu, *Langmuir*, **18** (2002), p. 1840.
10. Zhang, J., J.K. Whitesell and M.A. Fox, *Chem. Mater.*, **13** (2001), p. 2323.
11. Gu, T., T. Ye, J.D. Simon, J.K. Whitesell and M.A. Fox, *J. Phys. Chem. B*, **107** (2003), p. 1765.
12. Balasubramanian, R., B. Kim, S.L. Tripp, X. Wang, M. Lieberman and A. Wei, *Langmuir*, **18** (2002), p. 3676.
13. Manna, A., P.L. Chen, H. Akiyama, T.X. Wei, K. Tamada and W. Knoll, *Chem. Mater.*, **15** (2003), p. 20.
14. Ipe, B., S. Mahima and K.G. Thomas, *J. Am. Chem. Soc.*, **125** (2003), p. 7174.
15. Thomas, K.G. and P.V. Kamat, *J. Am. Chem. Soc.*, **122** (2000), p. 2655.
16. Lin, S.Y., S.W. Liu, C.M. Lin and C.H. Chen, *Anal. Chem.*, **74** (2002), p. 330.
17. Obare, S., R.E. Hollowell and C.J. Murphy, *Langmuir*, **18** (2002), p. 10407.
18. Kim, Y., R.C. Johnson and J.T. Hupp, *Nano Letters*, **1** (2001), p. 165.
19. Zhu, M.Q., L.Q. Wang, G.J. Exarhos and A.D.Q. Li, *J. Am. Chem. Soc.*, **126** (2004), p. 2656.
20. Everett, D.H., (1988), *Basic Principles of Colloidal Science*, Royal Society of Chemistry, London, pp. 26–27.
21. Murray, C.B., D.J. Norris and M.G. Bawendi, *J. Am. Chem. Soc.*, **115** (1993), p. 8706.
22. Murray, C.B., C.R. Kagan and M.G. Bawendi, *Annu. Rev. Mater. Sci.*, **30** (2000), p. 545.
23. Zaitseva, N., Z.R. Dai, F.R. Leon and D. Krol, *J. Am. Chem. Soc.*, **127** (2005), p. 10221.
24. Sandhyarani, N. and T. Pradeep, *Int. Rev. Phys. Chem.*, **22** (2003), p. 221.
25. Chaki, N.K. and K.P. Vijayamohanan, *J. Phys. Chem. B*, **109** (2005), p. 2552.
26. Wang, S., H. Yao, S. Sato and K. Kimura, *J. Am. Chem. Soc.*, **126** (2004), p. 7438.
27. Prasad, B.L.V., S.I. Stoeva, C.M. Sorensen and K.J. Klabunde, *Chem. Mater.*, **15** (2003), p. 935.

Additional Reading

1. Love, J.C., L.A. Estroff, J.K. Kriebel, R.G. Nuzzo and G.M. Whitesides, *Chem. Rev.*, **105** (2005), p. 1103.

Chapter 9

CORE-SHELL NANOPARTICLES

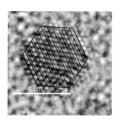

Core-shell nanoparticles are hybrid systems. They have a core and a shell. Various cores and diverse shells are available. The cores and shells can have distinct attributes such as metallicity, semiconductivity, magnetism, etc. Any combination of core and shell is possible. It is also possible to have one kind of core with the same kind of shell, namely a metallic core with another metallic shell, as Au@Ag, implying Au core and Ag shell. These systems are fascinating as they can protect the core from the chemical environment of the medium. A chemically active, biologically unsuitable core can be protected with an unreactive, chemically friendly shell. There are also several other possibilities. This chapter focuses on metallic cores.

Learning Objectives

- What are core-shell nanoparticles?
- Why do we need a core-shell system? What are the advantages of core-shell nanoparticles, in comparison to nanoparticles of core or shell?
- What are their optical, chemical and other properties? How can we understand these properties?
- What do we use them for?

9.1 Introduction

Controlled fabrication of nanomaterials has been one of the challenges faced by nanotechnologists and only limited progress has been achieved in this sphere so far. One of the fascinating characteristics of nanomaterials is that their properties are dependent on size, shape, composition and structural order. Therefore, it is imperative to develop effective and reliable methodologies to cater to the ever-increasing demands of tailored nanomaterials with the desired properties. Core-shell nanoparticles, i.e. particles with a well-defined core and a shell both in the nanometer range, have demanding applications in pharmaceuticals, chemical engineering, biology, optics, drug delivery and many other related areas in addition to chemistry.

During the past decade, there have been widespread research efforts to develop core-shell colloidal nanoparticles with tailored structural, optical, surface and other properties (Refs 1–3). Investigations on these types of materials have been catalyzed by their applicability in modern science and their technological edge over conventional materials. Such composite coatings are used in sensors (for protecting high-tech equipments from lasers in the form of optical limiters), nanoelectronics, catalysis and pharmaceuticals. These are also ideal systems used for probing the interfaces of the nanoparticle core and the shell, which is of fundamental relevance from the academic point of view. The term used to describe the synthesis of core-shell particles with well-defined morphologies and tailored properties is called 'particle engineering'. This is achieved by encapsulating the nanometal core within the shell of a preferred material, or by coating the nanoparticle core with the shell material. The shell protection imparts certain functional properties to the nanomaterial including: (1) monodispersity in size, (2) core and shell processibility, (3) solubility and stability, (4) ease of self-assembly, and (5) applications in nanoscale optics, nanoelectronics, as well as in magnetic, catalytic, chemical and biological fields. Shell protection is absolutely necessary for the following important reasons: (a) the shell can alter the surface charge, reactivity and functionality of the metal core thereby enhancing the stability and dispersibility of the colloidal materials; (b) by choosing a suitable shell-forming material, we can incorporate magnetic, optical and catalytic properties into the composite material; (c) encasing the metal core in a shell invariably protects it from physical and chemical changes; and (d) core-shells exhibit improved physical and chemical characteristics as compared to their single component counterparts. Various procedures have been employed for their synthesis, but the lack of suitable methodology for industrial production of core-shell nanomaterials has limited their applicability. A very critical aspect in the synthesis of such materials is the optimization of suitable/reliable synthetic parameters.

This chapter provides an overview of the various methods to synthesize core-shell nanoparticles, of their characterization using various spectroscopic and microscopic techniques, and their utility in material science, and relevance and future prospects, wherever necessary. The common characterization techniques involve UV-visible spectroscopy (UV-vis), transmission electron microscopy (TEM), powder X-ray diffraction (XRD), cyclic voltammetry (CV), etc.

The topic of core-shell nanoparticles can be divided into sub-sections depending on the nature of the core and the shell.

9.2 Types of Systems

9.2.1 Metal–Metal Oxide Core-shell Nanoparticles

These are among the most widely studied core-shell nanosystems. Nanosized metal clusters have intense colour, which can be tuned by varying the size of the clusters. One of the major problems associated with their handling is their vulnerability to aggregation. In order to avoid aggregation, various methodologies have been developed, with the most notable one being coating them with silica (Ref. 4), titania (Ref. 5), zirconia (Ref. 6) and maghemite (Ref. 7). Liz-Marzan, et al., have developed a synthetic procedure to prepare silica-coated nanosized metal clusters, and the same methodology has been applied to various metals like Au, Ag and CdS (Ref. 8). The methodology uses 3-aminopropyl trimethoxy silane (APS), the silane coupling agent, which can bind to the nanoparticle surface and can also function as an anchor point for the chemical deposition of active silica (SiO_3^{2-}). Briefly, the method adopted in the above procedure can be detailed as follows (Ref. 9): An aqueous dispersion of citrate-capped Au nanoparticles has been treated with amino propyl trimethoxy silane (APS). A thin layer of active silica is deposited onto the activated surface of Au clusters, which are thereby stabilized and can be transferred to ethyl alcohol. After the transfer into the ethanol medium, the silica shell can be grown on it by using the standard Stöber procedure (see chapter 10). The thickness of the silica shell can be adjusted by using this method and can be varied from 10–83 nm (Ref. 9). The absorption characteristics are tunable according to the shell thickness (Ref. 9). Figure 9.1 shows the transmission electron micrographs of Au@SiO_2 nanoparticles with different shell thicknesses synthesized by the above method. The shell thicknesses for a, b, c and d are 10, 23, 58 and 83 nm, respectively. Other inorganic coatings on nanocores include yttrium basic carbonate, titania, titanium nitride, zirconia and Fe_2O_3. In the inorganic coating procedures mentioned above, the sizes and shapes of the core particles as well as the relative ratios of the reactants influence the thickness of the shell. Hence careful and systematic experiments are usually required for optimizing the parameters for desired shell thickness. Compatibility between the metal core and the inorganic shell forming materials is

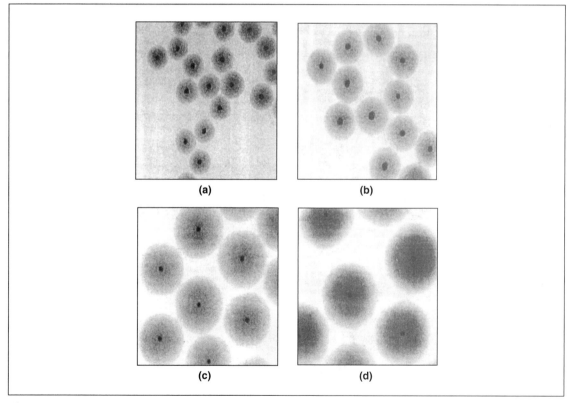

Fig. 9.1: *Transmission electron micrographs of silica coated gold nanoparticles. The shell thicknesses are (a) 10 nm, (b) 23 nm, (c) 58 nm and (d) 83 nm. Reprinted with permission from Liz-Marzan, et al. (Ref. 9). Copyright (1996) American Chemical Society.*

also a prerequisite for obtaining uniform coatings without aggregation (Ref. 9). Liz-Marzan, *et al.*, have demonstrated the synthesis of TiO$_2$-coated Ag nanoparticles by the simultaneous reduction of Ag$^+$ and condensation of titanium butoxide (Ref. 5). A recent study extended the methodology to the syntheses of ZrO$_2$ and TiO$_2$-coated Au and Ag nanoparticles through a one-step synthesis (Ref. 6). Figure 9.2 shows the TEM images of Au@ZrO$_2$ nanoparticles showing the core-shell geometry. Images A, B and C represent three different areas of the same sample. D is an expanded image of one nanoparticle, clearly depicting the core-shell morphology. In this case also, the absorption spectra are tunable depending upon the thickness of the shell. The synthetic insertion of Au nanoparticles into mesoporous silica was demonstrated by Konya, *et al.*, (Ref. 10) using the mesoporous silica materials MCM-41 and MCM-48. Another important approach towards the development of metal–metal oxide core-shell nanoparticles was that used by Teng, *et al.* (Ref. 7). Robust Pt-maghemite (Fe$_2$O$_3$) core-shell nanoparticles were synthesized by a one-pot method involving the reduction of platinum acetylacetonate in octyl ether-yielding Pt nanoparticles. Layers of iron oxide were subsequently deposited on their surface as a result of the thermal decomposition of iron pentacarbonyl (Ref. 7). These particles have potential applications in catalysis, high density storage media and as precursors for making tunable magnetic nanoparticles, thin films, etc. Figure 9.3 shows the TEM images of Pt-maghemite core-shell nanoparticles having two different shell thicknesses, synthesized by varying the concentration of the shell-forming precursor.

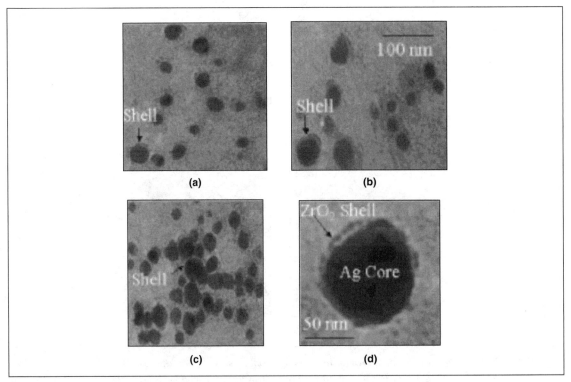

Fig. 9.2: *TEM image of ZrO$_2$ coated Ag nanoparticles. Reprinted with permission from Tom, et al., (Ref. 6) Copyright (2003) American Chemical Society.*

Plate 1

Fig. 1.1: *Faraday's gold preserved in Royal Institution. From the page, http://www.rigb.org/rimain/heritage/faradaypage.jsp.*

Fig. 1.2: *The Lycurgus Cup made from glass appears red in transmitted light and green in reflected light. The glass contains 70 nm particles as seen in the transmission electron micrograph. The cup itself is dated to 4th century AD, but the metallic holder is a later addition. From the site, http://www.thebritishmuseum.ac.uk.*

Plate 2

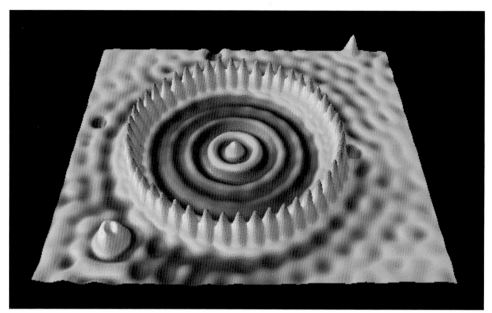

Fig. 2.17: Quantum corral made of 48 Fe atoms on a Cu(111) surface. Color image taken from the website, http://www.almaden.ibm.com/vis/stm/stm.html. Original image was published in Crommie, et al., 1993 (Ref. 7). Reprinted with Permission from Ref. 7. Copyright (1993) AAAS.

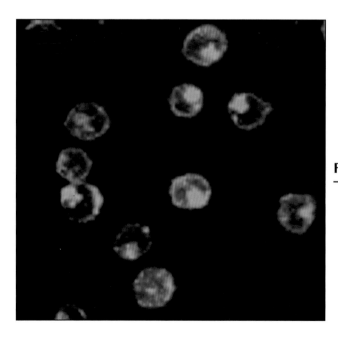

Fig. 2.22: Confocal fluorescence image of human promyelocytic leukemia (HL 60) cells after incubation with fluorescein isothiocyanate labeled single wall carbon nanotube solution for 1 h. The image shows that nanotubes have been delivered into the cells. Reprinted with permission from Nadine Wong Shi Kam, et al. Copyright (2004) American Chemical Society.

Plate 3

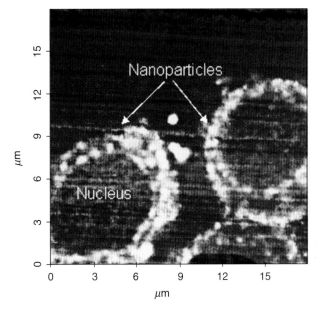

Fig. 2.24: *SNOM image of SiHa cells incubated with gold ions showing the formation of nanoparticles in the cytoplasm (From the author's work).*

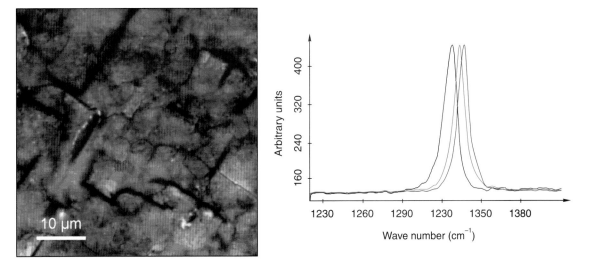

Fig. 2.31: *Raman micrograph of a CVD grown diamond film. The spectra from various regions are shown on the right. Spectra of different colours are used to construct the images of the same colours. Data and image are with the courtesy of www.witec.de.*

Plate 4

Fig: 4.6: *Picture of a 5" diagonal active display developed by Samsung Corporation. Reused with permission from Chung, et al. (Ref. 25). Copyright 2002, American Institute of Physics.*

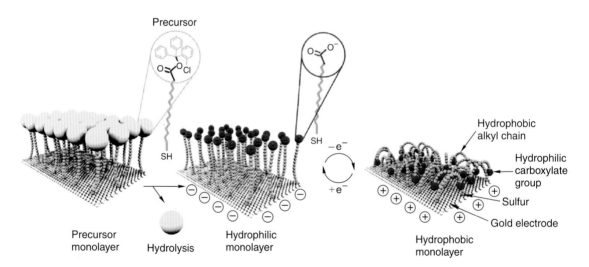

Fig. 5.19: *Schematic showing a monolayer which changes conformation with applied electric field. As a result of this, the polarity of the surface can be switched reversibly by changing the electric field. This allows one to control the wetting behavior of metal plates. Reprinted with permission from Lehann, et al. (Ref. 32). Copyright (2003) AAAS.*

Plate 5

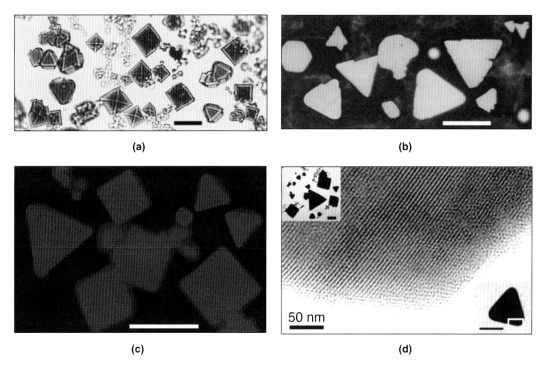

Fig. 8.14: Superlattices built from a solution of CdSe crystals in nonanoic acid. (a) Optical micrograph of the superlattice and (b), (c) fluorescence microscopic images of the superlattices made from 3.5 and 5.3 nm nanocrystals, respectively. (d) TEM image taken from the edge of a superlattice made from 5.3 nm CdSe nanocrystals with the inset showing a low magnification image. Reprinted with permission from Zaitseva, et al. (Ref. 23). Copyright (2005) American Chemical Society.

Fig. 9.15: Transmitted and reflected colors from Au@SiO$_2$ multilayer thin films with varying silica shell thickness. Reprinted with permission from Ung, et al. (Ref. 49) Copyright (2001) American Chemical Society.

Plate 6

Fig. 10.2: *Nanoshells designed to absorb various wavelengths of light (the six vials on the right), including infrared (vial at far right) compared to gold colloid (far left). Used with permission from www.ece.rice.edu/people/faculty/halas.*

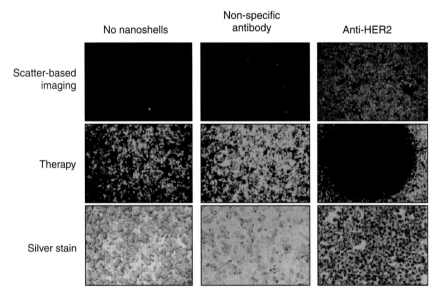

Fig. 10.11: *Combined imaging and therapy of SKBr3 breast cancer cells using HER2-targeted nanoshells. Scatter-based darkfield imaging of HER2 expression (top row), cell viability assessed via calcein staining (middle row), and silver stain assessment of nanoshell binding (bottom row). Cytotoxicity was observed only in cells treated with a NIR-emitting laser following exposure and imaging of cells targeted with anti-HER2 nanoshells. Increased contrast (top row, right column) and cytotoxicity (dark spot) are seen in cells treated with nanoshells as compared to others, called controls (left and middle columns). It is important to note that all experiments of this kind are done with appropriate controls. Reprinted with permission from Christopher, et al. (Ref. 10), Copyright (2005), American Chemical Society.*

Plate 7

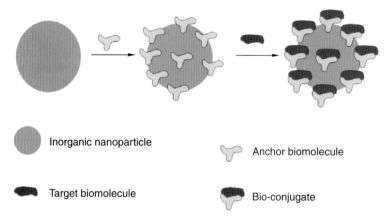

Scheme 11.1: This scheme represents hybridization of conjugate biomolecules on inorganic nanoparticle surfaces.

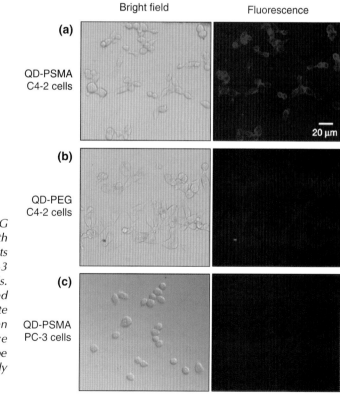

Fig. 11.5: Interaction of CdSe@ZnS@PEG (QD-PEG) quantum dots with cancerous cells. C4-2 represents prostrate cancer cells; PC-3 represents non-cancerous cells. QD-PSMA is functionalized with the antibody of prostrate selective membrane antigen (PSMA-Ab). The negative staining of (b) and (c) can be explained by antigen-antibody interaction. From (Ref. 26).

Plate 8

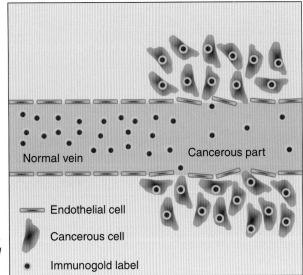

Scheme 11.2: *The scheme represents the mechanism of site-specific immunogold labeling.*

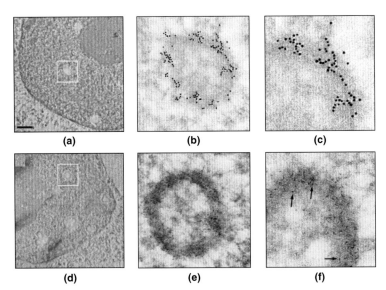

Fig. 11.6: *Target selective fluorescent and TEM imaging of PML bodies on cell nucleus using 10 nm gold and 10–15 nm CdSe nanocrystals. Figures (a) and (d) are fluorescence image of HEp-2 PML I cells labeled with 10 nm gold and CdSe particles, respectively. On gold-labeled sections, a Cy3 dye labeled secondary antibody was used, after incubation with gold, for generation of the fluorescence signal. (b) and (e) are the TEM images of the marked areas given in Figs (a) and (d), respectively. (c) and (f) are the enlarged views of (b) and (e), respectively. The dimension of the bar corresponds to 1000 nm, 100 nm and 50 nm for the set of Figs (a, d), (b, e) and (c, f), respectively. Sections were stained with uranyl acetate. The arrows in (f) mark CdSe nanocrystals. Reprinted with permission from Ho, et al. (Ref. 27). Copyright (2004) American Chemical Society.*

Plate 9

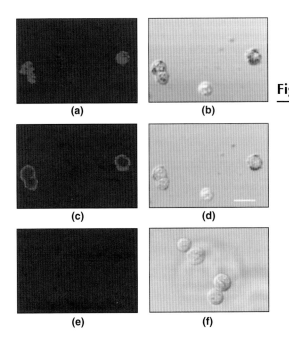

Fig. 11.8: *High resolution optical images of SiHa cells labeled with anti-EGFR/gold conjugates. Non-specific labeling using gold conjugates with BSA is shown in (e) and (f). Laser scanning confocal reflectance (a, c and e) and combined confocal reflectance/transmittance (b, d and f) images were obtained with 40 X objective. Scattering from gold conjugates is false-colored in red. In (a) and (b), the focal plane is at the top of the cells. The middle cross-section of the cells is in focus in images (c) and (d). The confocal reflectance and transmittance images were obtained independently and then overlaid. Reflectance images were obtained with 647 nm laser excitation. The scale bar is 50 μm (a–f). From Sokolov, et al. (Ref. 28). With permission from Cancer Research.*

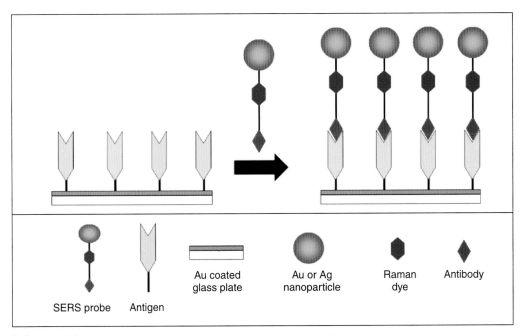

Scheme 11.3: *This scheme represents an immunoassay based on target selective surface-enhanced Raman scattering (SERS) probe.*

Plate 10

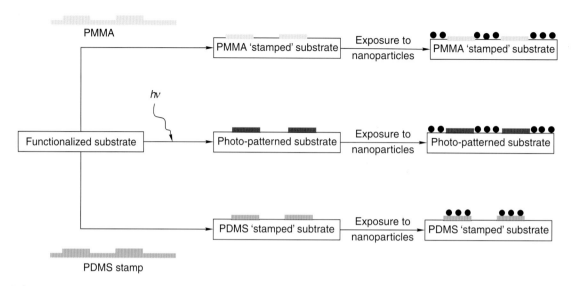

Scheme 12.3: *Some of the methods of template-assisted organization of nanoparticles.*

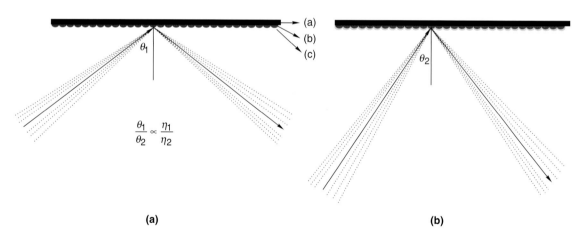

Scheme 12.4: *Schematic of SPRS: (a) Opaque substrate, (b) nanoparticle and (c) plasmon electric field in parent (a) and the modified (b) surfaces.*

Plate 11

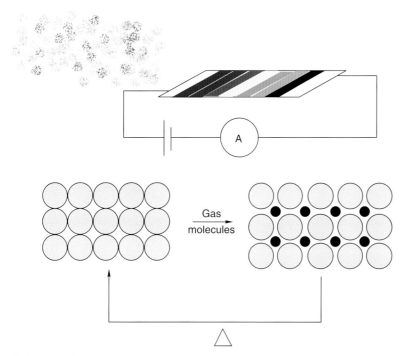

Scheme 12.5: *The working of an electronic nose. The diffused colors represent different gases while the solid colors represent the various nanoparticles that are sensitive to the specific gas molecules. The gases thus adsorbed on the films might be removed by suitably heating the array, thus making repeated use of the system possible.*

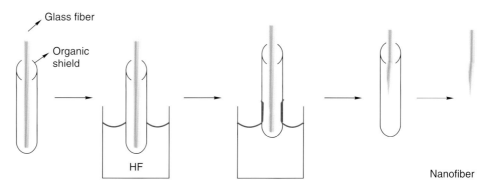

Scheme 12.7: *Cartoon representation of fabrication protocol of a nanofiber as a probe for biological sensing and imaging. As the optical fiber is withdrawn from the HF solution, surface tension causes the HF to initially rise along the organic cladding. Slow draining of HF from there causes a nanotip to be formed, after which the organic cladding is removed by treating with a suitable solvent.*

Plate 12

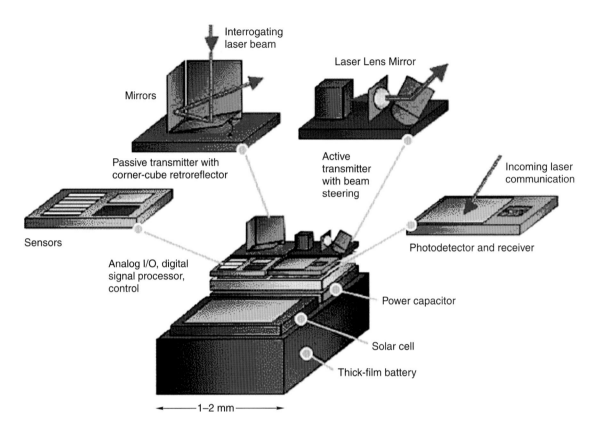

Scheme 12.9: *Proposed structure of a smart dust multifunctional sensor. From the Webpage, http://www.bsac.eecs.berkeley.edu/archive/users/warneke-brett/SmartDust/index.html. Used with permission.*

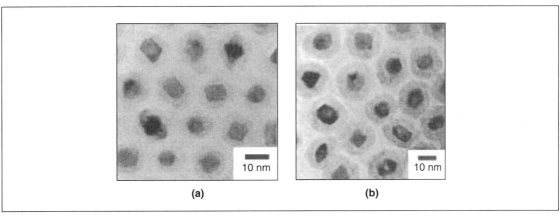

Fig. 9.3: *TEM image of Pt-maghemite core-shell nanoparticles having different shell thickness made with different shell-forming precursors. The shell thicknesses are 3.5 nm and 5.4 nm, respectively (left to right). Reprinted with permission from Teng, et al., (Ref. 7) Copyright (2003) American Chemical Society.*

Mayya, et al., (Ref. 11) demonstrated the coating of Au nanoparticles with titania by using a facile approach based on the complexation of a negatively charged titanium (IV) bis (ammonium lactate) dihydroxide with poly (dimethyldiallyl ammonium hydroxide). This method has an advantage over other reported methods in the fact that in it controlled hydrolysis and condensation reactions of titanium (IV) bis (ammonium lactate) dihydroxide are possible, thereby enabling controlled coating of the nanocore. Reverse micelle and sol–gel techniques are also employed in the synthesis of metal–metal oxide core-shell nanoparticles. The inorganic coatings around the nanoparticles modify the optical properties of the systems in addition to stabilizing them against coalescence. Core-shell geometry allows shell functionalization too (Ref. 12) for better re-dispersibility and ease of handling using appropriate organic monolayers. Figure 9.4 depicts the high-resolution TEM image of a stearate functionalized Ag@ZrO_2 core-shell nanoparticle. The distinct core-shell geometry is visible from the image, though the organic monolayer build-up around the ZrO_2 shell is not clearly seen.

The inorganics-coated particles can catalyze redox reactions on their surface, which will be discussed in detail in Section 9.5.3. Heat dissipation from Au@SiO_2 core-shell nanoparticles in both water and ethanol were studied by Liz-Marzan, et al., (Ref. 13) using time-resolved spectroscopy. The characteristic time constant for heat dissipation depends on the thickness of the silica shell and the solvent. Chen, et al., (Ref. 14) have done the synthesis and characterization of Au@SiO_2 core-shell nanoparticles by incorporating a mercaptosilane at the core-shell interface. The highest degree of functional group organization at the core-shell interface was achieved using this methodology which involves exploiting the strong interaction between thiols and gold. Recently Wang, et al., (Ref. 15) synthesized Au-SiO_2 inverse opals by colloidal crystal templating. Inverse opals represent an excellent class of materials for the manipulation of the flow of light when their lattice parameter is comparable to the wavelength of electromagnetic waves. Inverse opals are highly ordered 3D macroporous structures that can exhibit a bandgap at optical wavelengths.

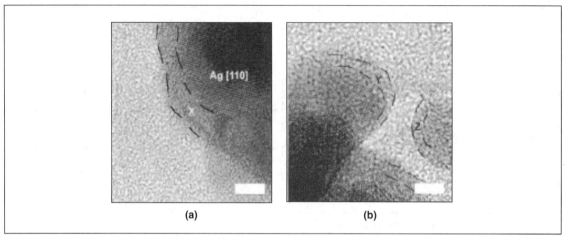

Fig. 9.4: *HRTEM images of Ag@ZrO$_2$ core-shell nanoparticles functionalized with a stearate monolayer. From Nair et. al. (Ref. 12). Reproduced by permission of the Royal Society of Chemistry.*

Figure 9.5 shows the transmission electron micrographs of Au@SiO$_2$ inverse opals described above. Images a, b and c have different shell thicknesses. Hodak, *et al.*, (Ref. 16) proved that the frequency of the breathing modes of core-shell nanoparticles strongly depend on the thickness of the shell. They found that a visible or near-UV pulse could be used to selectively excite small metal spheres, which coherently excites the acoustic vibrational modes as the photon energy is absorbed by the particles before any significant heat transfer to the dielectric shell. A theoretical model of the acoustic breathing modes of core-shell nanoparticles has recently been proposed by Sader, *et al.* (Ref. 17).

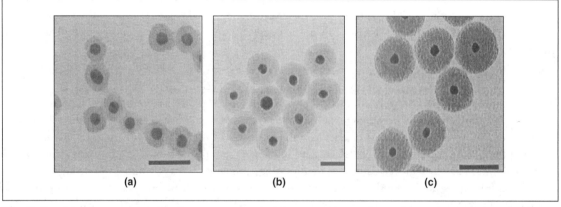

Fig. 9.5: *TEM images of Gold-silica inverse opals. The core dimension is ~15 nm and the silica shells are around 8, 18 and 28 nm, respectively. Scale bars are 50 nm in all cases. Reprinted from Ref. 15. Copyright (2002) Wiley-VCH.*

9.2.2 Bimetallic Core-shell Nanoparticles

Various bimetallic core-shell nanoparticles have been synthesized in the recent past because of their renewed interest in catalysis. Henglein (Ref. 18) synthesized core-shell alloys of Au-Pt through the simultaneous reduction of chloroauric acid and chloroplatinic acid. The TEM images of the Au@Pt core-shell nanoparticles (image on the right) are shown in Fig. 9.6, produced by the coating of Pt on Au nanoparticles (image on the left). Yonezawa, et al., (Ref. 19) reported the synthesis of Au core-Pt shell type nanoparticles through a single-step procedure. Schmid, et al. (Ref. 20) also reported the synthesis of the Au core-Pt shell nanoparticle by the simultaneous reduction of $PtCl_6^{2-}$ in an aqueous gold sol. γ-irradiation based synthesis and linear optical characterization of Ag/Cd, Ag/Pb and Ag/In core-shell nanoparticles were reported as early as 1980 by Henglein (Ref. 21). Mulvaney, et al. (Ref. 22) and Kreibig, et al. (Ref. 23) synthesized Ag core-Au shell nanosystems. Link, et al. (Ref. 24) synthesized Ag-Au alloy core-shell nanoparticles of 17–25 nm size. The synthesis of Au core-Ag shell nanoparticles and their optical characteristics like linear extinction and resonant hyper-Rayleigh scattering studies, were carried out by Kim, et al. (Ref. 25).

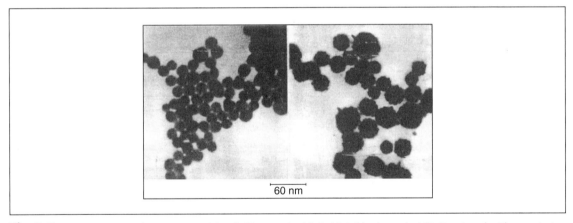

Fig. 9.6: *TEM images of Au nanoparticles (left) coated with Pt (right) in the ratio 1:2. Reprinted with permission from Henglin. (Ref. 18) Copyright (2000) American Chemical Society.*

Sobal, et al. (Ref. 26) showed that Ag core-Co shell exhibit optical behavior distinct from that of individual components. The presence of a noble metal also protects the Co shell against oxidation. The synthesis of Pt core-Co shell nanocrystals was also reported by Sobal, et al. (Ref. 27) by thermal decomposition of cobalt carbonyl in the presence of nanosized Pt dispersion. The thickness of the Co shell can be controlled by varying the amount of cobalt carbonyl. These types of systems are ideal for probing induced polarization of Pt at the core-shell interface. There are also several reports about the synthesis of various other core-shell nanoparticles in literature. A novel low temperature synthetic protocol for Cu@Au core-shell nanoparticles was developed by Cai, et al. (Ref. 28). These are extremely useful systems for the electrochemical DNA hybridization detection assay. The use of core-shell alloy nanoparticles combines the easy surface modification properties of Au with the good voltammetric response of Cu. The oxidative peak current of Cu colloid is much more than that of Au of the same size and quantity. Also, one layer of Au coated on the Cu core is sufficient for protecting Cu from oxidative degradation and can also provide an active surface for the immobilization of oligonucleotides.

The synthesis of dumb-bell shaped Au-Ag core-shell nanorods by seed-mediated growth under alkaline conditions was reported by Huang and Matijevic (Ref. 29). The method uses gold nanorods as seeds in the presence of Ag and ascorbate ions. The Ag ions that are being reduced by ascorbate ions become deposited on the surface of the Au nanorods to form dumb bell shaped Au-Ag core-shell nanorods. Recently, the synthesis of Au-Ag core-shell nanoparticles using tyrosine as a pH dependent reducing agent was reported by Sastry, *et al.* (Ref. 30).

9.2.3 Semiconductor Core-shell Nanoparticles

Semiconductor nanocrystals exhibit interesting size-dependent optical properties because of the confinement of electronic wavefunctions. Control of their surface is the key to highly luminescent nanocrystals. This is because of the presence of a large number of surface defects arising from nonstochiometry, unsaturated bonds, etc. Core-shell type composite quantum dots exhibit novel properties which make them attractive for chemists. Overcoating the nanocrystallites with higher band gap inorganic materials has been shown to increase the photoluminescence quantum yield due to the passivation of surface non-radiative recombination sites. Also, particles passivated with inorganic composite materials are much more reliable and robust than their corresponding organic analogues. The synthesis and characterization of strongly luminescent CdSe-ZnS core-shell nanocrystals were reported by Hines and Guyot-Sionnest (Ref. 31). Layered and composite semiconductor nanocrystals have been widely studied by several groups. Epitaxially grown CdSe/CdS core-shell nanocrystals with high luminescence properties and photostabilities were reported by Alivisatos, *et al.* (Ref. 32). Similarly the synthesis and characterization of highly luminescent CdSe-ZnS core-shell quantum dots were reported by Bawendi, *et al.* (Ref. 33). One pot synthesis of highly luminescent CdSe/CdS core-shell nanocrystals *via* organometallic and greener chemical approaches was reported by Weller, *et al.* (Ref. 34). Molecular nanocluster analogues of CdSe/ZnSe and CdTe/ZnTe core-shell nanoparticles were reported by DeGroot, *et al.* (Ref. 35). The off-resonance optical non-linearities of Au@CdS core-shell nanoparticles, embedded in BaTiO$_3$ thin films, were reported by Yang, *et al.* (Ref. 36). Semiconductor nanocrystallites are studied extensively because of their large third order non-linearities and ultra-fast non-linear optical response.

9.2.4 Polymer-coated Core-shell Nanoparticles

Polymer-coated core-shell nanoparticles have interesting applications ranging from catalysis to industry in making additives, paints and pigments and a host of other materials. The synthetic methodologies adopted for polymer capping of nanoparticles are of two main classes, namely: (a) polymerization at the nanoparticle surface, (b) adsorption of pre-formed polymer onto the nanoparticle cores. The monomer adsorption onto nanoparticles followed by polymerization (Refs 28, 34, 36–39) heterocoagulation-polymerization

(Ref. 40) and emulsion polymerization (Refs 41–44) are the most commonly adopted methods for the polymerization leading to core-shell nanostructures. Polymerization can be catalyzed by an initiator or by colloidal particles themselves. Huang and Matijevic (Ref. 29) reported the synthesis of silica particles coated with polydivinyl benzene layers by a pre-treatment method. Using similar approaches, poly (vinylbenzyl chloride), copolymers of polydivinyl benzene-poly (vinylbenzyl chloride) and double shells of polydivinyl benzene and poly (vinylbenzyl chloride) were also synthesized. Polymer coating of the particles allowed cores incorporating dyes to be retained on the nanocores as the polymer shells are permeable to small inorganic ions. Polymerization can also be achieved in the presence of catalytically active cores. This approach was utilized in the synthesis of poly (pyrrole) coated nanoparticles of catalytically-active cores like Fe_2O_3/ceria. Hematite (α-Fe_2O_3), silica-modified hematite and cerium (IV) oxide were coated with polypyrrole by using exactly the same procedures. Polypyrrole-coated α-Fe_2O_3 and CeO_2 are electrically conductive core-shell nanosystems. Uniform coatings can be achieved by this method as is shown in Fig. 9.7. Figure 9.7 shows the core-shell nanoparticles of SiO_2-polypyrrole with distinct and robust polypyrrole shell around SiO_2 core.

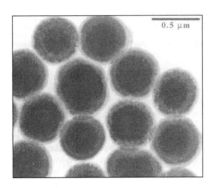

Fig. 9.7: *TEM of polypyrrole coated SiO_2 core-shell nanoparticles. Reproduced from Ref. 29, Copyright 1995 Materials Research Society, also published in F. Caruso, Adv. Mater., 2001, 13, 11. Copyright (2001) Wiley-VCH.*

The thickness of the polymer coating can be controlled by varying the contact time with the core. The polymer thickness is also dependent on the type of core employed and the nature of the polymer (Ref. 29). An excellent strategy for the synthesis of polymer coated nanoparticles was developed by Feldheim, et al. (Ref. 40). This is followed by trapping and aligning the particles in the pores of the membrane by vacuum filtration followed by polymerization of the monomer inside the pores. Au nanoparticles can be coated with polypyrrole by this method and the corresponding TEM images are shown in Fig. 9.8. The increased shell thickness of the polymer shell as a result of change in the monomer is evident from the figure on the right. The attractive feature of this methodology is that it facilitates controlling the thickness and composition of the polymer coating. Usually in almost all these polymerization reactions, the thickness of the coating depends on the polymerization time. Another widely used method for the polymer coating is the heterocoagulation of small particles with larger ones, followed by heating.

Polystyrene core coated with a uniform layer of polybutylmethacrylate is prepared by this method. Emulsion polymerization is another widely accepted strategy for the synthesis of polymer-capped core-shell nanoparticles.

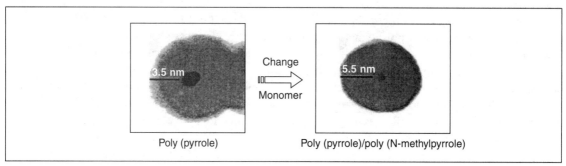

Fig. 9.8: *TEM of polypyrrole-capped Au nanoparticles (left) and with further increase in shell thickness by polymerization with poly (N-methylpyrrole). Reprinted with permission from Marinakos, et al. (Ref. 40) Copyright (1999) American Chemical Society.*

Sub nanometer- and micrometer-sized organic and inorganic particles can be coated with polymer by this method. The polymerization of styrene and/or methacrylic acid in emulsions of oleic acid resulted in the formation of a uniform polymer layer around the metal core. Unlike the uncoated particles, the polymer-coated particles can be easily centrifuged and re-dispersed. They usually exhibit strong resistance to etching. Polymer-coated nanoparticles are very easy to modify in the form of thin films by using self-assembly techniques. Layer by layer (LbL) templating strategy is commonly used for the same. Here a polymer solution (having an opposite charge to that on the particles) in excess to that required for saturation adsorption, was added to the colloidal dispersion. The polymer is adsorbing to the nanoparticles by electrostatic interactions. The LbL assembly of the polystyrene capped Au nanoparticles is shown in Fig. 9.9.

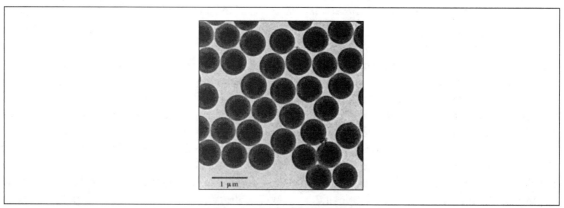

Fig. 9.9: *TEM of LbL assembled polystyrene-capped Au nanoparticles. Reproduced from F. Caruso, Adv. Mater., 2001, 13, 11. Copyright (2001) Wiley-VCH.*

9.3 Characterization

9.3.1 X-ray Diffraction (XRD)

Powder X-ray diffraction is one of the powerful techniques for the characterization of core-shell nanoparticles. In the case of materials where the core and the shell are crystalline, diffraction patterns from the prominent lattice planes can be seen in the diffractograms. A representative powder XRD pattern (Ref. 7) of Pt/Fe_2O_3 core-shell particles is shown in Fig. 9.10. The X-ray diffraction peaks at $2\theta = 39.8°$, $46.3°, 67.5°$ and $81.3°$ can be assigned to (111), (200), (220) and (311) planes of cubic phase of Pt particles. The Fe_2O_3 features are not seen because of the strong scattering from Pt nanoparticles. The XRD patterns of Ag@ZrO_2 and Ag@TiO_2 core-shell nanoparticles reveal the diffractions from the crystalline core and the shell, however (Ref. 6).

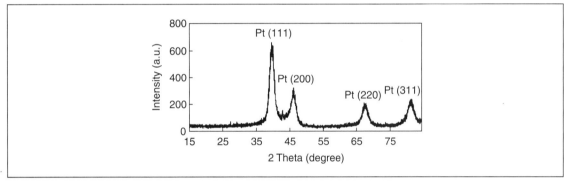

Fig. 9.10: *Powder XRD pattern of Pt@Fe_2O_3 core-shell nanoparticles. Reprinted with permission from Teng, et al. (Ref. 7) Copyright (2003) American Chemical Society.*

9.3.2 Optical Spectroscopy

Optical spectroscopy or absorption spectroscopy is another very important tool used for the characterization of nanomaterials. All the nanostructured materials exhibit unique and complex optical properties. We will confine ourselves to metal nanoparticles here. The most striking phenomenon encountered in these structures is the electromagnetic resonances resulting from the collective oscillations of conduction band electrons called plasmons. Plasmon modes vary depending on the geometry and are studied especially in noble metals like silver, gold and copper. The electrons in these metals originate from the completely filled d bands, which are relatively close to the Fermi energy. Since the diameter of the nanoparticle is of the order of the penetration depth of electromagnetic waves in metals, the excitation light is able to penetrate the particles. The field inside the particle shifts the conduction electrons collectively with respect to the fixed positive charge of the lattice ions. The electrons build-up a charge on the surface at one side of the particle. The attraction of this negative charge and positive charge of the remaining lattice ions on the opposite side results in a restoring force. If the frequency of the excitation light field is in resonance with the eigen frequency of this collective oscillation, even a small exciting field leads to strong oscillation. The resonance frequency mainly depends on the restoring force. This force, in turn, depends on the separation of the surface charges, i.e. the particle size, and the polarizability of the medium between and around the charges, which, in turn, depends on the embedding medium and the polarizability of the core electrons of the metal particle. The alternating surface charges effectively form an oscillating dipole, which radiates electromagnetic waves. The core-shell nanoparticles of noble metals and most of semiconductor nanoparticles are characterized by surface plasmon resonance, which results in strong absorption characteristics in the visible or UV region. The absorption characteristics are tunable depending on the nature and thickness of the shell material, which is described in detail in Section 9.4.2.

9.3.3 Zeta Potential

Due to dipolar characteristics and ionic attributes, the colloidal particles (including nanoparticles) suspended in solvents are electrically charged. For example, the surface groups of a colloid may be ionized (COO^-, NH_4^+, etc.). This leads to a net electric charge at the surface, which causes the accumulation of opposite charges (counter ions) around them. This, in turn, results in an electrical double layer. The ion (with positive or negative charge) with a set of counter ions form a fixed part of the double layer. The diffuse or mobile part of the double layer consists of ions of different polarities, which extends into the liquid phase. This double layer may also be considered to have two parts, an inner region which includes ions bound relatively strongly to the surface, and a diffuse region in which the ion distribution is determined by balance of electrostatic forces and random thermal motion. When an electric field is applied, the particles are attracted towards electrodes depending upon their polarity. The potential at which the fixed part of the double layer along with a part of the mobile layer move towards an electrode, is called Zeta potential or electrokinetic potential. It can also be defined as the potential at the shear plane of the particle when it moves in the medium.

The zeta potential depends on a number of parameters like surface charges, ions adsorbed at the interface and the nature and composition of the surrounding medium. The net charge in a specific medium depends on the particle charge and counter ions. The zeta potential is an index of interaction between the particles.

Zeta potential is calculated according to, Smoluchowski's Formula.

$\zeta = 4\pi\eta/\varepsilon \times U \times 300 \times 300 \times 1000$,

where ζ = zeta potential in mV, ε = dielectric constant of the medium, η = viscosity of solution, U = electrophoretic mobility ($v/V/L$), v = speed of the particles in the electric field in cm/s, V = applied voltage and L = distance of the electrode.

The measure of zeta potential throws light on the stability of colloidal/nanoparticle solutions. If all the particles in a suspension have large negative or positive zeta values, then they will repel each other and there will not be any tendency to flocculate. However, if the particles have low zeta potential values, then there is no force to prevent the particles from coagulating. The threshold of stability of colloidal/nanoparticle solution in terms of the zeta potential is ± 30 mV. The greater the zeta potential, the greater will be the stability. The value of the zeta potential is largely affected by pH.

The zeta potential is traditionally measured by using the 'micro electrophoresis' method, which needs extreme dilutions and hence stringent sample handling requirements. Microelectrophoresis is a technique based on light scattering by particles. In the case of nanoparticle solutions, however, the microelectrophoresis is not ideal due to the Doppler broadening of the scattered light from the fine particles. Modern methods used for zeta potential measurements are based on electro-acoustic methods based on electrokinetic properties. In this method, the application of a high frequency electric field sets in motion the eletrophoretic movements of the particles. This generates an alternating acoustic wave due to the density difference between particles and the medium. The velocity of the particles is measured by using laser Doppler electrophoresis. The velocity of these particles or mobility is converted into the zeta potential using Henry's equation:

$U = 2\varepsilon z f(ka)/3\eta$, where ε = dielectric constant, z = zeta potential, η = viscosity and $f(ka)$ = Henry's function. Zeta potential measurements in aqueous media and moderate electrolyte concentration generally employ $f(ka)$ value of 1.5 (Smoluchowski approximation). $f(ka)$ value is generally taken as 1 for the measurements of zeta potentials of small particles in non-aqueous media (Huckel approximation). The zeta potential measurement by microelectrophoresis is a passive technique as it does not alter the chemical properties of the systems.

9.4 Properties

9.4.1 Electrochemistry

The shells in core-shell nanoparticles are porous and hence permit electron transport through them. Ag@ZrO$_2$ shows a characteristic anodic peak at 0.310 V and a cathodic peak at 0.120 V in the solution phase containing 0.1M tetrabutyl ammonium hexafluorophosphate (TBAHFP)/CH$_3$CN on the Pt electrode surface (Ref. 45). The redox couple is centered at $E_{1/2}$ = 0.215 V vs. Ag/AgCl with a peak separation of 0.190 V at 25°C for the anodic curve (trace a in Fig. 9.11 A). The sharp and symmetrical anodic peak at 0.310 V with FWHM (full width-half maximum) of 60 mV suggests one electron transfer of silver nanoclusters. The quasi-reversible peaks corresponding to the oxidation and reduction of Ag clusters can be represented as:

$$Ag_n \rightarrow Ag_n^+ + e.$$

In the case of Au, the peak was observed at $E_{p1/2}$ = 320 mV (Ref. 45). The electron transport is sensitive to the shell matrix. With the adsorption of long chain fatty acids like stearic acid on the shell matrix of ZrO$_2$, electron transport through the shell is hindered. Traces (a-f) in Fig. 9.11 (A and B) show a decrease in the peak current of Ag@ZrO$_2$ and Au@ZrO$_2$, respectively (Ref. 6) with an increase in the stearic acid concentration. Similar trends occur with the adsorption of dyes on shell surfaces as well. The porosity of core-shell nanoparticles was extensively investigated by using cyclic voltammetry and absorption spectroscopy. Small molecules (whose sizes are equivalent to or less than the pore diameter of the shells) which react/interact with the metal cores, penetrate through the shell. The subsequent changes in the metal core caused by these interactions were manifested in the absorption as well as redox characteristics of the core (Ref. 45). Electro chemistry of core-shell particles has been investigated by Koktysh, et al. (Ref. 46) also.

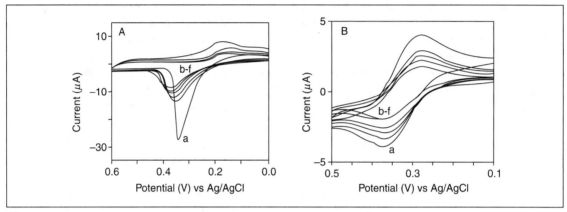

Fig. 9.11: *Cyclic voltammetry responses of core-shell nanosystems. Reprinted from Tom, et al. (Ref. 6). Copyright (2003) American Chemical Society.*

9.4.2 Optical Properties

Optical properties of nanoparticles

The optical properties of nanoparticles have been extensively investigated in recent years. When an electromagnetic wave passes through a metal particle, the electronic and vibrational states get excited. The optical interaction induces a dipole moment that oscillates coherently at the frequency of the incident wave. The frequency of this oscillation depends on the electron density, its effective mass, the shape and size of the charge undergoing oscillation. There can be other influences such as those due to other electrons in the system. The restoring force arises from the displacement of the electron cloud relative to the nuclei, which results in the oscillation of the electron cloud relative to the nuclear framework. The collective oscillation of the free conduction electrons is called 'plasmon resonance' or 'dipole plasmon resonance' of the particle, and is schematically depicted in Fig. 9.12 (Ref. 47). In this resonance, the total electron cloud moves with the applied field. There can be higher modes of plasmon resonance as well. In the quandrupole mode, half the electron cloud is parallel while the other half is anti-parallel to the applied field.

The dipole plasmon frequency is related to the dielectric constant of the metal. The frequency dependent dielectric constant of a bulk metal $\varepsilon(w)$ is measurable. In the discussion, in order to simplify matters, we consider a spherical particle whose diameter is much smaller than the wavelength of the electromagnetic radiation. As can be understood from Fig. 9.12, under such conditions, the electric field of light felt by the particles can be regarded as a constant. This reduces the interaction to be treated by electrostatics, rather

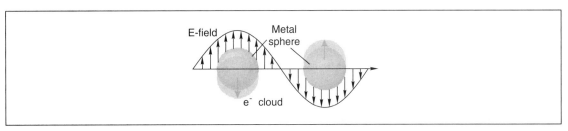

Fig. 9.12: *Surface plasmon resonance in metal nanoparticles in an electromagnetic field. The displacement of the conduction band electrons relative to the nuclei can be seen. Reprinted with permission from Kelly, et al. (Ref. 47) Copyright (2003) American Chemical Society.*

than electrodynamics. This treatment is referred to as the quasi-static approximation—'quasi' because we consider the wavelength dependent dielectric constant. In electrostatic theory when the incident electric filed of the radiation interacts with the electrons, we get a net field due to the applied field and its induced field. This field in reality is a radiating one and contributes to extinction and Rayleigh scattering by the particle. The strength of extinction and scattering can be given in terms of their efficiencies.

Extinction efficiency, $Q_{Ext} = 4 \times Im g_d$

Scattering efficiency, $Q_{scat} = (8/3) x^4 |g_d|^2$

Where $x = 2\pi R \varepsilon_m / \lambda$, $g_d = (\varepsilon_c - \varepsilon_m)/(\varepsilon_c + 2\varepsilon_m)$, ε_c and ε_m are the dielectric constants of the metal and the medium, respectively, R is the particle radius. Dielectric functions are complex quantities and Im refers to the imaginary part.

The efficiency = cross section/area (πR^2).

In particles of diameter, less than 10 nm light scattering does not make a significant contribution.

$Q_{Ext} \sim Q_{Abs} = [4(2\pi R \varepsilon_0^{1/2})/Im[(\varepsilon_c - \varepsilon_m)/(\varepsilon_c + 2\varepsilon_m)]$

When $\varepsilon_c = -2\varepsilon_m$, we get the resonance condition when Q_{Abs} goes to a maximum. Since the dielectric function is a complex quantity this equation can be given in terms of the real and imaginary parts of the metals dielectric function, ε' and ε'', respectively. There are two distinct size regimes of the particles; in both the plasmon resonance depends on size. For particles larger than 10 nm in diameter, the dielectric function itself is independent of size. The shape and size dependence of plasmon resonance in this regime is due to the dependence of electrodynamics on size and shape. This is called the extrinsic size regime. In the intrinsic regime, for particles less than 10 nm in diameter, the dielectric function itself changes with size.

For metals, the absorption characteristics depend, to a large extent, on the conduction band electrons. The spatial confinement of the free conduction band electrons results in plasmon excitations that are restricted to a small range of frequencies, usually in the UV-visible region. Bulk metals absorb very strongly in the IR or near IR region, but colloidal metals are transparent.

Assuming the metal to have a simple dielectric function, $\varepsilon'_c = \varepsilon_\infty - \lambda^2/\lambda_p^2$ where ε' refer to the real part (and ε'', the imaginary part) of the dielectric function. Then the peak position of the absorption obeys the equation, $\lambda_{peak}^2 = \lambda_p^2 (\varepsilon_\infty + 2\varepsilon_m)$, where ε_∞ is the high frequency value of the dielectric function of the metal and λ_p is the bulk plasma wavelength of the metal. (The bulk plasma wavelength of the metal is given by $\lambda_p^2 = 4\pi^2 c^2 m\varepsilon_0/Ne^2$, where N is the electron concentration, 'm' is the effective mass of the conduction band electrons and ε_0 is the vacuum permittivity.) ε_∞ is determined by all the transitions within the metal at UV and higher frequencies. It is clear from the equation that the peak maximum depends on ε_m (solvent dielectric constant) and changes in λ_p. The most important parameter which governs λ_p is N. Variations in electron charge density will alter the plasma frequency and results in shifts in peak maximum. Similarly the particles in the media of varying refractive index can also alter the absorption maxima.

A dispersion diagram showing the conditions of surface plasmon resonance as a function of wavelength of the incident light is shown in Fig. 9.13. The figure shows a plot of the real part of the dielectric function for two metals with two different values of λ_p, determined by $\varepsilon_c' = \varepsilon_\infty - \lambda^2/\lambda_p^2$, keeping a constant value of 1 for ε_∞. The variation of the dielectric function is shown by the curve for the two metals. When the dielectric constant of the medium is ε_{m1}, we get resonance at point A for metal 1, but at point B for metal 2. For the medium with dielectric constant, ε_{m2} the points are C and D. Also if the alloying of the metals 1 and 2 is assumed, the alloy tends to have a dielectric function intermediate between that of 1 and 2, and therefore, the plasmon band of the alloy will peak between that of 1 and 2 depending on the total electron concentration.

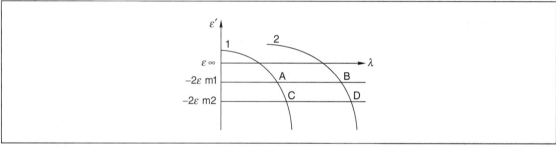

Fig. 9.13: *Dispersion diagram showing the conditions of surface plasmon resonance absorption as a function of the wavelength of the incident light. Reproduced from Mulvany, et al. (Ref. 8). Reproduced by permission of the Royal Society of Chemistry.*

Optical properties of core-shell nanoparticles

The optical characteristics of core-shell nanoparticles can be explained on the basis of the modified Mie's theory (Ref. 48), which considers the scattering of light from spheres coated with a homogeneous layer of uniform thickness. The absorbing medium of this layer has dielectric properties that are different from those of the core and the surrounding medium. The electromagnetic radiations scattered from the core-shell geometry can be described in the same form as that scattered from a homogeneous sphere by considering the influence of the radial variation of scattering coefficients on the extinction cross section. For the nanometer-sized objects of core-shell nanoparticles considered here, the particles may be described as dipole oscillators. The extinction efficiency can be calculated as:

$$Q_{ext} = 4x \, Im\left[\frac{(\varepsilon_s - \varepsilon_m)(\varepsilon_c + 2\varepsilon_s) + (1-g)(\varepsilon_c - \varepsilon_s)(\varepsilon_m + 2\varepsilon_s)}{(\varepsilon_s + 2\varepsilon_s)(\varepsilon_c + 2\varepsilon_s) + (1-g)(\varepsilon_s - 2\varepsilon_m)(\varepsilon_c - \varepsilon_s)}\right]$$ where the subscript 'c' refers to the core and 's' to the shell layer. The radius R refers to the coated particle radius and 'g' is the volume fraction of the shell layer. In the above expression, ε_c is a complex function, while ε_s and ε_m are real. The existence of shell layer modifies the surface plasmon resonance condition and the surface plasmon resonance occurs when the denominator of the above equation is zero, i.e. $\varepsilon_c' = -2\varepsilon_s [\varepsilon_s g + \varepsilon_m(3-g)]/[\varepsilon_s(3-2g) + 2\varepsilon_m g]$. For thin shells, $g \ll 1$ the condition of surface plasmon resonance becomes: $\varepsilon_c' = -2\varepsilon_m [2g(\varepsilon_s - \varepsilon_m)/3]$. This gets reduced to the usual resonance condition for small spheres without coating as $g \to 0$. Taking the metal to have a simple dielectric function, $\varepsilon_c' = \varepsilon_\infty - \lambda^2/\lambda_p^2$, where λ_p is the metal's bulk plasma wavelength, the peak position can be determined by:

$$\lambda_{peak}^2/\lambda_p^2 = \varepsilon_\infty + 2\varepsilon_m + [2g(\varepsilon_s - \varepsilon_m)/3].$$

A plot of λ_{peak}^2 vs. ε_m will now have a slope less than 2. Compare this with the case of isolated particles (without shell). Physically this is due to the fact that the shell layer is also polarizable by light and sets up a dipole. This would cause a red shift or it may act to reduce the core polarization thereby giving a blue shift as compared to the bare nanoparticle plasmon resonance. In the limit of very thick shells, $g \to 1$ the surface plasmon band becomes insensitive to the medium. The resonace is similar to a core immersed in a solvent made of the shell material. This becomes apparent when we consider the decreasing core volume fraction f. Since $f = 1-g$ and $f \ll 1$, the resonance condition can be written as,

$$\varepsilon_c' = -2\varepsilon_m [\varepsilon_s(1-f) + \varepsilon_m(2+f)/\varepsilon_s(1+f) + 2\varepsilon_m(1-f)].$$ For small f,

$$\varepsilon_c' = -2\varepsilon_m [(\varepsilon_s + 2\varepsilon_m) + f(\varepsilon_m - \varepsilon_s)/(\varepsilon_s + 2\varepsilon_m)].$$ As $f \to 0$, the surface plasmon resonance shifts towards a limiting value of $\varepsilon_c = -2\varepsilon_s$, similar to the case where the shell is the medium for the particles. The surface plasmon resonance peak position becomes:

$$\lambda_{peak}^2/\lambda_p^2 = \varepsilon_\infty + 2\varepsilon_s [(\varepsilon_s + 2\varepsilon_m) + f(\varepsilon_m - \varepsilon_s)/(\varepsilon_s + 2\varepsilon_m)].$$ Taking values of $\varepsilon_s = 2.25$, and $\varepsilon_m = 1.7$ to 3, and with $f = 0.05$, a plot of λ_{peak}^2 vs. ε_m will have a slope of less than 0.1, i.e. the surface plasmon mode will show no sensitivity to the solvent refractive index. The degree of scattering by the shell layer can be controlled by altering the refractive index of the solvent. This means that even if the particles are grown to micron size by silica deposition, the sol can still exhibit the optical properties of the nanoparticle cores. Figure 9.14 (top) shows the changes in the optical absorption spectra of Au@SiO$_2$ nanoparticles as a function of the increasing shell thicknesses of the SiO$_2$ shell. An increase in the background of the absorption and shift in the absorption maximum accompanied by broadening due to strong scattering are evident from the traces. Figure 9.14 (bottom) shows the absorption characteristics of Au@TiO$_2$ (a) and Ag@ZrO$_2$ (b) nanoparticles. Figure 9.14(a) shows the absorption spectral characteristics as a function of the increase in core dimension while Fig. 9.14(b) shows the corresponding changes as a function of shell thicknesses.

The color of the core-shell solutions of nanoparticles changes by varying the shell thickness, as shown in Fig. 9.15 (Plate 5). The transmitted and reflected colors of thin films of Au@SiO$_2$ colloids as a function of silica shell thicknesses are shown in the Fig. 9.15 (Plate 5) (Ref. 49).

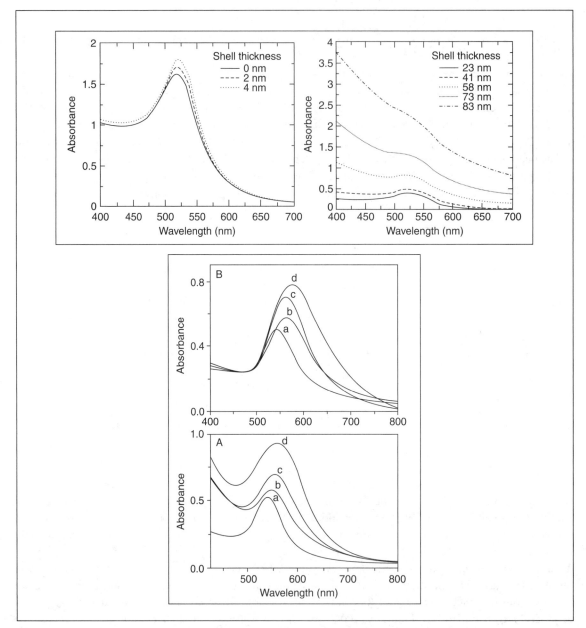

Fig. 9.14: *The absorption spectra of Au@SiO$_2$ colloids as a function of solvent refractive index (top figures) and Ag@TiO$_2$ (bottom (A)) and Ag@ZrO$_2$ (bottom (B)) as a function of increasing core dimension, a–d (A) and increasing shell thickness, a–d (B). Reprinted with permission from Liz-Marzan, et al. (Ref. 2) (top) and Tom, et al. (Ref. 6) (bottom). Copyright (1996 and 2003, respectively) American Chemical Society.*

9.4.3 Optical Non-linearity

There is a considerable interest in the role of non-linear optical materials in manipulating optical signals in optical communication and optical signal processing applications. Organic non-linear optical materials can be used for high density data storage, phase conjugation, holography, and spatial light modulation. Among the non-linear optical applications, optical limiting has been the most promising. Optical limiters have a transmission that varies with the incident intensity of light. Transmission is high at normal light intensities, but low for intense beams. The intensity-dependent transmission can limit the transmitted light intensity so that it is always below some threshold, and hence the name 'optical limiting'. This is useful in protecting elements that are sensitive to the sudden high intensity light, such as optical elements, sensors, the human eye and other sensitive devices. A non-linear absorption wherein the material's absorbance increases with the intensity of the incident light is obviously useful for good optical limiting. One mechanism that has been particularly useful for optical limiting is two-photon absorption and is shown by most of the materials. A simple energy level diagram where this can occur is shown in Fig. 9.16, depicting the 'reverse saturable absorption (RSA)' or excited state absorption (Refs 50–52).

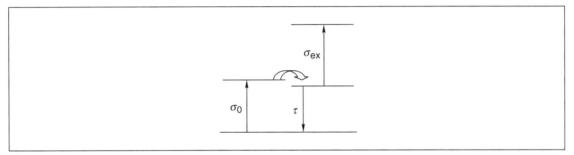

Fig. 9.16: *Schematic of two-photon absorption mechanism of optical limiting.*

In case when the incident light intensity is sufficiently high to allow a significant population to accumulate in the excited state and if the material has an excited state absorption cross section, σ_{ex} which is larger than the ground state absorption cross section, σ_o, the effective absorption coefficient of the material increases. The excited state may relax and subsequent absorption may occur from another state. In order to achieve largest non-linear absorption, both a large excited state absorption cross section and a long excited state life-time, τ are required, else the excited state population will not be high to increase net absorption. This sequential two-photon absorption is also called a large, positive non-linear absorption. A large σ_{ex} and a large difference between the ground and excited state absorption cross sections are essential for this to occur. The non-linear response should possess a low threshold value and remain large over a range of fluences before the non-linearity saturates. A high saturation fluence normally requires a high concentration of the non-linear material in the optical beam. Since no communication from the sensor or from any other device is required for the optical limiting devices to become active, they provide a type of 'smart' protection. Increased speed is particularly important for applications wherein the sensor may be exposed to sudden bursts of high intensity light. Optical limiting devices are used, for example, to protect optical sensors used in conjunction with pulsed laser systems. The most important difficulty with the usage of common optical limiters is their low damage threshold, that is, the materials themselves get damaged under the impact of high intensity laser pulses. Studies have shown that protected nanoparticles of silver, gold and Pd are good optical limiters in the nanosecond regime. They give materials with a high damage threshold when incorporated into polymer matrices. The utility of the core-shell nanoparticles in optical limiting is given in Section 9.5.6.

9.5 Applications

9.5.1 Biological Applications

Magnetic core-shell nanoparticles are used in clinical applications and biotechnology, as the shells can be made biocompatible. Such core-shell particles find use in hyperthermia as immunoassays. They also find extensive uses in the transportation of drugs to the sites of diseases, and function as magnetic carriers for the identification and isolation of blood cells and antibodies. Further they are used in imaging for the magnetic mapping of organs, tissue repair, detoxification of biological fluids, drug delivery and in cell separation. The biomedical applications of polymeric core-shell nanoparticles of poly(ethylene oxide-β-ε-caprolactone)(PEO-β-PCL) diblock copolymer have been demonstrated by Gan, et al. (Ref. 53). The engineering of polymeric materials at the nanometer scale holds great promise in various biological applications. Shell cross-linked nanoparticles consisting of a polyisoprene (PI) core and poly (acrylic acid-co-acrylamide) shell have been used in drug delivery (Ref. 54). Functionalized internal surface of the nanocage, obtained by the dissolution of the polystyrene cage by ozonolysis, which uses masked fluorescent compounds, has biomedical applications. The masked compound, conjugated to the nanocage becomes fluorescent, when treated with aminopeptidase. Core-shell nanoparticles serve as intriguing building blocks for designing chemically and biologically active materials for analytical applications. Nanostructured thin films of gold nanoparticles and molecular linkers of different binding properties, namely 1,9-nonanedithiol and 11-mercaptoundecanoic acid, act as vapor sensors since the vapor accumulation at the interface changes the electrical conductance. Shell cross-linked nanoparticles possessing novel surface functionalities are ideal candidates for probing multivalent contacts with bacterial moieties and cell membranes. Shell cross-linked nanoparticles, when developed as multifunctional delivery and sequestration vessels, find utility in the treatment of bacterial infections. Amphiphilic block copolymers, consisting of poly(γ-benzyl L-glutamate) (PBLG) as the hydrophobic part and poly(ethylene oxide) (PEO) [or poly(N-isopropylacrylamide) (PNIPAAm)] as the hydrophilic one, can self-assemble in water to form nanoparticles which show a slow release of hydrophobic drugs from these polymeric nanoparticles. The entrapment of proteins in biodegradable nanospheres of amphiphilic di-block copolymers obtained by the ring opening polymerization, have been widely investigated as sustained release formulations for proteins and anti-cancer drugs. Monodisperse core-shell maghemite nanoparticles protected with a thin layer of divinyl benzene cross-linked polystyrene molecules are used in biological applications. The formation of core-shell type nanoparticles of linear polymer-dendrimer block copolymer-plasmid DNA complexes are used in gene delivery applications *in vitro*. The super paramagnetic magnetite-silica core-shell nanoparticles find extensive uses in biology because of the presence of the inert silica shell (Ref. 55). The development of a novel biological reporter system based on the surface enhanced Raman scattering properties of silica-gold core-shell nanoparticles, was reported by Jensen, *et al.* (Ref. 56). The SERS activity of the nanostructures can be tuned by changing the dielectric constant and size of both the core and the shell.

9.5.2 Magnetism of Core-shell Nanoparticles

The magnetism of core-shell nanoparticles can be tuned by varying the shell thickness. The $Fe_{58}Pt_{42}/Fe_3O_4$ core-shell nanoparticles are ferromagnetic at low temperatures, but super-paramagnetic at room temperature. Such materials find applications in ultra high density magnetic storage media, biological labeling and detection and drug delivery. The bimetallic core-shell nanoparticles exhibit new properties of the core-shell magnetic systems which are often distinct from those of the corresponding monometallic properties. Fe-Pt, Pt-Co and Pt_3-Co alloy core-shell systems show high magnetic anisotropy and chemical stability. Bimetallic colloids of this type are expected to reveal interesting magnetic behavior like an induced polarization at the bimetallic interface or a giant magnetoresistance effect. The super paramagnetism of γ-Fe_2O_3 nanoparticles has been utilized in making facile recyclable catalysts in Suzuki cross-coupling reactions (Ref. 57).

Core-shell type magnetic nanosystems are also of great interest from the technological point of view. Magnetic nanoparticles are used in various areas such as bearing, seals, lubricants, heat carriers, printing, recording and polishing media. One of the rapidly emerging applications of magnetic nanoparticles is in biologically labeled areas including MRI, drug delivery, rapid biological separations and therapy. The synthesis of magnetic nanoparticles is also generating interest for the purpose of research. Polymeric shells have some unique advantages because of the flexibility in the control of chemical compositions and functions of the polymers. Magnetic nanoparticles are being tested as contrast-enhancing agents for cancer imaging and therapy. Aqueous magnetic fluids composed of small magnetic particles, covered with biocompatible functionalized shells, find use in biology as mentioned in Section 9.5.1.

9.5.3 Catalysis

Catalysis by core-shell nanoparticles is a very important, active and emerging area, which can have a tremendous impact in the chemical industry, pharmaceutical products and the fuel sector. More than 90 per cent of all industrial processes are catalytically controlled. Enhancement in specific catalytic activity and selectivity coupled with a reduction in the cost of catalysts would benefit the industrial sector in a big way. Catalysis of core-shell particles can either be due to the core or due to the shell.

The specific features which were seen to enhance the utility of the nanoparticles in the catalytic industry were their large surface-to-volume ratio and the specific binding sites on them (Ref. 58). Nanosized gold has high catalytic activity in the oxidation and reduction of hydrocarbons. Monolayer-encapsulated Au and alloy core-shell nanoparticles are model building blocks for devising nanostructured catalysts.

Tremendous variation can be achieved in the catalytic activity by changing the composition of the core, size, shape and surface properties.

Catalysis by shell includes asymmetric catalysis and mediated electrotransfer. In catalysis by nanoparticle cores, three types of nanoparticles are of relevance. These are: (a) nanoparticles supported on oxides including core-shell nanoparticles or polymers, (b) nanoparticles encapsulated in dendrimers and (c) nanoparticles encapsulated in alkane thiolate monolayers. Nanosized Au supported on oxide surfaces showed a high degree of catalytic activity as compared to bulk gold, which is a poor catalyst. A typical example is the oxidation of CO. It is observed that the catalytic activity of Au is high when its size is around 5 nm. Partial electron transfer from the Au clusters' surface is believed to play an important role in the activation of nano gold clusters as catalysts. A decrease in the mean coordination number and the ready mobility of Au atoms on the surface could lead to greater chemisorptivity, larger coverage of O_2 and a stronger interaction with the support to create special gold sites near the support. The oxidation of CO was considered to occur preferentially on the peripheral edges between Au nanoparticles and oxide support. However, the exceptionally high catalytic activity of the nanosized gold in the restricted size range is still a puzzle for chemists. The porosity of the shells facilitates size-selective accessibility of molecules to the metal cores.

It is well known that nanoparticles incorporated in the dendrimer cavities exhibit remarkable catalytic activity and selectivity towards hydrogenation and C-C coupling reactions like Suzuki, Heck, and Sonogashira reactions. Such nanoparticles are prepared by trapping the metal ions within the dendrimer cavities followed by a chemical reduction to yield metal nanoparticles of desired size range, usually 1–5 nm (Ref. 59). Nanoparticle encapsulation in dendrimer cavities possess advantages like uniformity in structure, protection against agglomeration, and the access of only small molecules to the nanoparticles' surface.

In order to exploit core-shell nanoparticle-based catalysis, the most important issue that must be addressed is the surface passivation of nanoparticles. This can be enhanced by place exchange, interparticle linking, size processing, electrochemical, thermal and photochemical annealing.

9.5.4 Sensing

The most important challenge in the filed of chemical/biological sensors is the rational design of materials with high sensitivity and selectivity. Nanostructured materials provide challenging opportunities for addressing problems because of their new and unique interparticle spatial and chemical properties that can be fine-tuned with various parameters. The sensing properties are highly dependent on several design parameters such as particle size, interparticle distances and the dielectric constant of the surrounding medium. Sensors in the field of biology are in great demand now because of their potential uses in detecting the mutations of genes. Biosensor technologies for DNA are also in great demand because of the sensitivity and selectivity they offer. The construction of nanoarrays on the basis of design parameters facilitates real-world sensing applications like detection of toxic gases, explosives and toxins. Alternative layer-by-layer assemblies of core-shell nanostructured materials have found interesting new applications in sensors and actuators on a wide scale. Dye-embedded core-shell nanoparticles with surface Raman enhancement are excellent spectroscopic tags for detection protocols. These core-shell nanoparticles have a metallic core as an optical enhancer, a reporter molecule as a spectroscopic tag, and an inert shell for stabilization and conjugation. These find better applications in the sensing of biological warfare agents. Core-shell nanoparticles having a metallic core and an organic monolayer shell in ionic liquids have interesting applications such as optical sensors for anions, which work on the basis of aggregation-induced color changes. Thin films/membranes of pre-engineered gold nanoparticles derivatized with thiolate shells with carboxylate endings and a polymer, poly(2-hydroxyethyl methacrylate), are novel systems for chemical/biological sensing applications. Core-shell nanomaterials with gold/alloy/semiconductor nanocrystal cores and functional polymer shells, in solution state and thin films, have been studied as model systems for chemical and electrochemical sensing. Systems with the above-mentioned metal cores with thiols, thioethers, carboxylic acids, polymeric matrix, etc. as monolayers are excellent materials for the sensing of nitro aromatics and for the electrochemical detection of metal ions and biologically relevant molecules. The sensitivity, selectivity, detection limit and response time depend on a number of parameters like electronic conductivity, interfacial mass flux, binding specificity of the ligand and catalytic properties of the nano-framework. Such systems involve inter-core or inter-shell covalent linkages or non-covalent hydrogen bonding. The responses of material to volatile organic compounds, toxic gases and explosive vapors have been discussed by Han, *et al.* (Ref. 60). Organic-inorganic network assembly comprising gold nanoparticle core and organic linkers such as 1, 9-nonanedithiol and 1, 5-pentadithiol is responsive towards organic vapors and their nature of response depends on the chain length of the linker. These are chemically sensitive interfacial materials. Core-shell nanoparticle arrays cross-linked by the molecular receptor species act as sensing interfaces. Layered metal and semiconductor nanoparticle systems cross-linked by nucleic acids find extensive uses in the optical, electronic and photoelectrochemical detection of DNA. Similarly, nanoparticles incorporated in hydrogel matrixes yield new composite materials with novel sensing properties.

Bacteria, viruses and other micro-organisms responsible for diseases are also detectable via their unique and specific nucleic acid sequences. Metal nanoparticles are excellent candidates for the purpose because of their unique optical and electrical properties. Mirkin, *et al.* (Ref. 61) used hybridization-induced optical changes of oligonucleic acids-modified gold nanoparticles for devising DNA sensors. The electrochemical stripping detection of DNA by gold nanoparticles and further enhancement in signal amplification with silver have been reported by Cai, *et al.* (Ref. 28) and Wang, *et al.* (Ref. 15), respectively. The detection of glucose (as a sensor for blood glucose) with metal core-polymer shell system in an aqueous solution using surface-enhanced Raman scattering was reported by El Khoury, *et al.* (Ref. 62).

9.5.5 Chemical Reactivity

The chemical reactivity of core-shell nanoparticles is fascinating because they are excellent substrates for the conduct of C-C coupling reactions. Also, their specific chemistry leads to the metal core getting leached away by suitable molecules and thereby results in oxide shells or oxide nanobubbles (see Chapter 10) (Ref. 63). The synthesis, characterization and use of highly crystalline γ-Fe_2O_3 nanoparticles capped with a very thin polymer shell of polystyrene for the loading of Pd catalyst to facilitate Suzuki cross-coupling reactions was recently demonstrated by Stevens, *et al.* (Ref. 57). The most important aspect of the above method is that the catalyst can be recovered to the extent of 97 per cent by a permanent low-field magnet since the catalyst-loaded substrate is magnetic. Dendrimer-encapsulated Pd core-shell nanoparticles are used as catalysts for the Stille coupling reactions of aryl halides and organostannanes. The nanoparticle Pd catalyst brings about catalysis at room temperature very efficiently even with 0.10 per cent Pd incorporation. Core-shell nanoparticles of Pd embedded into aluminum hydroxide act as dual catalysts for alkene hydrogenation and aerobic alcohol oxidation. The construction of core-shell nanoassemblies on gold nanostructures via Sonagashira coupling reactions was reported by Xue, et al. (Ref. 64) Suzuki coupling reactions using colloidal Pt nanocrystals with polyvinyl pyrrolidone(PVP) capping were recently investigated by El-Sayed, et al. (Ref. 65). There is a marked difference in the catalysis of tetrahedral PVP-Pt nanoparticles versus spherical PVP-Pt nanoparticles towards Suzuki reactions. The effect of molecular mass of the capping agent on the reactivity and the shape transformation of Pd nanoparticles during catalysis was also investigated in detail by the authors. Polyethylene glycol-capped Pd catalyst is an ideal catalyst for Heck coupling reactions. The utility of core-shell nanoparticles of gold stabilized by poly (N-vinyl-2-pyrrolidone) shows a high efficiency in catalyzing C-C bond formation in the aerobic homocoupling of aryl boronic acid in water.

The reactions between core-shell nanoparticles and reactive molecules present an entirely different chemistry leading to new systems. The most widely studied core-shell nanosystems in this category are Ag@TiO_2, Au@TiO_2, Ag@ZrO_2, Au@ZrO_2 and Au@SiO_2, respectively. The oxide shells in the above core-shell nanoparticles are porous and therefore, allow ionic and molecular transport through the shell. The dissolution of the core in Ag@TiO_2 by NH_3 resulting in TiO_2 nanoshells is shown in Fig. 9.17(b). The core-shell geometry of Ag@TiO_2 nanoparticles is shown in Fig. 9.17(a). Reactions of Ag@ZrO_2 and Au@ZrO_2 core-shell nanoparticles with halocarbons also result in the selective leaching of the metal core to give oxide nanobubbles of ZrO_2 (Ref. 63). The time-dependent reaction between Ag@ZrO_2 core-shell nanoparticles with CCl_4 and benzylchloride results in nanobubbles (see Chapter 10). The surface

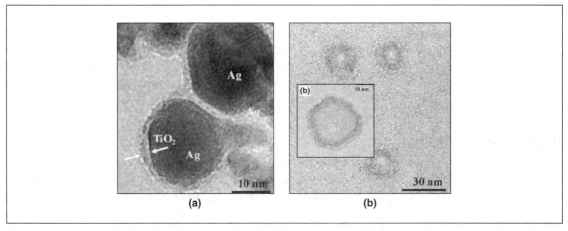

Fig. 9.17: *TEM images of Ag@TiO_2 core-shell nanoparticles (a) and the TiO_2 nanoshells formed from the same after leaching the core with NH_3. The inset of Fig. 9.17. (b) shows an expanded view of a TiO_2 shell. Reprinted from Ref. 46. Copyright (2002) Wiley-VCH.*

plasmon resonance of the Ag nanoparticles steadily decreases in intensity with the addition of the halocarbon, signifying the conversion of the Ag nanoparticles into silver halide. The dissolution of the metal core was also evident from cyclic voltammetry measurements. With the addition of the halocarbon, both the anodic and cathodic peak current intensity decrease and finally become flat. The decrease in peak current occurs because of the removal of the electro-active Ag nanoparticles in the form of AgCl from the system by the degradation of the halocarbon. The reactivity difference between CCl_4 and benzylchloride towards Ag@ZrO_2 is also evident from the absorption spectra. The ZrO_2 nanobubbles resulting from the reaction of CCl_4 with Ag@ZrO_2 core-shell nanoparticles can be seen in TEM images. Various types of carbonaceous materials were seen in the above reactions. The chemistry works with a series of other halocarbons like benzylchloride, $CHBr_3$, CH_2Cl_2, $CHCl_3$ and CCl_2F_2. The halocarbon reaction between Au@SiO_2 core-shell nanoparticles and CCl_4 results in carbon onions within SiO_2 nanoshells (Ref. 66). Increasing the shell thickness results in a decrease in the core-removal kinetics.

9.5.6 Optical Limiting

Optical sensors are important, for the light-sensitive devices that are used in the detection of light. However, they may be susceptible to damage if exposed to a high intensity of light, especially as in the case of a laser. Therefore, a proper protective device is essential for protecting the sensor at high light intensities. Nanomaterials constitute one of the best optical limiters discovered so far. The optical limiting characteristics of bare nanoparticles of silver and gold have been studied in detail by Mostafavi, et al. (Ref. 67). The optical limiting mechanisms of carbon nanotubes, carbon nanoparticles and carbon black suspensions have already been studied in detail. The biggest problem with bare nanoparticles is that they are prone to laser-induced damage at higher laser fluences. Therefore, the relevance of core-shell nanomaterials for this purpose has been considered as an alternative. The presence of inert shells around the core of nanoparticle protects them from higher laser powers as illustrated in Fig. 9.18. The optical limiting characteristics of thiol protected Au, Ag and Au/Ag alloy core-shell nanosystems have been investigated in detail. The core-shell nanomaterials are very good optical limiters and a comparative study of their optical limiting characteristics was attempted by Anija, et al. (Ref. 68). The most prominent mechanism observed in the case of core-shell nanoparticles for optical limiting is non-linear scattering, like in the case of bare metal nanoparticles or carbon nanotubes or carbon black suspensions. The optical limiting characteristics, as revealed by Z-scan experiments, of Au@ZrO_2 and Au@TiO_2 core-shell nanomaterials are shown in Fig. 9.19. The inset of the figure shows the Z-scan curve of Ag@ZrO_2. The z-scan technique is a method

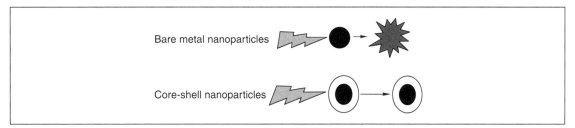

Fig. 9.18: *Schematic showing the stability of core-shell nanoparticles with intense laser fluences.*

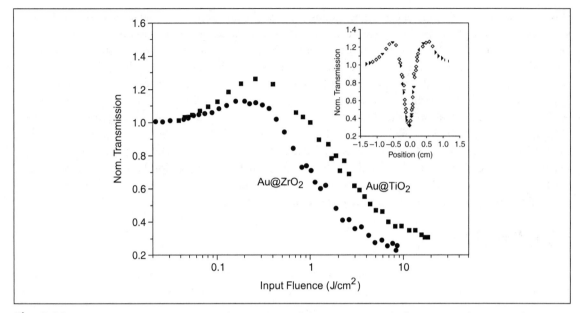

Fig. 9.19: *Optical limiting responses of Au@TiO$_2$ and Au@ZrO$_2$ core-shell nanoparticles using the Z-scan technique. Inset shows the Z-scan curve of Ag@ZrO$_2$ system. Reprinted with permission from Tom, et al. (Ref. 6). Copyright (2003) American Chemical Society.*

by which the sample is scanned along an axis (z-axis) while the laser is focused on it by a lens. At $z = 0$ the light intensity falling on the sample is a maximum. The intensity of the transmitted light is measured, which shows a minimum at $z = 0$, when light intensity is maximum. The optical limiting response of Au-Ag alloy systems with varying mole fractions of Au and Ag has also been investigated in detail by Nair, et al. (Ref. 69). The core-shell nanosystems incorporated into polymer matrices signify new approaches to help increase the limiting performance and damage threshold of optical limiting materials (Ref. 70). In the case of oxide-protected nanoclusters, no signs of damage were seen with laser pulses of fluences up to 20 J/cm^2 and intensities up to 2.8 GW/cm^2. However, the exact mechanism of energy transport, i.e. how the electrons communicate with the insulator shell, is not completely understood. Plasmon damping and electron-phonon scattering at the interface, however, may be important in these cases and more studies are being carried out to understand the electron transport mechanism in detail.

Review Questions

1. What are the principal core-shell systems?
2. How do we understand the optical properties of core-shell systems starting from the core?
3. What are intrinsic and extrinsic size regimes?
4. What are the essential aspects of Mie theory?

5. What are the principal applications of core-shell systems?
6. Why are they important in biology in contrast to simple particles?
7. Why core-shell particles are important in optical limiting applications?

References

1. Davies, R., G.A. Schurr, P. Meenan, R.D. Nelson, H.E. Bergna, C.A.S. Bervett and R.H. Goldbaum, *Adv. Mater.*, **10** (1998), p. 1264.
2. Liz-Marzan, L.M., M. Giersig and P. Mulvaney, *Langmuir*, **12** (1996), p. 4329.
3. Hofman-Caris, C.H.M., *New. J. Chem.*, **18** (1994), p. 1087.
4. Liz-Marzan, L.M., M. Giersig and P. Mulvaney, *Chem. Commun*, (1996), p. 731.
5. Pastoriza-Santos, I., D.S. Koktysh, A.A. Mamedov, M. Giersig, N.A. Kotov and L.M. Liz-Marzan, *Langmuir*, **16** (2000), p. 2731.
6. Tom, R.T., A.S. Nair, N. Singh, M. Aslam, C.L. Nagendra, R. Philip, K. Vijayamohanan and T. Pradeep, *Langmuir*, **19** (2003), p. 3439.
7. Teng, X., D. Black, N.J. Watkins, Y. Gao and H. Yang, *Nano Lett.*, **3** (2003), p. 261.
8. Mulvaney, P., L.M. Liz-Marzan, M. Giersig and T. Ung, *J. Mater. Chem.*, **10** 2000, p. 1259.
9. Liz-Marzan, L.M., M. Giersig and P. Mulvaney, *Langmuir*, **12** (1996), p. 4329.
10. Konya, Z., J. Zhu, A. Szegedi, I. Kiricsi, P. Alivisatos and G.A. Somorjai, *Chem. Commun.*, (2003), p. 314.
11. Mayya, K.S., D.I. Gittins and F. Caruso, *Chem. Mater.*, **13** (2001), p. 3833.
12. Nair, A.S., T. Pradeep and I. MacLaren, *J. Mater. Chem.*, **14** (2004), p. 857.
13. Hu, M., X. Wang, G.V. Hartland, V. Salgueirino-Maceira and L.M. Liz-Marzan, *Chem. Phys. Lett.*, **372** (2003), p. 767.
14. Chen, M.M.M. and A. Katz, *Langmuir*, **18** (2002), p. 8566.
15. Wang, D., V. Salgueirino-Maceira, L.M. Liz-Marzan and F. Caruso, *Adv. Mater.*, **14** (2002), p. 908.
16. Hodak, J.H., A. Henglein and G.V. Hartland, *J. Chem. Phys.*, **111** (1991), p. 8613.
17. Sader, J.E., G.V. Hartland and L.M. Liz-Marzan, *J. Phys. Chem. B*, **106** (2002), p. 1399.
18. Henglein, A., *J. Phys. Chem. B*, **104** (2000), p. 2201.
19. Yonezawa, T. and N. Toshima, *J. Mol. Catal.*, **83** (1993), p. 167.
20. Schmid, G., A. Lehnert, J.-O. Malm and J.-O. Bovin, *Angew. Chem. Int. Ed. Engl.*, **30** (1991), p. 874.
21. Henglein, A., *J. Phys. Chem.*, **83** (1979), p. 2209.
22. Mulvaney P., M. Giersig and A. Henglein, *J. Phys. Chem.*, **97** (1993), p. 7061.
23. Sinzig, J., U. Radtke, M. Quinten and U. Kreibig, *Zeitschrift Fur Physik D-Atoms Molecules and Clusters*, **26** (1993), p. 242.

24. Link, S., C. Burda, Z.L. Wang and M.A. El-Sayed, *J. Chem. Phys.*, **111** (1999), p. 1255.
25. Y. Kim, R.C. Johnson, J. Li, J.T. Hupp and G.C. Schatz, *Chem. Phys. Lett.*, **352** (2002), p. 421.
26. Sobal, N.M., M. Hilgendorff, H. Mohwald, M. Giersig, M. Spasova, T. Radetic and M. Farle, *Nano Lett.*, 2 (2002), p. 621.
27. Sobal, N.S., U. Ebels, H. Mohwald and M. Giersig, *J. Phys. Chem. B*, **107** (2003), p. 7351.
28. Cai, H., N. Zhu, Y. Jiang, P. He and Y. Fang, *Biosensors and Bioelectronics*, **18** (2003), p. 1311.
29. Huang, C.L. and E. Matijevic, *J. Mater. Res.*, **10** (1995), p. 1327.
30. Sastry, M., A. Swamy, S. Mandal and P. R. Selvakannan, *J. Mater. Chem.*, **15** (2005), p. 3161.
31. Hines, M.A. and P. Guyot-Sionnest, *J. Phys. Chem.*, **100** (1996), p. 468.
32. Peng, X., M.C. Schlamp, A.V. Kadavanich and A.P. Alivisatos, *J. Am. Chem. Soc.*, **119** (1997), p. 7019.
33. Dabbousi, B.O., J. Rodriguez-Viejo, F.V. Mikulec, J.R. Heine, H. Mattoussi, R. Ober, K.F. Jensen and M.G. Bawendi, *J. Phys. Chem. B*, **101** (1997), p. 9463.
34. Mekis, I., D.V. Talapin, A. Kornowski, M. Haase and H. Weller, *J. Phys. Chem. B*, **107** (2003), p. 7454.
35. DeGroot, M.W., N.J. Taylor and J.F. Corrigan, *J. Mater. Chem.*, **14** (2004), p. 654.
36. Yang, Y., J.I. Shi, H. Chen, S. Dai and Y. Liu, *Chem. Phys. Lett.*, **370** (2003), p. 1.
37. Sprycha, R., H.T. Oyama, A. Zelenev and E. Matijevic, *Colloid Polym. Sci.*, **273** (1995), p. 693.
38. Marinakos, S.M., L.C. Brousseau, A. Jones and D.L. Feldheim, *Chem. Mater.*, **10** (1998), p. 1214.
39. Marinakos S.M., D.A. Shultz and D.L. Feldheim, *Adv. Mater.*, **11** (1999), p. 34.
40. Marinakos, S.M., J.P. Novak, L.C. Brousseau, A.B. House, E.M. Edeki, J.C. Feldhaus and D.L. Feldheim, *J. Am. Chem. Soc.*, **121** (1999), p. 8518.
41. Ottewill, R.H., A.B. Schofield, J.A. Waters and N.S.J. Williams, *Colloid Polym. Sci.*, **275** (1997), p. 274.
42. Hergeth, W.D., U.J. Steinau, H.J. Bittrich, K. Schmutzler and S. Wartewig, *Prog. Colloid Polym. Sci.*, **85** (1991), p. 82.
43. van Herk, A.M., *NATO ASI ser. E*, **335** (1997), p. 435.
44. Quaroni, L. and G. Chumanov, *J. Am. Chem. Soc.*, **121** (1999), p. 10642.
45. Nair, A.S., R.T. Tom, V. Suryanarayanan and T. Pradeep, *J. Mater. Chem.*, **13** (2003), p. 297.
46. Koktysh, D.S., X. Liang, B.-G. Yun, I. Pastoriza-Santos, R.L. Matts, M. Giersig, C. Serra-Rodriguez, L.M. Liz-Marzan and N.A. Kotov, *Adv. Funct. Mater.*, **12** (2002), p. 255.
47. Kelly, K.L., E. Coronado, L.L. Zhao and G. C. Schatz, *J. Phys. Chem. B*, **107** (2003), p. 668.
48. Mie, G., *Ann. Phys.*, **25** (1908), p. 377.
49. Ung, T., L.M. Liz-Marzan and P. Mulvaney, *J. Phys. Chem. B*, **105** (2001), p. 3441.
50. Pittman, M., P. Plaza, M.M. Martin and Y.H. Meyer, *Opt. Commun.*, **158** (1998), p. 201.
51. Chen, P.L., I.V. Tomov, A.S. Dvornikov, M. Nakashima, J.F. Roach, D.M. Alabran and P.M. Rentzepis, *J. Phys. Chem.*, **100** (1996), p. 17507.

52. Philip, R., G. Ravindra Kumar, N. Sandhyarani and T. Pradeep, *Phys. Rev. B*, **62** (2001), p. 13160.
53. Gan, Z.H., T.F. Jim, M. Li, Z. Yuer, S.G. Wang and C. Wu, *Macromolecules*, **32** (1999), p. 590.
54. Liu, J.Q., and K.L. Wooley, *Abstr. Paper Amer. Chem. Soc.*, **221** (2001), U 439.
55. Sun, Y.K., L. Duan, Z.R. Guo, D.M. Yun, M. Ma, L. Xu, Y. Zhang and N. Gu, *J. Mag. Mag. Mater.*, **285** (2005), p. 65.
56. Jensen, T., L.A. Kelley, A. Lazarides and G.C. Schatz, *J. Cluster Sci.*, **10** (1999), p. 295.
57. Stevens, P.D., J. Fan, H.M.R. Gardimalla, M. Yen and Y. Gao, *Org. Lett.*, **7** (2005), p. 2085.
58. Templeton, A.C., M.J. Hostetler, E.K. Warmoth, S. Chen, C.M. Hartshom, Y.M. Krishnamurthy, M.D.E. Forbes and R.W. Murray, *J. Am. Chem. Soc.*, **120** (1998), p. 4845.
59. Zhao, M.Q. and R.M. Cooks, *Adv. Mater.*, **11** (1999), p. 217.
60. Han, L., J.M. Kneller, D.R. Daniel, S.R. Kowaleski, F.L. Kirk, J. Luo and C-J. Zhong, *Mater. Res. Soc. Symp. Proc.*, **710** (2002), DD 6.4.1.
61. Mirkin, C.A.; R.L. Letsinger, R.C. Mucic and J.J. Storhoff, *Nature*, **382** (1996), p. 607.
62. http://www.ohiolink.edu/etd/view.cgi?arkon1123637252
63. Nair, A.S., R.T. Tom, V.S. Suryanarayanan and T. Pradeep, *J. Mater. Chem.*, **13** (2003), p. 297.
64. Xue, C., G. Arumugam, K. Palaniappan, S.A. Hackney, H. Liu and J. Liu, *Chem. Commun.*, (2005), p. 1055.
65. Narayanan, R. and M.A. El-Sayed, *Langmuir*, **21** (2005), p. 2027.
66. Rosemary, M.J., I. MacLaren and T. Pradeep, *Carbon*, **42** (2004), p. 2352.
67. Francois, L., M. Mostafavi and J. Belloni, *J. Phys. Chem. B*, **104** (2000), p. 6133.
68. Anija, M., J. Thomas, N. Singh, A.S. Nair, R.T. Tom, T. Pradeep and R. Philip, *Chem. Phys. Lett.*, **380** (2003), p. 223.
69. Nair, A.S., V. Suryanarayanan, T. Pradeep, J. Thomas, M. Anija and R. Philip, *Mater. Sci. Engg. B*, **117** (2005), p. 173.
70. Porel, S., S. Singh, S.S. Harsha, D.N. Rao and T.P. Radhakrishnan, *Chem. Mater.*, **17** (2005), p. 9.

Additional Reading

1. Link, S. and M.A. El-Sayed, *Ann. Rev. Phys. Chem.*, **54** (2003), p. 331.
2. Hanack, M., T. Schneider, M. Barthel, J.S. Shirk, S.R. Flom and R.G.S. Pong, *Coord. Chem. Rev.*, (2001), pp. 219–221, 235.
3. Bond, G.C. and D.T. Thomson, *Catal. Rev.*, **41** (1999), p. 319.
4. Kolmakov, A. and M. Moskovits, *Ann. Rev. Mater. Res.*, **34** (2004), p. 151.
5. Kreibig, U. and M. Vollmer, *Optical Properties of Metal Clusters*, (1993) Springer, Berlin.

Chapter 10

NANOSHELLS

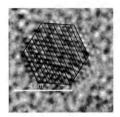

Nanoshells are important from the point of view of cancer therapy and spectroscopic applications. As we have seen, their method of synthesis is quite simple and includes one-step and two-step approaches. Silica nanoshells are used mainly in the field of molecular encapsulation while metal nanoshells are used for cancer therapy. Principal tools for their characterization are absorption spectroscopy, fluorescence spectroscopy and transmission electron microscopy. Even though intense research is taking place in this area, many questions remain unanswered regarding the biodegradability of these shells. The questions that need to be answered are, what the byproducts of these materials will be and what changes they are likely to create in the body.

Learning Objectives

- What are the various nanocavity systems?
- How do we study them?
- What are their properties?
- Are there any immediate applications for them?

10.1 Introduction

Nanoparticles are stabilized by different types of ligands such as organic molecules, polymers, surfactants, etc. The advantage of this kind of stabilization is that both the properties of the metal (or other) core as well as the stabilizing ligand can be made use of, in addition to the hybrid properties (properties due to both the core and the shell). Nanoparticles, whose surface is passivated by a shell with its own distinct properties other than the core, are called *core-shell particles*. Depending upon the use, this shell can also be made of metals, oxides, etc. Such coatings not only stabilize colloidal dispersions but also allow modification and tailoring of the particle properties (e.g. optical, magnetic, catalytic, etc.). Oxide-protected nanoparticles are found to be more stable even in extreme conditions such as exposure to intense lasers, which normally degrades materials. The shell makes the metal nanoparticle inert to chemical reagents. Nano-sized objects which have only the shell and are devoid of the core, are called nanoshells. They are also called by other names such as nanocapsules and nanobubbles.

There are some other types of nanoshells which are made of metals such as gold, silver, etc. and have a dielectric core. Nanoparticles of metallic, semiconducting, and magnetic materials have recently generated interest in terms of research because of their potential uses in optoelectronics, reprography, catalysis, chemical and biological sensing, etc. Among the metallic particles, the study of colloidal gold particles particularly stands out. It is one of the most widely studied systems, scientific interest in which can be traced back to the time of Faraday (see Chapter 1). This nanostructure has a unique optical property in that by changing the relative sizes of the core and the shell, its surface plasmon resonance can be tuned in a broad spectrum of wavelength (see Chapter 9). Halas and co-workers (Ref. 1) have developed a procedure to make gold nanoshells on silica treated with (aminopropyl) triethoxysilane. They have also tuned the properties of this system in such a way that it has been used for biological imaging and the therapy of cancer cells. There are reports of gold shells with polystyrene cores (Ref. 2). These gold nanoparticle-decorated silica cores can be modified for further stability by using self-assembled monolayers of alkane thiols. Other kinds of nanospheres are formed by the deposition of silica on biological systems such as liposomes.

The nanoshells which are obtained in the case of oxide nanoshells have diameters in the range of 10–20 nm. However, this is a variable and can be changed depending upon the required size. In the case of metal nanoshells like gold with silica shell, the thickness of the shell can be up to around 20 nanometers. It has been found that these nanoshells are highly porous in nature. One of the main uses of these hollow silica shells is in the form of containers of drugs. The outer surface of these shells can be used for attaching antibodies so that the silica shell-antibody complex can be used to bind to a specific antigen in fluid systems.

This chapter deals with different types of nanoshells, principally nanoshells of silica and gold. The applications of these nanoshells in different fields are also discussed. In the case of nanoshells made of silica, encapsulation of different analyte molecules has been done. Different kinds of methods used for the synthesis of coreshell particles as well as nanoshells are listed and the different techniques used for their characterization are discussed in this chapter.

10.2 Types of Nanoshells

It is possible to obtain nanoshells by a variety of methods. Given an application and size, one can select a method of synthesis. But most of these methods have the disadvantage of non-uniform coating which limits their applications. Hence it is very important to decide on the method of synthesis as it is difficult to obtain controlled and uniform coating. The preferred method involves surface precipitation of inorganic molecular precursors on particles and removal of the core by thermal or chemical means. However, the method used also depends on the type of the shell.

The two principal methods available for the synthesis of oxide-protected core shell particles, are 'layer-by-layer precipitation' of the inorganic precursor from solution and 'one-pot synthesis' carried out in the presence of several chemical ingredients. The multi-step synthesis involves making the core particle first and then covering it by using the shell forming precursor (e.g. TiO_2, SiO_2) and one-pot synthesis entails reduction of the metal in the presence of the shell-forming precursor, typically an organometallic species. In some other cases, dielectric substances such as silica or polymers are covered with gold or silver, thereby forming metal-coated particles. Nanoshells are obtained when the core of these particles is removed by chemical or physical means.

Most of the nanoshells are made of oxides, which include silica, zirconia and titania and oxides of a few other metals. As mentioned earlier, their method of synthesis varies from one another. Silica nanoshells, which are formed by two-step synthesis, are mainly used for studies related to the encapsulation of molecules, which may be dyes or drugs. While the titania shell is used for sensing purposes, new materials can also be formed from it.

For synthesizing metal nanoshells, first the dielectric material was made followed by the adsorption of the small metal nanoparticle over it, that was synthesized separately. Gold nanoshell is especially important in bio-imaging because gold is biocompatible. Another type of silica shell system with a liposome core, which is useful for drug delivery has also been reported. This is also highly biocompatible and biodegradable. These nanoshells can basically be divided into groups: (a) those that are formed from the oxide core shell particles, with a hollow core, and (b) those having a dielectric core and a metallic shell. Both these shells have applications in the fields of drug delivery, sensing and catalysis as also in the study of the micro-environment in which they are placed.

10.2.1 Oxide Nanoshells

This group includes silica, titania and zirconia nanoshells. The unique feature of this group is that they have a hollow core and a covering made of oxide. Their main applications are in the fields of encapsulation of molecules and in spectroscopy.

Hollow silica nanoshells

Silica has certain advantages when used as a protecting agent. It is chemically inert and therefore, does not affect the reactions which are taking place with the core. Also, since it is optically transparent, the spectroscopy of the system can be easily investigated. Most of the silica nanoshell systems are found to be useful in the study of the photochemistry of molecules, mainly fluorophores.

In the typical synthesis of silica-covered gold and silver core-shell particles (Refs 3, 4), organic molecules were used first to stabilize the metal nanoparticles. These ligands have the dual role and are used also as the primers for the growth of shells on the core. This means that the first layer of oxide is grown with the help of these molecular linkers. First, gold and silver silica colloids are prepared. 3-aminopropyltrimethoxysilane (APS) is used as the primer (and stabilizer) to make the gold surface vitreophilic and it also allows the formation of a thin layer of silica. The silica layer is formed by the hydrolysis of the methoxy groups. After this, another precursor of silica, i.e. sodium silicate solution is added at the appropriate pH to get a thicker layer of shell followed by the Stöber process (Ref. 5), which is a method to make larger particles of silica. This process is achieved by transferring the particles to an alcohol medium and creating the silica monomer *in-situ* by adding tetraethoxy silane, which leads to a slow homogeneous growth of silica around the particles. The reactions involved in the Stöber process are,

$$Si(OC_2H_5)_4 + 4H_2O \rightarrow Si(OH)_4 + 4C_2H_5OH$$

$$Si(OH)_4 \rightarrow \text{(in presence of } NH_3 \text{ and alcohol)} \ SiO_2 + 2H_2O$$

Hydrolysis of the tetraethoxysilane as well as the condensation of silica are base catalyzed events. This also provides the particles with a negative charge, thereby stabilizing the surface. It is possible to get a silica shell covered gold and silver core shell particles by using this method. Since standard methods of synthesis of gold and silver particles having a definite size are available in literature, it is possible to obtain core shell particles with a fixed core size.

Nanoshells can be obtained from these coreshell particles through the removal of the core material. It is important to use a suitable procedure for removing the core without disturbing the shell structure. The most common methods involve the use of cyanide ion in the case of gold and ammonia in the case of silver. Another method using halocarbons, mainly CCl_4 and benzyl chloride, is also found to be useful for making nanoshells.

In order to obtain nanoshells from Au@SiO_2, sodium cyanide solution was added to it and the core of the particle was dissolved. The dissolution of the core was monitored by using absorption spectroscopy from the disappearance of the surface plasmon peak of the gold nanoparticle (see Chapter 8). In the case of Ag@SiO_2, ammonia solution was used to remove the core.

The reactions can be represented as follows:

$$4Au + 8NaCN + 2H_2O + O_2 \rightarrow 4NaOH + 4NaAu(CN)_2$$

$$4Ag + 8CN^- + O_2 + 2H_2O \rightarrow 4Ag(CN)_2^- + 4OH^-$$

$$4Ag + 8NH_3 + O_2 + 2H_2O \rightarrow 4Ag(NH_3)_2 + 4OH^-$$

Zirconia nanoshells

Zirconia nanoshells also constitute an important class of materials that can be made of core-shell particles of silver and gold. Even though there are no reports of any specific application of this material nor has any study of molecular encapsulation been carried out, this material is important, especially form the point of view of catalysis and sensing. The shells are made (Ref. 6) from core-shell particles synthesized by two-step or one-step routes.

Synthesis of core-shell particles

1. Two-step synthesis: The first step is the synthesis of the core-shell particles. Silver nanoclusters are made first as in the case of monolayer protected clusters (Chapter 8). The monolayer made has a carboxyl functionality. The method used is called the Brust method (see Chapter 8). In a standard synthesis, 0.01g of the as prepared cluster is dispersed in a toluene-methanol mixture. To this mixture, zirconylchloride octahydrate is added in a 1:10 (weight) ratio. This solution is equilibrated and then 5 ml of diethylamine is added slowly over a period of 15 minutes. 50 ml of water is then added. The solution is stirred for over 24 hours. The mixture is centrifuged, washed with plenty of methanol and air dried. The powder obtained is zirconia coated nanoparticles and the shell thickness is several nanometers (Ref. 7).

2. One pot synthesis: In this method of making core-shell nanoparticles, the method adopted is the reduction of noble metals by dimethylformamide in the presence of oxide forming precursors. It is possible to make silver- and gold-covered zirconia as well as titania in this manner. A typical procedure is as follows: A solution containing equimolar (19.9 mM) amounts of titanium isopropoxide (or zirconium (IV) propoxide) and acetylacetone (a complexation agent for metal ions, to increase the stability of the alkoxide used) in 2-propanol is prepared. A clear solution is formed upon mild sonication. Another solution of 8.80 mM $AgNO_3$ (or $HAuCl_4 \cdot 3H_2O$) and 13.88M H_2O in DMF is prepared. A 40 mL sample of the first solution and 20 mL of the second solution are mixed and stirred for about ten minutes. The mixture is transferred to a heating mantle and refluxed for 45 minutes. The solution becomes pink in the case of Au and green-black in the case of Ag. The color change is abrupt in the case of Au and more gradual in the case of Ag. Further refluxing of the solution results in the formation of a precipitate, which can be dispersed by sonication. The colloidal material is precipitated by the addition of toluene. The precipitate is washed repeatedly with toluene and re-dispersed in 2-propanol. The cleaning procedure is important for obtaining well-defined absorption spectra and electrochemical properties. This one-pot synthesis can be used to make coreshell particles, especially in the case of zirconia, titania, etc. (Refs 6, 8).

Formation of nanoshells from core-shell particles prepared by one pot synthesis

The core shell particles which are synthesized by the above methods are subjected to different reactions which result in nanoshells. For getting nanshells, 4 ml of the core-shell particle dispersion was mixed with 1 ml each of CCl_4 or benzyl chloride separately. The reaction took place a period of several hours and the solution turned colorless. Silver ions were leached out and precipitated as AgCl during the reaction.

Titania nanoshells

Titania nanoshells can also be prepared by the dissolution of the cores of Ag@TiO_2 nanoparticles by using ammonia solution (Ref. 8). Ammonia complexes with the surface silver atoms, oxidizing the silver metal to silver ion complexes which diffuses from the interior of the core-shell particle through the pores in the titania shell. The shells have diameters of 10-30 nm and thicknesses of 3-5 nm. The diameter of the channels in the titania shells were comparable to the thickness of the electrical double layer (0.3-30 nm). The permeation of ions through these channels can be tuned by parameters such as pH and ionic strength. This has applications in the biomedical field (Ref. 8).

Zinc oxide nanoshell having silica core

Almost the same method that was used for the synthesis of zinc oxide covered silica (Ref. 9) which is photoluminescent. Simultaneous introduction of triethanolamine and $Zn(Ac)_2$ into the SiO_2 ethanol-water dispersion yields this composite. About 0.2 g of SiO_2 was dispersed in 30 mL ethanol-water (2:3 in volume) mixture and heated to 90 °C. After 10 minutes, 1.6 M triethanolamine and 0.02 mol/L Zn(Ac)$_2$ were dropped simultaneously through latex tubes into the previous mixture at a constant flow rate. The mixture was stirred for 1 h at 90°C. The resulting white powders were recovered by centrifugation, washed repeatedly with water, and dried in vacuum. Finally, the powders were sintered at 700°C for 3 hours.

10.2.2 Metal Nanoshells

Metal nanoshells are different from oxide nanoshells in their structure, which has a dielectric core made of silica. The main advantage of such a system is that the optical properties can be tuned by changing the thickness of the shell. Among this group, gold nanoshells find application in the field of cancer therapy.

Gold nanoshells

The novelty of this material is that its properties can be tuned to scatter or adsorb light over a broad spectral range including the near IR, a wavelength region that provides maximal penetration of light through the tissue. This property offers the opportunity to design nanoshells which can be used for diagnostic and therapeutic applications and also can be used to monitor different analytes such as immunoglobulin in blood, saline, etc. This system is found to have sensitivity in the subnanogram per mL range within 10-30 min. Gold nanoshells have many advantages over silica nanoshells. Apart from the fact that their optical properties, especially absorption, can be tuned as a function of the thickness of the gold layer, they have large optical cross section as compared to the conventional near-IR (NIR) dyes, such as

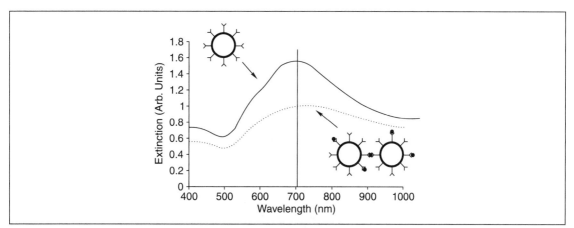

Fig. 10.1: UV-vis spectrum of dispersed nanoshells fabricated with a 96-nm-diameter core and 22-nm-thick gold shell (—); spectrum of nanoshells/antibody conjugates following addition of analyte (- - -). Extinction reduction upon aggregation in the presence of analyte was monitored at 720 nm, as indicated. Reprinted with permission from Hirsch, et al. (Ref. 1), Copyright (2003), American Chemical Society.

indocyanine green. They also show improved photostability. Another advantage is that the gold surface can be used to anchor different molecules, especially proteins and other biomolecules. The gold surface is generally considered to be biocompatible. It can be modified according to the type of anchoring needed by using molecules like polyethylene glycol. Alkanethiols can be used for functionalization to facilitate increased solubility in the organic media, see Fig. 10.2 (Plate 6).

Synthesis of gold nanoshell having silica as the core

Method I Nanoshells with extinction maximum at ~800 nm were synthesized as follows (Ref. 10). Using the Stöber method, 120 nm diameter silica nanoparticles were made first. The surface of these particles were functionalized using (aminopropyl) triethoxysilane (APTES). Small gold colloid was grown as per the method of Duff and Baiker (Ref. 11) and these were adsorbed on aminated silica nanoparticles. More gold was reduced into these nucleation sites using potassium carbonate and formaldehyde.

Method II The original method is due to Sun and Xia (Ref. 9). Ag nanoparticles were prepared by injecting $NaBH_4$ (50 mM, 2 mL) into 0.2 mM, 100 mL aqueous solution of $AgNO_3$, in the presence of sodium citrate (0.5 mM). The Ag colloid formed was kept at 70 °C for 2 hours. $HAuCl_4$ (0.1 M, 0.68 mL) was added to 100 mL of the Ag colloid. A series of color changes from yellow to red to dark blue were observed during the course of the replacement reaction (Refs 12, 13). After 1 hour of stirring, the particles were purified by gradient centrifugation, washed using aqueous sodium citrate (0.3 mM), re-dispersed in 5 mL of sodium citrate (0.3 mM), and finally kept at 4°C.

Silver nanoshell

In another method, the seed growth approach was used to prepare to silver nanoshells on silica nanoparticles (Ref. 14). In this case also, the optical properties could be changed by changing the thickness of the shell. Since its Mie resonance occurs at energies distinct from any bulk interband transition, a silver colloid is expected to have a stronger and sharper plasmon resonance than gold. Another advantage of a silver colloid is that the plasmon resonance of a solid silver nanoparticle appears at a shorter wavelength than that of gold.

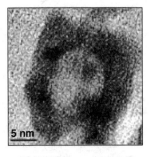

Fig. 10.3: *High-resolution TEM photograph of an individual Au nanoshell. Reprinted with permission from Hao, et al. (Ref. 12), Copyright (2004), American Chemical Society.*

Preparation of a silver nanoshell on a silica sphere

Scheme 1 shows the typical strategy used to prepare a silver nanoshell (Ref. 14). First, the silica sphere core is treated with an amine terminated surface silanizing agent (step I, see Fig. 10.4). The resultant terminal amine groups act as attachment points for small colloidal silver particles, which is used for the growth of a silver nanoshell overlayer (step II). This was followed by the growth of silver particles (using the standard sodium citrate route, step III). By repeating step III, the thickness of the silver shell can be tailored, as expected. Reduction of silver ions can also be done by other reducing agents such as $NaBH_4$.

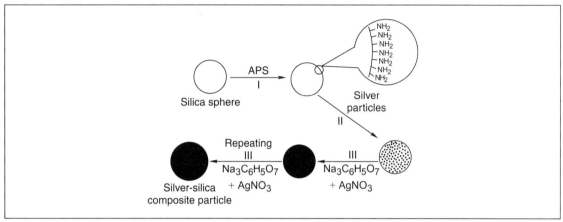

Fig. 10.4: *Fabrication procedure of a silver nanoshell on the silica sphere. Reprinted with permission from Jiang and Liu (Ref. 14). Copyright (2003) American Chemical Society.*

10.2.3 Nanoshells from Liposomes

Another method followed for the synthesis of silica nanoshells involves the use of liposomes as the templates.

It was found that liposomes which are covered with silica are nonporous in nature. Hence, the exact properties of the lipid system are retained. This is important as liposomes are very sensitive to parameters such as pH and ionic strength, and also to the presence of organic solvents in the reaction mixture. Hence this kind of a system can find use in the case of drug delivery mainly because liposomes are biological systems and are also biodegradable.

Synthesis of Liposomes Covered with Silica

Unilamellar liposomes are prepared according the procedure of Bangham (Ref. 15). A suitable amount of L-α-dipalmitoylophosphatdylcholine (which forms the liposome) was dissolved in chloroform. Solvent was removed, a phosphate buffer solution was added and a 10 mg ml^{-1} lipid suspension was made. This contains multi lamellar vesicles. This suspension was then extruded above the transition temperature of

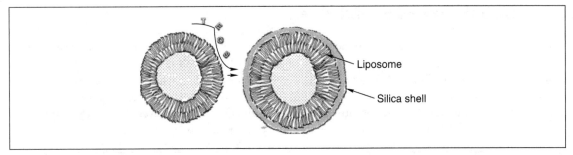

Fig. 10.5: *Schematic of liposome covered with silica. Taken from the Graphical Content of Begu, et al. (Ref. 15). Reproduced by permission of the Royal Society of Chemistry.*

the lipids (41°C) using an extruder having polycarbonate membranes. The mean size of the vesicles was 108 nm, from quasi-elastic light scattering (QELS) measurements. QELS allows one to study the dynamics of a system in real-time (see chapter 9). The silica shell is grown by adding tetraethylorthosilane (TEOS) to the diluted suspension of liposomes at room temperature with stirring, followed by the addition of NaF (4% molar ratio with respect to TEOS). The reaction mixture was again stirred for 48 h and dried at 40°C to obtain silica covered liposomes (Ref. 16).

10.3 Properties

10.3.1 Reactions Inside the Silica Nanoshell

Colloidal Alloy Formation Inside Silica Shells

$AuCl_4^- + 3Ag \text{ (core)} \rightarrow Au \text{ (core)} + 3AgCl \text{ (core)} + Cl^-$

Reactions like colloidal alloy formation are possible inside the silica shell. It is found that the silver core of the core-shell silica particle can be converted to gold core via oxidation with $AuCl_4^-$. Generally this process is expected to be very fast. But since a shell is present outside and the silver surface is passivated by the precipitated AgCl, the reaction is bound to be slow. The AgCl precipitated in this case is very thin, which is why the shell doesn't beak or expand. It is found that there is a definite rate dependence on the rate of diffusion of the gold ions through shell, clear from the fact that the reaction is very fast in the case of bare silver particles compared to that of core shell particles.

These processes are taking place because the silica shell is porous in nature and also catalytically active, even though the core is separated by a shell. Thus the shell acts as a selective membrane which controls the rate of the chemical reactions with the metal cores. In some cases, almost complete inhibition of the reactions happens while in some others, the rate is retarded. The rate can be controlled by varying the thickness of the silica shell.

Formation of Carbon Onions Inside Silica Nanoshells

Nanoshells can be formed from Au@SiO$_2$ by leaching gold using carbon tetrachloride. It is found that the amorphous carbon produced in the leaching process gives rise to carbon onions inside the shell.

This is an important reaction since this is the first report of formation of carbon onions at room temperature in a solution. Figure 10.6 shows a high resolution image of silica shell containing a carbon onion. It gives important information regarding the nature of the shell. The silica shell stretches itself so that it can accommodate carbon onions formed inside implying the plastic nature of the shell. The cavity left behind after the removal of gold is smaller than the dimension of the leached out metal implying that catalytic destruction of the halocarbon occurs and more carbon per metal ion is generated (Ref. 17).

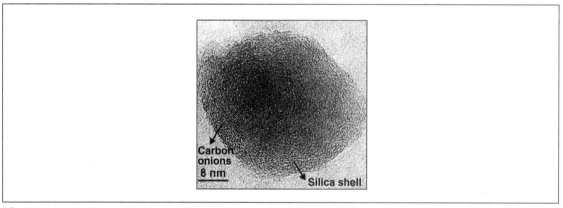

Fig. 10.6: *High resolution TEM images of carbon onion containing shells. The outer few nm shows amorphous contrast and is the silica shell, inside this the typical concentric ring structure of a carbon onion is seen, in this case more than 20 concentric graphitic planes are visible. Reprinted from Rosemary, et al. (Ref. 17), Copyright (2004), with permission from Elsevier.*

10.3.2 Incorporation of Molecules Inside the Nanoshells

One of the main advantages of nanoshells is that they can be used as carriers of molecules. This allows nanoscopic containers to be used for controlling the environment of the molecule. This helps protect the molecule from unwanted chemical reactions and the cavity also provides a rigid environment for adsorbed molecules. Colloidal dispersions of silica shells are optically transparent, providing an opportunity to study the behavior of the molecules incorporated without excessive light scattering problems. Imhof, *et al.* have studied the incorporation of fluorescein isothiocyanate inside silica spheres where the objective was to increase the photostability of the dye molecule (Ref. 18). The results of this study seemed to suggest an inhomogeneous distribution of the molecules inside the shell. Studies have also been carried out on the excited state reactions of the photochemically important molecule, ruthenium tris (bipyridyl) dye inside silica shells, wherein the excited Ru (II) shows a significant enhancement of the phosphorescence yield and lifetime. Moreover, the dye reacts with molecules such as methylviologen. Bosma, *et al*. (Ref. 19) have synthesized colloidal poly(methyl)methacrylate (PMMA) particles wherein fluorescent dyes are incorporated into the polymer network. In these studies, there has been a significant focus on the photochemistry and spectroscopy of the adsorbed molecules, possibly to the neglect of other interesting properties of the incorporated molecules. Since these molecules are isolated, they may show different properties in contrast to the free ones when confined within a shell.

Since most of these nanoshells are used as containers for molecules, they can also be used for drug delivery as well as for studying the photochemistry of these molecules.

Incorporation of Dye Molecules Inside Silica Nanoshells

Most of such molecules have been incorporated inside silica shells. This is because the formation of silica-covered core-shell particles provides an easy method for the incorporation of these molecules. The formation of a silica shell includes the usage of aminopropylsilane as a precursor for the shell formation. Molecules which have amino or thiocyanato functional groups can also be used along with aminopropylsilane, while the silica shell formation takes place. In order to incorporate any molecule, the first step is to functionalize it (the molecule) to have an amino or thiocyanto group. The next step is to add the functionalized molecule together with aminopropylsilane, do further polymerization using tetramethyl orthosilicate and allow further shell growth in ethanolic medium (Ref. 3).

The subsequent removal of the core can be done by any of the methods discussed above using NaCN, NH_3 or CCl_4.

Fluorescein Isothiocyante@SiO_2

Fluorescein isothiocyante is a very important dye molecule, with applications in the field of protein labeling. But one disadvantage of this dye is that it undergoes photobleaching very fast. Hence, it is important to make this dye inside a shell which can be studied by spectroscopic techniques as well as used for different applications. The disadvantage of this system is that self-quenching occurs in it because of which there is a decrease in the quantum yield of the encapsulated dye. It was found that the incorporation of this dye on the shell surface decreases the porosity of the nanoshell. It takes more time for mineralization using cyanide ion when the shell has the dye.

10.3.3 Modification of Silica Shell for Immunoassays

Binding of immunoreagents to different particles is important from the point of view of many applications like site-specific drug delivery to the targeted portions of the body. The success of targeted drug delivery depends upon two facts: (1) attachment of antibody to the silica surface, and (2) the ability of the antibodies to bind their specific antigen while they are attached to the shell. In the case of the first criterion, the molecule which has a silanol group at one end and an organic group at the other has been found to be useful. In most of the cases, APS is used for this purpose. Since APS doesn't have any functional group to make covalent bond with the molecules, another molecule, N-5-azido-2-nitrobenzoyloxysuccinimide (ANB-NOS), which can form a strong covalent bond with the antibody, is often used. About 10 percent of the bound antibodies on the APS and 12 percent on the ANB-NOS surface were found to be active (Ref. 20).

10.4 Characterization

Nanoshells can be characterized in terms of many spectroscopic techniques, especially transmission electron microscopy, atomic force microscopy, etc.

Transmission Electron Microscopy

Transmission electron microscopy is the main technique used for the characterization of nanoshells as we can see the nano structures. As shown in Fig. 10.7, we can clearly see approximately 15 nm-sized silica nanoshells.

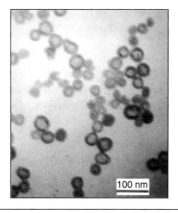

Fig. 10.7: *A TEM micrograph showing nano-sized silica shells with an average inside diameter of ~15 nm containing 195 Cascade Blue dye per particle. Reprinted from Ostafin et al. (Ref. 4), Copyright (2003), with permission from Elsevier.*

Optical Spectroscopy

There are many methods by which one can use optical spectroscopy to characterize nanoshells. In the case of silica nanoshells formed by the leaching of gold or other nanometal core, it is possible to monitor the disappearance of the plasmon peak characteristic of it (Fig. 10.8). In the case of shells incorporated with molecules, it is possible to get the absorption characteristics of that molecule from absorption

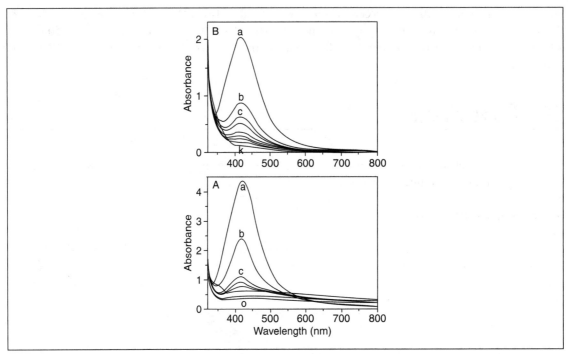

Fig. 10.8: *Time-dependent UV–visible spectra of the reaction of Ag@ZrO$_2$ with (A) CCl$_4$ and (B) benzyl chloride indicating the selective leaching of the metal core. Trace a corresponds to the parent material. In (A) the traces were recorded after every 30 minutes and in (B) after every 10 minutes (after addition of CCl$_4$ and benzyl chloride, respectively). From Nair et. al. (Ref. 6). Reproduced by permission of the Royal Society of Chemistry.*

spectroscopy. For metal nanoshells, the formation is monitored principally using the absorption spectroscopy. For the different applications of metal nanoshells, it is possible to tune its surface plasmon resonance by changing the layers of the metal shell.

Fluorescence Spectroscopy

Apart from absorption spectroscopy, fluorescence spectroscopy is also used to characterize nanoshells. Here the emission spectrum of the incorporated molecule can be used for its characterization (Ref. 3).

Cyclic Voltammetry

Cyclic voltammetric studies provide valuable information on the molecules present inside the nanoshell, thus probing the shell indirectly. Apart from the usual characterization, it also provides information regarding the environment of the molecule inside the nanoshell. It is a very useful tool in cases where one is looking at the reactions taking place inside the nanoshell. For example, ciprofloxacin-incorporated silica nanoshell shows two irreversible cathodic reduction peaks at −0.75 V and −1.18 V (Ref. 18). In the case of pure ciprofloxacin molecules, however, the peak potentials are at −0.81 and −1.25 V, respectively, with a shift of 60 mV towards the cathodic region with respect to ciprofloxacin@SiO_2. This may be due to the presence of the SiO_2 shell outside the ciprofloxacin molecule which retards the electrochemical reduction.

Infrared Spectroscopy

Infrared spectroscopy provides information about the kind of linkages present in nanoshells. In the case of silica nanoshells, the Si-O-Si linkage can be characterized by using stretching modes which appear at 801 and 1100 cm^{-1}. Zirconia shell shows two features at 534 and 725 cm^{-1} which are analogous to the 530 cm^{-1} main feature and 725 cm^{-1} shoulder seen for cubic zirconia (Ref. 6).

10.5 Applications

Ion Selective Films

Strong ion sieving propertied of nanoshell films can be used for applications. Dopamine can be detected electrochemically by carbon fiber electrode, but the major problem in this is the interference from ascorbic acid, which also falls in this electrochemical window. The layer-by layer assembly of (TiO_2 nanoshells/poly acrylic acid) acts as an excellent detection tool for dopamine without any interference from ascorbic acid as shown in Fig. 10.9.

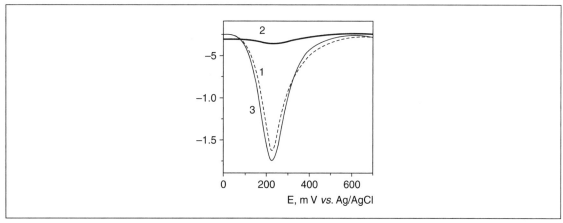

Fig. 10.9: *Electrochemical dopamine sensing using an LbL of TiO_2 nanoshells. Reprinted from Ref. 8. Copyright (2002) Wiley-VCH.*

At pH = 7, the TiO_2 nanoshells are negatively charged and therefore, the diffuse part of the electrical double layer is composed primarily of cations, an ideal condition necessary for the selective detection of positively charged dopamine over negatively charged ascorbic acid. The ratio between the dopamine and ascorbic acid signals changes from 1:3 for a native glassy carbon electrode surface to 9:1 for a nanoshell modified surface, which gives an overall 27 fold enhancement of the selectivity between these substances. The signal from the mixture of ascorbic acid and dopamine (trace 1 in the figure) is virtually equal to that from 1mM dopamine (trace 3). The ascorbic acid peak current (trace 2) was negligible under these conditions.

Gold Nanoshells for Blood Immunoassay and Cancer Detection and Therapy

For conventional blood immunoassays, optical tests are performed at visible wavelengths. Since it is necessary to separate out several unwanted biomaterials which absorb visible light, the whole procedure can take long time, of the order of several hours or days. In the immunoassay procedure proposed by Halas and West (Ref. 1), nanoshells are conjugated with antibodies that act as recognition sites for a specific analyte. The analyte causes the formation of dimers (Fig. 10.10), which modify the plasmon-related absorption feature in a known way. A fast absorption measurement can determine the presence of the molecule, avoiding the purification step.

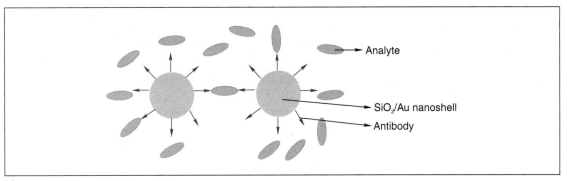

Fig. 10.10: *Schematic of the interaction of a gold nanoshell modified using an antibody with an analyte. Adapted from Ref. 22. Used with permission from the author, Copyright (2003) Nature Publishing Group.*

Since these nanoshells have a large optical scattering cross section, they can be used as potential contrasting agents for photonics-based imaging modalities. Among the methods, reflectance confocal microscopy (RCM) and optical coherence tomography (OCT), which facilitate early cancer detection are important. Optical properties of the nanoshells can be tuned in such a way that that they can be used for both imaging and therapy. The colloidal regime allows controlling both scattering and absorption properties simultaneously by changing size. Selective accumulation of the nanoshell can be used to image the tumor by using the high permeability and retention properties of the cancer cells.

Figure 10.11 (Plate 6) shows SKBr3 breast cancer cells imaged by using targeted HER2 nanoshells (Ref. 10). HER2 is an acronym and refers to human epidermal growth factor receptor, a protein. HER2 receptors are over-expressed in the case of a cancer cell. Among the three figures in the first row, the one with nanoshells shows a better image. After the therapy, it is also found that the cells treated with nanoshells are dead. This effect was not observed in the case of non-cancer cells. This is also found to have applications in the case of silver staining also were better imaging was possible. The cytotoxicity of the nanoshells was checked by treating cancer cells with the nanoshells and taking the image without NIR radiation. It was found that there was no death of cells due to the nanoshells alone.

Even though these kinds of gold nanoshells are found to have applications in many biological fields, no one really knows what will happen to the material itself after treatment with the tumor cells. The fact that these are non-biodegradable and also catalytically active which may create problems for the biological systems.

Gold Nanoshells for Enhancing the Raman Scattering

Raman scattering is a phenomenon which is used to study different surfaces. But this technique has a very weak sensitivity. However, roughened surfaces can be used for enhancing the Raman signal (Ref. 23). Interactions of light with individual nanoshells allowed a 10 billion-fold increase in the Raman effect.

Review Questions

1. What are nanoshells?
2. What are the principal kinds of nanoshells? How to make them?
3. What makes gold nanoshells attractive in biology? What are their applications?
4. What are the major breakthroughs in this area which make the application of nanoshells promising?
5. What are the essential features of nanoshells useful in biology?
6. How would one investigate the porosity of the shells?
7. Why nanoshells are made over dielectric materials and not over metallic particles?

References

1. Hirsch, L.R., J.B. Jackson, A. Lee, N.J. Halas, J.L. West, *Anal. Chem.*, **75** (2003), p. 2377.
2. Graf, C., A.V. Blaaderen, *Langmuir*, **18** (2002), p. 524.
3. Makarova, O.V., A.E. Ostafin, H. Miyoshi, J.R. Jr. Norris, D. Meisel, *J. Phys. Chem. B*, **103** (1999), p. 9080.
4. Ostafin, E., M. Siegel, Q. Wang, H. Mizukami, *Micropor. Mesopor. Mater*, **57** (2003), p. 47.
5. Stöber, W., A. Fink, E. Bohn, *J. Colloid Interface Sci.*, **20** (1968), p. 62.

6. Nair, A.S., R.T. Tom, V. Suryanarayanan, T. Pradeep, *J. Mat. Chem.*, **13** (2003), p. 297.
7. Eswaranand, V., T. Pradeep, *J. Mat. Chem.*, **12** (2002), p. 2421.
8. Koktysh, D.S., X. Liang, B. Yun, I.P. Santos, R.L. Matts, M. Giersig, C.S. Rodriguez, L.M.L. Marzan, N.A. Kotov, *Adv. Funct. Mater.*, **12** (2002), p. 255.
9. Sun, Y., Y. Xia, *Anal. Chem.*, **74** (2002), p. 5297.
10. Christopher, L., L. Amanda, N. Halas, J. West, R. Drezek, *Nano. Lett.*, **5** (2005), 709.
11. Duff, D.G., A. Baiker, *Langmuir*, **9** (1993), p. 2301.
12. Hao, E., S. Li, R.C. Bailey, S. Zou, G.C. Schatz, J.T. Hupp, *J. Phys. Chem. B*, **108** (2004), 1224.
13. Xia, H.L., F.Q. Tang, *J. Phys. Chem. B*, **107** (2003), p. 9175.
14. Jiang, Z.J., C. Liu, *J. Phys. Chem. B*, **107** (2003), p. 12411.
15. Bangham, A.D., *J. Mol. Biol.*, **13** (1965), p. 238.
16. Begu, S., S. Girod, D.A. Lerner, N. Jardiller, C.T. Peteilh, J.M. Devoisselle, *J. Mater. Chem.*, **14** (2004), 1316.
17. Rosemary, M.J., I. MacLaren, T. Pradeep, *Carbon*, **42** (2004), p. 2352.
18. Imhof, J.A., M. Megens, J.J. Engelberts, D.T.N.D. Lang, R. Sprik, W.L. Vos, *J. Phys. Chem. B*, **103** (1999), p. 1408.
19. Bosma, G., C. Pathmamanoharan, E.H.A. de Hoog, W.K. Kegel, A.V. Blaaderen, H.N.W. Lekkerkerker, *J. Colloid Interface Sci.*, **245** (2002), p. 292.
20. Wong, C., J.P. Burgess, A.E. Ostafin, *Journal of Young Investigators*, issue 1, volume 6, (2002).
21. Rosemary, M.J., V. Suryanarayanan, P.G. Reddy, I. MacLaren, S. Baskaran, and T. Pradeep, *Proc. Indian Acad. Sci.*, **115** (2003), p. 703.
22. Brongersma, M.l., *Nature Materials*, **2** (2003), p. 296.
23. Jackson, J.B., S.L. Westcott, L.R. Hirsch, J.L. West, N.J. Halas, *Appl. Phys. Lett.*, **82** (2003), p. 257.

Additional Reading

1. Nalwa, H.S. (ed.) (2002), *Nanostructured Materials and Nanotechnology*, Academic Press, New York, 2002.
2. New, R.R.C. (ed.) (1990), *Liposomes: A Practical Approach*, Oxford University Press, Cambridge, 1990.

PART FOUR

Evolving Interfaces of Nano

Contents:
- Nanobiology
- Nanosensors
- Nanomedicines
- Molecular Nanomachines
- Nanotribology

Chapter 11

NANOBIOLOGY

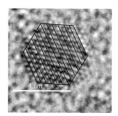

Biological functions take place in nanometer-sized objects. Therefore inorganic nanoparticles can be integrated into biological molecules such that one can understand, control and manipulate biological processes. This has been made possible now through the utilization of several properties of nanosystems, many of which have been discussed in earlier chapters. Here the objective is to appreciate the science of bio-nano hybrid systems with selected examples.

Learning Objectives

- What is nanobiology?
- How can one use nanomaterials for biological applications? What are the examples?
- What are the applications of nanomaterials in biology?
- What are the promising areas for the immediate application of nanomaterials in biology?
- What are the outstanding issues in nanobiology?

11.1 Introduction

With the evolution of science, all scientific phenomena have been investigated in progressively microscopic dimensions. Thus we see that science has progressed from big to small. Optical microscopy helped the scientist to look at phenomena in the regime of the micrometer. Later on, the invention of electron microscope helped the scientist to examine phenomena in the regime of nanometer. Due to the electronic confinement of nano objects, their chemical and optical properties are different from those of larger objects. As we look at life forms in smaller and smaller dimensions, we end up with biological objects which are involved in fundamental life processes. The key molecules in biology such as DNA, enzymes, receptors, antigens, antibodies and oxygen carriers can be included in the dimension of nanometers. Thus it is clear that all fundamental processes in biology are taking place in the dimension of 1–100 nm. Molecular self-organization around nanoparticles, utilizing the tools of surface science, can be of use in the development of probes for understanding fundamental life processes. The synergy of surface science and molecular biology has given birth to a new subject called 'nanobiology', which symbolizes a path-breaking evolution in the progress of biology. This new subject is capable of unveiling many fundamental secrets of life forms.

The science of nanobiology has progressed significantly and can be branched into several disciplines of life science such as cellular biology, genomics, proteomics, oncology, immunology, diagnosis, targeted drug delivery, etc. The idea of creating the hybrid systems of inorganic nanoparticles with biological moieties has helped solve several technical difficulties in medical and biosciences. Using nanomaterials, instead of conventional materials, in life science applications has increased efficiency while decreasing cost. Thus the hybridization of nano-bio-technologies has laid the ground for many novel methodologies, which are capable of solving several technical difficulties in bio-analysis. The ideas and innovations of nanobiology can be extended to the frontiers of biotechnology. In this chapter we discuss the following four major aspects of nanobiology:

1. Interaction between biomolecules and nanoparticle surfaces
2. Biological imaging using nanoparticles
3. Analytical applications of nanobiology
4. Medical diagnosis and targeted drug delivery
5. Biosynthesis of nanomaterials

11.2 Interaction Between Biomolecules and Nanoparticle Surfaces

One of the objectives of nanotechnology is the packing of nanoparticles in ordered arrays with the ability to tailor the size and inter-particle distance. While the assembly of nanoparticles from solution into hexagonally close-packed monolayers and superlattice structures on solid surfaces has been fairly successful (see Chapter 8), the controlled assembly of nanoparticles in solution had hitherto remained a relatively unexplored area. The construction of three-dimensional arrays of nanoparticles in the aqueous phase can be achieved by bioconjugation, the phenomenon in which intermolecular interactions lead to assembly. The interaction of nanoparticles functionalized with conjugate biomolecules, can lead to the formation of desired superstructures in the aqueous phase. The first steps in this direction leading to the construction hybrid bio-nano assemblies, were taken by the groups of Mirkin (Ref. 1) and Alivisatos (Ref. 2), who demonstrated that DNA-modified nanoparticles could be assembled into superstructures by the hybridization of complementary base sequences of the surface-bound DNA molecules. From a fundamental point of view, Mirkin, et al. have used this strategy to critically study the role of inter-particle separation (Ref. 1) and aggregate size on the optical properties of DNA-modified colloidal gold solution. Other interactions such as the biotin-avidin molecular recognition process, hydrogen bonding between suitable terminal functional groups bound to the nanoparticle surface, electrostatic assembly on DNA templates, and control over electrostatic interactions stabilizing nanoparticles in the aqueous phase have been used to construct the nanoparticle assembly in solution (Ref. 3). Sastry, et al. (Ref. 4) described the surface modification of aqueous silver colloidal particles with the amino acid, cysteine and the cross-linking of the colloidal particles in solution. Capping of the silver particles with cysteine was accomplished by a thiolate bond between the amino acid and the nanoparticle surface. The silver colloidal particles were stabilized electrostatically by ionizing the carboxylic acid groups of cysteine. The amino acid, cysteine (H_2N–CH(CH_2SH)–CO_2H) plays an important role in defining the tertiary structure of proteins through disulfide (cystine) bridges (Ref. 5).

11.2.1 Influence of Electrostatic Interactions in the Binding of Proteins with Nanoparticles

It has been suggested that each gold nanoparticle has an Au (0) core and an Au (I) surface, as a result of the preferential adsorption of Au (I) ion on the surface at the time of formation of the nanoparticle. Citrate ions co-ordinate to the Au (I) atoms on the surface, resulting in an overall negative charge for each particle. Nanoparticles are capable of binding with oppositely charged species in aqueous solutions through electrostatic interactions based on the ionic characteristics of their surfaces (Ref. 6). Amphiprotic species, such as peptides and proteins, have unique isoelectric points (pI). When the pH of a protein sample solution is below the value of the pI of the protein, the protein molecules have a net positive charge. This means that by increasing the pH from 0 to 14, the molecule will get more and more negatively charged by the consequent removal of protons. At a particular pH, the molecule will have a net charge zero. This point is called the isoelectric point (pI). Therefore, below the pI value, the molecule is positively charged. Negatively charged gold nanoparticles tend to attract positively charged protein molecules to their surfaces through electrostatic interactions. On the other hand, if the pH of a protein solution is above the value of the pI of this protein, the protein molecules are negatively charged and repel any negatively charged gold nanoparticles. With regard to the nature of interaction of the biomolecule and nanoparticle, it is clear that they are interacting through electrostatic attraction. Thus by considering the attraction between two opposite charges,

Electrostatic force $F = (kq_+q_-/r)$

F is the force in Newton and $k = 8.9874 \times 10^9$ Nm^2C^{-2}, r is the distance in meter (m), q_+ and q_- are the positive and negative charges in coulomb (C), respectively. For simplicity we can consider, r and q_- (the charge on the nanoparticle surface) are constant. Thus the electrostatic force between the biomolecule and nanoparticle,

$F \alpha\, q_+$

i.e. the electrostatic force is proportional to the net charge on the biomolecule (q_+).

From the above equation, it is clear that the pI is the key factor in the binding of the biomolecule on the nanoparticle. Above the pI value, there is no attraction between the biomolecule and the nanoparticle because the biomolecule becomes negatively charged. The selective binding of proteins on the nanoparticle surface as a function of the pH is illustrated by matrix-assisted laser desorption ionization mass spectrometry (MALDI-TOF MS). The pI values of the proteins, cytochrome c and myoglobin are 10.6 and 7, respectively (Ref. 7).

The absence of the adsorption of proteins, cytochrome c and myoglobin on the gold nanoparticle surface above their pI value is reported by Teng, et al. (Ref. 8). In Fig. 11.1, aqueous solutions of cytochrome *c* and myoglobin have been employed to demonstrate this. From a solution at pH 6, both cytochrome *c* and myoglobin were detected in MALDI-TOF MS analysis (Fig. 11.1(a)). At pH 8, only cytochrome *c* was detected in the MALDI-TOF mass spectrum (Fig. 11.1(b)), and signals for neither analyte are observed in the MALDI-TOF mass spectrum at a pH 12 (Fig. 11.1(c)) (Ref. 8). These results demonstrate that isoelectric points play an important role in the binding of nanoparticles with proteins. Thus, gold nanoparticles can be used to bind specific proteins selectively by adjusting the pH of their sample solutions.

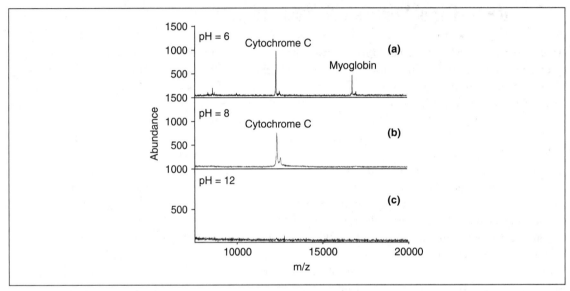

Fig. 11.1: *MALDI-TOF mass spectra obtained from a mixture of cytochrome c and myoglobin at: (a) pH 6, (b) pH 8 and (c) pH 12. Reprinted with permission from Teng, et al. (Ref. 8). Copyright (2004) American Chemical Society.*

11.2.2 The Electronic Effects of Biomolecule-Nanoparticle Interaction

When a biomolecule has interacted with the surface of a noble metal nanoparticle, its surface charge density is deformed by the electrostatic attraction. The extent of deformation of surface charge density depends on the following factors:

1. The size of the noble metal nanoparticle
2. Molecular diameter of the biomolecule
3. pH of the medium
4. Ionic strength of the medium

As a result of the interaction, the surface charge density of both the nanoparticle and the biomolecule get perturbed. This perturbation of the surface electron density is manifested in the change in the electronic spectrum of both the nanoparticle and the biomolecule. The functional groups of the molecules which are close to the vicinity of the nanoparticle surface get perturbed more. The binding of the biomolecule can affect the surface plasmon resonance of the nanoparticle. Sometimes, this interaction can be manifested in the assemblies of nanoparticles and biomolecules. The surface electron density of the nanoparticle is spherically symmetric. The perturbation to charge density of the nanoparticle induced by the interaction with the biomolecule is isotropic. But in the case of noble metal nanorods, their charge density is anisotropic. Thus the interaction with the biomolecule can lead to anisotropic perturbation of electron density of the nanorods. The difference in the interaction of the biomolecule to the nanorod and nanoparticle is manifested in the UV-visible spectrum. The surface plasmon resonance (SPR) of noble metal nanoparticles is due to polarized oscillation of the electron cloud, induced by the electric vector of the electromagnetic wave according to Mie's theory. This phenomenon is highly sensitive to the dielectric constant of the micro-environment around the nanoparticle. Thus, such an interaction with a biomolecule can be manifested in the form of alterations in the SPR band. The molecular diameter of a biomolecule is a key factor which decides the biomolecule–nanoparticle interaction. Bigger molecules will form stable nano-bio conjugates. They will form thicker shells above the surface of the nanoparticle. Such shell formation can prevent the aggregation of nanoparticles. As an example we can consider the interaction between an antibiotic vancomycin and gold nanoparticle (Ref. 9).

The changes in the SPR band of the nanoparticle and the dimension of the interacting molecule are directly related. In Fig. 11.2, the dashed line represents the free nanoparticle. After the capping of vancomycin, the SPR band is broadened and shifted slightly from 520 to 528 nm. This indicates a change in the dielectric constant of the micro-environment around the gold nano-core. The formation of aggregated structures can be observed from the SPR features of Au@cysteine, including the shifting of the SPR band

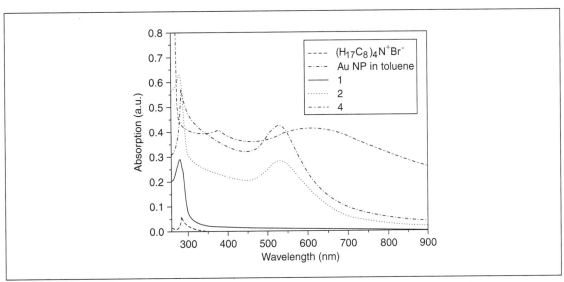

Fig. 11.2: *UV-visible spectrum of: (1) vancomycin, (2) Au@vancomycin and (4) Au@cysteine. Reprinted with permission from Gu, et al. (Ref. 9). Copyright (2003) American Chemical Society.*

to 631 nm as well as the increase in background intensity. A similar kind of an aggregation is reported for cystiene-capped silver nanoparticles (Ag@Cysteine) by Sastry, et al (Ref. 4). The bare nanoparticles (AuNP) stabilized with tetra N-octyl ammonium bromide show the SPR band at 520 nm. After the vancomycin capping of the nanoparticle surface, the SPR feature has no significant shift. This indicates that due to the bigger molecular diameter of vancomycin, the nanoparticle aggregation is absent.

The TEM image Fig. 11.3(a), (b) clearly illustrates the difference between the interaction of a bigger and smaller biomolecule with the nanoparticle. In the TEM image, Au@cysteine shows the aggregated structure, while the Au@vancomycin shows well-separated particles. This indicates that there is a thicker shell of vancomycin around the gold core, which prevents the aggregation of the nanoparticle. As mentioned earlier, similar kinds of observations are reported in the case of cytochrome c capped nanoparticles also (Ref. 10). The absence of nanoparticle aggregation is due to electrostatic repulsion between the thicker cytochrome c shells of the neighboring nanoparticles. Also, the formation of aggregates or scaffolds in the bio-nano hybrid system, may bring two nanoparticles closer. So their electron clouds get perturbed. This kind of perturbation can be manifested in the form of a coupling between the SPR bands. In the case gold nanoparticle–biomolecule hybrid systems, this inter-plasmon coupling is isotropic. But in the case of gold nanorod–biomolecule hybrid systems, this inter-plasmon coupling is anisotropic. These kinds of electronic interactions are manifested in the form of a color change in both gold nanoparticles and nanorods after binding with biomolecules. This observation can be used for developing cheaper and efficient technologies for medical diagnosis, proteomics, genomics and biotechnology. In 2005, Thomas, et al. (Ref. 11) reported the inter-plasmon coupling of cystiene and glutathione-capped Au nanorods. Gold nanorods possess two plasmon absorption bands. In the case of both cysteine and glutathione, a dramatic decrease in the intensity of the longitudinal surface plasmon absorption band, with a concomitant formation of a new band was observed (Fig. 11.4(a), (b)). The appearance of a new red shifted band at 850 nm in the presence of cysteine/glutathione results from the coupling of the plasmon absorption of Au nanorods assisted through self-assembly. The longitudinal alignment of nanorods after the binding of cysteine and glutathione has resulted in the anisotropic coupling of the SPR band.

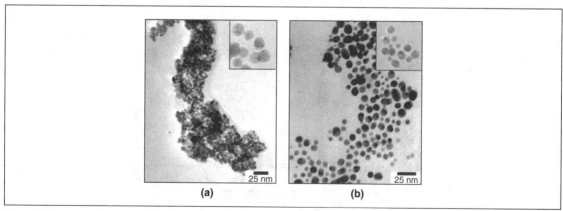

Fig. 11.3: *TEM image of cysteine (a) and vancomycin (b) capped gold nanoparticles, in the aggregated state after cryodrying at concentrations of 6.7 and 50 µg/mL. Reprinted with permission from Gu, et al. (Ref. 9). Copyright (2003) American Chemical Society.*

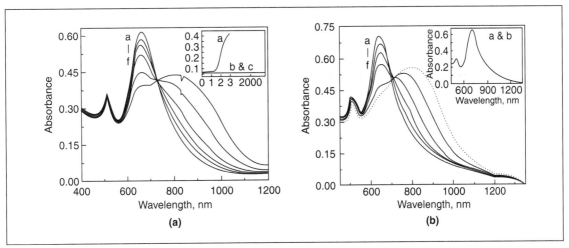

Fig. 11.4: *(A, B) Absorption spectral changes of Au nanorods (0.12 nM) in acetonitrile/water (4:1) on addition of (A) cysteine at: (a) 0 (b) 1.75 (c) 2.0 (d) 2.25 (e) 2.5, and (f) 3 µM, or (B) glutathione at: (a) 0, (b) 7, (c) 9, (d) 11, (e) 13, and (f) 14 µM. Figure 11.3(a) (inset): Changes in optical density at different concentrations of: (a) cysteine, (b) tyrosine, and (c) leucine. Figure 11.3(b) (inset): Effect of addition of 1-hexylmercaptan at (a) 0, and (b) 10 µM. Reprinted with permission from Thomas, et al. (Ref. 11). Copyright (2005) American Chemical Society.*

11.3 Different Types of Inorganic Materials Used for the Synthesis of Hybrid Nano-bio Assemblies

11.3.1 Noble Metal Materials

Normally nano materials made of silver and gold are used for biological applications. There are various types of noble metal nanomaterials such as noble metal nanoparticles, nanorods, metal nanoshells, nanocages, etc. Nanoshells (Chapter 10) constitute a novel class of optically tunable nanoparticles that consist of a dielectric core surrounded by a thin gold shell. Antibodies can be immobilized to the gold surface of nanoshells (Ref. 12). The gold surfaces of nanoshells are considered to be biocompatible and polymers such as polyethylene glycol (PEG) may be attached to nanoshell surfaces to further enhance biocompatibility and improve circulation in the bloodstream (Ref. 13). Silver (Ref. 14) and gold (Ref. 15) nanoparticles are used as intracellular SERS (surface-enhanced Raman scattering) probes. The advantages of noble metal nanoparticles are that they offer and/or function as:

1. Ideal immuno labels for transmission electron microscopy
2. Efficient contrast agents for optical microscopy

3. Powerful surface-enhanced Raman scattering (SERS) probes
4. Inert to physiological conditions
5. Most biocompatible nanomaterials
6. Ideal systems for immobilizing biomolecules.

11.3.2 Semiconductor Nanocrystals (Quantum Dots)

Quantum dots are small (<10 nm) inorganic nanocrystals that possess unique luminescent properties; their fluorescence emission is stable and can be tuned by varying the particle size or composition. Quantum dots are generally composed of atoms from groups IIB–VI or III–V of the periodic table and are defined as particles with physical dimensions smaller than the exciton Bohr radius (Ref. 16). This size leads to a quantum confinement effect, which endows nanocrystals with unique optical and electronic properties. Semiconductor nanocrystals can be capped with inert oxides or sulphides and can be attached to a biomolecule with a specific function. The salient features of semiconductor nanocrystals are as follows:

1. Size-tunable emission (from the UV to the IR) of quantum dots
2. Narrow spectral line widths
3. High luminescence
4. Continuous absorption profiles
5. Stability against photobleaching
6. Ideal immuno-labels for *in vitro* and *in vivo* fluorescent imaging

11.3.3 Magnetic Nanoparticles

The phenomenon of magnetism has received significant attention in life sciences and medical sciences. Intracellular ferrimagnetic nanocrystals of greigite (Fe_3S_4) magnetite (Fe_3O_4) have been found in magnetotactic bacteria (Ref. 17). These ferrimagnetic nanocrystals are aligned in a linear fashion in the intracellular part called 'magnetosome', which helps the bacteria in its alignment and motion parallel to the geomagnetic field. A variety of remarkable applications in biology such as in the field of biosensors, magnetic resonance imaging, drug delivery and magnetic fluid hyperthermia can be attained by the synergy of magnetic nanocrystals and biomolecules. Magnetic nanocrystals are size-compatible, competent enough to inter-relate with biological entities, detectable in an applied magnetic field and capable of conveying energy from an alternating magnetic field. Magnetic nanomaterials have a variety of applications in the following sphers:

1. Proteomics
2. Molecular cell biology
3. Medical science

4. Analytical biochemistry
5. Clinical diagnostics
6. Microbiology
7. Immunology
8. Biotechnology
9. Targeted drug delivery

11.4 Applications of Nano in Biology

Accurate and sensitive detection of water-soluble analytes such as toxins, carbohydrates, ionic species, and various biomolecules including DNA, proteins and peptides is a highly sought after scientific goal with implications in healthcare and industrial applications (Ref. 18). The interaction of a target molecule with a protein receptor in a biological recognition process is often associated with a change in the protein conformation as a response to the binding event. Many scientists are focusing on the design and development of recognition-based sensing assemblies that can account for such changes via signal transduction in a medium of interest (Ref. 19). The above drive leads us to the key issue concerning the fabrication of nano-bio hybrid systems. First, we have to think about inorganic nanoparticles in aqueous medium. These nanoparticles have either a positive or a negative charge on their surface depending on the synthetic methodology adopted for their preparation. A majority of biomolecules have electrostatic charge due to the presence of acidic and basic functional groups. Thus electrostatic self-assembly of biomolecules around the nanoparticle can be used to fabricate nano-bio hybrid systems. Another protocol used for the fabrication of assemblies is the immobilization of biomolecules above the surface of inorganic nanoparticles by using covalent bonding with the aid of an anchor molecule, see Scheme 11.1 (Plate 7). Both the methodologies can be used according to the requirement of the desired application. The tactics used in the construction of nanoprobes for a specific purpose can be adopted from the logic of bio-conjugation. This means that we are utilizing the principle of molecular interactions in the construction of nanoprobes. Immobilization of the molecular conjugate of the target molecule on the nanoparticle surface is the most efficient protocol. For example, antibody–nanoparticle hybrid systems can be used for the detection of antigens and vice versa. Such nanoprobes are now available in the market. The kind of nanoprobes used for optical microscopy depends on the nature of the technique. For instance, semiconductor quantum dots are used for confocal fluorescence microscopy of biological samples while noble metal nanoparticles are used for transmission electron microscopy of biological samples. Among the metal nanoparticles, gold nanoparticles are widely used due to their unique properties such as strong optical absorption, chemical inertness and ease of surface functionalization.

Inorganic nanoparticles such as noble metal nanoparticles, magnetic metal oxide nanoparticles and semiconductor nanocrystals are used for bio-conjugation. They are further utilized in different analytical applications. In normal protocols, an antibody of biologically important molecule is immobilized on the surface of the nanoparticle which, in turn, is used for applications such as molecular detection, targeted drug delivery and biological imaging. Sensing studies utilizing Fluorescence Resonance Energy Transfer (FRET) between a fluorescent donor molecule bound to the target and an acceptor attached to a receptor protein, have been widely used to study receptor–ligand interactions and changes in protein conformation upon binding to a target analyte. It has also been used to study changes in the solution conditions (e.g., temperature, pH conditions, etc.). FRET is extremely sensitive to the separation distance between the donor and acceptor, and is ideal for probing such biological phenomena. For example, Clapp, *et al.* (Ref. 20) used luminescent CdSe-ZnS core-shell quantum dots (QDs) as energy donors in FRET assays. The interface of biological systems and inorganic nanomaterials has recently attracted widespread interest in biology and medicine. Nanoparticles are believed to have potential as novel intravascular probes for both diagnostic (e.g., imaging) and therapeutic purposes (e.g., drug delivery) (Ref. 21).

11.4.1 Biological Imaging Using Semiconductor Nanocrystals

In vivo targeting by using semiconductor quantum dots has proved to be highly feasible, as found out by Akerman, *et al.* (Ref. 22). It was found that ZnS-capped CdSe quantum dots coated with a lung-targeting peptide accumulate in the lungs of mice after injection, whereas two other peptides specifically direct quantum dots to blood vessels or lymphatic vessels in tumors. It was also seen that on adding polyethylene glycol to the quantum dot, coating prevents the non-selective accumulation of quantum dots in reticuloendothelial tissues. These results encourage the construction of more complex nanostructures with capabilities such as disease sensing and drug delivery. ZnS-capped CdSe quantum dots emitting in the green and the red (550 nm and 625 nm fluorescence maxima, respectively) were coated with peptides by using a thiol-exchange reaction. These peptide-coated quantum dots were injected into the tail vein of a mouse and allowed to circulate for a given time. Then it was frozen, sectioned and examined under an inverted fluorescent microscope or confocal microscope. The use of peptides to target drug carrying nanostructures such as those composed of dendrimers or stabilized drug nanocrystals is also possible (Ref. 22).

11.4.2 Immuno Fluorescent Biomarker Imaging

A critical requirement in molecular cell biology is the localization of specific biomolecules in cells and tissues. Immuno-fluorescent labeling is the standard approach, but fluorescence-based imaging is limited in spatial resolution by the wavelength of light. Hence, the application of transmission electron microscopy (TEM) in conjunction with immuno-labeling has proved to be advantageous for high-resolution structural studies (Ref. 23). The prevalent strategy for immuno-localization via TEM is to employ antibodies conjugated with colloidal gold of various dimensions (Ref. 24). The recent development of luminescent semiconductor nanocrystals, also termed as quantum dots (QDs), for immuno detection raises the possibility of their use as probes (Ref. 25). For instance, we can consider target selective staining by using CdSe@ZnS@PEG core shell quantum dots (see Chapter 9 for a discussion on core-shell nanoparticles). This quantum dot is red emitting and its solubility can be increased by capping it with PEG (polyethylene glycol). We can denote these quantum dots as QD-PEG and QD-PSMA for the one without and with antibody anchoring, respectively.

Target-specific fluorescent imaging is shown in Fig. 11.5 (Plate 7). The methodology adopted in fluorescent imaging is the same as that explained in Scheme 11.1 (Plate 7). The anchor molecule used here is an antibody of prostrate selective membrane antigen (PSMA-Ab) (Ref. 26). The target molecule is antigen present in the cell membrane. This antigen is present only on the cell membrane of a prostrate cancer cell (C4-2). The mechanism of target-specific fluorescent cellular imaging entails bio-conjugate formation by the hybridization of the antibody on the nanoparticle surface with the antigen present of the cell wall of the cancer cell. Fluorescent staining happens only in the case of (a). This is due to the presence of the antigen containing cancer cells (C4-2) and antibody-capped nanoparticles (QD-PSMA). In the case of (b), the quantum dots are unable to attach on the cancer cell wall surface because they are not anchored with the antibody. In case (c), the quantum dots are unable to attach on the cell wall surface even though they are anchored with the antibody because of the absence of the target antigen on the cell wall of non-cancerous PC-3 cells. The above-mentioned cases illustrate the target specificity of immuno-fluorescent imaging and unveil the potential of semiconductor nanocrystal-based labels for medical diagnosis. The next example illustrates the imaging of the nucleus of a cell by using fluorescence microscopy and transmission electron microscopy. The nuclear promyelocytic leukemia (PML) protein was chosen as the target biomolecule (Ref. 27). The localization of PML protein in discrete sub-nuclear bodies has been well characterized with both fluorescence and electron microscopy. The inorganic nanoparticles used for fluorescence microscopy and transmission electron microscopy are antibody-anchored CdSe 10–15 nm nanocrystals and Au 10 nm nanoparticles, respectively.

Figure 11.6 (Plate 8) clearly explains the potential of inorganic nanoparticles for target selective imaging of intracellular parts. From this figure it is clear that both fluorescence and transmission electron microscopies are used for imaging the neucleus of the HEp-2 PML I cells. The mechanism of the imaging is the same as showed in Scheme 11.1. It also illustrates the potential of immuno-targeted gold nanoparticles (immunogold) for high resolution target selective imaging of intracellular parts with the aid of transmission electron microscopy. The schematic representation of target-selective binding of immunogold labels on cancerous cells is shown in Scheme 11.2 (Plate 8).

11.4.3 Immunogold Labeling

The term 'immunogold' indicates an immuno-targeted gold nanoparticle, which is functionalized with an antibody of a specific biomolecule of interest. In this protocol, gold nanoparticles are used as labels for imaging cell lines and tissues. Immunoglobulin G-capped gold nanaoparticles are used to image pathogenic organisms like *Staphylococcus aureus*, *Staphylococcus pyrogenes* and *Staphylococcus saprophyticus*. Ig G can bind specifically to the pathogens created by the bacteria (Ref. 27). Thus Ig G-capped gold nanoparticles are used to label the bacteria specifically.

Figure 11.7 illustrates a TEM image of the selective interaction of Immunoglobulin G-capped gold nanoparticles to the bacterium, *Staphylococcus saprophyticus*. From Fig. 11.7, it is shown that targeted gold nanoparticles can be used for the immuno-targeted imaging of bacteria. Figure 11.7(a) represents the selective binding of Ig G-capped gold nanoparticles. There is no labeling when citrate capped (Fig. 11.7(b)) and BSA-capped (Fig. 11.7(c)) gold nanoparticles are used. Gold nanoparticles have a good optical absorption in the visible region. Thus they can also be used for optical microscopy. Many companies are selling immunogold probes in the market.

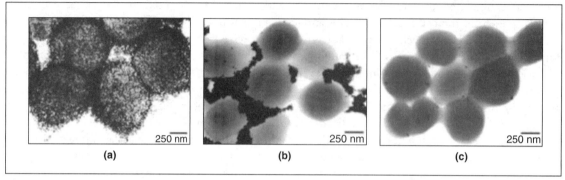

Fig. 11.7: *TEM images of Staphylococcus saprophyticus obtained after incubating these bacteria with: (a) Au-IgG nanoparticles, (b) unmodified gold nanoparticles and (c) Au-BSA nanoparticles. Reprinted with permission from Ho, et al. (Ref. 27). Copyright (2004) American Chemical Society.*

11.4.4 Diagnostic Applications of Immuno-targeted Nanoparticles

One of the major applications of immunogold labeling is that it facilitates the early detection of cancer. Gold nanoparticle-antibody bioconjugates are used for the vital imaging of cervical cancer cell suspensions as well as for fresh cervical biopsies. The principle applied for imaging the pre-cancerous tissues by reflectance imaging is based on the attachment of gold nanoparticles to probe molecules with high affinity for specific cellular biomarkers. Gold nanoparticle-antibody bioconjugates can be used as vital reflectance agents for the *in vivo* imaging of cancer affected parts for laser scanning confocal reflectance microscopy (Ref. 28). In cervical cancer cells, the epidermal growth factor receptor (EGFR) is over-expressed. The target selective imaging of cervical cancer cells (SiHa) was carried out by using 70 nm-sized gold nanoparticles anchored with monoclonal antibodies against EGFR. The mechanism of nanoparticle staining is similar to that cited in previous examples. This methodology, can deliver a cheaper technology for the early detection of cancer for the people in the Third World countries.

Anti-epidermal growth factor receptor antibody-conjugated gold nanoparticles (anti-EGFR/gold conjugates) were used to image the suspended SiHa cell lines as well as the biopsy of the cervical tissue. The scanning reflectance microscopic image shows efficient staining of SiHa cells by gold nanoparticles. The cross-sectional images (Figs 11.8 (c) and (d), (Plate 9)) show that the nanoparticles are staining only the outer surface of the cell. The target selectivity is clear from the absence of staining for the BSA-capped gold nanoparticles (Figs 11.8 (e) and (f), Plate 9).

11.4.5 Targeted Drug Delivery Using Nanoparticles

By definition, targeted drug delivery implies the slow and selective release of drugs to the targeted organs. Nanoparticles are 100–1000 times smaller than human cells. A drug-carrying magnetic nanocrystal or a drug-carrying fluorescent particle can be synthesized. We can fabricate the desired nano-sized drug delivery vehicle with the required properties. Two different kinds of nanomaterials are used for drug delivery applications. One kind is organic, while the other constitutes organo-inorganic hybrid systems. Each kind of vehicle has its own advantages and disadvantages. Here we illustrate an example, namely vancomycin-capped gold nanoparticles on different kinds of micro-organisms (Ref. 9).

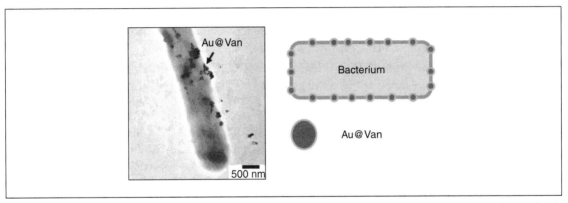

Fig. 11.9: *TEM image of E. coli after being treated by Au@van at minimum inhibition concentration (50 μg/mL). The scheme represents a cross section of E-coli after Au@van immobilization. The scheme shows that Au@van is selectively sitting on the cell membrane of the bacterium. Reprinted with permission from Gu, et al. (Ref. 9). Copyright (2003) American Chemical Society.*

The antibacterial activity of Au@van can be checked with different kinds of bacteria. The minimum inhibitory concentration (MIC) of vancomycin-capped nanoparticles is much lower than that of pure vancomycin (only one-fiftieth of pure vancomicin). Au@van efficiently binds with both gram-positive and gram-negative bacteria. It also acts effectively against vancomycin resistant *E. Coli* (VRE). This suggests that nanoparticle-based drugs have great potential for the treatment of drug-resistant microbes.

The key point in the fabrication of such a vehicle is related to its mode of usage, i.e. whether it is delivered orally or directly injected into the blood. A drug delivery vehicle should be chemically stable at the locations to which it is transported. Thus key aspects in such fabrication are as follows:

1. Lack of cytotoxicity
2. Absence of side effects
3. Chemical stability at transported locations
4. Target selectivity

5. Good performance
6. Cost-effectiveness.

From the above examples, it can be safely concluded that nanobiology can lead to better drug delivery systems, and better technologies for bioimaging and synthetic viruses for gene delivery.

11.5 Nanoprobes for Analytical Applications—A New Methodology in Medical Diagnosis and Biotechnology

Several probes have been used during the past decades to understand the mechanisms of biological processes. The probes are normally molecular in nature, which are either incorporated with a fluorescent tag or labeled with a radioactive isotope. In order to understand the biological activity of a target molecule, the same is labeled with either a fluorescent tag or radioactive isotope and injected into the biological system for tracing the pathway of the molecule of interest. As a result of the malignancies or infections in the biological systems, some of the proteins are over-expressed. These over-expressed proteins are called 'biomarkers' for each malignancy or disorder. The antibody of the biomarker is used as a molecular probe with appropriate labels. As described earlier, the inorganic nanoparticles exhibit size compatibility and have excellent optical features in the visible region. The trend of using nanoparticles for probing biological processes is very recent. The nanoparticles exhibiting different properties such as light scattering, fluorescence or magnetism are used for particular applications. Ferrimagnetic Au-Fe_3O_4 nanocomposite particles are used as pre-concentration probes for peptides at extremely low concentrations (10^{-8}–10^{-7}M), which are formed by the digestion of cytochrome c by enzymes (Ref. 8). The mass spectrometric analysis of the enzyme digests of proteins is a methodology in proteomics for the elucidation of amino acid sequence. But the chances of obtaining the mass signature of these peptides in MALDI-TOF MS analysis are slim due to the large concentration of ionic impurities in broth obtained after enzyme digestion. The digest contains several impurities such as phosphate, chloride, sodium dodecyl sulphate and urea. These impurities are able to suppress the molecular ion peak in MALDI-TOF MS analysis. The electrostatic adsorption of each peptide on Au-Fe_3O_4 seed is driven by the surface positive charge of the peptide. This positive charge density depends on the pH of the medium and the intrinsic pI value of each peptide, as discussed earlier (Ref. 8). These seeds are incubated overnight in the digest and separated with the aid of applied magnetic field. The ionic impurities and urea are removed by washing with water. The MALDI-TOF MS analysis shows mass signatures of each peptide component. Another report describes the utility of magnetic nanoparticles for the detection of pathogenic bacteria in biological fluids by the selective adhesion of immunoglobulin G (IgG)-capped Fe_3O_4 nanoparticles on the cell wall of the bacterium (Ref. 27). The MALDI-TOF MS analysis of the residue obtained by magnetic separation shows the mass signatures of the pathogens secreted by the bacteria. This protocol shows the efficient detection from urine samples of pathogens at very low concentrations. The detection protocol of pathogens using Ig G-capped magnetite (Fe_3O_4) nanoparticles can be used in the clinical detection of many diseases in their early stages (Ref. 27).

The surface enhancement of Raman signals by noble metal nanoparticles has been known for several years. The use of noble metal nanoparticles as surface-enhanced Raman scattering (SERS) probes in biological applications was first investigated by the research group of Mirkin (Ref. 29). Their salient features are given below.

1. Metal nanostructures are useful for fabricating sensitive optical probes, based on enhanced spectroscopic signals.
2. The remarkable effect is connected with the confined optical fields is SERS.
3. SERS enhanced features can attain larger intensity, which produces signals to a level comparable to or even better than fluorescence.
4. In contrast to fluorescence, which creates comparatively broad bands, Raman scattering creates a distinctive spectrum composed of a number of narrow spectral lines, ensuing in well-noticeable spectra.
5. SERS probes offer elevated spectral selectivity, and are resistant towards photo-bleaching.

Metal nanostructures capped with Raman active molecules can be used as intracellular SERS probes. Silver nanoparticles capped with 4–mercapto benzoic acid were used as SERS probes for sensing the pH of the cytoplasm of Chinese hamster ovary cells (Ref. 30). Gold nanoparticle capped with indocyanine green is used as an intracellular SERS probe for rat prostrate carcinoma cell line (Ref. 15).

The SERS probe for immunoassay is fabricated by the integration of a noble metal nanoparticle, a Raman active dye and a target selective biomolecule. The target selective biomolecule is anchored to the nanoparticle through the dye molecule by covalent bonding. The analyte molecule is adsorbed on the gold coated glass plate functionalized with dithiobis (succinimide undecanoate) (DSU). The succinimide moiety of DSU reacts with the analyte molecule forming an amide linkage with the mercapo undecanoyl group on the gold coated glass plate. Thus the covalently fixed analyte molecule is marked by the bioconjugate formation as shown in Scheme 11.3 (Plate 9). The Raman dyes used for SERS experiments are bifunctional, which include disulphides for chemisorption to the nanoparticle surface and succinimides for forming amide linkage with monoclonal antibodies. In this way probe fabrication will confine the dye molecules in the optical field of the noble metal nanoparticles and confinement of the analyte for the enhancement of Raman modes occurs. Thus the hybridization of the monoclonal antibody tail of SERS probe with the analyte molecule on the Au coated glass substrate will result in the adhesion of probe on the substrate. The substrate is washed and analyzed for the SERS intensity from the dye. There are earlier reports similar to the antigen antibody hybridization protocol; a DNA strand with a known sequence can be used instead of a monoclonal antibody in the SERS probe, in order to detect target DNA (Ref. 29). The detection of extremely small quantities of biomarker is seldom achieved by using the normal methodologies of immunoassays.

Prostate-Specific Antigen (PSA), a 33-kDa glycoprotein, a biomarker for prostate cancer, is present in blood plasma, in concentrations ranging between 4 and 10 ng/mL for a healthy adult male. As a result of an increase in the occurrence of prostrate cancer, the chances of the complex formation of PSA and α-antichymotrypsin are elevated. This phenomenon decreases the amount of free PSA in cancer patients. Thus the concentration of free PSA is important for the determination of the extent of prostate cancer.

A protocol has been developed recently for using SERS-based immunoassay using Au nanoparticle probe for femto molar detection of PSA (Ref. 31). 5, 5′-dithiobis (succinimidyl-2-nitrobenzoate) (DSNB) was used as the bifunctional Raman dye to connect Au (32 nm) nanoparticle and the monoclonal antibody of PSA. The immobilization of PSA was done on an Au-coated glass surface which is functionalized with dithiobis (succinimide undecanoate) (DSU). The PSA solution in human serum was taken in a concentration between 1 μg/mL (30 nM) to 1 pg/mL (30 fM) and was used for fixing on the DSU fuctionalized Au-coated glass plate. After the fixation of PSA at different concentrations on the substrate, it was allowed to hybridize with SERS probes as shown in Scheme 11.3. The SERS probe consists of a self-assembled monolayer of 5-merapto 2-nitro benzamide attached to a monocolonal antibody. The antibody was anchored to the 5-meracapto 2-nitro benzoyl group with a peptide linkage, which was formed by the condensation of the amino group of the antibody and the succinimidyl 2-nitro benzoate group. The Raman scattering intensity of the symmetric stretching of the nitro group of the 5-meracapto 2-nitro benzoyl moiety at 1338 cm^{-1} was taken as the maker signal for the immunoassay. In Fig. 11.10(a), the spectra were acquired using 60-s integration of SERS-based immunoassay. As mentioned above, the Raman features of the DSNB-labeled nanoparticles at 1338 cm^{-1} demonstrate a gradual enhancement as a function of the concentration of PSA. The concentration level shown in Fig. 11.10 is critical for prostate cancer diagnosis (Ref. 32). The trace labeled blank represents the Raman spectrum of the sample in the absence of PSA on the substrate. Additional information can be obtained from the dose-response curve shown in Fig. 11.10(b). This curve was created by plotting the intensity of the symmetric stretching of the nitro group at 1338 cm^{-1} versus logarithm of the concentration of PSA in pg/mL used for analysis. This report demonstrates the potential of the gold nanoparticle-based SERS probe for the early detection of prostrate cancer in a quantitative fashion. This also unveils the utility of nanoparticle-based SERS probes in a variety of applications in the fields of medical diagnosis and biotechnology.

11.6 Current Status of Nanobiotechnology

An overlap of nanotechnology, biotechnology and bio-informatics has resulted in the advent of a new technology called 'nanobiotechnology' (see Scheme 11.4). It is defined as a field that applies the principles of nanotechnology and bio-informatics to probe and modify biological systems or to apply the principles of life sciences and surface sciences to develop novel devices and systems for biocatalysis. During the past decade, this technology has reached a level, which enabled it to develop probes for understanding the intracellular and intercellular biological process, novel targeted labels for biological systems, and efficient nanoparticle-based immunoassays for the detection of biomarkers at extremely low concentrations and pre-concentration probes for selectively adsorbing the analyte from crude samples.

As compared to other branches of nanotechnology, nanobiotechnology has a high potential for catering to societal needs. The application of this technology in medical diagnosis and therapy is well known. There are reports highlighting the use of nanoceramic materials based supports for enzyme catalysis. This demonstrates the potential of nanomaterials as supports for several processes in industrial biotechnology. The application of nanomaterials for protein separation and analysis is also well-known. Many companies

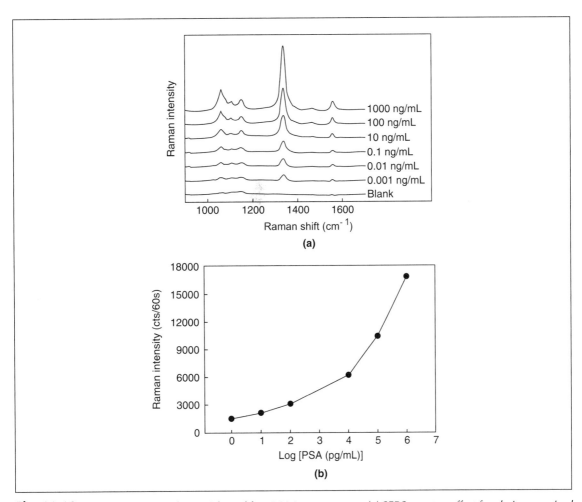

Fig. 11.10: *Demonstration of a SERS-based free PSA immunoassay. (a) SERS spectra, offset for clarity, acquired at various PSA concentrations. (b) Dose-response curve for free PSA in human serum. Reprinted with permission from Grubisha, et al. (Ref. 31). Copyright (2003) American Chemical Society.*

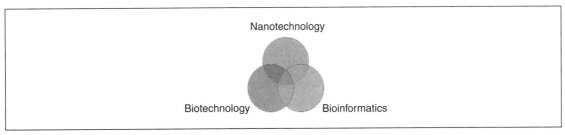

Scheme 11.4: *Schematic representation of the evolution of nanobiotechnology.*

are marketing nanoparticle-based labels in the market. A few of them are SPI-Mark™, Nanoink Inc, Nanoprobes, Biocompare Inc. and Tedpella Inc. Plenty of nanoprobes are available in the market for various applications in life sciences. Thus nanobiotechnology can be used in common applications such as clinical assay and medical diagnosis, and can provide cheaper and efficient solutions.

11.7 Future Perspectives of Nanobiology

Nanobiology can provide better methodologies for medical diagnosis and therapy. It can also provide better tools for biochemical analysis.

The future directions of nanobiology will be seen in the following areas:

- *Proteomics*: The methodology based on magnetic nanoparticles can be used for protein purification and separation. The magnetic nanoparticles can act as pre-concentration probes for MALDI TOF analysis. These techniques can be selectively used to adsorb each peptide formed by the enzyme digestion of a particular protein.
- *Medical sciences*: Nanoparticle-based targeted imaging of tissues and cell lines can be utilized for the early detection and screening of diseases. Also antibody-capped nanoparticles can be used for the selective destruction of tumors.
- *Biotechnology*: Nanoceramics can be used as inert supports for enzyme catalysis. Thus enzymes supported on nanometer-sized inert metal oxides have potential applications in industry. Magnetically triggered drug delivery and gene delivery vesicles are also promising areas. The development of nanoparticle-based SERS and fluorescent probes for immunoassays and DNA finger printing is a novel methodology in biochemical analysis.
- *Microbiology*: Nanobiology can be used to efficiently understand the pharmacological responses of different micro-organisms. The mechanism of drug resistance, drug action and drug uptake by the micro-organism can be probed by nanoparticle-based spectroscopic techniques. The selective imaging of intracellular parts is another salient application of nanoprobes. The SERS probes can deliver chemically selective vibrational information of intracellular parts. Finally nanobiology can offer target selective probes to examine fundamental issues such as cell division and membrane transport.

Review Questions

1. How do we probe the nature of molecular interactions between nanoparticles and biomolecules?
2. What makes the fusion of nanotechnology and biology possible?
3. What are the specific properties of metal nanoparticles in understanding biology?

4. What are the analytical applications of nanoprobes in biology? How single molecule detection become possible?
5. How can biomolecules be used for nanomaterial assembly?

References

1. Mirkin, C.A.; R.L. Letsinger, R.C. Mucic and J.J. Storhoff, *Nature*, **382** (1996), pp. 607–608.
2. Alivisatos, A.P.; K.P. Johnsson, X.G. Peng, T.E. Wilson, C.G. Loweth, M.P. Bruchez and P.G. Schultz, *Nature*, **382** (1996), pp. 609–611.
3. Rosi, N.L. and C.A. Mirkin, *Chem. Rev.*, **105** (2005), pp. 1547–62.
4. Mandal, S., A. Gole, N. Lala, R. Gonnade, V. Ganvir and M. Sastry, *Langmuir*, **17** (2001), pp. 6262–8.
5. Brennan, L., D.L. Turner, P. Fareleira and H. Santos, *J. Mol. Biol.*, **308** (2001), pp. 353–65.
6. Storhoff J.J., A.A. Lazarides, R.C. Mucic, C.A. Mirkin, R.L. Letsinger and G.C. Schatz, *J. Am. Chem. Soc.*, **122** (2000), pp. 4640–50.
7. Pan, P., H.P. Gunawardena, Y. Xia and S.A. McLuckey, *Anal. Chem.*, **76** (2004), pp. 1165–74.
8. Teng, C.H., K.C. Ho, Y.S. Lin and Y.C. Chen, *Anal. Chem.*, **76** (2004), pp. 4337–42.
9. Gu, H., P.L. Ho, E. Tong, L. Wang and B. Xu, *Nano Lett.*, **3** (2003), pp. 1261–3.
10. Jiang, X., J. Jiang, Y. Jin, E. Wang and S. Dong, *Biomacromolecules*, **6** (2005), pp. 46–53.
11. Sudeep, P.K., S.T.S. Joseph and K.G. Thomas, *J. Am. Chem. Soc.*, **127** (2005), pp. 6516–7.
12. Chen A.M. and M.D. Scott, *Bio Drugs*, **15** (2001), pp. 833–47.
13. Talley, C.E., L. Jusinski, C.W. Hollars, S.M. Lane and T. Huser, *Anal. Chem.*, **76** (2004), pp. 7064–8.
14. Kneipp, J., H. Kneipp, W.L. Rice and K. Kneipp, *Anal. Chem.*, **77** (2005), pp. 2381–5.
15. Chan W.C., D.J. Maxwell, X. Gao, R.E. Bailey, M. Han and S. Nie, *Curr. Opin. Biotechnol.*, **13** (2002), pp. 40–6.
16. Dunin-Borkowski, R.E., M.R. McCartney, R.B. Frankel, D.A. Bazylinski, M. Posfai and P.R. Buseck, *Science*, **282** (1998), pp. 1868–70.
17. Iqbal, S.S., M.W. Mayo, J.G. Bruno, B.V. Bronk, C.A. Batt and J.P. Chambers, *Biosens. Bioelectron.*, **15** (2000), pp. 549–78.
18. Looger L.L., M.A. Dwyer, J.J. Smith and H.W. Hellinga, *Nature*, **423** (2003), pp. 185–90.
19. Clapp, A.R., I.L. Medintz, M.J. Mauro, B.R. Fisher, M.G. Bawendi and H. Mattoussi, *J. Am. Chem. Soc.*, **126** (2004), pp. 301–10.
20. Niemeyer, C.M., *Angew. Chem.*, **40** (2001), pp. 4128–58.
21. Akerman, M.E., W.C.W. Chan, P. Laakkonen, S.N. Bhatia and E. Ruoslahti, *PNAS*, **99** (2002), pp. 12617–21.

22. Boisvert F.M., M.J. Hendzel and D.P. Bazett–Jones, *J. Cell Biol.*, **148** (2000), pp. 283–92.
23. Jensen H.L. and B. Norrild, *Histochem. J.*, **31** (1999), pp. 525–33.
24. Chan C.W. and S. Nie, *Science*, **281** (1999), pp. 2016–18.
25. Gao, X., Y. Cui, R.M. Levenson, L.W.K. Chung and S. Nie, *Nature Biotechnology*, **22** (2004), pp. 969–76.
26. Nisman, R., G. Dellaire, Y. Ren, R. Li and D.P. Bazett–Jones, *J. Histochem. Cytochem.*, **52** (2004), pp. 13–18.
27. Ho, K.C., P.J. Tsai, Y.S. Lin and Y.C. Chen, *Anal. Chem.*, **76** (2004), pp. 7162–68.
28. Sokolov, K., M. Follen, J. Aaron, I. Pavlova, A. Malpica, R. Lotan and R. Richards-Kortum, *Cancer Research*, **63** (2003), pp. 1999–2004.
29. Cao, Y.W.C., R.C. Jin and C.A. Mirkin, *Science*, **297** (2002), pp. 1536–40.
30. Talley, C.E.; L. Jusinski, C.W. Hollars, S.M. Lane and T. Huser, *Anal. Chem.*, **76** (2004), pp. 7064–68.
31. Grubisha, D.S., R.J. Lipert, H.Y. Park, J. Driskell and M.D. Porter, *Anal. Chem.*, **75** (2003), pp. 5936–43.
32. Polascik, T.J., J.E. Oesterling, and A.W. Partin, *J. Urol.*, **162** (1999), pp. 293–306.

Additional Reading

1. Jones, Richard A.L., *Soft Machines: Nanotechnology and Life*, (2004) Oxford University Press.
2. Goodsell, D.S., *Bionanotechnology: Lessons from Nature*, (2004), John Wiley and Sons Inc.
3. Robert A. and Jr Freitas, *Nanomedicine, Vol. IIA: Biocompatibility*, (2003), Landes Bioscience.
4. Caruso, Frank, *Colloids and Colloid Assemblies: Synthesis, Modification, Organization and Utilization of Colloid Particles*, (2004), Wiley-VCH.
5. Prasad, Paras N., *Introduction to Biophotonics*, (2003), Wiley-Interscience.

Chapter 12

NANOSENSORS

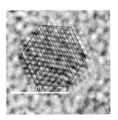

Sensors can be developed by using the properties of matter at nano dimensions. In such devices, the functional units will be nano entities though the material itself may not be nano in dimension. The properties of nanomaterials such as the large surface to volume ratio and photophysical properties will be used in generating a signal when analyte molecules interact with it. Several such sensors have been developed. Our objective here is to categorize these systems and put them in the perspective of materials discussed so far.

Learning Objectives

- What are nanosensors?
- What are the typical things one can sense with nanodevices?
- What are the properties used for sensing?
- What is the status of current research in this area?
- How useful can a smart dust be?

12.1 Introduction

One unifying property for all life forms is their ability to sense, perceive and react to situations. This characteristic is exhibited, albeit to different extents, by the most developed living organisms down to the lowest organisms. What is surprising, however, is that man, having proclaimed himself as the most advanced and developed of all species, looks at other 'lower' species for their exceptional ability to sense and react to their surroundings. They provide him with ideas and inspirations on how to develop and acquire an improved sense of perception. Since science has still not allowed man to tamper with and modify his own self to the extent he would like to, he attempts to develop external sensors and devices to obtain a better knowledge of his surroundings. This understanding equips him with the ability to react and eventually assume better control of his surroundings. Therefore, in his never-ending quest for controlling his locality, he tries to constantly innovate and invent devices equipped with better sensory perceptions than what has been bestowed on him naturally.

A careful analysis would reveal that the development of sensory devices has been vital for ushering in various technological advancements that have revolutionized civilizations. After all, every experiment, entails a measurement, which involves sensing something. In the modern day world, our ability to point to the exact location of a person owes its existence to a satellite that can sense the time delay between the signals it receives from various other satellites and translate it into distance on a three-dimensional space. Perhaps, it would be difficult to believe that this complex concept is also employed by a bat to sense its environment!

12.2 What is a Sensor?

In the context of this book, the term 'sensor' can be referred to any device that uses an active chemical species or component that generates a signal in presence of an analyte molecule. The signal, in turn, is used either directly or after suitable amplification to trigger a suitable detector.

Thus, the three essential components of a sensor device are:

1. *The responding element*, which recognizes the presence of the analyte species and generates a signal. This is the principal component which 'sees' the molecule, ion or process. In general, a sensor element has to satisfy certain requirements like:
    (a) It should be capable of detailing the analyte in a qualitative and quantitative manner.
    (b) It should be able to detect even very small amounts of the analyte.
    (c) The signal it generates should be reproducible. This implies that the sensor should not have a strong affinity for the analyte, in which case the sensor will be passivated with the analyte after some time, leading to an irreproducible response. The affinity should be optimal so that the analyte is disengaged from the sensor after a short interaction time period, during which the sensor transfers its response to the next stage.
    (d) The sensor should be very selective and specific in its response towards a target molecule.
2. *An amplifier*, which receives the signal of the sensor as an input and amplifies it to a level that is acceptable for processing by the detector, and
3. *A detector*, which receives the output of the amplifier as an input and converts it, in a pre-programmed manner, to a parameter which represents either the analyte species or its concentration or both. In most cases, the detector is equipped with a feedback control through which it signals to the sensor that the input has been received and processed. This acts as a trigger for the sensor to release the analyte molecule. This process causes an induction time period during which the sensor is not available for response. Therefore, the induction period should be as small as possible. An overall schematic of a sensor device is represented in Scheme 12.1.

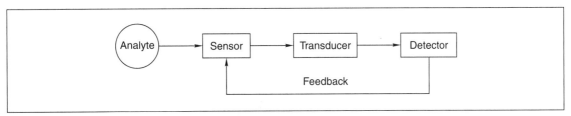

Scheme 12.1: *Schematic of the design of a sensor.*

12.3 Nanosensors—What Makes Them Possible?

So where do nanomaterials, and the properties that arise thereof, stand in the realm of things? As has been discussed in all the previous chapters, the size regime of nanomaterials gives rise to an entire gamut of physical and chemical properties, which are being currently explored by numerous research groups worldwide. The main properties of nanomaterials that are utilized for conceiving and designing a viable sensor are as follows:

1. Properties derived from the high surface to volume ratio
2. Optical properties
3. Electrical and electrochemical properties
4. Physical properties.

Apart from these broadly classified attributes, some specific properties also find application in the development of sensors. Such properties will be briefly discussed as and when required.

12.4 Order from Chaos—Nanoscale Organization for Sensors

A translation of liquid state properties to solid state is of utmost importance in the development of sensors and devices. One of the major hurdles in achieving this is that this transfer has to be done without altering the liquid-phase properties. In this process, assembling of the particles and an enhanced ordering have to be achieved to enable uniform and isotropic detection protocol. A variety of substrates are employed for carrying the nanoparticles, with the most common among them being metals like Au, Pt, etc. and conducting glass substrates. Metallic surfaces are utilized for the spectroscopic probing of the molecules using Surface Enhanced Raman Spectroscopy (SERS) that gives information about the alignment and the nature of interaction existing between the molecule and metal. On the other hand, transparent conducting glass substrates lend themselves to a variety of electrical and microscopic (as well as spectroscopic) probing, besides being technologically more attractive. Many methods are used to achieve the assembly and further ordering of the nanoparticles. The most important and relevant techniques are discussed here, with examples.

12.4.1 Self-assembly

This is one of the most widely used and simplest methods of organizing nanoparticles onto substrates. Normally, this method is adopted for attaching the nanoparticles (usually metallic) on surfaces like conducting glass or metal films. The substrate is first cleaned and activated suitably, with the process depending on the nature of the substrate and the nanoparticles. The surface is made either hydrophobic or hydrophilic as is necessitated by the type of the nanoparticle. In the case of metallic substrates, the cleaned surface might be sufficiently active for the attachment of the nanoparticles. Otherwise, the activated surface is functionalized with anchor molecules to facilitate the binding of the nanoparticles. These anchor molecules are characterized by the presence of certain functional groups to which the nanoparticles will adhere. A silane containing thiol (Ref. 1) or an amino functional group, capable of forming a covalent link, is best suited for the adhesion of gold nanoparticles. The type and extent of functionalization is controlled by the nature of application for which it is fabricated. For example, for electrochemical studies, a thick polymeric layer of alkoxysilane will hinder the transport of ions and electrons between the nanoparticles and the substrate. A representative schematic of the procedure is shown in Scheme 12.2.

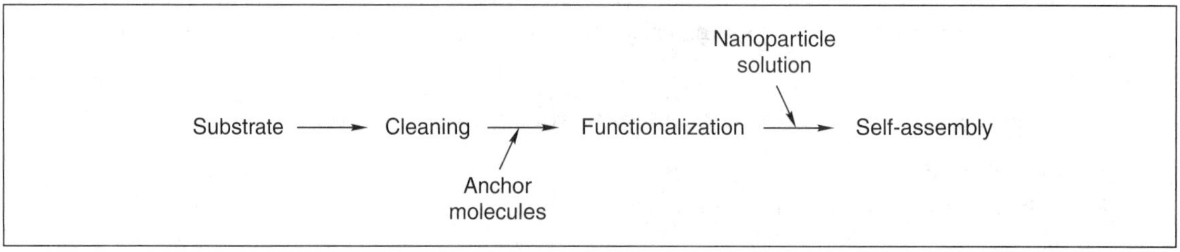

Substrate → Cleaning → Functionalization → Self-assembly
(Anchor molecules; Nanoparticle solution)

Scheme 12.2: *Schematic representation of self-assembly.*

The assembly thus achieved is referred to as a 'self-assembled layer' (SAL). More specifically, if a single monolayer adsorption is achieved, it is known as a 'self-assembled monolayer' (SAM). This is more relevant with respect to the organization of molecules on a metal or nanoparticle surface. It is thus known as there is no external driving force other than the mutual affinity between the nanoparticles and the functional groups on the substrate. This plays a crucial role in deciding the specificity of the nanoparticles that goes on to form the SAL or SAM. To know more about self-assembled monolayers, see Chapter 5.

12.4.2 Template Method

One of the main disadvantages of the self-assembly method is its sensitivity in terms of external parameters. It is almost impossible to obtain two similarly arranged surfaces, with the adhesion and the organization depending on various factors like solvent, concentration, temperature, type of substrate, and so on. In fact, even on a given surface, there is a high possibility of the existence of a coverage gradient, leading to inconsistency in results.

One way to overcome this is to fabricate a template with regions wherein the possibility of the adsorption of a nanoparticle is artificially enhanced. The most common example is the use of polymethylmethacrylate (PMMA) (Ref. 2) to create the necessary patterns on the substrate. On activating the plate, the portions containing the PMMA remain passive and hence the nanoparticles are not able to bind to the substrates at those places. Conversely, if a suitably charged microstructured polymer is chosen, instead of PMMA, the nanoparticles adhere on the substrate to the regions containing the polymer. An example in this case would be the patterning of Pd nanoparticles using a polydimethylsiloxane (PDMS) template.

An advancement of this technique is the use of light to create the patterns on a functionalized surface. The silanization is done with a suitable group such as amine or thiol in the case of gold nanoparticles. The substrate is then exposed to light in certain areas leading to oxidation of the groups. Exposure to nanoparticles makes the light-exposed regions passive. A schematic overview of the various methods is given below.

Although only a few of the template-based techniques have been discussed here, the possibilities with this methodology are innumerable, with creativity being the only limitation. See also Chapter 5 on self-assembled monolayers.

12.4.3 Biological Assembling

With assembling constituting the *raison d'etre* for the existence of life, it is hard not to think of strategies based on biological entities for assembling and patterning nanoparticles. Accordingly, many approaches have been developed by combining these two areas. Since an exhaustive discussion on this has already been provided in Chapter 5, a more detailed picture is beyond the scope of this book. Thus only a few of the interesting approaches utilizing biological entities for assembling and patterning are discussed here, see Scheme 12.3 (Plate 10).

Complimentary DNA strands can be used to self-assemble nanoparticles in the form of wires and helical structures (Ref. 3). Going further, the recent trend has been to immobilize DNA, either as oligonucleotides or as duplex strands, on substrates. This is followed by the deposition of nanoparticles to create specific patterns. These patterns arise due to the enhanced affinity of the nanoparticles towards certain residues of DNA. For example, it has been established that DNA strands can act as stabilizers for CdS nanoparticles and that the content of adenine influences its morphology and photophysical properties (Ref. 4). Accordingly, duplex DNA are immobilized on substrates and Cd^{2+} ions are assembled on it, with high degree of specificity. Following exposure to H_2S, these are converted into CdS nanoparticles. These methodologies offer a high degree of flexibility in generating various patterns of arrangement.

The recognition-based organizational ability of biomolecules has led to their qualification as ideal candidates for organizing and patterning nanoparticles. One of the strategies adopted for generating nanoparticle arrangements is through the interaction between antigens and antibody. Besides their high specificity, these interactions are chemically stable and offer a wide range of equilibrium constants, thereby allowing fine control over the organizational behavior and pattern. The most commonly exploited biomolecule for achieving this is the biotin-sav (See Box 12.1) combination. Nanoparticles are configured by tagging them to one of the biomolecules and arranged by using the other molecule. An extrapolation of this approach allows one the flexibility to order different types of nanoparticles. This can be achieved by capping each type of nanoparticles to one specific type of biomolecule, and then bringing about regularity by utilizing the recognition between the biomolecules involved in the capping.

Box 12.1
Biotin-Sav

The interaction between biotin, a member of the Vitamin B family, and streptavidin (sav), a tetrameric protein, is of great interest due to many significant reasons. Biotin, besides playing a part in the synthesis of fatty acids (like L-leucine and L-valine), is also believed to take part in gene expression and DNA replication. The binding constant between biotin and sav is exceptionally high and is stable over a wide range of pH and temperature. These aspects have made this system the focus of a number of studies with particular focus on drug designing and targeted drug delivery.

12.4.4 Lithographic Techniques

Lithography, which has its origin in the Greek words—*lithos* meaning stone and *graphy* meaning writing, has developed into one of the most sophisticated and accurate techniques for creating molecular and nano-architectures. Lithography can be classified into various types based on the type of probe used for accomplishing the arrangement. Some of these important classifications are Atomic Force Microscopy (AFM), Scanning Probe Microscopy (SPM), Focused Ion Beam (FIB) lithography and e-beam lithography. Besides providing lithographic resolution to a few nanometers, these techniques are also used in the characterization of surfaces. These procedures have been used for various advanced accomplishments such as creating a nanopen with a molecular 'ink' to develop patterns on surfaces. This device works by making the ink flow through micro-fluidic channels to an AFM tip, which then 'writes' on a suitable substrate. The basis and the use of these techniques are best discussed separately as an in-depth discussion is beyond the scope of this chapter. While these techniques hold the promise of taking nanoscience to a new level, their only drawback is their limited affordability and high sensitivity to handling. Some of the lithographic techniques along with their working principles are summarized in Table 12.1. The interested reader is urged to look up the relevant literature for further details (See Chapter 2 for references).

12.5 Characterization—To Know What has been Put In

The surface coverage and its uniformity are decisive factors in the functioning and reliability of the sensor or device. These, along with other information about the surface, can be measured with the help of a Quartz Crystal Microbalance (QCM) or investigated with various microscopic techniques like AFM and SPM. Other methods of analysis are discussed in the following sections, as and when the relevant properties are being used for detection and quantification.

Table 12.1: *Summary of some of the lithographic techniques and their working principles*

Technique	Working Principle
Atomic Force Microscopy	Oscillating tip which contacts the surface at the atomic level resulting in variation of the oscillation characteristics.
Scanning Tunneling Microscopy	A bias voltage is applied between the tip and the substrate and the variation in tunneling current is used to study surface morphology.
Dip Pen Nanolithography	An AFM tip which delivers molecules with atomic precision, onto a surface through a solvent meniscus.
Electron Beam Lithography	Electron beam exposes the resist causing physical or chemical changes at the exposed position, modifying the surface at those positions.
Focused Ion Beam Lithography	Lenses focus the metal ion beam resulting in deposition, etching or imaging of the surface.

After digressing to discuss a few existing procedures for the realization of nanoscale order and architectures, we will proceed with the main focus of this chapter—the development of sensors based on nanosystems. An attempt has been made here to classify the nanosensors according to the properties of the nanomaterials being utilized in the sensing process.

12.6 Perception—Nanosensors Based on Optical Properties

One of the most easily noticeable features, common to almost all types of nanoparticles, is their color. This intense, characteristic quality arises because of the excitation of the electron cloud present at the surface, in the case of metal nanoparticles. This feature in the Ultraviolet–Visible absorption spectroscopy (UV–Vis) is known as 'Surface Plasmon Resonance' (SPR). For semiconductor nanocrystals, quantum size effects operate resulting in an enhanced overlap of the electron and hole wave functions.

The occurrence of SPR provides information about the size of the nanoparticles. This is truer in the case of semiconductor nanoparticles, wherein there is a non-linear increase in the band gap with sizes approaching the Bohr exciton radius of the materials (See Box 12.2). Apart from this, it also provides information about the local environment of the nanoparticle. The position of SPR in UV-Vis is known to be a reflection of the local environment of the particle. Even a small chemical change in its surroundings (typically about a few hundred square nanometers) causes a monitorable shift in the occurrence of SPR. This property is exploited to track chemical changes taking place in the vicinity of the nanoparticles. If monitored in the liquid state, it might lead to: (a) a perceivable shift in SPR, (b) complete disappearance of the SPR band, sometimes accompanied by a simultaneous appearance of another band, or (c) change in the structure or intensity of SPR. Inter-particle interaction leads to a coupling of the SPR resulting in the emergence of another band, which is usually red-shifted as compared to the original feature. These types of interactions arise in solid-state immobilization of the particles, or due to analyte-mediated agglomeration. An analysis of the coupled plasmon band, which belongs to the latter category, has been used for sensing biomolecules like DNA (Ref. 5). Here, single strands of DNA act as capping agents for the nanoparticles. Recognition between the bases of one nanoparticle, to the complimentary bases on another particle, leads to duplex formation in suitable conditions. This draws the nanoparticles together, thereby inducing an aggregated state. This is reversed by altering the conditions (like temperature, pH, etc.), which causes the duplex DNA to unwind, leading to a disaggregated state of the nanoparticles. Sometimes, a metal ion can chelate with the functional groups on the nanoparticle surface, thereby bringing the nanoparticles within the interaction range (Ref. 6). This causes changes in the SPR of the nanoparticles; the effect can be reversed by adding a more powerful chelating agent of the metal ion, whereby the metal–ion bound nanoparticles are released back into the solution.

The sensing protocol involves the binding of one component of a recognition pair on the surface of a nanoparticle, which is present either in the solution or immobilized on a surface, and allowing the other component to interact with it. The recognition, interaction and binding events give rise to a change in the chemical environment of the nanoparticle. The nanoparticle responds to this modification, either by altering its SPR feature to a different frequency or by losing its resonant frequency altogether. Both these actions can be detected through UV-Vis spectroscopy. A knowledge of the kind and amount of change produced in the SPR helps in making an accurate qualitative and quantitative estimate of the analyte.

An alternative means of following these changes is through Surface Plasmon Resonance Spectroscopy (SPRS). This is widely used for opaque substrates, wherein the absorbance cannot be monitored. In such cases, light of a known intensity is shone on the substrate containing the nanoparticles and the intensity of the reflected light is monitored. The angle of incidence is varied and at a specific angle, the intensity of the reflected light falls to a minimum. This is because at a specific angle, which is greater than the total internal reflection angle, the electron oscillations of the nanoparticles resonate with the frequency of the incident light. The plasmon oscillations create an electric field which pervades to about 100 nm around the particle.

Box 12.2
Exciton Radius

An exciton, in the context of semiconductors, refers to the combination of an excited electron in the conduction band and the corresponding hole thus produced in the valence band. The electron and the hole constituting the exciton, always possess a mean separation distance, dependent on the nature and the size of the material. In the nanoscale size regime, the size of the semiconductor becomes small enough to be comparable to the exciton radius. At this limit, the electron energy levels can no longer be treated as a continuum, but are discrete. As a result of this discretization of the energy levels of the band gap of the semiconductor material, minor variations in the number of atoms constituting the semiconductor or its local environment have immediate consequences in the absorption and emission characteristics.

Any change in the environment of the particles within this boundary results in shifting of the angle at which maximum absorbance occurs. In a physical sense, the incident angle at which the maximum absorbance occurs is mainly controlled on the refractive index of the medium surrounding the nanoparticles. Thus SPRS is a powerful and sensitive analytical tool for comprehending binding events occurring at the nanoparticle surface. A schematic representation of the working of the SPRS is shown in Scheme 12.4 (Plate 10).

Nanoparticles cast either separately as free standing films or incorporated into film-forming polymeric matrices and gels, are being developed as sensors for gas molecules. Depending on the size of the gas molecules, they are either physisorbed or chemisorbed in the voids between the nanoparticles. This causes changes in the interparticle distance, which results in the decoupling of the interparticle plasmon. An alternative approach is to form films or gels of materials that are sensitive to the analyte molecules of interest, and then incorporate nanoparticles in them. The interaction between the analyte and the film/gel forming materials leads to a change in its physical structure, which is picked up by the nanoparticle. This gives rise to changes in the SPR features, sometimes leading to a complete disappearance of the interparticle interaction feature. The advantage of such systems lies in their high sensitivity and reproducibility. Efforts are on to optimize the residence time of the gas molecules in the matrix, which will result in increased efficiency of detection and also reusability of the material.

12.7 Nanosensors Based on Quantum Size Effects

The electrical, and consequently electrochemical, properties of materials in nanosize are entirely different from those exhibited by the bulk state of the corresponding substances. These variations stem from the fact that the electronic state of nanomaterials lies somewhere in between that of the bulk and atomic states. In bulk materials, there are distinct continua of valence and conduction bands. On the other hand, the electronic states in atoms and molecules are distinct with discrete energy levels. In comparison to these two, nanosystems possess a quasi-continuum state, wherein the bands are distinct, but with discrete energy levels. They exhibit size- and material-dependent spacing of the energy levels. This offers us a great handle to tune the electronic properties, by regulating the size of the particles.

The most unique property that arises in nanoscale materials, which is currently being developed further, is known as 'Coulomb Blockade' (CB). In the nanoregime, due to high surface/volume ratio, charge and energy quantization become the dominant forces in deciding electron transport. The charge quantization dictates the capacitance of the material and restricts it to the nanodimension. This implies that the energy that has to be supplied to add an electron to an uncharged nanoparticle varies inversely with the dimension of the particle and so far exceeds the thermal energy available at room temperature. This provides us with a handle to regulate and manipulate charges flowing through nanosystems. This phenomenon is known as Coulomb Blockade and has been detailed in both theory and experiments. Coulomb Blockade is discussed in more detail in Chapter 7.

Of late, sensor devices called electron turnstiles (Ref. 7) have been developed on the basis of this concept. These are being tried out for their applicability to count electrons. It consists of a number of islands interconnected by insulting barriers, through which charges can tunnel. By controlling the gating voltage of individual islands, one can regulate the number of electrons fed into the system.

With the rapid advancement of lithographic techniques and achievement of maneuverability of a single nanoparticle, research in single electron tunneling in nanoparticles has also gained prominence in recent years. Conventional electronics relies on the transport of charge carriers. In nanodevices, the transfer of the charges has to take place in a controlled manner. This is realized by tunneling of the charges. The set-up consists of two electrodes separated by a mesoscopic island of nanoparticles. The transfer of charges in such islands is largely governed by a minimum capacitance of the island. If that is the case, then the creation of a very small charge excess in the island affects its potential significantly. This acts as a feedback system and prevents further charging of the island, until the excess charge is dissipated to the other electrode. Essentially, this methodology enables us to control the transport of a small but definite amount of charge carriers. As is evident, these single electron tunneling events are greatly dependent on the capacitance of the island. If the nanoparticle comprising the islands has imbibed some of the analyte species present either in gaseous or liquid state, then the capacitance of the island is modified. This is reflected on its step value of the coulomb staircase in the I–V characteristics. If the affinity of the analyte to the nanoparticles is low, then the I–V characteristics would be similar to the nascent case. Thus, all it takes for a signal variation is a few atoms of the analyte. The unprecedented detection capability and sensitivity of single electron sensors (SES) make them very advantageous in specialized applications wherein detection is of utmost importance.

Semiconducting metal oxide nanoparticles are being widely researched as chemoresistive gas sensors (Ref. 8). Single gas detection, leakage detectors, and fire detectors, and humidity sensors fall under the category of gas sensors. The operating principle of these electronic noses is that of variation in the resistance of these materials, when exposed to certain gases. When the gas comes into contact with the semiconducting metal oxide nanoparticles, it results in structural variations of the particles. These variations might occur either at the surface or in the bulk of these particles. As a consequence, electrical conductivity of the material varies. The variation in the conductivity can be monitored and tabulated for a series of gases over a range of partial pressures. This is then analyzed to provide information about the best sensor for a particular gas over a given range of partial pressures. The important considerations to be taken into account while designing the sensor are its cross-sensitivity and detection capability.

The commercial production of some materials has also been started and they are more commonly known as electronic noses (Ref. 9). Essentially, an electronic nose functions just like a human nose and hence its name. These consist of an array of varying nano-metal oxides, each of which has a selective and specific response to certain gas molecules. While the changes in conductivity in a single type of nanoparticle film might not be sufficient to identify an analyte, the varied changes in the array of films produce a distinctive, identifiable pattern. Together, they are able to respond to mixtures of gases, providing both quantitative and qualitative information about them. The working is better shown as a schematic in Scheme 12.5 (Plate 11).

12.8 Electrochemical Sensors

A discussion of sensors based on their electrical properties will not be complete without detailing some of the aspects of electrochemical sensors. Normally, multi-layers of metal nanoparticles find extensive use as electrodes, since they are known to retain the electrochemical activity of the analyte species. The electrodes also mediate charge transfer between the bulk electrode and the analyte molecules. The single electron events discussed earlier induce changes in the double layer charging of the nanoparticle, which, in turn, modifies the redox potential of the system. This gives us an opportunity to quantify the charge transfer occurring at the electrode surface. The deviation from the peak voltages of the nascent electrode helps us to qualitatively arrive at the chemical nature of the analyte. The sensitivity of the electrode can be tuned by controlling the number of layers fabricated on it. The use of nanoelectrodes has fuelled an enormous increase in the surface area which causes a multifold increase in the sensitivity to the analyte molecules. As in the previously discussed techniques, a similar strategy of binding suitable receptor molecules on the surface of the nanoparticle is followed. On coming into contact with the analyte molecules, a host–guest interaction occurs, which is manifested in the I–V characteristics of the electrode. The specificity is best illustrated in the detection of dopamine (DA) in the presence of ascorbic acid (AA), using a cysteine-capped gold nanoparticle electrode (Ref. 10, Fig. 12.1). This drives home the point of specificity, since dopamine and ascorbic acid are not distinguishable through their I–V characteristics using a normal bulk electrode.

The enhanced sensitivity in this case arises because of the preferential catalytic oxidation of ascorbic acid at the surface of the nanoelectrode, which shifts the anodic current for oxidation of ascorbic acid to

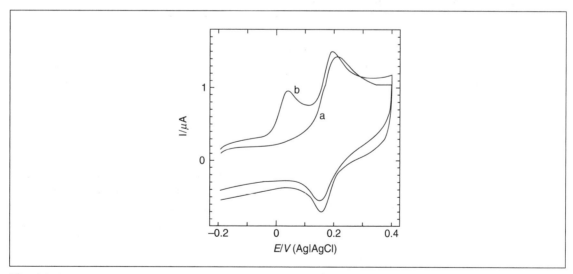

Fig. 12.1: *Cyclic voltammograms of a binary mixture containing equimolar concentrations (50 mM each) of AA and DA at the bare Au (a) and nano-Au (b) electrodes in 0.1 M PBS (pH 7.2). Scan rate: 100 m Vs^{-1}. Reprinted from Raj, et al. (Ref. 10). Copyright (2003), with permission from Elsevier.*

less positive values. Thus the oxidation of ascorbic acid is finished well before the onset of oxidation of dopamine.

In summary, an ordered nanoparticle array shows environment–dependent electrical properties (such as conductivity). These properties are modified by the chemical species present in its vicinity. The conductivity of nanoparticles is believed to occur due to:

1. Tunneling of electrons through the metal core.
2. Hopping of the electrons along the atoms constituting the chain of the ligand molecule encapsulating the nanoparticle.

By changing the parameters of the nanoparticle-modified electrode such as its particle diameter, space between the particles and the number of layers, the conductivity of the system can be altered. The analyte can be made to interfere with any one of the processes and hence can help vary the conductivity. This could lead to a sensing of the analyte.

The use of silicon nanowires (SiNWs) as gas sensors has been reported (Ref. 11). The I–V characteristics of bundles of SiNWs have been investigated as a function of exposure for a range of concentrations of various gas analytes (Fig. 12.2). They are found to possess high sensitivity to humidity and can act as water vapor sensors. The sensitivity is shown up in the form of resistance changes upon exposure to water vapors. Interestingly, the process is found to be perfectly reversible, with the resistivity reaching the original value on removing the gas analyte.

12.9 Sensors Based on Physical Properties

A new field of research has been emerging due to the utilization of physical properties at the nanoscale. This field is known as 'Nano-electro-mechanical-systems' (NEMS). This comprises a class of devices that relies on the mechanical properties at the nanoscale to power them. The fabrication of such devices is attracting heavy attention with newer strategies being developed everyday.

Under this category, cantilevers fabricated in the nanodimension are gaining prominence for use in sensor gadgets. The discussion on this subject will first focus of some of the known methods to fabricate these nanocantilevers followed by their applications and advantages.

The usual methods of nanolithography like AFM, STM and electron beam lithography can reach up to tens of micrometer dimensions, but are not economically viable for large-scale production. Therefore, FIB is a more compatible technique. Here a Ga ion beam is used to physically scoop out nanodimensional cantilevers from silicon substrates. Going a step further, FIB can be complemented with wet etching, wherein the FIB is used not to etch the surface but rather to dope the wet-etched surface selectively. The primary requirement for these techniques is that one has to start with a single crystal. Although widely done on a silicon substrate, it can also be extended to other surfaces.

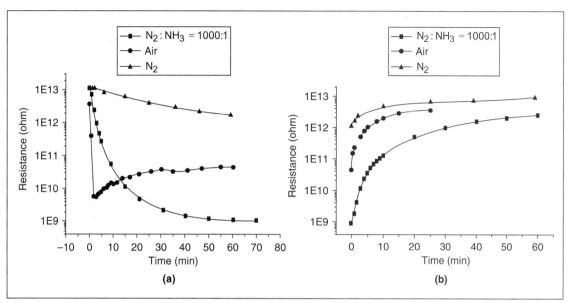

Fig. 12.2: *Electrical responses of the Si nanowire bundle to N_2, a mixture of N_2, NH_3 (NH_3 concentration: 1000 ppm), and air with a relative humidity of 60%; (a) When the gases were introduced into the chamber, and (b) when the gases were pumped away. Reprinted from Zhou, et al. (Ref. 11). Copyright (2003), with permission from Elsevier.*

The normal techniques of detecting cantilever motion, such as piezoelectricity and optical beam, are not useful in the case of nanosize cantilevers (Ref. 12). This is because the nanosize causes greater scattering of the light beam, thus making it difficult to exactly fix the resonant frequency. Therefore, techniques like electron tunneling or shuttling which are compatible with the dimension of the cantilever, are used for their detection. A preliminary set-up consists of two micro-machined electrodes, between which a nanocantilever is fixed. At the resonant frequency of the cantilever, one of the electrodes is made to act as a source with the other functioning as the drain. As the cantilever oscillates, it makes electrical contact with either of the electrodes and thus acts a gate for the flow of charges. This is determined by monitoring the current generated, which is typically of the order of pico amperes, for a bias voltage of few millivolts.

When the cantilever is exposed to analyte gas vapors, its resonant frequency is bound to vary, with the oscillations becoming more damped (Scheme 12.6). This will be discernible in the plot of bias voltage versus current. By comparing the voltage at which the peak current is obtained in both the damped and undamped cases, the effective mass of the cantilever before and after exposure can be calculated. The difference between the two readings will point to the quantity and nature of the gas to which the cantilever was exposed. Phenomenal mass sensitivity of a few femtograms has been achieved with the use of nanocantilevers. This corresponds to a multifold increase in sensitivity of nanocantilevers as compared to normal cantilever sensors. At the masses of these ranges, it can be safely assumed that the detected mass is proportional to the original mass of the molecule. Therefore, a very efficient way of calculating the mass, which is a characteristic property of every matter, is developed.

Moreover, recent research has been concentrating on developing techniques to capture the phase changes and the amplitude of vibration of the nanocantilever sensors. Achieving this would only result in unprecedented ability to image and view, as though through a microscope, the analyte molecules. One main limitation of this sensor is its inability to function in environments wherein a leakage current is possible, since this will obscure the actual current from the oscillating cantilever. Its use in biological areas is also restricted due to the increased presence of particulate matter. Liquid state applications are also constrained due to large-scale damping. However, research is underway to overcome these shortcomings, mainly inspired by its increased sensitivity.

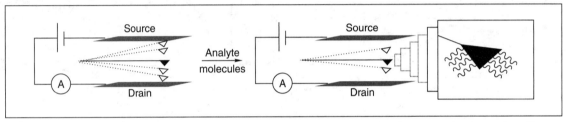

Scheme 12.6: *Graphical representation of the working of a nanocantilever sensor. The damping of oscillations after exposure to analyte molecules is observed. The cantilever tip is enlarged to show the presence of the analyte (indicated by curved lines).*

Another detection protocol, utilizing the weight of the nanoparticles as amplifying agents for the detection of complementary DNA pairs and misfits, is Quartz Crystal Microbalance (QCM). The inverse piezoelectric effect is the key to the operation of the QCM. The application of an electric field to the quartz crystal causes a shear deformation of the order of few nanometers. Initially, the gold electrode is modified with single stranded DNA. When complementary-DNA functionalized gold nanoparticles are allowed to interact on the electrode, the vibrating frequency of the quartz crystal varies and therefore, the mass change can be detected. When the same process is carried out without the complementary DNA functionalizing the nanoparticles, the weight change is not significant enough for the QCM to register. Therefore, in this case the nanoparticles do not influence or participate in the detection process, but are merely involved in amplifying the signal.

12.10 Nanobiosensors—A Step towards Real-time Imaging and Understanding of Biological Events

The evolution of nanoscience and that of biology have complemented each other beautifully. There is widespread interest in combining both the fields, so much so a new field known has nanobiotechnology is one of the frontier areas of research. In keeping with the time at which this book is being written, an attempt has been made to emphasize the role of nanomaterials in developing sensors for biological applications. Any of the protocols detailed in the previous sections can be extended to biosensors. Besides those techniques, a few techniques specific for following bio-related events have been developed, which will be the focus of this section. In this section also, like in other parts of the book, there will be greater focus on the current research directions, with a pointer to the future prospects.

A nano-biosensor concept has to contain a biological recognition component, which is involved in interacting with the analyte molecule. This interaction causes reversible changes in both the bioreceptor and the analyte. These chemical changes have to be transformed into recordable signals. In most of the cases, the transducer has to convert the chemical modification into an electrical signal. The signal from the transducer is analyzed and interpreted to provide details about the analyte. Therefore, the bioreceptor controls the selectivity and specificity while the transducer determines the sensitivity of the sensor.

The nature of the bioreceptor depends on the type of analyte molecule. Since the binding and recognition events in biology are highly specific, the choosing of the bioreceptor should be done keeping in mind the kind of application the device is going to be used for. Since the biological medium is mostly aqueous, Surface Enhanced Raman Spectroscopy (SERS) is a widely used tool to bring out the maximum information from biological sensing events. However, due to limited low spatial resolution, near field spectroscopy and associated microscopy have been receiving attention. With near field SERS (NF-SERS), single molecule detection is now a reality.

Two main approaches are widely followed for the fabrication of nanofiber probes, and are needed to sense the plethora of processes occurring inside individual cells. The first approach is known as the 'heat and pull' method (Ref. 13). A glass fiber is spot-heated by using a focused heating source such as a laser. In the hot condition, the fiber is pulled apart, yielding probes the morphology of which depends heavily on the experimental parameters. The second, more systematic procedure is known as 'Turner's method' and entails chemical etching of the surface to yield nanoscale probes (Ref. 14). The etching, in case of glass probes is done with HF. A fiber with a silica core and an organic shielding material is exposed to HF. The slow etching of the inner silica core by HF takes place. This is followed by the controlled emergence of the silica core, during which HF rises up the fiber due to capillary action. On complete emergence, the HF drains of the glass fiber, etching a nanotip in the process. A schematic procedure of the etching process is shown in Scheme 12.7 (Plate 11).

The nanofiber tip thus fabricated is functionalized with an appropriate bioreceptor. The functionalization is carried out by suitably activating the nanofiber tip and coating silver over it. The silver coating is done by evaporating the metal under a rotating tip, which results in a uniform layer of silver on the tip. The tip is used for immobilization of the bioreceptor, which can be anything ranging from antigens to biomimetic molecules.

The use of this functionalized nanotip is clearly demonstrated by its capability to penetrate the cells and thus study the actual locations inside a cell wherein the analyte molecules are accumulated or processed. Benz[a]pyrene (BaP) is a chemically generated carcinogenic compound and there has been a keen interest in tracking its chemistry inside the cells. Benzopyrenetetrol (BPT) has been used as an antibody probe for this chemical and to study its metabolic pathway inside the cells and the mechanism by which it causes mutations. After penetrating the cell with a bioreceptor bound nanotip, the BaP and BPT form a receptor-ligand complex. The fluorescent signal form this complex is monitored for detailing the concentration of BaP and its mechanism of action (Ref. 13).

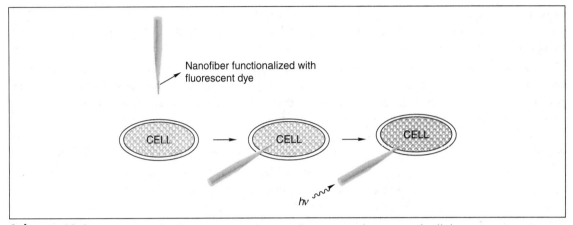

Scheme 12.8: *Diagrammatic representation of in-situ detection and imaging of cellular components.*

Another interesting protocol involves incorporation of the dye molecules inside the cells. Then a nanotip is carefully inserted inside the cell. Light of suitable frequency is passed through this fiber optic nanotip to cause fluorescent emission from the dye molecules. This makes possible a chemical photograph of the cell, indicating the regions where the dye molecules are present (Scheme 12.8).

Sönnichsen and Alivisatos, have utilized the light scattering property of gold nanorods (and metallic nanorods in general) as a means of sensing its motion (Ref. 15). A similar protocol can be extended to detect orientation changes and track the movement of large biomolecules. The strategy adopted is to attach gold nanorods loosely on the surface of a glass flow cell. On this surface, under the influence of a small flow current, the rotational behavior of the nanorods can be mapped. The detection protocol consists of a dark field set-up, with an illumination source and an objective to collect the scattered light. Due to their anisotropy, nanorods are the best suited candidates for observing rotational dynamics, the resolution of which can be brought down to as far as 20 nm.

12.11 Smart Dust—Sensors of the Future

In the constant urge to innovate and improvise, man has been breaking many barriers. This trend has been observed, in the field of nanosensors also, giving rise to fierce competition for developing detection capabilities and sensor characteristics. Towards this end, 'stand-off detection' has been undertaken in a big way. This pertains to the detection of harmful and toxic chemicals which cannot be analyzed under laboratory conditions. Paradoxically, the detection of these chemicals is much more important and relevant as compared to the detection of normal chemicals. This is achieved by the 'smart dust' approach, wherein small particles resembling dust are carried by air currents to the area under investigation. In one such study, porous silicon was etched out from crystalline silicon by a combination of anodic and galvanic currents. These were ground or ultrasonicated to form the materials for smart dust. Porous silicon used here exhibits visible photoluminescence. When it comes into contact with explosive chemicals like dinitrotoluene (DNT) or trinitrotoluene (TNT), its photoluminescence is quenched due to NO_2 produced by catalytic oxidation. The signal can be monitored from a safe distance and provides information about the nature and quantity of the chemical present in the ambience. Detection capabilities of parts per billion (ppb) have been achieved using this system.

Taking this development further, researchers at UC, Berkeley have embarked on an ambitious project, the outlay of which is shown in Scheme 12.9 (Plate 12) (Ref. 16). It entails building an integrated sensor array, complete with power supply, receivers and transmitters. Each little sensor has the capability of analyzing a multitude of parameters ranging from humidity to explosives. Every 'dust' particle can communicate with another through a distance of about thousand feet, thus enabling three-dimensional concentration profiling. Although questions are being raised about the environmental and health safety aspects of these systems, it is likely that these problems will be solved in the coming years.

Review Questions

1. What are sensors?
2. What is the ultimate size of a sensor? What decides it?
3. What is the requirement of a nanosensor?
4. What are the current nanotemplating methods?
5. If the sensor is nano in dimension, how can one measure the property, as the magnitude of which will be reduced by that order?
6. What properties of the sensor are modified during sensing?
7. What are the physical properties used for sensing?

References

1. Chumanov, G., K. Sokolov, B.W. Gregory and T. M. Cotton, *J. Phys. Chem.*, **99** (1995), p. 9466.
2. Hidber, P.C., W. Helbig, E. Kim and G.M. Whitesides, *Langmuir*, **12** (1996), p. 1375.
3. Mitchell, G.P., C.A. Mirkin and R.L. Letsinger, *J. Am. Chem. Soc.*, **121** (1999), p. 8122.
4. Storhoff, J.J., C.A. Mirkin, *Chem. Rev.*, **99** (1999), pp. 1849–62.
5. Boal, A.K., F. Ilhan, J.E. Derouchey, T.T. Albrecht, T.P. Russell and V.M. Rotello, *Nature*, **404** (2000), p. 746.
6. Kim, Y., R.C. Johnson and J.T. Hupp, *Nano Lett.*, **4** (2001), p. 165.
7. Keller, M.W., A.L. Eichenberger, J.M. Martinis and N.M. Zimmerman, *Science*, **285** (1999), p. 1706.

8. Savage, N., B. Chwieroth, A. Ginwalla, B.R. Patton, S.A. Akbar and P.K. Dutta, *Sensors and Actuators B*, **79** (2001), p. 17.
9. Dibbern, U., "Sensory and Sensory Systems for an Electronic Nose" in J.W. Gardner and P.N. Bartlett (eds), *NATO ASI Series,* (1992).
10. Raj, C. Retna, Takeyoshi Okajima and Takeo Ohsaka, *Journal of Electroanalytical Chemistry*, **543** (2003), pp. 127–133.
11. Zhou, X.T., J.Q. Hu, C.P. Li, D.D.D. Ma, C.S. Lee and S.T. Lee, *Chem. Phys. Lett.*, **369** (2003), p. 220.
12. Bottomley, L.A., M.A. Poggi and S. Shen, *Anal. Chem.*, **76** (2004), p. 5685.
13. Vo-Dinh, T., J.P. Alarie, P.M. Cullum and G.D. Griffin, *Nature Biotechnol.*, **18** (2000), p. 76.
14. Turner, D.R. and U.S. Patent, **4** (1984), pp. 469, 554.
15. Sönnichsen, C. and P. Alivisatos, *Nano Lett.*, **5** (2004), p. 301.
16. http://robotics.eecs.berkeley.edu/~pister/SmartDust/.

Additional Reading

1. Rao, C.N.R., A. Müller and A.K. Cheetam, *Chemistry of Nanomaterials Vol 1 and 2*, (2004), Wiley-VCH.
2. Schmid, Günter, *Nanoparticles: From Theory to Application*, (2004), Wiley-VCH.
3. Shipway, N., E. Katz and I. Willner, *Chemphyschem*, **1** (2000), 18–52.
4. Nalwa, H.S., *Encyclopedia of Nanoscience and Nanotechnology*, (2005), American Scientific Publishers.

Chapter 13

NANOMEDICINES

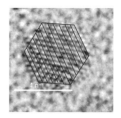

Nanomaterials are finding their way into biology in the form of drug carriers. This is probably the most important application of nanomaterials right now. The property utilizes the large surface area available to load materials. Due to their small size, nanomaterials can be transported into cells and nuclei. Specificity to the target can be achieved by appropriate labeling. The materials put in can be subjected to magnetic fields, photons, etc. and can respond to all these situations. The diagnostic and therapeutic applications of such systems are being suggested. Here we present an overview of this area.

Learning Objectives

- Why should nanomaterials be used in medicine?
- How do we apply them?
- What are the materials currently used in this area?
- What are the medical applications currently being investigated?
- What is the future of nanomedicine?

13.1 Introduction

The objective of nanotechnology is to gain atomic and molecular control over matter. It involves the creation of functional materials with control over their physical sizes, which exhibit novel physical and chemical properties that are drastically different from the corresponding bulk forms. The physical sizes of these materials create a strong possibility for their interactions with biological systems (see Chapter 11). Biological systems themselves contain various components which are essentially in the nanometer dimensions (proteins, nucleic acids, membranes); a fact implying possible synergies between nanosystems and biological components. This can have implications for the understanding of biology. Such an understanding can be achieved through the use of nanosensors or probes for disease detection, all of which will ultimately offer robust solutions for the well-being of all.

The concept of the effective use of nanotechnology in disease treatment was suggested as early as 1959 by Nobel Laureate Richard Feynman in his famous talk on, "Plenty of room at the bottom" (Ref. 1).

Feynman provided insights into how nanomedicines could be developed as effective solutions for heart disease: "A friend of mine (Albert R. Hibbs) suggests a very interesting possibility for relatively small machines. He says that, although it is a very wild idea, it would be interesting in surgery if you could swallow the surgeon. You put the mechanical surgeon inside the blood vessel and it goes into the heart and looks around (of course the information has to be fed out). It finds out which valve is the faulty one and takes a little knife and slices it out. Other small machines might be permanently incorporated in the body to assist some inadequately functioning organ."

Today nanomedicines are being developed to have accurate, controllable, reliable, economic and rapid responsive diagnostic and treatment solutions for various kinds of diseases. With advancements in drug discovery processes, stress is on effective drug delivery to the affected organ. It is well-known that many therapeutic agents have intracellular compartments as their site of action. For example, the nucleus is the site of action for anti-cancer intercalating agents whereas cytoplasm is the centre for a number of steroids. Accordingly, the efficacy of drug depends on its sustained availability at the targeted point of delivery. As such, drug administration is affected by the inability of the drug molecule to effectively escape the endosomal/lysosomal pathways, get transported across the membranes and reach the intended location of delivery inside the cell. Even though liposomes have been tried as potential drug carriers because of their unique abilities to avoid drug degradation, reduction in side effects and targeted delivery, their effective usage has been limited due to their low encapsulation efficiency, rapid leakage of water-soluble drugs in the presence of blood components and poor storage stability (Ref. 2). This fact emphasizes the desired attributes of an effective drug carrier system. The use of nanoparticles for drug delivery purposes becomes important because of their high surface-to-volume ratio, enhanced detection features, easier transport across the membrane and possible protection of drug molecules. A high proportion of the atoms in small metal nanoparticles will be present at the surface. The surface-to-bulk ratio bears a strong inverse dependence on particle size. A high surface-to-bulk ratio ensures strong interaction between nanoparticles and the reacting species. Additionally, there is a need to develop a molecular tool for disease detection and treatment because of the uniqueness of each individual's response to therapeutic intervention. This uniqueness is the result of differences in the interaction of therapeutic tools and biological processes, which means that an individualized approach to this problem can lead to a dramatic improvement in results.

Various studies confirm the fact that particle size should be sufficiently small for it to get transported across the membrane and this transport occurs more readily for nanoparticles rather than for microparticles (Refs 3, 4, 5).

Here we discuss the various approaches that are currently being researched for developing nanomedicines. Additionally, we also discuss various nanomaterials which are strong candidates for use in nanomedicines.

13.2 Approach to Developing Nanomedicines

Depending on the method of preparation and the capping agent present, nanoparticles vary in size from 10 to 1000 nm. Drugs can be associated with the nanoparticles in entrapped, encapsulated or attached form. Nanodrugs are being synthesized in various forms such as nanospheres (drug present on the nanoparticle as the capping agent), nanocapsules (drug confined in a cavity surrounded by a polymeric layer, see Chapter 10 on nanoshells) (Refs 6, 7, 8), nanopores (nanoparticle surface perforated with holes, holes contain drug molecules) (Refs 9, 10), dendrimers (Refs 11, 12), etc. The purpose of encapsulation or entrapment is to gain a better degree of control over the drug release process. This approach finds favor for effective and steady drug delivery over conventional drug due to kinetic behavior observed during drug release. Encapsulated nano-system based drugs are observed to show nearly zero-order kinetic profile whereas conventional oral drugs follow first-order kinetics leading to unsteady drug release at the location of drug delivery (Ref. 13).

Recently, attempts have also been made towards developing biodegradable polymeric nanoparticles as potential drug delivery devices. In addition to the inherent property of reduced cytotoxicity, biodegradable polymeric nanoparticles have been found to be extremely effective in controlled and targeted drug release, even through administration is oral (Refs 8, 14). The phenomenon of zero-order kinetics has been observed predominantly for polymeric nanoparticles. Additionally, various research groups have also established the use of polymeric nanoparticles for nasal (Ref. 15) and ophthalmic delivery of drugs (Refs 16, 17). This group of nanoparticles has also shown prominence for use in neuro-disorders, in which case a large number of other drugs fail (Refs 18, 19). Furthermore, nanosize carriers of vitamin molecules such as vitamin A and E, have potential applications in dermatology and cosmetics (Refs 20, 21).

Various kinds of approaches can be used to attach drugs to nanosystems. There can be electrostatic interaction or covalent binding between the nanoparticle and the drug. The nanoparicle surface can be made electrically neutral or charged, depending on the functional group present on the surface. The surface properties can be tuned depending on the drug-nanoparticle interaction required.

13.3 Various Kinds of Nanosystems in Use

Metal nanoparticles themselves are used as drug delivery vehicles (see Chapter 8). However, there are several other systems for this application which are briefly reviewed here.

13.3.1 Nanoshells

Nanoshells (Chapter 10) represent a unique class of medically prominent nanoparticles. These are made of drug-coated metal nanospheres/dielectric metal nanospheres (e.g. gold-coated silica nanoparticle). Typical metals include gold, silver, platinum and palladium. It is quite evident that the response of these nanoshells is a function of the thickness of the shell/capping agent. When these nanoshells are irradiated with a laser of known intensity, it causes release of the drug coat present on the nanoparticle surface. The release process can be accomplished with the use of an alternating magnetic field as well (Ref. 11).

This approach to the release of capping agent can have implications in cancer treatment. A high surface-to-volume ratio for nanoparticles enables large quantity of drugs to be transported into the affected region.

Attempts have also been made to coat nanoparticle surfaces with antibody molecules, specific to a particular protein present in the human body. This can have profound implications in cancer detection, protein immunoassay and biosensing (Ref. 22).

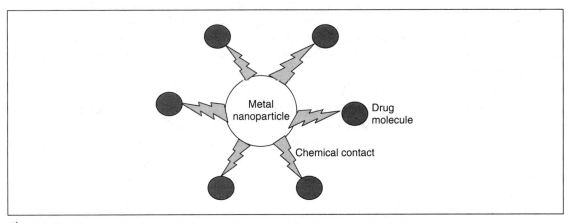

Fig. 13.1: *Drug molecule attached to metal nanoparticle via covalent/ionic interaction.*

13.3.2 Nanopores

Nanopores are essentially nanoparticles whose surface contains pores, which can be used for containing drugs. Uniformly spaced holes are created on the surface in which a drug molecule is contained. The pore size imposes a restriction on the size of the biomolecules present. This means that small molecules like oxygen, glucose, insulin, neurotransmitters, etc. can move across the pore surface while large immune system molecules like immunoglobulin cannot. The released molecule can therefore be used in disease treatment, e.g. the use of insulin in diabetes treatment, use of neurotransmitters in neural disorders, etc. (Refs 9, 10).

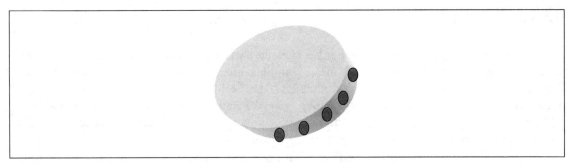

Fig. 13.2: *Cross section of nanopore with drug molecule contained inside the pore.*

Attempts are also being made to control the flow of the molecules across the pore for highly specific molecular transport capabilities, by the use of voltage gating (Ref. 23) and immobilized biochemical molecular-recognition agents (Ref. 24). The first attempt to create a voltage gated nanopore involved an array of cylindrical gold nanotubules with inside diameters as small as 1.6 nm. On developing a positive charge on the tubule, positive ions were not transported inside the nanopore to undergo a reaction with the drug molecule trapped inside. Similarly, only positive ions could pass on applying a negative voltage. The aim is to gain a significant improvement in isolating the targeted molecule with which drug molecules have to interact, by the use of combinatory tools such as voltage gating, pore size and shape.

13.3.3 Tectodendrimers

Dendrimers are branched tree-shaped nanoparticles, which have an immense potential for use in clinical diagnostics and therapeutics. Various research groups (Refs 11, 12) have also synthesized multi-component nanodevices called 'tectodendrimers' which are formed by attaching different types of dendrimers with each other through their branches. These smart nanodevices have been synthesized for applications ranging from the detection to treatment of diseases.

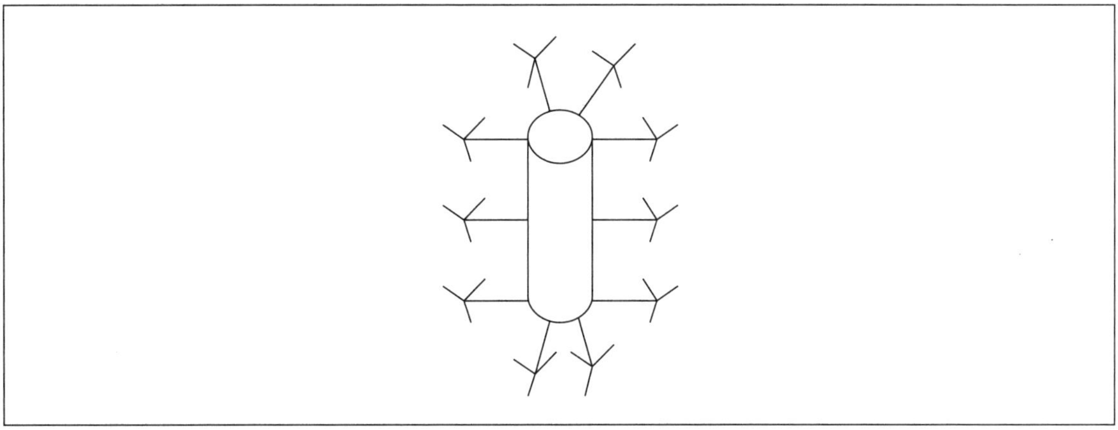

Fig. 13.3: *Dendrimer (tree-shaped nanoparticle), nanometers in dimension with branches projecting out.*

13.4 Protocols for Nanodrug Administration

13.4.1 Nanoparticle-drug System for Oral Administration

Various kinds of approaches are being attempted for the delivery of nanoparticle–drug complex to target particular locations in the human body. An analysis of conventional oral administration indicates that the basic requirements for the successful delivery of a nanoparticle-drug system via oral administration are:

1. The complex should be stable in the gastrointestinal tract.
2. Digestive system enzymes should act on the complex and digest it, and the product should subsequently get transported across the intestinal epithelium.
3. Products from the digestion of nanoparticle system complex should not be cytotoxic for the human body.

In order to avoid the disintegration of the complex before the digestive enzymes start interacting with it, a hybrid system of hydrophobic core–hydrophilic shell has been designed which acts as a carrier for drug molecules (Ref. 25). The core is made of hydrophobic material such as oils or lipids whereas the shell is hydrophilic in nature and composed of polyethylene glycol (PEG which protects against protein adsorption) or chitosan (a known permeability enhancer). Chitosan is a naturally occurring substance (shown in Fig. 13.4, chemically similar to cellulose, $-NH_2$ group in chitosan is replaced by $-OH$ group in glucose) with the ability to significantly bind fat without itself being digested. The various reported applications for chitosan are that it:

1. Absorbs and binds fat/promotes weight loss.
2. Promotes healing of ulcers and lesions.
3. Is used as an anti-bacterial agent and antacid.
4. Inhibits the formation of plaque/tooth decay.
5. Helps control blood pressure and prevent constipation.
6. Has an anti-tumor action.

The drug molecules present inside the core of the nanoshell are protected against degradation by the shell. This system is created by forcing the orientation of the PEG segment towards the surface and concentrating the hydrophobic polymer to core (Fig. 13.4). This nanosystem proved its efficacy in drug delivery when it safely transported the tetanus toxoid protein to the blood stream. Similar successful results were reported for salmon calcitonin peptide-nanoparticle complex when used in rats (Ref. 26). It has also been proved that nanocapsules made of poly alkylcyanoacrylate are able to increase the absorption of insulin when administered orally (Ref. 27).

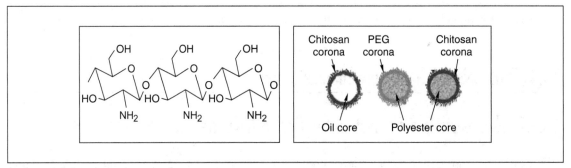

Fig. 13.4: *Structure of chitosan and various kinds of core-corona (core-shell) nanosystems.*

13.4.2 Nanoparticle-drug System for Nasal Administration

It has been established that the nasal route of drug delivery is more effective (especially for small peptides) due to a better transport process and lower enzymatic activity for nasal mucosa. Studies show that the nanoparticle–drug system is capable of crossing the nasal epithelium with the strong influence of nanoparticle surface composition on transport rates (Ref. 28). Bare nanoparticles aggregated on the mucus layer and hence were not transported with the model protein tetanus toxoid whereas a significant increase in the absorption was observed when PEG was used as a surface cover on the bare nanoparticle. PEG-covered nanoparticles were observed to be circulating in the blood stream and the associated protein was delivered at an appropriate location. Similar results have also been observed for various other kinds of vaccines (Ref. 29).

Chitosan-coated nanoparticles have also been observed to respond in a similar way. When they were tested for the nasal absorption of insulin in rabbits, it was observed that the absorption of the drug was significantly higher (Ref. 30). Similarly, improved nasal transport of the tetanus toxoid protein was observed when it was encapsulated in a chitosan-coated nanoparticle. This is because of the facilitated interaction and internalization of these nanosystems in the nasal epithelium.

13.4.3 Nanoparticle-drug System for Ocular Administration

It was observed that polyalkylcyanoacrylate nanoparticles were able to enter the well-organized corneal epithelium though it caused a slight damage to the epithelial cells. Due to better organization of cells in the corneal epithelium, the dimension of the carrier must be in the sub-micron region. It has also been established that the coating present on the surface of the nanoparticle has an important effect on drug transport through the corneal epithelium. When the experiments were conducted with ^{14}C-indomethacin-chitosan-coated nanosystems, it was observed that the complex penetrated to the superficial layers of the epithelium through a trans-cellular pathway and chitosan-coated systems had a good ocular tolerance (low ocular lesion index). This nanoparticle complex was also able to provide a selective and prolonged delivery of cyclosporine A to the ocular mucosa without compromising the inner ocular tissues by avoiding systemic absorption. This prolonged delivery was attributed to the ocular retention of chitosan nanoparticles (Refs 31, 32).

13.5 Nanotechnology in Diagnostic Applications

Research efforts are also being driven in the direction of using nanotechnology for molecular diagnostic purposes such as biological research, clinical diagnostics, detection of biomolecules and drug discovery. The focus of this section is on understanding the use of nanosystems for clinical diagnostics, especially in the early diagnosis of various forms of cancer.

The main drawback of the laboratory tests currently used for the detection of cancer is that they try to identify visible changes in cell morphology through microscopy. Evidently, this practice cannot state with 100 per cent specificity and sensitivity the true cases of disorder (identified by the clinical criterion) in the malignancy stage of cells. Unfortunately, detection of many cancers at the microscopic level often takes place when it is too late for successful intervention and these techniques suffer from intra-observational subjectivity (Ref. 34).

The development of tumors is a complex process, requiring the co-ordinated interactions of numerous proteins, signal pathways and cell types. As a result of extensive studies of the molecular pathogenesis of cancer, several novel regulatory pathways and networks have been identified. The delineation of these pathways has revealed several unique events, marked by morphological and histological changes of cells, and the expression of genes and proteins that accompany oncogenic transformation. Thus, the cell signature changes during cancer development. If these changes are read accurately, there is a strong likelihood of improving the early detection and diagnosis capabilities for various forms of cancer. These early changes in cell signature are reflected in the state of biomarkers. Biomarkers are measurable phenotypic parameters that characterize an organism's state of health or disease, or its response to a particular therapeutic intervention (Ref. 35).

Various approaches are being followed for improving clinical diagnostic capabilities. Essentially these approaches have to be molecular in nature. Here we discuss two approaches to develop molecular-based diagnostic tools, which can detect cancer in the very early stages.

Several groups have reported intra- as well as extra-cellular synthesis of metal nanoparticles using bacteria, fungi and viruses. This process involves the reduction of metal ions added to cells under suitable conditions, and the resultant appearance of nanoparticles or their aggregates, both inside as well as outside the cell boundary (Refs 36, 37).

Recent attempts to synthesize gold nanoparticles using human cells indicate that the cellular response towards the reduction of chloroaurate ions is different for normal and malignant cells. The growth of gold nanoparticles is confirmed by TEM images presented in Fig. 13.5. The cells shown here are HEK

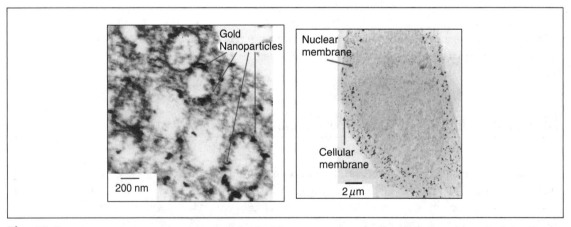

Fig. 13.5: *Electron micrographs of the growth of gold nanoparticles in HEK 293 (left) and SiHa (right) cells after incubation with 1 mM chloroaurate ions for 96 h. The left side image shows nanoparticles around cell bodies (From the author's work).*

(normal cell) and SiHa (cervical cancer) cells. The images show the presence of Au nanoparticles as black dots. The particles are in the size range of 20–100 nm and are distributed throughout the cytoplasm. Particles are found in the nucleus too, which are much smaller in diameter. The ability with which nanoparticles are synthesized by cells in different stages of infection can be quantified and may be used to pinpoint the level of infection in various kinds of cancer. The results indicate that there is a significant difference between the responses of cancer cells and those of non-cancer cells towards the bio-reduction process, which may be attributed to differences in the metabolism and kinetics of nanoparticle formation in the cells being investigated.

When the growth of gold nanoparticles was observed over a duration of 96 hours after incubating various cell lines with 1 mM chloroaurate ion solution, it was found that there existed a difference in the UV-visible feature for the cancer and non-cancer cells (Fig. 13.6). What one observes is that as a function of incubation time, the peak at 560 nm, corresponding to the plasmon excitation of gold nanoparticles, increases in intensity. Further, the curves for cancerous cells broadened after the cells were lysed, which suggests that the nature of the nanoparticles present inside and outside the cells was different. Lysing releases the nanoparticles into the solution, as a result of the breaking of the cell membrane. During the course of incubation, the supernatant solution above the cell line shows the gradual evolution of nanoparticle signature as mentioned before, because part of the nanoparticles can be leached of the cell. However, it is important to note that since most of the particles grown are large, they cannot diffuse out of the cell membrane and are seen in the absorption spectrum only after lysing. The quantification of the cellular response towards the nanoparticle synthesis of cells during various stages of cancer can then be utilized to develop a protocol for the early diagnosis of cancer.

In the second approach, research attempts are being made to read the state of various kinds of biomarkers for different types of cancer. The known biomarkers which can be utilized for this purpose are changes in protein concentration, genetic mutations, etc. For example, the concentration of protein p16, which is known to over-express itself during the Human Papilloma Virus (HPV) infection, can be used as a biomarker for cervical cancer detection. Hence, when the antibody specific to protein p16 is bound to nanoparticles and used for reading the p16 concentrations, the absolute concentration level can be ascertained. When this level is quantified for various stages of cervical cancer it can help in early diagnosis. The use of nanosystems for reading the state of biomarkers becomes important because of the high surface area offered by the nanoparticle surface (which facilitates the reading of even small changes in the biomarker state) and the ease with which nanosystems can interact with biomolecules. The feasibility of reading the state of biomarker (though without the use of nanosystems) has been demonstrated by the use of a molecular-based tool developed by Digene Corporation, The Hybrid Capture System. It is a signal amplification assay that uses antibody captures and signal detection for cervical cancer diagnosis. This protocol involves releasing the target DNA of HPV and combining it with an RNA probe. The resultant DNA–RNA hybrid is captured by using the antibody specific to the hybrid. The antibodies are bound to alkaline phosphatase which assists in the detection by the use of a luminometer (Ref. 38).

Various kinds of nanoparticles have been used for diagnostic applications. These include gold nanoparticles, quantum dots and magnetic nanoparticles, which are described in detail below.

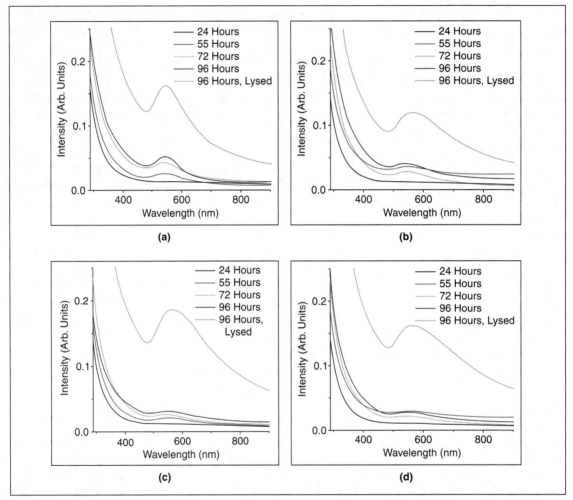

Fig. 13.6: *UV-visible spectrum for: (a) HEK 293 cells with 1 mM chloroaurate ion solution. (b) HeLa cells with 1 mM chloroaurate ion solution. (c) SiHa cells with 1 mM chloroaurate ion solution. (d) SKNSH cells with 1 mM chloroaurate ion solution. Except HEK 293, all others are cancer cells. Note the increase in width of the lyseel sample in b, c and d. All other measurements correspond to the supernatant solution in the cell culture well (From the author's work).*

13.6 Materials for Use in Diagnostic and Therapeutic Applications

13.6.1 Gold Nanoparticles

Gold nanoparticles are extraordinarily efficient for clinical diagnostic purposes as they give strong signatures in optical absorption and fluorescence spectroscopy, X-Ray diffraction and electrical conductivity.

Additionally, gold nanoparticles interact strongly with biomolecules containing thiol or amine groups and can be suitably modified with a number of small molecules, proteins, DNA and polymers. Various biomolecules bound to the gold nanoparticle surface can be detected by using various analytical measurement tools such as MALDI-TOF MS and confocal Raman spectroscopy. Gold can be synthesized routinely in sizes varying continuously from 0.8 to 200 nm with <10 per cent size dispersity.

Gold nanoparticle characterization can be done by UV-visible spectroscopy in which a surface plasmon band appears at 500–700 nm. This happens because of electronic oscillations in the conduction band of metals, on exposure to electromagnetic waves (see Chapter 9, for a detailed description). The surface plasmon resonance phenomenon occurs due to a matching of the frequency of the oscillation of the electron cloud and that of the incident light. For gold nanoparticles, the frequency is in the visible region which imparts intense color to the nanoparticle solution.

13.6.2 Quantum Dots

Quantum dots (QDs) are nanoscale crystals synthesized with semiconductor materials (see Chapter 7). QDs are generating strong research interests in biology due to their fluorescence property seen when they are excited by a laser. Their fluorescence intensity is also significantly higher and are more stable as compared to conventional fluorescent markers. QDs have fairly broad excitation spectra which can be tuned by varying the physical size and composition. Additionally, QDs have narrow emission spectra, which means that it is possible to resolve the emissions of different nanoparticles simultaneously and with minimal overlap. Finally, QDs are highly resistant to degradation.

QD technology holds special promise for use with biomolecules. QDs can be specifically attached to molecules like proteins and nucleic acids. Some of the value additions that QDs can bring are:

1. It is easy to excite QDs which means that the possibility of drug degradation due to a high intensity excitation beam is discounted.
2. Whole blood assay can be done with the use of QDs because they can emit light throughout the electromagnetic spectrum.
3. This technology has high sensitivity and is easy to use.
4. Photobleaching does not occur with QDs which is a serious limitation in the case of fluorescent dyes.

The difficulty encountered with QDs is that their surface is hydrophobic in nature which makes its interaction with water-friendly molecules like proteins and DNAs extremely difficult. Research is being carried out to modify the surface of QDs so as to make them more biocompatible. For example, short size peptides have been used to coat QD surfaces and make them interact with cells. A unique new coating has been developed for inorganic particles at the nanoscale that may be able to disguise QDs as proteins—a process that allows particles to function as probes which can penetrate the cell and light up individual proteins inside, thereby creating the potential for a wide range of applications including cell imaging and clinical diagnostics (Ref. 39). Highly luminescent and stable QD bioconjugates are constantly being evaluated for use in cell imaging, which will help in developing better tools for clinical diagnostics. QDs, coated with a polyacrylate cap and covalently linked to antibodies or to streptavidin, have been used for the immunofluorescent labeling of breast cancer biomarker HER2 (human epidermal growth factor receptor 2). The advantages of using QD are that it facilitates highly specific labeling and makes the label brighter and more stable than that obtained with the use of conventional fluorescent markers.

13.6.3 Magnetic Nanoparticles

Another special class of nanoparticles, which is being intensively researched for use in biological systems is magnetic nanoparticles. These particles are superparamagnetic, i.e. they do not possess any magnetism in the absence of an applied field. They are being used for the detection of various kinds of biomolecules. Research efforts have proven the potential of cellular organisms to synthesize various magnetic nanoparticles. Like gold nanoparticles, they can also be used for cancer diagnostics through two approaches discussed in Section 13.5. Magnetic immunoassay techniques have been developed in which the magnetic field generated by the magnetically labeled targets is detected directly with a sensitive magnetometer. The binding of the antibody to the target molecules or to the disease-causing organism forms the basis of several tests. Antibodies labeled with magnetic nanoparticles emit magnetic signals on exposure to a magnetic field. Antibodies bound to targets can thus be identified as unbound antibodies are dispersed in all directions and produce no net magnetic signal. Another challenge in cancer diagnosis is the detection of circulating cancer cells in the blood. Magnetic nanoparticle-based tests are being developed to screen, diagnose, stage and monitor cancer on the basis of the circulating cancer cells in the blood.

13.7 Future Directions

Nanosize materials have found practical implementation in the field of medical diagnoses with the proper and efficient delivery of pharmaceuticals. While on the one hand, attempts are being made to develop accurate and versatile biomarkers for various kinds of diseases, on the other hand, research is constantly being driven to develop new protocols for creating synergies between bio- and nano-systems. However, there are several other intriguing proposals for the practical applications of nanomechanical tools into the fields of medical research and clinical practice. Such nanotools still await construction, but they may become a reality in the near future.

Newer avenues are being developed for using nanodevices in the field of clinical diagnostics and therapeutic applications. Currently, research efforts are being directed to develop suitable nano-bio-systems which can successfully replace defective/incorrectly functioning cells in various parts of the body. Attempts are being made to create artificial red blood cells which will provide oxygen more effectively as compared to the natural ones, deliver them into the human body and monitor their flow by using onboard nanosensors (Ref. 40). An onboard nanocomputer and numerous chemical and pressure sensors would facilitate complex device behaviors that will be remotely reprogrammable by the physician via externally applied acoustic signals. The ultimate aim is to develop a robust mechanism whereby the therapeutic intervention is able to locate the site of infection/disorder, diagnose the level of infection, deliver the drug in case therapeutic treatment is required (or kill the cells if necessary) and in the meantime, provide metabolic support in the event of impaired functioning. Attempts are being made to develop a feedback system integrated to therapeutic intervention which will constantly update the physician about the action of the drug and possibly make it feasible for the physician to change the guidelines for drug delivery as he/she may deem it better for more precise treatment. Such devices would have a small computer for information analysis, several binding sites to determine the concentration of specific molecules, drug molecules for treatment, and a supply of some 'poison' that could be released selectively. Similar machines equipped with specific 'weapons' could be used to remove obstructions in the circulatory system, or to identify and kill cancer cells. The use of such nanorobots would enable medical specialists to ascertain the level of infection after examination of the tissue location and variations in its biochemistry and biomechanics. Accordingly, a programmable algorithm would initiate therapeutic intervention for accurate and reliable drug delivery (Refs 41, 42).

These steps for reaching the ultimate objective of gaining control over various kinds of diseases that humans suffer from, will involve a three-dimensional approach, i.e. development of a better understanding of biological systems, creation of nanosystems and integration of nano-bio systems. It may seem impossible to develop a kind of nanomachine which when injected into the human body, will itself do the job of finding the location of the disease, treating/killing it and providing transitory support for the metabolic processes but a step-by-step approach to nano-bio integration could make this possibility real in the distant future.

Review Questions

1. Will nanomedicines be used in nanoscale? How do we apply them?
2. What are the systems used and how nano-properties are useful?
3. What are the properties useful for diagnostic applications?
4. What are the advantages of magnetic nanoparticles in nanomedicine?
5. What are the properties of nanoparticles themselves (as opposed to the molecules anchored on them) useful for therapeutic applications?

References

1. Feynman, R.P., "There's Plenty of Room at the Bottom". *Engineering Science* (Cal Tech), **23** (1960), pp. 22–36.
2. Knight, C.G., *Liposomes from Physical Structure to Therapeutic Applications*, (1981), Elsevier, 1981.
3. Desai, M.P., V. Labhasetwar, G.L. Amidon and R.L. Levy, *Pharm Res.*, **13** (1996), pp. 1838–45.
4. Jani, P.U., G.W. Halbert, J. Langridge and A.T. Florence, *J. Pharm. Pharmacol.*, **42** (1990), pp. 821–26.

5. Alexander, G. Tkachenko, Huan Xie, Donna Coleman, Wilhelm Glomm, Joseph Ryan, Miles F. Anderson, Stefan Franzen, and Daniel L. Feldheim, *J. Am. Chem. Soc.*, **125** (2003), pp. 4700–4701.
6. Kumar, M.N.V., *J. Pharm. Pharmaceut. Sci.*, **3** (2000), pp. 234–58.
7. Lambert, G., E. Fattal and P. Couvreur, *Adv. Drug. Deliv. Rev.*, **47** (2001), pp. 99–112.
8. Soppimath, K.S., T.M. Aminabhavi, A.R. Kulkarni and W.E. Rudzinski, *J. Control. Release*, **70** (2001), pp. 1–2.
9. Tao, S.L. and T.A. Desai, *Adv. Drug Delivery Rev.*, **55** (2003), pp. 315–328.
10. Leoni, L, D. Attiah and T.A. Desai, *Sensors*, **2**, pp. 111–120.
11. Sun, Y., B.T. Mayers, Y. Xia, *Nano Lett.*, **2** (2002), pp. 481–485.
12. Quintana, A., E. Raczka, L. Piehler, I. Lee, A. Mue and I. Majoros, *et al.*, *Pharmaceutical Res.*, **19** (2000), pp. 1310–16.
13. Baker, J.R. Jr, A. Quintana, L. Piehler, Holl M. Banaszak, D. Tomalia and E. Raczka, *Biomed. Microdevices*, **3** (2001), pp. 61–69.
14. Langer, R., *Acc. Chem. Res.*, **33** (2000), pp. 94–101.
15. Illum, L., I. Jabbal-Gill, M. Hinchcliffe, A.N. Fisher and S.S. Davis, *Adv. Drug. Deliv. Rev.*, **51** (2001), pp. 81–96.
16. Bourlais, C.L., L. Acar, H. Zia, P.A. Sado, T. Needham and R. Leverge, *Prog. Retin. Eye Res.*, **17** (1998), pp. 33–58.
17. de Campos, A.M., A. Sanchez and M.J. Alonso, *Int. J. Pharm.*, **224** (2001), pp. 159–168.
18. Kreuter, J., *Adv. Drug. Deliv. Rev.*, **47** (2001), pp. 65–81.
19. Schroeder, U., P. Sommerfeld, S. Ulrich and B.A. Sabel, *J. Pharm. Sci.*, **87** (1998), pp. 1305–1307.
20. Jenning, V., A. Gysler, M. Schafer-Korting and S.H. Gohla, *Eur. J. Pharm. Biopharm.*, **49** (2000), pp. 208–211.
21. Dingler, A., R.P. Blum, H. Niehus, R.H. Muller, and S. Gohla, *J. Microencapsul.*, **16** (1999), pp. 751–67.
22. Hirsch, L.R., J.B. Jackson, A. Lee, N.J. Halas and J.L. West, *Anal. Chem.*, **75** (2003), pp. 2377–2381.
23. Trofin, L., S.B. Lee and D.T. Mitchell, *et al.*, *J. Nanosci. Nanotechnol.*, **4** (2004), pp. 239–244.
24. Martin, C.R. and P. Kohli, *Nature Rev. Drug Discovery*, **2** (2003), pp. 29–37.
25. Alonso, M.J., *Biomedicine Pharmacotherapy*, **58** (2004), pp. 168–72.
26. Prego, C., E. Fernandez-Megia, R. Novoa-Carballal, E. Quiñoá, D. Torres and M.J. Alonso, (2003), Proceedings of the 30th International Symposium on Controlled Release of Bioactive Materials, Glasgow.
27. Damgé, C., C. Michel, M. Aprahamian, P. Couvreur, *Diabetes*, **37** (1988), pp. 246–51.
28. Tobío, M., R. Gref, A. Sánchez, R. Langer, M.J. Alonso and Stealth, *Pharm Res.*, **15** (1998), pp. 270–75.

29. Vila, A., A. Sánchez, C. Pérez and M.J. Alonso., *Polym. Adv. Technol.*, **13** (2002), pp. 1–38.
30. Fernandez-Urrusuno, R., C. Calvo, C. Remuñan-López, J.L. Vila-Jato and M.J. Alonso, *Pharm. Res.*, **16** (1999), pp. 1576–1581.
31. Calvo, P., J.L. Vila-Jato, M.J. Alonso, *Int. J. Pharm.*, **153** (1997), pp. 41–50.
32. De Campos, A., A. Sanchez and M.J. Alonso, *Int. J. Pharm.*, **224** (2001), pp. 159–168.
33. Gareth, A. Hughes, *Dis. Mon.*, **51** (2005), pp. 342–361.
34. Fong, K.M., *et al.*, *Molecular Genetic Basis for Early Cancer Detection and Cancer Susceptibility*, (1999), (IOS press, Amsterdam) pp. 13–26.
35. Robert, S. Negm, Mukesh Verma and Sudhir Srivastava, *Trends in Molecular Medicine*, **8** (2002), No. 6, June.
36. Nair, B., and T. Pradeep, *Crystal Growth and Design*, **2** (2002), pp. 293–298.
37. Mukherjee, P., A. Ahmad, D. Mandal, S. Senapati, S.R. Sainkar, M.I. Khan, R. Parishcha, P.V. Ajaykumar, M. Alam, R. Kumar and M. Sastry, *Nano Lett.*, **1(10)** (2001), pp. 515–519.
38. www.digenecorp.com.
39. Jaiswal, J.K., E.R. Goldman, H. Mattoussi, S.M. Simon, *Nature Methods*, **1** (2004), pp. 73–78.
40. Freitas, R.A., *Artif. Cells, Blood Subtit. and Immobil. Biotech*, **26** (1998), pp. 411–430.
41. Deo, Sapna K., Moschou A. Elissavet, Serban F. Peteu, Leonidas G. Bachas, Sylvia Daunert, Patricia E. Eisenhardt and Marc J. Madou, *Analy. Chem.*, (2003), pp. 206–213.
42. Freitas, R.A., *Analog*, **116** (1996), pp. 57–73.

Additional Reading

1. Robert, A. Freitas Jr., *Nanomedicine, Volume I: Basic Capabilities,* (1999), Landes Bioscience
2. Robert, A. Freitas Jr., *Nanomedicine, Volume IIA: Biocompatibility,* (2003), Landes Bioscience

Chapter 14

Molecular Nanomachines

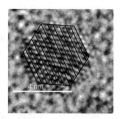

The designing of molecules performing mechanical functions is one of the larger goals of nanoscience. The development of such molecular systems would lead to the creation of integrated nanodevices. Several molecular systems have been designed to perform translations and rotations. These have also been used as switches and logic gates. In this chapter, we provide an introduction to this area, purely from the point of view of synthetic molecular machines, while focusing on one class of chemical systems, namely rotaxanes. We also highlight the problems that occur in this area when such systems are used as device components.

Learning Objectives

- What are molecular machines?
- What are the various functions achieved by these systems?
- What are rotaxanes and catenanes?
- How do we construct molecular shuttles with them?
- What are the difficulties in developing molecular devices?

14.1 Introduction

Nanoscience and technology are concerned with the manipulation of objects of nanometer dimensions. This entails playing with atoms and molecules in a programmed fashion. The motion of objects as per instructions (or stimuli) can be used to store and retrieve information, construct switches, and deliver materials of molecular dimensions among other things. This branch of science, which is concerned with the motion of molecules and atoms as a result of external stimuli, is called 'molecular nanomachines'. In brief, it deals with molecular objects performing mechanical functions. All these functions are similar to those being performed by biology in a highly precise manner for millions of years. The current interest in such processes stems from the fact that device dimensions are shrinking, and it is necessary to rely on to the molecular regime ultimately to perform functions which have hitherto been undertaken with macroscopic objects. Obviously in this size regime, systems have to obey the rules of atoms and molecules, namely quantum mechanics. Newton's equations of motion have to be satisfied by a macroscopic object in motion, and the former give the position at every stage of motion. In the case of molecular objects, the forces and quantum states of the system are related by the Schrödinger equation of motion. Due to the abstractness of the wave equation, the macroscopic analogies of motion may not be accurate enough to describe molecular processes. In this chapter, we shall examine some of the examples of molecular nanomachines constructed in the laboratory so that the reader can appreciate the challenges and prospects related to these machines. The discussion will not include the theoretical aspects of molecular motion for which the reader may consult the additional reading material listed at the end of the chapter.

14.2 Covalent and Non-covalent Approaches

As mentioned in the Introduction, Feynman postulated the possibility of atomically manipulated matter in 1959. He said, "The principles of physics do not speak against the possibility of maneuvering things atom by atom." However, the manipulation of atoms is difficult as isolated atoms are extremely reactive and form bonds instantaneously. The construction of an architecture arranged by atoms, one by one, therefore, may be possible in an inert atmosphere, but it is practically impossible in the laboratory ambience, unless, of course, it is used in synthesis in condensed media as in the case of nanoparticles. However, such approaches are important even for devices and we have outlined atomic manipulations in vacuum in Chapter 2 of this book. Thus the 'bottom-up' methodology is unlikely to be realized by starting with atomic building blocks.

The manipulation of molecules and implementation of their functions requires the use of synthetic chemistry. In the covalent approach of synthesis or routine chemistry, covalent bonds are added in a series to assemble larger structures. These approaches have yielded phenomenal successes in the past such as advances in pharmaceuticals, in which complex structures with structural and steriochemical specificity are synthesized in large quantities. However, this approach necessitates precise control of synthetic parameters to assemble units one after another. There can be minor errors in the construction and the synthesized structure can have a smaller fraction of the unwanted structures. This requires purification, which involves additional effort. Thus a typical covalent synthesis requires precise control of parameters, longer time and more effort.

The alternate approach used for making larger structures is non-covalent self-assembly. In this case, smaller building blocks are added together to form larger structures that utilize weak non-covalent interactions between the molecular units. These interactions ultimately create a structure which is thermodynamically stable. This process makes the assembly reversible, implying that dismantling and reconstruction are possible. In biology, all the structures are made by self-assembly, and the DNA double helix is a classic example wherein non-covalent interactions make and reform structures. This branch of chemistry, called 'supramolecular chemistry', was recognized with a Nobel prize awarded to C.J. Pedersen, D.J. Cram and J.–M. Lehn in 1987.

The approaches of supramolecular chemistry are better for making bottom-up architectures, especially those intended to function as machines. This is because molecular building blocks have definite shapes, sizes and properties. Developments in areas such as scanning probe microscopy have made it possible to manipulate molecules effectively. These modified structures can be studied at the single molecule level by advanced spectroscopic techniques such as single molecule spectroscopy. Today, it is possible to make electrical contacts with single molecules so that single molecule conductivity can be investigated. All these advances have propelled device structures at the single molecule level to the forefront of research.

We must also remember that self-assembly can also occur with covalent modifications. Here, a self-assembled superstructure goes through a covalent modification. As a result of this change, the system, as a whole, achieves functional competence. Several of the structures discussed in this chapter are formed as a result of this process. Here kinetic control of the process makes the system.

Numerous kinds of structures can be made by using the above approaches. All of them are directed towards making 'functional structures'. In the context of a molecular system, a functional assembly constitutes chemical architecture in which responsive modules can be made to perform specific functions by activation by external stimulus including photons, electrons, ions, heat, etc. Such structures can be used for a variety of functions such as chemical switching, logic gates, molecular shuttles, etc. In the following sections, we shall study some functional entities. It is important to realize that a functional system will have several components and all those components have to be linked in a co-ordinated manner in order to form a working machine.

14.3 Molecular Motors and Machines

Terms such as 'molecular motors' and 'molecular machines' are commonly used in literature. Before discussing specific examples, it is important to understand what their macroscopic analogues are. A macroscopic motor is a device which converts energy into mechanical work, often in the form of displacement. A machine is an assembly of devices designed to perform functions by consuming energy. Thus machine is a larger entity in comparison to a motor in its complexity. Molecular analogues of motors and machines facilitate the displacement of molecular co-ordinates by external energy stimulus. Although a single bond allows free rotation resulting in atomic displacements in most molecules in the gaseous and liquid phases, they are not referred to as molecular motors because designed molecular motion is not possible. In a molecular motor or machine, displacement can be controlled by a chemical or physical stimulus. A physical stimulus is far better than a chemical one as a chemical reaction leads to by-products which may hinder the performance of the motor or machine on a larger scale. The displacements themselves will be only of the order of molecular dimensions, but when scaled to the size of macroscopic objects, the distances involved become very large.

Molecular machines are characterized by the same kind of parameters that characterize macroscopic motors. They could involve: (i) a type of energy stimulus (chemical reactions, photons), (ii) a type of mechanical transformation (which could be translational or rotational), (iii) repeatability of the event as and when required (controllability) and (iv) time scale needed for the process to occur (as all these machines depend on a change in the nuclear co-ordinates of atoms). Ultimately, the motor results in a function which can also be used to categorize it. The resultant motions should be monitored, which makes it mandatory to bring about changes in the spectroscopic/molecular properties of the system while the change occurs. The changes should be monitored with the help of experimental techniques which become the readout when the device is used for a read-write operation. Obviously, the reading and writing mechanism should be able to pick the molecular scale objects. Ultimately, the molecular motion devices made should be usable somewhere. While the functions of macroscopic analogues are performed in day-to-day life, the functions that molecular motors are likely to perform are still being discussed. In several cases, such operations can be used to construct logic gates and transistors, and may be useful in molecular computing. These ideas have been proven in the simplest cases, but no functional architectures have been made. For a recent review of the topic, see Ref. 1.

The molecular machines discussed here are distinctly different from the molecular motors investigated in biology. Biological systems work through the co-operation of several molecules with different functions. Here, the motor proteins bind to a cytoskeletal filament and move along its length by repeated adenosine triphosphate (ATP) hydrolysis. Several kinds of motor proteins co-exist in the cell and they differ in various aspects such as the binding filament, direction of motion and the cargo carried. The protein contains a head region which binds and hydrolyses ATP, and a tail region which recognizes the cargo. The protein motion occurs in steps wherein the protein is bound and unbound to the filament, and in between the steps, conformational changes occur, which trigger motion. There are broadly three kinds of motor protein families: myosin, kinesin and dynein. While the myosin proteins move on actin filaments, the kinesin and dynein proteins move on microtubules. Within a family, the motor head is shared and can be attached to a variety of tails, as a result of which they perform different functions. Myosin and kinesin walk along different tracks but they share a structural core suggesting that they have common ancestry. A discussion of molecular motors of biology can be found in Ref. 2. The motion of single actin filaments supported on myosin heads, which in turn, are adsorbed on nitrocellulose-coated glass cover slips, has been imaged. The filaments are fluorescent-labeled in this kind of study, which is carried out by optical microscopy. Although the filament has a ~9 nm diameter, the fluorescence technique suffers from the diffraction limit and the single filaments appear much larger in the optical images. In contrast, the molecular motors of chemists, are single molecules in which one part undergoes motion upon excitation by a stimulus.

14.4 Molecular Devices

The most important aspect of nanoscale devices is the incorporation of molecular building blocks. As we have seen in several chapters (Chapters 4, 5, 7 and 8) in order to make architectures efficiently, it is necessary to use the approaches of self-assembly as the serial arrangement of units will be practically impossible (to make structures on a large scale). The approaches must be similar to those used in biology. In such structures, the molecular systems used are functional in nature. This functionality is triggered by activation involving photons, electrons and ions. Any form of effect seen in a material made of molecular components is caused by an external trigger, as very few systems change without any stimulus. An exception would be phenomena such as radioactivity. In a molecular system, there can be many responses to a stimulus. However, here we will consider only molecular level changes, which are easier to probe by using the techniques of spectroscopy. Properties such as magnetism or conductivity are also observed when an assembly of molecules is considered, though there are also molecular analogues of such phenomena. As a result of the stimulus, several processes can occur, including changes in the volume or shape (photoresponsive polymers, actuators), in color (photochromic), in molecular order (liquid crystal), in sensing (molecular recognition), in movement (molecular machines), etc. The nanosystem or device can be categorized as being photoactive, redoxative or ionactive depending on the nature of the stimulus, namely photons, redox chemistry or ions, respectively.

14.5 Single Molecule Devices

14.5.1 Switches

The most important aspect of a switch is that it has the ability to exist in two states, which can be interconverted. This property is called 'bistability'. The manner in which the change is brought about is referred to as 'writing' and the process through which these states are identified constitutes 'reading'. Thus the switch can be used as a memory element in binary logic (Fig. 14.1).

Several molecular systems are used as switches (Ref. 3). These conform to several broad categories, namely conformational change, configurational change and constitutional change. In conformational change, the conformational property of the system is altered when the stimulus is applied. In configurational change, the typical approach is to change *cis-trans* isomerization resulting in a system which has different properties from the earlier one. In configurational change, the chemical entity formed as a result of external stimulus is different as in the case of photoinduced cyclization. Here, we shall discuss one example from each of these categories. The objective is not to discuss the entire literature, but to present the concepts.

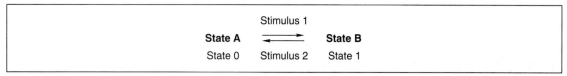

Fig. 14.1: *Concept of bistability. The system is stable in two states and they can be interconverted by the action of external stimulus.*

Conformational change The process shown in Fig. 14.2 can be illustrated with the example of a reversible molecular brake. Here, the triptycene unit undergoes free rotation around the single bond which connects it to the bipyridine unit. This occurs in solution. Upon the addition of the Hg salt, a conformational change occurs and the free rotation around the single bond is hindered. In this process, the bipyridine unit acts as a brake. The removal of the Hg^{2+} ion by the addition of EDTA restores free rotation. Thus the system is reversible. A motion of this kind is observable in terms of the NMR signatures of the system (Ref. 4).

Configurational change   Azobenzenes exist in two forms (*E* and *Z*) and these are interconvertible by light irradiation. The state in which azobenzenes exist can be probed by absorption spectroscopy. When incorporated in matrices such as polymeric liquid crystals, such systems induce observable changes.

Constitutional change   The photochromic behavior of syropyrans is one example of a light-induced structural change. Here the closed colorless form (spiropyran) gets converted into an open colored form (merocyanine) upon light irradiation. The open form goes back to the closed form photochemically or thermally. The lifetime of the zwitterionic form can be increased by the introduction of substituents that have an electron withdrawing character. The incorporation of these units into materials can result in observable changes.

14.5.1.1 Supramolecular systems

Supramolecular structures can also be effective in switching. In fact, it is possible to make switches easily with weak non-covalent interactions as they are reversible. Numerous kinds of such systems are known. The following case is an example of such systems. The incorporation of azobenzene unit into a crown ether can make the complexation ability of the latter photoswitchable (Ref. 5). The (*E*)-isomer of the azacrown molecule does not have space to accommodate the alkali metal cations, but the (*Z*)-isomer shows a high affinity. The (*E*) to (*Z*) conversion is both photoswitchable and reversible. Thermal re-isomerization, as it occurs in azobenzenes, is inhibited due to the formation of the complex. The switching is complete as the (*E*)-isomer does not bind at all. The examples listed above are presented in Fig. 14.2.

There are several other ways in which supramolecular structures can show switching behavior. The switching on and off of fluorescence is a common approach. Changes in the ionic concentration and pH can be used to effect this change. The switching of ligands and the resultant change in the properties of a metal containing system is another method used to bring about switching. The photoswichable complexation of metalloporphyrins is another example of such switching. The spectroscopic signature of the complexed and uncomplexed states will be different which can be used for identification.

14.5.2 Molecular Ratchet

Ratchets allow motion in one direction only. The simplest devices of this kind would have a toothed ratchet wheel, a pawl that prevents rotation in the unintended direction and a spring to hold the pawl in position, as shown schematically in Fig. 14.3. One can think of designing a ratchet with these components in a molecular sense. A candidate for the same is shown in Fig. 14.3, which has a [4]helicene unit. Helicenes are polycyclic aromatic compounds which possess a helical structure as a consequence of the steric repulsion of the aromatic nuclei at the ends. The structure has a helicity, which is evident from Fig. 14.3. Rotation of the triptycene unit will make the helicene more aplanar so that an energy maximum is reached. This rotation in the clockwise direction is preferred over the anti-clockwise rotation. The NMR investigations of this system have indicated that the rotation is possible in both the directions and has no specificity (Ref. 6). This is because the molecular system obeys the principle of microscopic reversibility, i.e. in a system at equilibrium, the reverse process is equally probable as the forward process. What this means is that every position of the molecular conformation is possible and the system has an equal probability of

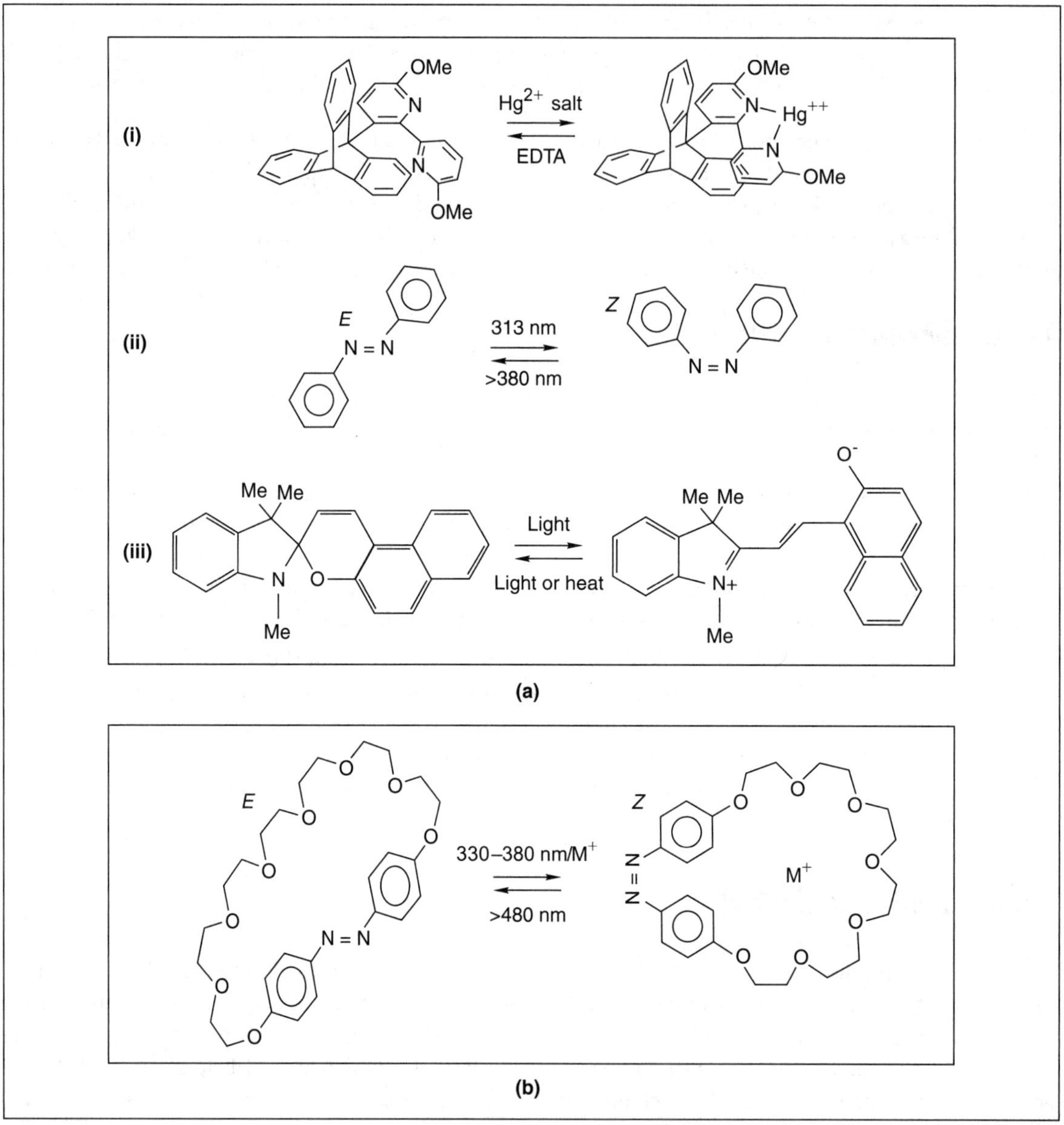

Fig. 14.2: *Various kinds of molecular switches. The state on the left can be considered as '0' and that on the right as '1'. Various kinds of stimuli are indicated. Cases listed in (a) correspond to molecular systems while (b) corresponds to a supramolecular system. In (b), the metal ion is released out of the crown ether cage in the reverse direction.*

going along either of these directions, when it is at a given state. This shows that in isothermal conditions, systems such as the one shown here cannot be used for unidirectional motion.

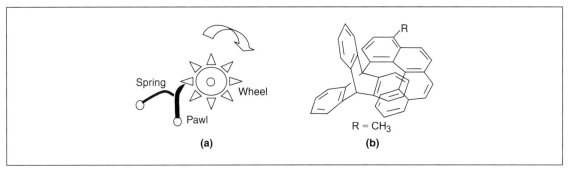

Fig. 14.3: *Schematic illustration of a macroscopic ratchet and a proposed molecular analogue. The helicene acts as both the pawl and the spring, while the triptycene acts as the wheel.*

14.5.3 Molecular Shuttles

Several chemical systems may be classified as molecular machines which utilize one of the above stimuli for activation. Among these, two broad categories of systems have been generally noticed due to their versatility. They are 'rotaxanes' and 'catenanes', on which we will focus in the rest of this chapter. These names are derived from the Latin words *rota*, *axis* and *catena* (which mean wheel, axis and chains, respectively). These are mechanically interlocked systems which utilize the molecular recognition of the components. A rotaxane contains one or more bead-like components which can be threaded into a rod. The rod has stoppers at both the termini so that the bead is not free to get out. Without the stopper, the bead is free to get out of the rod and such a structure is termed as a 'pseudorotaxane'. The rod can have multiple locations for the bead to reside in due to preferential molecular interactions called 'stations'. Catenanes are interlocked rings. In many a case, the rod is a polyether chain with aromatic units and the bead is cyclobis (paraquat-*p*-phynylene) tetracation (Fig. 14.4). Crown ether and cyclophane derived systems are used for assembling both rotaxanes and catenanes. Both these systems show translational isomerism. Since it is possible to control the kind of isomer used by external stimuli, and this property becomes the central aspect of designing molecular switches using rotaxanes and catenanes.

The cartoon representations of rotaxanes and catenanes are given in Fig. 14.4. The numerical prefix refers to the number of crossing points in the molecular graph of the structure. This number pertains to the molecular components present and not the number of beads or stoppers or the stations. In the case of catenanes, the nomenclature is evident.

A synthesis of these systems offers great challenges to organic chemists. However, over the years, efficient methods have come into existence. The very early approaches are based on chance encounters of the reacting partners. This leads to very poor yields as the dominant product is always without restrictions. In the case of [2]rotaxane synthesis, it can be easily seen that a mixture of rod, bead and stopper is likely to yield a rod with stoppers and not a rod with the bead in the middle and stoppers at the end. This is a matter of probability. However, if the interacting units can be self-organized to some smart reacting system forming a pre-assembly, the reaction may become more feasible. It can be seen that a rod and bead can pre-assemble and the stopper is added later. The routes used for the synthesis are listed in Fig. 14.5,

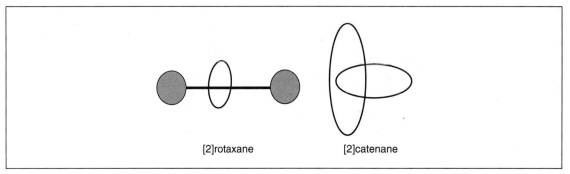

Fig. 14.4: *Schematic representations of a rotaxane and a catenane. There are two interconnected units in both and that makes the prefix [2].*

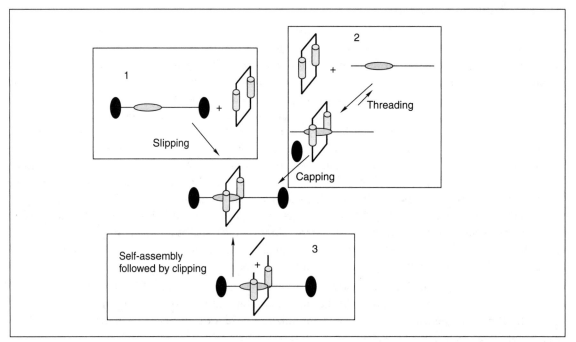

Fig. 14.5: *Synthetic approaches used for rotaxanes. In step 2, the de-threaded state exists in equilibrium with the threaded state. Three strategies are shown of which 3 involves proper design of components so that the starting material is assembled to some extent.*

while specific chemical examples are not given. Each of the specific methods is named and the process is self-explanatory. In the case of clipping, part of the structure is self-assembled utilizing preferred interactions and the last unit is integrated to the assembled structure.

There are several ways whereby mechanical motions can be introduced in this kind of systems. In the case of psueodorotaxanes, wherein the molecules are not physically restricted, two kinds of processes, namely 'threading' and 'de-threading', can occur without external stimulus. The two states are, in fact, in dynamic equilibrium. In rotaxanes and catenanes, two (or more) possible arrangements can also occur as a result of thermal activation. In Fig. 14.6, we show the schematic of a [2]rotaxane in which two states are possible. These correspond to the residence of the bead at the two possible stations. If the energy difference between the states is not large, and is within thermal energy, the states can co-exist and interconvert at room temperature. The equilibrium between various structures can be shifted by temperature. However, in the case of a molecular machine, we are interested in programmed conversion between structures by a stimulus. Ideally only one state should exist (corresponding to occupancy of station 1), which will get converted completely to the other (corresponding to occupancy of station 2) as a result of the stimulus. The state 1 will convert back to 2 and this process is reversible.

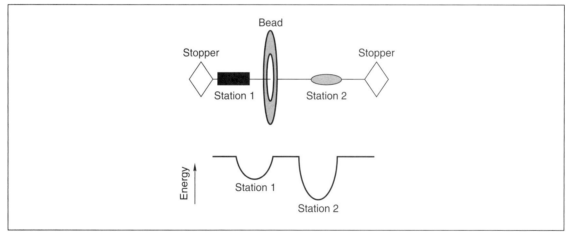

Fig. 14.6: *Schematic illustration of a [2]rotaxane structure in which two stations are incorporated on the rod. The bead can slide over the thread and can be held at station 1 or station 2, depending upon the strength of the interaction as shown in the potential energy diagram. Deeper potential well at station 2 makes that isomer to form preferentially. When the potential is unfavorable at that position, due to repulsive interaction between the station and the bead (as in the case of an electrochemical oxidation), the potential energy surface gets modified and the minimum becomes a maximum. As a result the bead slides over to the other station. When the oxidized state is reduced, the original situation returns and the bead returns.*

The first molecular shuttle (Ref. 7) was reported in 1991. It has a rod with two hydroquinone rings separated by a polyether chain and two stoppers at the end. The bead is the cyclobis(paraquat-*p*-phynylene) tetracation. The bead shuttles between the two degenerate positions approximately 500 times a second in solution, at room temperature. The movement of this kind of a unit shows no switching properties. In order to incorporate switching properties, the bead has to preferentially reside at one place and has to move to another location by the use of an external stimulus which may be chemical, photochemical or electrochemical. The effect of the stimulus has to be reversible so that the system can work as a switch.

One of the systems (Ref. 8) in which such a switching bahavior is manifested is shown in Fig. 14.7. The stations consist of benzidine and biphenol units. The bead preferentially lies on the π-electron rich benzidine unit. This is inferred from the proton NMR spectrum measured in CD_3CN. The two possibilities, namely occupancy of station 1 (benzidine) is 84 per cent in comparison to station 2 (biphenol). However, upon electrochemical oxidation of the benzidine unit to the radical cation, the cyclophane faces repulsion and moves to the neutral biphenol unit. When the benzidine unit is reduced back to the neutral form, the system comes back to the original situation. The same switching behavior is produced by protonating the bezidine unit by treatment with trifluoroacetic acid (TFA) and subsequent treatment with pyridine to get back the neutral form. Therefore, this reversible character makes it possible to form a binary molecular switch.

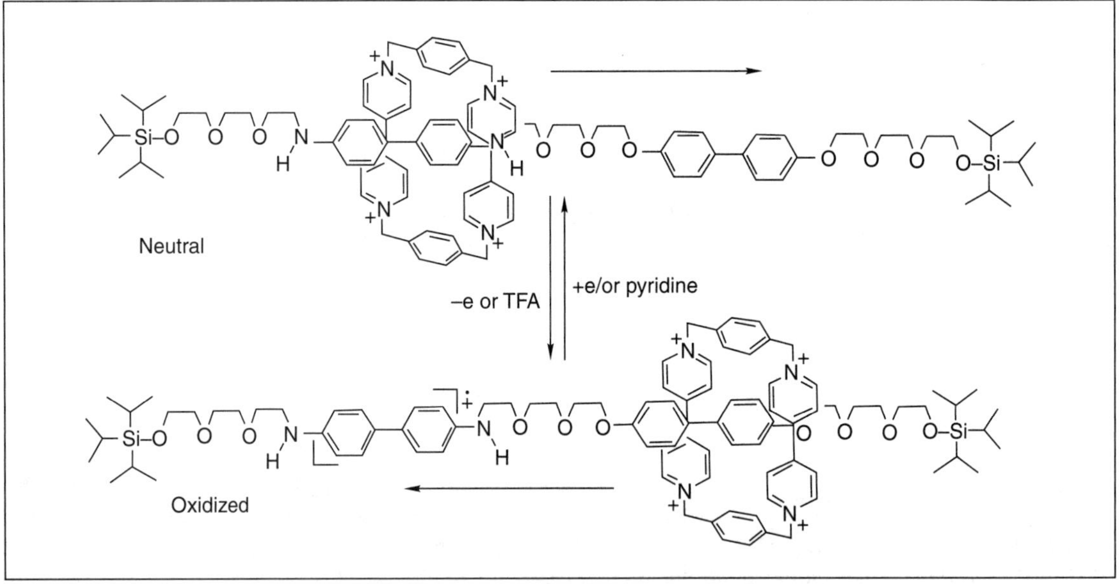

Fig. 14.7: *A rotaxane in which molecular motion has been demonstrated. The neutral molecule predominantly exists in the structure shown (86 per cent occupancy) and when the benzidine unit is oxidized either electrochemically or chemically, the bead shifts to the other station.*

The two isomeric structures have their characteristic charge-transfer bands. The differences are quite distinct and the color changes can be observed even by the naked eye. The charge transfer of benzidine occurs at 690 nm whereas that of biphenol occurs at 480 nm. As a result of the different charge transfer bands, the color changes from deep green to light red when acid is added, and the green color is regenerated by the addition of base. Thus the read-write property of the system can be demonstrated.

The principal advantage of the rotaxane- or catenane-based system is the confinement of the moving part within the molecule. It is possible to have a system in which the electron acceptor is in equilibrium with the two possible donor molecules. The binding with one of the donors can be stronger than that of the other but a complete switching from one to the other donor is not possible as the trimolecular system will always be in thermodynamic equilibrium between the various possibilities. It is important to recall that complete reversible switching is required for the system to be used as a device. The co-existence of other structures would lead to information scrambling in a storage device.

Similar kinds of molecular switches have been made with stimuli such as pH and light. In the case of light, photoisomerization is often utilized. The light-induced structural change of the rod makes the bead move in one direction. This process is reversed by a photon of another frequency thereby effecting recovery of the initial state. The binding of ions at specific locations and reversible electrochemistry of such systems can lead to interesting changes such as color (light absorption) or emission.

14.6 Practical Problems with Molecular Devices

It would be interesting to use any device of this kind for technological applications only if it has specific advantages. These advantages should pertain to device density, switching speed and device stability. A brief discussion of the important parameters is given below. The switching rate between the two states is an important parameter. In the specific case of the rotaxane discussed earlier, the measured shuttling time is of the order of milliseconds (Ref. 9). This means that one can have 1000 operations per second. In a semiconductor device such as MOSFET, the switching speed is of the order of 10 MHz, while the switching time is in the range of nanoseconds. What is the best time that can be achieved in a rotaxane? In the most simplistic model, the motion of the bead along the thread can be treated as a random walk in the absence of any interaction. The displacement, d after a time t is characterized by a diffusion coefficient, D. One can write $d = \sqrt{(2Dt)}$. In the case discussed above, d is 16 Å and if one takes the typical diffusion coefficient of 10^{-5} cm^2/s in condensed media, the shuttling time is of the order of nanoseconds. However, this is not achieved experimentally. This indicates that the interaction of the bead with the thread and the station limits its motion. Faster shuttling motion will be difficult to attain in these molecular systems and it is not possible to achieve the motion as fast as electrons or holes. This is because it will be practically impossible to eliminate interactions.

However, it may be possible to increase storage density by using rotaxanes. Substantially smaller number of atoms are needed to store per bit of information in rotaxanes in comparison to silicon. In the case discussed, one bit of information is stored with 274 atoms while typical electronic circuits require about 10^6 atoms to do the same job, though this number is shrinking. In real devices, it may not be possible to pack the molecules close together to get larger storage densities as intermolecular interactions would cause scrambling. In a device using the rotaxane discussed, the information is written in terms of the oxidized or neutral form. The oxidized form has to be stable for the period of storage for the information to be safe. This would necessitate the inclusion of spacers in order to avoid intermolecular electron transfer processes, which would reduce storage density.

When the device dimension shrinks, it is important to address single molecules with electrical connections. Although molecular wires have been developed, facilitating single molecule electrical contacts is a problem, which is yet to be solved. It may be necessary to develop devices with scanning probe microscopies to read and write the information. In a recent demonstration, a L-B film of rotaxane was made on a conducting glass surface and a potential was applied to it by using a scanning tunneling microscope. The potential switches the position of the bead to another area of the thread, changing the conformation and part of the molecule stick out from the surface by a small distance. This makes it possible to record information. However, erasing the information has not been possible (Ref. 10). The stability of such devices for repeated usage is an important issue to be considered.

Review Questions

1. What are the differences between molecular machines and macroscopic machines?
2. What are molecular switches?
3. What are the molecular properties used to achieve mechanical functions?
4. What are the functions achieved by rotaxanes and catenanes?
5. How fast molecular shuttles are?
6. What are the limitations of molecular devices?
7. What are the possible device applications?
8. What are molecular logic gates?

References

1. See for a recent review, Tian, H. and Q.C. Wang, "Recent Progress on Switchable Rotaxanes", *Chem. Soc. Rev.*, **35** (2006), p. 361.
2. Alberts Bruce, Alexander Johnson, Julian Lewis, Martin Raff, Keith Roberts and Peter Walter, *Molecular Biology of the Cell*, (2002), 4th edition, Garland Science, New York.
3. For a discussion, see Lopez, M.G., and J.F. Stoddart, *Molecular and Supramolecular Nanomachines*, (2000), Chapter 14, *Handbook of Nanostructured Materials and Nanotechnology*, H.S. Nalwa (ed.), Academic Press.
4. Kelly, T.R., M.C. Bowyer, K.V. Bhaskar, D. Bebbingto, A. Garcia, F. Lang, M.H. Kim and M.P. Jette, *J. Am. Chem. Soc.*, **116** (1994), p. 3657.
5. Shinkai, S., Y. Kusano, O. Manabe, T. Nakaji and T. Ogawa, *Tetrahedron Lett.*, **20** (1979), p. 4569.
6. Kelly, T.R., I. Tellitu and J.P. Sestelo, *Angew. Chem. Int. Ed. Engl.*, **36** (1997), p. 1866; Kelly, T.R., J.P. Sestelo, I. Tellitu, *J. Org. Chem.*, **63** (1998), p. 3655.

7. Anelli, P.L., N. Spencer and J.F. Stoddart, *J. Am. Chem. Soc.*, **113** (1991), p. 5131.
8. Bissel, R.A., E. Cordova, A.E. Kaifer and J.F. Stoddart, *Nature*, **369** (1994), p. 133.
9. Liu, J., M.G. Kaifer and A.E. Kaifer, *Struct. Bond*, **99** (2001), p. 141.
10. Feng, M., X. Guo, X. Lin, X. He, W. Ji, S. Du, D. Zhang, D. Zhu and H. Gao, *J. Am. Chem. Soc.*, **127** (2005), p. 15,338.

Additional Reading

1. Schalley, C.A., K. Beizai and F. Vögtle, "On the Way to Rotaxane-based Molecular Motors: Studies in Molecular Mobility and Topological Chirality", *Acc. Chem. Res.*, **34** (2001), p. 465.
2. Sauvage, J.P., "Transition Metal-containing Rotaxanes and Catenanes in Motion: Toward Molecular Machines and Motors", *Acc. Chem. Res.*, **31** (1998), p. 611.
3. Pease, A.R. and J.F. Stoddart, "Computing at the Molecular Level", *Struct. Bond*, **99** (2001), p. 189.
4. For a discussion of the theoretical aspects of single molecular rotors see Joachim, C. and J.K. Gimzewski, *Struct. Bond*, **99** (2001), p. 1. Note that the entire volume 99 of *Structure and Bonding* is on molecular machines and motors.
5. Lehn, J.M., *Supramolecular Chemistry*, (1995), VCH, Weinheim.

Chapter 15

NANOTRIBOLOGY

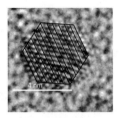

Nanotribology is concerned with the interactions of objects at the nanoscale. These interactions are important as device structures are shrinking in size and information is compressed in space. Reducing friction of interacting surfaces will be very important in the area of micro and nano electromechanical devices. As one studies these interactions at nanometer precision, we see that they have a molecular basis. Thus the study of nanotribology is to do with understanding macroscopic processes such as friction at the molecular level. This is possible with the application of modern techniques to engineering problems.

Learning Objectives

- What is nanotribology?
- Why are interfacial problems important at nanodimensions?
- How do we study interfacial properties with atomic precision?
- What are the typical applications of nanotribology?

15.1 Introduction

Tribology is one of the oldest sciences, but is still not well understood. 'Tribology', like many other technical terms, has a Greek origin. The two Greek words 'tribo' and 'logy' refer to 'rubbing' and 'knowledge', respectively, defining tribology as 'the knowledge of rubbing'. The Greeks applied it to understand the motion of large stones across the earth's surface. However, today tribology plays a crucial role in a number of technological areas with the main ones being polishing and lubrication of substrates for electronic applications (MEMS/NEMS) and in increasing the lifespan of mechanical components.

With advances in technology, the size of mechanical, electrical and optical components is reducing rapidly. Today, we require rapid actuation, which requires fast moving interacting surfaces. Extraordinary advances in MEMS/NEMS, data storage devices and micro-machines have been observed in recent times. Aerospace and automotive industries require materials with low friction and adhesion. Lubrication has served as a solution to face these needs, but the development of lubricants in the automobile industry depends on the adhesion of monolayers (see Chapter 5) to the material surface and hence a need to study tribology at a proportionate scale is being felt.

Technically nanotribology can be defined as the investigations of interfacial processes, on molecular length scales, occurring during adhesion, friction, wear, nanoindentation and thin-film lubrication at sliding interfaces. Nanotribology studies reveal behavior that can be quite different from that observed at macroscopic levels. A study of tribological behaviors can help us manipulate matter at the nanoscale. We can take advantage of the new electronic and atomic interactions as well as the new mechanical and magnetic properties observed at the nano levels to understand the synthesis, assembling and processing of nanoscale building blocks, composites, coatings, and smart materials with reduced and controlled friction, wear and corrosion. Ultimately these interactions will decide the applicability of nanomaterials in emerging areas (Refs 1, 2, 3).

15.2 Studying Tribology at the Nanoscale

Several instruments are being used nowadays to study tribology at this length scale. Some of them have been described below. Some of these techniques are discussed in detail in Refs 4–8.

15.2.1 Nanotribometer

In nano-tribometer (Fig. 15.1) a flat, a pin or a sphere is pressed onto the test sample with a precisely known force. These are mounted on a stiff lever, which acts as a transducer to measure the friction force. The friction coefficient is determined by measuring the normal and lateral forces as the pin moves by

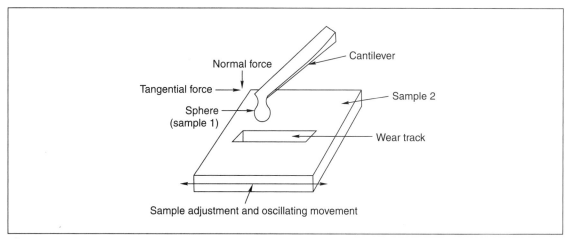

Fig. 15.1: *A nanotribometer being used to study the wear of sample 2 by sample 1, spherical in shape. The tangential force causes a deflection in the cantilever (attached to the ball), which gives the measure of the friction force (by using a traducer). By studying the wear tracks, wear coefficients are obtained.*

keeping track of the deflection in the cantilever. Wear coefficients for the pin, disc, sphere or plate are calculated from the amount of material lost during the test. The method facilitates the study of friction and wear behavior of almost every solid combination with or without the presence of a lubricant in between.

15.2.2 Surface Force Apparatus (SFA)

SFA or the surface force apparatus was developed in the 1960s and has been commonly used to study the static and dynamic properties of molecularly thin films sandwiched between two molecularly smooth surfaces. The SFA has a pair of atomically smooth surfaces (mica sheets) mounted on crossed cylinders (cylinder axis 90° to each other). This pair of crossed cylinders of the same radius of curvature, is the geometrical equivalent of a sphere in interaction with a flat surface. Molecules of our interest can be attached to these mica surfaces, and then surfaces may be immersed completely within a liquid, or maintained in a controlled environment. Actuators attached to either or both of the surfaces' supports are used to apply a load or shear force and used to control the distance of separation between them. Load and friction forces are measured with the help of the sensors. The contact area and relative separation of the surfaces can be measured with optical (as shown in Fig. 15.2) or capacitance measurements. The separation distances can be measured and controlled to the angstrom levels.

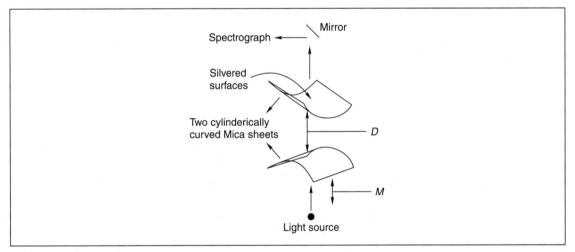

Fig. 15.2: *Schematic of a SFA (top view).*

The distance D between the surfaces is varied by using a piezoelectric actuator (which moves by a distance M). When the two surfaces are sufficiently far apart, the motion of the actuator will result in an equal change in the surface distance. But when the surfaces are close to each other, due to the interactions, the actuator motion will not be equal to the distance between the surfaces. Hence, by calculating the actual distance between the sheets by interferometry, we can learn about the forces of interaction. The difference in the actual distance and that detected by the device is positive if there is repulsion between the sheets and vice versa.

The lateral resolution is limited to the range of several micrometers. The instrument is thus a model contact wherein the contacting geometry is known. Therefore, by varying the material between the surfaces, the interaction forces can be controlled and measured. The drawbacks of the instrument are that the lateral resolution is limited, molecular smoothness is required to obtain meaningful results, and so usually, the substrate is restricted to mica.

15.2.3 Quartz Crystal Microbalance (QCM)

QCM is used for monitoring thin film growth with sub-monolayer sensitivity, since the shift in resonance frequency (of the sensing surface) is proportional to the mass of the absorbed film. It is nothing but an ultra-sensitive mass sensor. It consists of a quartz crystal sandwiched between two electrodes, as shown in Fig. 15.3. These electrodes are further connected to an oscillator, which make the quartz crystal oscillate with a stable frequency of f_o. If a rigid layer is evenly deposited on one or both of the electrodes (sensing surface), the resonant frequency will decrease proportionally to the mass of the adsorbed layer according to the Sauerbrey equation:

$$D_f = -[2f_o^2 D_m]/[A(r_q m_q)^{1/2}],$$

where

D_f = measured frequency shift,

f_o = resonant frequency of the fundamental mode of the crystal,

D_m = mass change per unit area (g/cm²),

A = piezo-electrically active area,

r_q = density of quartz, 2.648 g/cm³,

m_q = shear modulus of quartz, 2.947×10^{11} g/cm. s².

The typical surface used for QCM studies is a thin film gold coated over a quartz crystal (Fig. 15.3). The electrode surfaces are deposited by using vapor deposition.

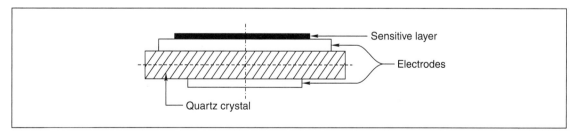

Fig. 15.3: *Schematic of a quartz resonator working in shear mode. The standard quartz crystal has a gold sensor surface. Other surfaces are available such as metals, polymers and chemically modified surfaces.*

In other words,

$$D_f = C \times D_m/A,$$

where C directly depends on the quartz crystal properties. C is typically around 10^8 and hence even a slight change in mass during the tribological action can be detected easily by keeping note of the QCM frequency. Thus a quartz crystal microbalance can be helpful in studying tribology at the nano-level.

15.2.4 Atomic Force Microscope

For friction force calculations, we slide the cantilever orthogonal to the long axis of the cantilever (in the contact mode). Here the changes in the intensities in the right and left halves of the four quadrant photodiode (because of twisting of the cantilever due to friction) help us calculate the friction parameters. Please see Chapter 2 for a discussion of the principle of AFM.

Figure 15.4 gives the images after scanning on the surface along the long axis of the cantilever, (a), and perpendicular to it, (b). Note how different the two scans are. Actually, Bhushan (Refs 6, 7) has shown that the friction force scan images are found to be more similar to the 3D plot of slope of the roughness scan, (a), versus distance. For example, the initial peak in (b) is similar to the sudden increase in slope of the roughness scan in (a) at the same location.

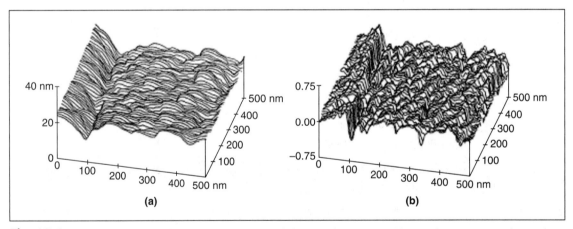

Fig. 15.4: *Comparison between the roughness and friction force scan. (a) Roughness scans of a surface, (b) friction force scan of the same surface. Reprinted from Bhushan, (Ref. 7). Copyright (1999), with permission from Elsevier.*

For studying the adhesive properties of one material with another, a tip of one material (generally a micropsphere on the usual tip) and the sample of another are used. The tip is brought closer and closer to the sample till it comes into contact with it. It is then pulled back. However, due to adhesive forces, it stays stuck to the surface beyond the point it came into contact with it. The extra displacement given to the piezo drive to loose contact is noted. Multiplying this value with the cantilever stiffness directly gives the adhesion force value. For example, consider the deflection *Vs* distance (moved by the piezoelectric) in Fig. 15.5. While bringing the tip closer to the surface (marked 'extending' in the figure) the contact occurs at B. From A to B, attractive van der Walls forces come into play. After contact, the cantilever movement is directly proportional to the distance that the piezoelectric drive moves. However, in retracting, the contact is lost at C, way beyond point B due to the presence of adhesive forces from B to C.

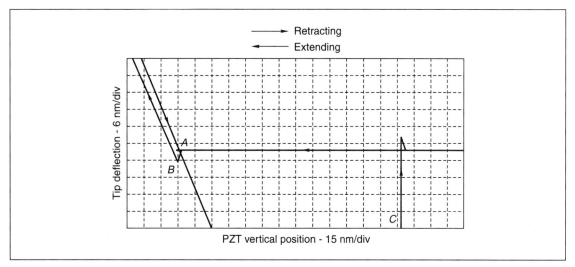

Fig. 15.5: *A deflection vs distance curve obtained to measure the adhesive force. Reprinted from Bhushan, (Ref. 7). Copyright (1999), with permission from Elsevier.*

The adhesive Force, $F = BC \times k$,

where, BC is the distance from the graph = the extra distance that the piezoelectric tube had to travel to loose contact and k is the stiffness of the cantilever.

For scratching and wear, a single crystal diamond (of high hardness) tip is used in the contact mode. After scratching, the surface around the scratch is scanned to study the effects more elaborately. Nanofabrication and nanomachining refer to scratching and wearing the surface at intended locations and in an intended fashion, with the mechanism being the same.

Nanoindentation is another important measurement in nanotribology. It can help us get the hardness and elasticity modulii values of a localized point. It is done in the normal mode of AFM, by indenting the surface with the tip after setting the scan size to zero. After indenting, a finite size scan is done around the indent to calculate the projected area of indent. The hardness (H) is then given by,

H = (load applied)/(projected area of indent),

where the projected area of indent is the area of the indent projected on to a plane perpendicular to the tip (Fig. 15.6(b)).

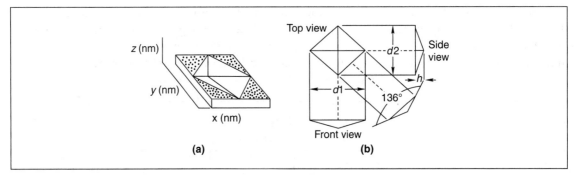

Fig. 15.6: *(a) A schematic showing an indent made by the pyramidal (136°) AFM tip. (b) The projected area of the indent.*

Young's modulus is obtained from the slope of the force-displacement curve while unloading. Using the AFM technique, we can investigate surfaces of interest at the atomic scale. The AFM relies on a scanning technique to produce three-dimensional images of sample surfaces of high resolution. AFM can be used to measure ultra small forces (<1 nN) present between the cantilever tip and a sample surface. These small forces are measured by tracking the motion of a highly flexible nano-sized cantilever (having a very small mass), by various measurement techniques like optical deflection, optical interference, capacitance and tunneling current. The deflection can be measured to the very low limits of 0.02 nm. For a typical cantilever having a force constant of 10 N/m, a force as low as 0.2 nN can be detected. In the operation of high-resolution AFM, it is the sample that is moved rather than the cantilever, as the movement of the cantilever may cause vibrations thereby affecting the measurements. AFMs are now available for large samples too, wherein the tip is scanned and the sample is stationary. In order to obtain good atomic resolution, the spring constant of the cantilever should be less than the equivalent spring constant between the atoms. As per the experimental results, a cantilever beam with a spring constant of about 1 N/m or lower is desirable. Tips have to be as sharp as possible. Tips with a radius ranging from 10 to 100 nm are commonly available.

Surfaces in contact will generally have a thin layer of liquid (and sometimes a solid) lubricant at the interface. Tribological studies under such environments can lead to completely different results. John Pethica and co-workers have made useful contributions to the study of nanoindentation of liquid environments (Ref. 9), comparing these data to the conventional results reported. The same group has also reported useful results based on the mechanical deformation of nanocontacts due to the size and structure of the asperities at the point of contact (Ref. 10).

15.2.5 Friction Force Microscope

Subsequent modifications to AFM led to the development of the friction force microscope or the lateral force microscope (LFM), designed for atomic-scale and micro-scale studies of friction and lubrication. This instrument measures the lateral or frictional forces (in the plane of sample surface and in the direction of sliding). By using a standard or a sharp diamond tip mounted on a stiff cantilever beam, AFM is also used in the investigations of scratching and wear, indentation, and fabrication/machining. Surface roughness, including atomic-scale imaging, is routinely measured by using the AFM. Adhesion, friction, wear and boundary lubrication at the interface between two solids with and without liquid films have been studied by the using AFM and FFM. Nanomechanical properties are also measured by using an AFM.

15.3 Nanotribology Applications

When the devices scale down from 1mm to 1nm, the surface area decreases by a factor of 10^{12}. At the same time, the volume decreases by a factor of 10^{18}. Therefore, the surface-to-volume ratio of the device increases a billion times. This results in an increase in the surface forces (proportional to the surface area) like friction, adhesion, meniscus forces and surface tension, by the same factor. However, studying tribology at a proportionate scale has helped us overcome this problem to a great extent, as described by the following applications.

15.3.1 Superlubricity

Hirano and Shinjo (Ref. 11) in 1990 showed the origin of the ultra low friction of graphite between two incommensurable (having an incommensurate contact geometry, which prevents collective slip-stick motion of all atoms in that contact) graphite layers rotated with respect to each other. Since then, many research groups have used the Frictional Force Microscope to study this further. Recently, Martin Dienwiebel, et al. (Refs 12, 13) have revealed that when two parallel surfaces slide over each other in an incommensurate contact, superlubricity has been observed in certain orientations. Superlubricity is defined as a phenomenon wherein when two parallel surfaces slide over each other in incommensurate contact, they experience non-existent or negligible friction. In such geometry, the lattice mismatch prevents a collective slip-stick motion of all atoms in the contact together, and hence the kinetic friction force can be vanishingly small. The atoms in graphite are oriented like an egg crate, forming an atomic hill and valley pattern. When the two graphite surfaces are in registry (every 60 degrees), the friction force is high. When rotated out of registry, the friction is largely reduced, just like two egg crates can slide over each other easily when twisted with respect to each other. In order to understand this better, consider Fig. 15.7. The maximum friction force is observed when the tungsten tip-surface orientation angle is 60 degrees (in registry) and decreases as the graphite surface is rotated about an axis perpendicular to the surface and parallel to the tip (out of registry). Although the superlubricity phenomenon was discovered during the 1990s, it has received little attention even though it drastically reduced friction in between dry, unlubricated surfaces, making it relevant to various applications.

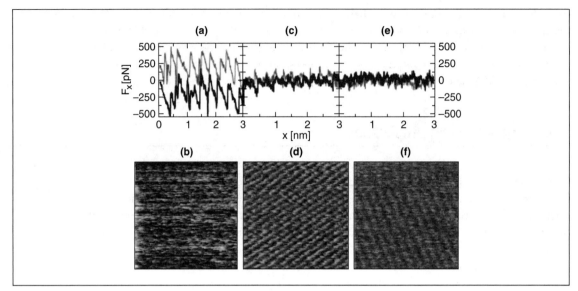

Fig. 15.7: *Friction loops (black, forward and grey reverse) and lateral force images (forward) measured along scanning direction at tip-surface orientation angles of: (a), (b) 60 degrees; (c), (d) 72 degrees; (e), (f) 38 degrees. Reprinted with permission from Dienwiebel, et al. (Ref. 13). Copyright (2004) by the American Physical Society.*

15.3.2 Head Disk Capacity

Technology today demands ultra low flying head-disks. The aerial densities of these head disks strongly depend on the distance maintained between the slider and the hard disks. This distance (known as the flying height) is generally kept in the nanometer range and hence the probability of contact is high. In order to decrease the chances of contact, the roughness values of the slider and disk surfaces are kept within molecular dimensions (see Fig. 15.8). However, at such low flying heights between the molecularly smooth surfaces, the surface forces (adhesion) are very strong. Due to this reason, modern hard disks are found to have a fly height of about 20–30 nm. A fly height of around 15 nm is known to give an aerial density of about 10 Gbit/in^2. However, technically it is desired to have an aerial density of about 100 Gbit/in^2, for which the head disks are required to operate at a flying height of 6–7 nm.

Fly heights of this order require the roughness of the disk to be of the order of 1nm peak to valley. Such smooth disk surfaces are undesirable because of elevated stiction. Also, at such low heights, the potential for discontinuous contacts between the slider and disk cause vibrations in slider that may lead to track mis-registration. The use of textured sliders (Fig. 15.9) has helped in reducing stiction at the head/disk interface for low flying heights with super smooth disk surfaces. The use of textured sliders is also helpful in reducing in-plane and out-of-plane vibrations at low flying heights. Also, textured sliders show less lubricant depletion on disk surface as compared to the untextured sliders.

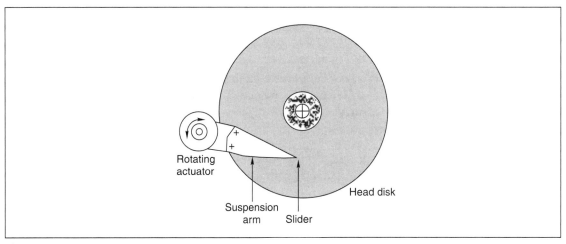

Fig. 15.8: *The image of a typical PC hard disk. The head (slider) and the disk can be clearly seen.*

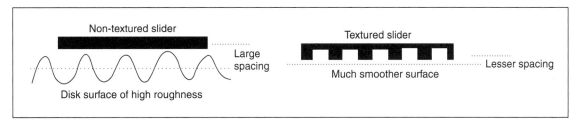

Fig. 15.9: *Reducing stiction by texturing slider rather than compromising for rough disk surface.*

The slider and the magnetic recording layer on the disks are generally given a carbon overcoat. This facilitates sliding between the two surfaces without any cohesive failure in the magnetic layer. The addition of a carbon overcoat also improves the slider wear durability. These coatings also prevent damage to the magnetic layers in corrosive environments. These carbon overcoats are lubricated by perfloropolyethers (PFPE) with polar end groups that enhance their attachment to the surface and limits spin-off. The commonly used PFPE has two –OH functional groups at the ends of the molecules. The film is deposited by dip coating and the thickness is generally kept to 1 ± 0.1 nm (Refs 14, 15).

15.3.3 Nanolubrication

One of the major aspects of tribology is lubrication. It is the most common way to prevent wear. The idea is to maintain a liquid or grease layer between the two solids and the compressive stresses generated in this layer keep the two solids from coming in contact. Theoretically, a good lubricant is supposed to:

1. Generate fluid pressures to keep the two surfaces separate.

2. Sacrificially wear off to protect the surface.
3. Redistribute the stresses at the contact.
4. Increase the contact area (lowering the contact pressure).

Another important property of a good lubricant is its ability to produce boundary lubricating films when needed—in situations when the film thickness is not sufficient to avoid contact (if the film thickness is lesser than the roughness of the surface), then the highest asperities of the two surfaces come into contact with each other. As the pressure increases, the deformation of these asperities becomes more and more plastic. This situation is referred to as 'boundary lubrication'. However, in this situation (under the conditions of high temperature of the asperity tips), the lubricant and its additives react with the solid surface forming a protective chemical film. This film is sheered away, thus protecting the surface from any damage.

The effectiveness of a film depends on several factors like its adhesive and cohesive strength, density, its thickness, etc. Most of these properties can be measured by using the techniques mentioned above to study nanotribology. It is desirable to use a film that is thicker than the surface roughness which is capable of generating fluid pressures to avoid contact and which also has good adhesion and cohesion properties.

However, most of the time at the nanoscale, there is only a limited supply of lubricant and only one or two monolayers are available to do the job. In such a situation, even a small damage to the film (by shear or oxidation, which is why the lubricating film should be oxidation-resistant and non-volatile) can continue to get aggravated, exposing the surface to contact and hence strong frictional forces. This is very common with the Langmuir-Blodgett films. Therefore, it is highly essential for these films to have self-repairing properties. This is possible only if the molecules from another location can move in to cover the exposed surface. Thus, the molecules should be free to move around on the surface, and should not be chemically absorbed on the surface, which would result in low bonding strength. In order to overcome these two contradicting problems, we use mixed molecular films, wherein one species bonds to the surface, while the other is free to move. Thus a carefully designed molecular assembly, wherein each molecule improves a certain property of the film, is required (Refs 1, 3, 16).

15.3.4 Micro-electro Mechanical Systems (MEMS)

Micro-Electro-Mechanical Systems (MEMS) signify the integration of mechanical elements, sensors, actuators and electronics on a common silicon substrate through microfabrication technology. While the electronics are fabricated by using integrated circuit (IC) process sequences, the micromechanical components are fabricated by using compatible 'micro-machining' processes that selectively etch away parts of the silicon wafer or add new structural layers to form the mechanical and electromechanical devices.

Most MEMS are made by using single crystalline silicon (generally doped). But bare silicon exhibits an inadequate tribological performance and needs to be coated with a solid and/or liquid overcoat to lower the coefficient of friction and wear factor at the interface. However, due to their dimensions, normal lubricants cannot be used. Instead ultrathin liquid films are deposited at the interfaces. These are generally the Langmuir-Blodgett (L-B) films or the self-assembled monolayers (SAMs). The L-B films are bonded to the substrate by weak van der Waals forces while the SAMs are bonded covalently. Therefore, SAMS exhibit better wear and hence better durability than the L-B films, thus proving a better choice for lubrication.

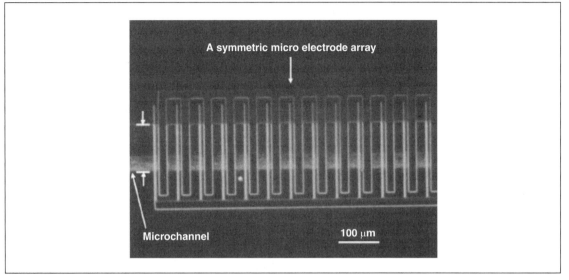

Fig. 15.10: *A microscopic picture of fabricated micro-fluidic devices (a micro-electro mechanical system) showing a 100 mm width channel and asymmetric micro-electrode array. Reprinted from Dhayal, et al. (Ref. 17). Copyright (2005), with permission from Elsevier.*

15.4 Outstanding Issues

Although progressive research has been conducted in the field of nanotribology since the 1970s, it is still a new field with a lot yet to be discovered. Some interesting work has been done and there is a lot more to do. The atomic-scale understanding of complex phenomena will help us design better engineering products with greater efficiency and durability. Several such efforts have been made and the interested reader may consult Ref. 5. Following are some of the important outstanding issues, which need to be addressed to facilitate progress in the sphere of nanotribology.

- Bridging the gap between macroscopic tribology and nanotribology. Atomic scale stick-slip behavior must be examined in more detail.
- Modifying the existing theories or proposing new continuum theories to accommodate nanoscopic experimental observations.

- Selective lubrication of the lubricant as it determines the effectiveness of lubricant films to protect the materials in contact. Molecular basis of lubrication: concept, design and principle of nanolubrication.
- Investigation of friction in low speed and moderate load conditions.
- Electronic and photonic contribution to friction.
- Modeling on the atomic-scale related to dissipation force microscopy and tunneling microscopy.
- Wear prediction and wear process must be considered in detail. The wear prediction of ceramics requires immediate attention in view of the extensive use of ceramics in industry.
- Applying what we understand to frictional processes in biology and medicine.

It is important to mention that several of the nanomaterials such as C_{60} are shown to have applications in lubrication. Studies (Ref. 18) have shown that the inorganic fullerenes of WS_2 are potential solid state lubricants.

Review Questions

1. What is the molecular basis of friction?
2. What are the methodologies to study interfacial properties at nano dimensions?
3. Where do we apply such molecular understanding?
4. State a few examples where nanotribology is important.

References

1. Hsu, S.M. and C.Z. Ying, *Nanotribology: Critical Assessment and Research Need*, (2002), Kluwer Academic Publishers, Netherlands.
2. Sherrington, I., *et al.*, *Total Tribology: Towards an Integrated Approach,* (2002), Professional Engineering Publishing Ltd., UK.
3. Bhushan, B., *Fundamentals of Tribology and Bridging Gap between Macro- and Micro/Nanoscales*, (2001), Kluwer Academic Publishers, Netherlands.
4. K. Jacqueline, *Surface Science*, **500** (2002), pp. 741–758.
5. Carpick, R.W. and M. Salmeron, *Chem. Rev.*, **97** (1997), pp. 1163–94.
6. Bhushan, B., *Wear*, **259** (2005), pp. 1507–31.
7. Bhushan, B., *Wear*, **225–229** (1999), pp. 465–92.
8. Dedkov, G.V., *Materials Letters*, **38** (1999), pp. 360–66.

9. Mann, A.B. and J.B. Pethica, *Langmuir*, **12** (1996), pp. 4583–86.
10. Mann, A.B. and J.B. Pethica, *Applied Physics Letters*, **69** (1996), pp. 907–09.
11. Hirano, M. and K. Shinjo, *Phys Rev.*, **B41** (1990), p. 11837.
12. Dienwiebel, M., P. Namboodiri, S.V. Gertjan, W.Z. Henny and W.M.F. Joost, *Surface Science*, **576** (2005), pp. 197–211.
13. Dienwiebel, M., P. Namboodiri, S.V. Gertjan, W.Z. Henny and W.M.F. Joost, *Phys. Rev. Lett.*, **92** (2004), p. 126101.
14. Zhou, L., K. Kato, N. Umehara and Y. Miyake, *Nanotechnol*, **10** (1999), No. 4, pp. 363–372.
15. Zhou, L., M. Beck, H.H. Gatzen, K. Altshuler and F.E. Talke, *IEEE Tansactions on Magnetics*, **39** (2003), No. 5.
16. Hsu, S.M., *Tribology International*, **37**, **7** (2004), pp. 537–545.
17. Dhayal, M., H.G. Jeong and J.S. Choi, *Appl. Surf. Sci.*, **252** (2005), pp. 1710–1715.
18. Rapoport, Y., Yu. Bilick, Y. Feldman, M. Homyonfer, S.R. Cohen and R. Tenne, *Nature*, **387** (1997), p. 791.

Additional Reading

1. Bhushan, B., *Principles and Applications of Tribology*, (1999), John Wiley and Sons, USA.
2. Bhushan, B., *Introduction to Tribology*, (2002), John Wiley and Sons, USA.
3. Bhushan, B., *Modern Tribology Handbook*, Vol. 1 (2000), CRC Press, USA.
4. Stachowiak, G.W. and Andrew W. Batchelor, *Engineering Tribology*, (2001), Butterworth-Heinemann, USA.
5. Moore, D.F., *Principles and Applications of Tribology*, (1975), Pergamon Press Ltd., Oxford.
6. Scherge, M. and Gorb Stanislav, *Biological Micro- and Nanotribology*, (2001), Springer-Verlag, Germany.
7. Bhushan, B., *Handbook of Micro-Nanotribology*, (1999) CRC Press, USA.

PART FIVE

Society and Nano

Contents:

- Societal Implications of Nanoscience and Nanotechnology

Chapter 16

Societal Implications of Nanoscience and Nanotechnology (in Developing Countries)

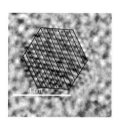

The phenomenal expansion and growth of nanoscience and nanotechnology has been historically unprecedented. Further the convergence of these two with the growth of information science and molecular biology has heralded new challenges for society and natural environment. Technological revolutions have shown that pioneering scientific discoveries have the potential to pave the way for radically innovative and integrated approaches and for providing new solutions to the international community's most pressing problems. In order to enable decision-makers to devise policies in keeping with the needs of the society, communities and nations, it is important to understand the societal implications of these newly emerging fields. Nanotechnology, unlike any other technology, can find application in virtually all areas of human life. Because of its distinctiveness and pervasiveness it has the potential to revolutionize the way we live, think, behave and act. In spite of the fact that it is an infant in terms of its evolution, some of the issues related to nanotech indicate a wide spectrum of potential societal impacts. These need to be studied further, especially in the context of the developing countries, where nano commodities, devices and services have the potential to make a significant difference, especially with regard to economic and social development. The current public nano-discourse offers sociology a unique opportunity to switch from a merely passive, observational role to an active participating one, especially wherein the key players involved meet to find joint and concerted solutions for the development of advanced sciences and technologies. Nanotechnology's unique and distinct features have the potential to bridge the technological gap between the developed and the developing worlds, if designed and implemented to serve the needs of people who were left out of the ambit of previous technological evolutionary processes. This larger objective necessitates the fusion of several ideas and experiences, from abstract science to the realities of the downtrodden, from sociology to materials science; nevertheless the realm of nano appears to possess the necessary ingredients to attain these goals.

Learning Objectives

- Are there specific implications for society when nanotechnologies are applied?
- What is unique to nanotechnology in comparison to the recently pioneered areas such as biotechnology?
- What are the parallels between nano and other technologies?
- How did technology impact society? How will its impact be different in the case of nano?

- Are we spending enough on nano and are we gaining enough from it?
- Does the use of nanotechnology signify specific areas of economic benefit for the developing world?
- How can we harness these benefits?

16.1 Introduction

The objective of this chapter is to address and outline the wider societal implications of nanosciences and its deriving technologies on society within the context of the developing world. In order to prepare the reader for the novel scientific and technological terrain, we begin with an introductory historical retrospective of the scientific and technological evolution, from the first Industrial Revolution to the third and onwards to the newly emerging technologies of the 21st Century to illustrate the influence of technological progress on the evolution of the society. The rapid pace of discoveries and development in the realm of nano indicates that this newly emerging field is different from others which gives rise to specific questions, even though the problems encountered in this field are common to other advanced technologies. The aim of exploring the world of nano is to discover new properties and to translate the new knowledge into the manufacturing process for obtaining enhanced structures and components with novel chemical, physical or biological properties. This futuristic manufacturing method can virtually invade and pervade all areas of human life, since it modifies the identity of all matter, animate and inanimate. It will definitely revolutionize human society in an ever unprecedented way. This new scientific branch and associated technology will doubtless have both desirable and undesirable repercussions, which will impact the society at large and change its structure, organization and functioning in the longer term. Since different aspects related to nanoscience and nanotechnologies need to be woven into the nano-discourse, we are examining these specific aspects with a special focus on the developing world. We try to provide a comprehensive framework on the issues relevant to nanoscience and technology, and map the social, economic, political, legal and ethical aspects, which are applicable to both the developed as well as the developing worlds, with particular attention to the latter. With nanoscience and nanotechnology being hailed as the science of the future and the technology of the next generation, respectively, along with their infinite market potential, the key focus lies on the control, manipulation and construction of matter at the atomic and molecular level. Recent studies have revealed the significance of nanotech for fostering economic growth, human health and increasing wealth in the developing world. The possible application of nanotech in the fight against poverty will be addressed and illustrated with practical examples. In the past, we have experienced the consequences of technology and society drifting apart. In order to prevent a further gap—the nano-divide—cross-cutting ties, and institutional linkages along with innovative, integrated and original solutions are needed.

16.2 From the First Industrial Revolution to the Nano Revolution

16.2.1 A Historical Retrospective

When James Watt invented the steam engine in 1765, little could he have imagined that his invention would unleash the first Industrial Revolution and transform society in a way only the invention of the wheel did in 3500 BC. The steam engine entered the history of technological evolution as a symbol of the era of mechanization. Watt's invention of steam-powered machinery revolutionized the manufacturing process and almost all areas related to human and social life in the course of the 18th and 19th Centuries in a massive and irreversible way. A new era for science, technology, economy, society and culture had just begun—called the "Modern Times". That period is characterized by the transformation of the transportation sector, the advent of capitalism and rapid urbanization, and the gradual rise of the modern industrial society. The rapid expansion of the industrial sector, successive inventions and technological breakthroughs enabled the extension of the factories to technology parks leading to increased production. All these developments laid the ground for the next revolution. Industrial Automation, also known as the second Industrial Revolution, started around the year 1870. Industrial robotics transformed the technological manufacturing process since it enabled automated large-scale production at reduced costs. The introduction of the assembly line on the verge of the 20th Century changed the entire organization of the manufacturing process and labour. Improved and more efficient production methods facilitated mass production at a global scale. Mass production at an ever-increasing productivity rate and speed became a must imperative in order to satisfy the rising demands of post-modern and upcoming consumer society. The invention of the transistor and the gradually growing semiconductor industry paved the way for the third revolution, the Digital Revolution. Computerization, starting in the early 1960s led to large-scale production and higher cost-efficiency all over the world and fostered flexibility in the manufacturing process. The era of post-modernism had just begun when in 1959, Richard Feynman gave his speech, "There is Plenty of Room at the Bottom". Most likely, he had imagined that his prognostics would lead to further transformations of the manufacturing process but he could not have thought that a new nano-revolution, would manifest itself at the threshold of the new millennium. The technologies of the late 20th Century and newly emerging technologies of the 21st Century, namely information and communication technologies, and biotechnology, have notably altered the industrial and service sectors, the production methods and society. They heralded the information age and the biotech era, and have ushered in a new kind of society, the knowledge society. Since nano-science and technology as well as manufacturing of nanomaterials are no longer futuristic fantasies but reality, we should assess how the 'disenchantment of the atomic world' will change our present societies. What comes next, the nano-society?

16.2.2 The Milestones on the Trajectory of Nanotech

The cornerstone of nanoscale science, engineering and technology was laid when in 1959, Richard Feynman envisaged the possibility of arranging atoms to create new matter at the atomic and molecular level. In 1964, the Nobel Laureate in Chemistry, Glenn T. Seaborg patented two of the elements he had synthesized—Americium and Curium. This was the beginning of patenting atomically and molecularly engineered matter. The term 'nanotechnology' was coined in 1974 by Norio Tanigutchi, professor at Tokyo Science University who pointed out the trend of precision manufacturing at the scale of nanometres. In his MIT doctoral thesis of 1981, Eric Drexler extended the term to a wider area and studied the subject in depth. During the same year, Gerhard Binning and Heinrich Rohrer, who were awarded the Nobel Prize in Physics in 1986, invented the scanning tunnelling microscope, a novel measurement tool facilitating the sensing of matter at the nanometre scale. This was a significant technological breakthrough and had a great impact on the future development of nanoscale science because every new science requires new instruments and equipment. In the early 90s, Warren Robinett of the University of North Carolina and R. Stanley Williams of the University of California established a virtual reality system and linked it with the aforementioned scanning tunnelling microscope to see and touch atoms. When D.M. Eigler placed xenon atoms to a shape, reflecting the logo of IBM in 1990, he showed the possibility and feasibility of atomic-scale manipulation, which marked a further pioneering breakthrough in this newly emerging field (Ref. 1). In 1993 the first academic research centre dedicated to nanotechnology was institutionalized in the USA, namely at the Rice University. Only five years later, Zyvex, the first molecular nanotechnology company, was established in the USA, which marked the beginning of nanotechnology private venture capital companies. In 2000, another step forward was taken, when Lucent and Bell Labs, along with Oxford University, created the first DNA motor, the first nano-biotechnological artefact. Since the beginning of the 21st Century, discoveries and new technological facilities, molecularly precise manufacturing, nanofactories, public and corporate investments in nano-research and public-private venture partnerships are rapidly expanding and increasing. The development in this newly emerging area suggests that the time span between milestones on nanotech's trajectory is reducing considerably, as we move further on. It is obvious that nano has gathered momentum and the nano-revolution has just begun.

16.2.3 The Distinctness of Nanoscience, Engineering and Technology

We have seen how the scientific discoveries and technological inventions of the past three centuries have revolutionized the manufacturing sector and society. The common traits of the new technologies that transformed human and social life in-between the first Industrial Revolution and the Digital Revolution are that they were designed for large-scale production which was capital- and energy-intensive in consonance with the needs of large manufacturing infrastructures, and were designed to compete in the global markets. Big became beautiful and the manufacturing process and methods were aligned along the macro-principle. However, the transistor and semiconductor industry ushered in a change in that trend and small became beautiful, as a result of which manufacturing began to focus on micro and then on nano. Nanoscience and technology require nanomaterials. The manufacturing of nanomaterial is not just another step in the growth ladder but is also about using the knowledge of the atomic realm to produce novel artefacts in a cheaper and cleaner way, with reduced capital and energy inputs, and with more precision. The unparalleled development of nanotech and the dissimilar preconditions for nano show that nanoscale science, engineering and technology are different from other newly emerging sciences and technologies. Here we highlight a few of the most salient peculiarities that explain the distinctness of nanoscience and manufacturing of nanomaterials. Firstly, at the nanometre scale, sciences and technologies converge and therefore go beyond the traditional boundaries of disciplines. Nanoscience is of a trans-disciplinary nature since it involves chemistry, physics, biology, mathematics, cognitive science and life sciences, in particular genomics and proteomics. Nanotechnology fuses with other recent technologies like information and communication technologies, and biotechnology. Secondly, control and manipulation of the very elementary building blocks of all objects of the living and non-living world—atoms and molecules—enable modifications of the same, which can influence each and every area of life. In other words, the core novelty in science and technology on the scale of the nanometre is that scientists and technologists do not invent the world *ex novo*, as in past, but *de novo* since the new artefacts are made of components which have no natural analogues. Thirdly, the term 'nano' refers to measurement, the nanometre, as it indicates the size of the matter being observed and manipulated and the term does not refer to any object *per se*. This explains the unlimited spectrum of nano since all physical matter, irrespective of its nature, can be measured, and the only condition is that measurement facilities for that size regime should exist. Fourthly, the manufacturing of nanomaterials does not entail enormous initial capital outlays for industrial infrastructure that other technologies require. This has made gigantic technological parks obsolete, at least in cases wherein controlling the shape of the nano object is not a stringent condition. There are factual examples substantiating the assumption that production can be cost-effective and tailored to local needs—either large or small. These examples include the production of nanomaterials through biology. We will come back to this argument later. A further novel aspect with regard to nanotechnology that gives a practical expression to the pace of technological change, is the reduction in time from the scientific discovery to the application of the new knowledge.

16.3 Implications of Nanoscience and Nanotechnology on Society

16.3.1 Science and Technology Change Society

In order to better understand the nexus between nanoscience, nanotechnology and society in the context of social change, a few preliminary concepts relevant to both nanotechnology and other technologies are briefly outlined here. Science, technology and society are intrinsically interlinked and characterized by mutual interdependency. Since technology is firmly imbedded in society, it cannot be looked at in isolation. The application of scientific knowledge and associated developments are two of the major factors determining social progress and prosperity. Social change is as dynamic and complex as social systems and is both an essential ingredient and the driving force of social evolution. A myriad other factors determine, shape and characterize the technological, social and cultural evolutionary processes of society. Advances achieved in scientific and technological knowledge in any discipline or branch inevitably lead to changes in social relations, meanings and societal patterns. Earlier, we have seen how scientific discoveries and technological inventions had literally revolutionized societies with time. Considering the time factor, technological and social changes may not occur contemporaneously since the social system takes its own time to respond to changes and to find its new equilibrium. The coexistence of several forms of societies and the analogous coexistence of technologies in the world today, from pre-industrial to state-of-the-art technologies, illustrates the above.

Progressive technological and social changes do not necessarily eradicate previous societal structures and historically, science and technology have been used by all kinds of societies, irrespective of their stage of development, instead of just being restricted to the most advanced societies. As regards social and technological change, as far as nanoscience and its deriving technologies are concerned, it is highly likely that their potential impacts will be stronger, because nano has the incomparable force to pervade all societies and economies, from the pre-industrial to knowledge societies, from ancestral to highly industrialized economies, and is not necessarily subjected to a nation's current development stage and/or geographical location. Nanotechnology can also bridge yesterday's missing technological link between the developed and the developing worlds.

16.3.2 Society and the Scientific and Technological Innovation Process

Protecting, improving and preserving life by using new scientific knowledge and technical findings is intrinsic to scientific and technological research since the very premise of science is to serve humanity. Society is reacting to technological change with new forms of institutions and develops its own responses to technological innovation, which is valid for nanoscience and technology as well. As in all other cybernetic systems, a change in one of the systems will generate changes in the others because these complex organic systems are linked by multiple feedback loops and therefore the social world always responds to scientific and technological innovations. Discoveries in nanoscale science and innovations in molecular manufacturing will certainly have an impact on society because technology has never been and will never be neutral, as knowledge itself is not neutral. The scientific and technological innovation process shapes the evolution of society, and therefore it is essential to understand the societal implications of nanoscale science, engineering and technology in order to understand the direction in which society is advancing. The current infant stage of this technology limits reliable or accurate prognostics but in spite of these limitations, a historical retrospective on how technologies of the previous centuries revolutionized the social organization, structure and value systems can help understand and predict the potential impacts of nanoscience and nanotechnology.

16.3.3 Forecasting the Nanofuture

How will society respond to the distribution and diffusion of engineered nanomaterials, including commodities, instrument facilities and services and how will these change society? The advancements in nanoscience and nanotechnology, necessitate these questions. It is commonly believed that the social implications in the developed as well as in the developing worlds will be similar to the other newly emerging technologies of the 21st Century like biotechnology and information and communication technologies. This assumption, however, ignores two facts: first, nanotechnology is a fusion technology and therefore incorporates, for instance, bio and information technologies. The synergy effects, resulting from the interface of two or more systems, will amplify the complexity and inevitably exceed the hypothetical consequences of one single technology and secondly, the world is entering the sphere of nano, even where information and communication technologies have not yet pervaded society at large. In the developing countries, where pre-industrialized and post-modern technologies coexist with the newly emerging technologies, nano-engineered commodities and services can be designed for the needs of people belonging to pre-industrialized, post-modern or knowledge societies since no preclusions apply. As far as the predictions of nano's future are concerned, global trends suggest that nano is gathering momentum. Expansion in scientific research and development, public and corporate investments, public-private partnerships, media coverage, patents, services and devices clearly indicate that nanotechnology is growing rapidly. From these positive projections, one can conclude that nano has the potential to become the flagship of the new millennium's industrial production methods in the developed as well as in the developing worlds. Nanotech will certainly not replace all other technologies, but will coexist with and borrow from the technological inventions of the past. It is thus unlikely that the nano era will replace the digital. Instead, the digital age will converge with the nano, and their synergy effects will lead to fundamental and irreversible alterations in the existing, cultures and institutions of society, societal organization, and various mechanisms and patterns, including the demographic structure of society. Its all pervasiveness and the magnitude of this new technology will exceed those of the precedent technologies because the intensity of the impact of any phenomenon's is positively correlated with its pervasiveness. The circumstances indicate that the possible impacts of nanotech will exceed even those of the first Industrial Revolution.

16.4 Issues—An Outlook

16.4.1 Artificial Evolution—How Green is Green Nanotechnology?

The term 'Green Nanotechnology' apparently seems a paradox *per se*, as it challenges both nature and the ecosystem, because the control and manipulation of matter at its very elementary level leads to the creation of new matter not present in the realm of nature. Since the concept 'green' refers primarily to environmental protection and not to the evolutionary process, nanoscience and technology are not inconsistent with 'green'. 'Green Nanotechnology', in fact, has the potential to play a pivotal role in the struggle against the world's most pressing environmental problems. Bio-nanotechnology, for instance, offers a wide spectrum of possibilities for mitigating the adverse effects of environmental degradation, regardless of its causes and sources. In particular, bio-nanotech can provide viable solutions to soil, air and water pollution, and the unsustainable exploitation of natural ressources. These solutions include the support of cleaner production methods, provision of alternative and renewable energy sources, reduction of input into the manufacturing process and purification of water. The interface of bio and nanotech, however, does not only generate positive results. The potential risks inherent in the convergence of life sciences and nanoscience and its deriving technologies need to be addressed and understood. Threats may arise from the increased chemical reactivity of materials at the nanoscale, the toxicity of nanoparticles and the yet unknown side effects of the atomic and molecular engineered materials. The release of atomic and molecular engineered matter into the biosphere poses additional problems to human beings and nature, since test results obtained in the laboratory may differ from those carried out in an open environment. It is still unknown how humans and the environment will respond with regard to the distribution and accumulation of novel materials. The lifecycle of products containing nanoparticles is difficult to establish since the degradation process of nanomaterials and components can only be estimated. The hazards identified with regard to health are chiefly related to the absorption of nanoparticles by the human body and their distribution as well as the risk of accumulation in organs. It is also unknown, as to how the human (and animal) metabolism will react to the intake of nano-engineered food and nanoparticles, which once introduced in the ecosystem, will enter the food chain. This necessitates research into the possible negative impacts of bio-nanotech, and transparency in the results in view of the credibility and plausibility of green nanotechnology. However, recent results obtained suggest that the benefits far outweigh the risks (e.g., applied nanotech techniques for water purification systems).

16.4.2 Crossing Land—Melting of the Traditional Boundaries of Natural and Human Sciences

The realization of the possibility to explore and control the world at the nanometre scale has given life to new scientific fields, and blurred the traditional boundaries of disciplines, and even led to fusion and convergence amongst natural sciences. These new circumstances present a unique chance for the sciences, both hard and soft, to meet and to overcome C.P. Snow's paradigm of the two cultures, dividing the scientific from the human sciences (Ref. 2). Despite the scientific tradition of hard and soft sciences using different methodologies and jargons, at the nanoscale there seems to be a trend of convergence of the two apparently opposing scientific cultures. Scientists belonging to the first category have recognized that at the atomic and sub-atomic level, an organic world view becomes extremely useful since the observation, control and manipulation of matter in realms inaccessible to human's ordinary senses demand not only new instrumental facilities but also novel approaches (Ref. 3). This revolutionary shift in the scientific mentality of natural scientists has the potential to overcome both the two-culture paradigm postulated by Snow and the long-lasting *Methodenstreit* (disputes about methodologies) between the natural and human (cultural) sciences. This new situation frees the various disciplines from their life in isolation, and offers natural and human scientists a common meeting ground. This, of course, does not mean that a physicist becomes an ethicist or vice versa, but that both are needed for finding solutions that contribute to the qualitative enhancement of human life and responsible scientific development. Recent conferences held on nanoscience and technology bear witness to the paradigm shift and have revealed that the scientific fraternity has recognized the importance of bridging the cultural gap. Today's scientists seem to be prepared and willing to cross the borders of their own disciplines.

16.4.3 Education and Training in Nanoscale Science and Technology

Frontier sciences and technologies represent a true opportunity to enhance a country's qualitative and quantitative level of human capital since new ways of manufacturing require the development of new capabilities and skills, which creates new job opportunities. Enhanced human capital strengthens a country's competitiveness, spurs economic growth and prosperity, all of which are essential ingredients for a more sustainable economic, human and social development. These factors are of paramount importance to developing countries since the latter are rich in human capital. In recent years, the academic world has begun to react to the upsurge of the nano-phenomenon and started preparing the future workforce for the emerging opportunities arising in the nano realm by offering multi-disciplinary curricula that complement basic natural science education with specialized courses in nanoscience, materials science and molecular biology as well as by sponsoring continuing education and training. Since the socio-economic situation of a country conditions its public expenditure in education, research and training, most educational and scientific activities in the nanoscale science, engineering and technology are offered in highly industrialized countries. Since nano-manufacturing is also possible in developing countries, it is vital to develop the necessary human intellectual resources for nano-manufacturing in the developing world. The development of educational, research and training programmes in nanoscience and its related fields of application, analogous to those of highly industrialized economies, has become a necessity. That this is possible in the developing world as well, can be the illustrated by existing strategic partnerships between government agencies, industries and businesses established for starting nanotech research centres of excellence. It is widely believed that scientific education, training and research have the potential to narrow the gap between the developed and the developing world, even if not within the society itself. It is a matter of fact that if the developing world does not catch up with the scientific progress of the industrialized world, the risk of being further marginalized becomes real and it would make real the public fear of a new gap between the haves and have-nots: the nano divide.

16.4.4 The Nano Economy

Technology has always played a central role in wealth generation and the emerging nanotech market has the potential to transform and reshape all economic sectors, from the primary to the tertiary, and this applies to nano as well. The often-advocated assumption that nanoscience and its deriving technologies will change the world must stem from a basic premise: namely that novel findings and innovations on the nanometre scale will have a visible and significant impact on productivity. The projected consequences of rapidly advancing technologies on the nanometer scale will see the rise of new industries and the fall of those stuck to the conventional or sustaining technologies of the past (Ref. 4). The marketing of scientific discoveries at the frontiers of science and technological innovations in the nano realm will become the driving force of the nanomarket. We can assume that in the years to come, this relatively new economic phenomenon will contribute to an increase in the global Gross Domestic Product (GDP) as projected by reliable global economic institutions like the World Bank (WB) and the International Monetary Fund (IMF). A comparison of the economic performance of the nano industry and business with the development of the national and global GDP could reveal possible correlations between the two and provide an answer to the question as to whether nanotech truly contributes to economic growth in quantitative terms.

16.4.5 The Nanobusiness and Finance

Since nanobusiness is still in its infancy even in the developed world, it is difficult to assess the possible economic impact it will have in the developing world. In highly industrialized economies, like the USA, the commercial spin-offs of nanoscientific research, start-ups and venture capital enterprises, together with multinational companies reaching out for nanotechnology and competing for market shares of the newly emerging nanotech market, will shape the national and world economies. Nanobusiness, which is still in its nascent phase but growing rapidly, encompasses a wide spectrum of well-established and solid manufacturing branches, from biotech, materials, electronics, energy, healthcare, textiles, sensors and many others, and does not preclude any business sector. The ongoing globalization process tends to amplify the importance of nanotech use, since the manufacturing of nanomaterials provides the link with international capital markets and global technology, and production networks. Not only is the future nano-market impressive in size, but so also is the potential clientele of nanotech derivates, products, devices and services. Since all forms of matter are ultimately composed of atoms and elements, theoretically all people could become exposed to the nearly unlimited nanomarket and become its benefactors. To put this into perspective, the US National Science Foundation (NSF) estimates the total global market for nanotechnology-related products and services to reach US$ 1 trillion by 2015 (Ref. 5). Based on these projections, the Nanobusiness Alliance forecasts the global nanotechnology market to reach US$ 225 billion by 2005 and expects it to touch US$ 700 billion by 2008 (Ref. 6). When these figures are seen in consonance with the projected world GDP, the share of the nanomarket can be expected to exceed one per cent by 2008 (Ref. 7). The ever-increasing number of private nanotechnology companies and the sharp rise in patent filings from public and private institutions testify the upward trend in the commercial innovations made in the field of nano-manufacturing and the run into nanobusinesses. To illustrate what is stated before, the US Patent and Trademark Office issued, until the first half of 2005, 3,818 patents with reference to nano and 1,777 patent applications were handed in and were waiting to be transformed into lucrative licences (Ref. 8).

Private nanobusiness was born only in 1997, when nano pioneers started the first venture capital company in the USA. While large and well-established trans-national companies (TNCs) followed Zyvex and started diversifying and integrating attributes of nanotech techniques into their manufacturing activities, small and medium-sized enterprises (SMEs) too entered the arena of the upcoming nanobusiness. The preliminary findings of business experts suggest that the latter have already conquered niche markets and by consolidating their positioning at the global level, they are challenging the big market leaders. A look at the companies and sponsors present at the Eighth Nanotech Conference and Trade Show in May 2005 in Anaheim, California reveals who the contenders in the newly emerging market are. More than 100 corporate companies, both TNCs and SMEs, are in some way or other involved in nanomaterial manufacturing (Ref. 9). As far as the capital market is concerned, nano-stocks are already traded at the world's major stock exchanges and financial institutions as well as insurance companies nurture high interest in this novel technology and its derivates. As regards private investments and risk management, they already play a significant role. At the micro-economic level, it is still difficult to express the prognostics because of two major reasons, namely, the absence of hard data about the cost structure of nano-engineered products, devices and services, and secondly, the fact that nanotech commodities are still not an integral part of everyday life in any part of the world (primarily because there are only a few commodities for mass consumption).

16.4.6 Public and Private Investments in Nano Research and Development

Public and private research funding in nanoscale science and technology has progressively increased over the past years in both the developed and the developing worlds. Research and development activities are generally segregated in relatively large industrial, government and academic laboratories but the latest trend suggests, to a smaller but not less remarkable extent, that the private sector, and in particular SMEs, are investing into the nanometer scale technology. The public sector still holds the lion's share of research and has been growing at an unprecedented rate during the last few years. The global public spending in nanoscience and technology exceeded US$ 3 billion in 2003 and it will increase further since more countries, including the developing countries, are planning to or have already launched national nano-initiatives (Refs 5, 10). A look at the government funding of the next five years of the USA confirms the progressive trend in public investments into nanoresearch and development. For the year 2005, a sum of US$ 809.8 million has been approved and the figure for 2008 exceeds an annual spending of US$ one billion (Ref. 11). The developing as well as the newly emerging economies have started realizing the inherent opportunities of nanoscience and nanotechnologies, and the importance to compete in these areas from the beginning and not, unlike in the case with information and communication technologies, wherein they could afford, to wait for these to be transplanted from the developed world at a later stage. For a good performance in the global research and development arena, public and private investments into nano are a precondition. The trend of increasing public spending into nano in the developing world is parallel to that seen in the highly industrialized world. However, investments in the development world in nano are lower than those of the developed world, in terms of their respective GDPs.

16.4.7 Reliability, Safety and Risks: Assessment and Management

Systematic identification and assessment of the risks posed by any new technology are essential. The idea of a system getting out of control is the nightmare of every conscientious scientist and technologist and not only of technophobes and those predicting a cataclysmic end to humanity because of the release of toxic nano-engineered artefacts from the laboratory into the environment. The potential risks might arise as a result of the characteristics of the nanoparticles themselves, as also of the properties of products manufactured with nanoparticles along with the manufacturing process itself. Systematic research into the flip side of this new technology, especially the risks that it entails, is essential, chiefly because of two reasons. Firstly, because if unintended consequences, including negative side effects and counter intuited results are known, it is possible to calculate the risk and to take precautions. The presentation of worst-case scenarios, which include the necessary mechanism and measures for managing such striking situations, serves to limit or mitigate the adverse effects on society and its natural habitat. Several existing instruments and methods used in technology management, like the Environmental Impact Assessment (EIA) help in analyzing, assessing and evaluating the reliability, safety and risks of new products, services and devices using atomic or molecular engineered matter. Secondly, the fear of derailing nanoscale science and technology, because it is evolving faster than the researchers' ability to keep pace with its development, is undermining the research and development process and gives rise to anti-nanotechnology feelings and attitudes. Therefore, it is imperative to foster methodical research in the fields of safety, reliability and risk management of nanocommodities, devices and services. The fantasy apocalyptical worst-case scenario involving nanotechnology, named by Eric Drexler as the "Grey Goo", has contributed little towards broadening our societal understanding and knowledge of the potential risks of molecular engineered artefacts (Ref. 12). Since people are receptive to fictional representations as they shape their imaginaries, the "nanomania" has rather alienated the entertainment and leisure industry, and given rise to misconceptions, misunderstandings and distorted views regarding this particular technology rather than contributed to a rational and genuine discussion.

16.5 Nano Policies and Institutions

When we look at the nano-phenomenon from the political perspective, in general, we look principally at issues relating to long-term strategic technology policies, including intellectual property reforms, international co-operation, monitoring and the regulation of research and development. Because of its unlimited potential, nanoscale science, engineering and technology necessitates the adoption of national and international nano policies. The management of research and development in nanoscience and technology, in terms of administration and control by the public authorities, has to guarantee the responsible use of the potentials that both offer. The claim for formalizing a regulatory regime over nanoscience and technology indicates that a minimum level of coercive intervention is needed by the state, especially vis-à-vis responsible nanotech. Scientists, politicians, the private sector and representatives of the civil society have opposing views as far as the intensity of the public interference into scientific and technological research and development is concerned and even within these groups, their views do not converge. The opponents, i.e. the nano sceptics, a heterogeneous flock, call for a regulatory and institutional framework that limits the scope of scientific activities, and of atomically and molecularly engineered commodities and services. Despite the clash of interests and opinions, the two blocks agree on three central issues in the public nano discourse, namely the non-interference into privacy, preservation of human dignity, and the protection of the society and natural environment from hazards. The first two, since they have strong ethical connotations, will be addressed separately while the third has already been addressed. In order to counter misuse, including activities and actions that are in contradiction with the universally shared ethics and principles, a minimum regulatory framework has become a necessity to guarantee the responsible use of nanoscale science and its deriving technologies. The current scenario suggests that the institutionalization of a supranational body with a standing committee exercising minimum international control, monitoring the technological development in the atomic realm and providing a legal framework to which the countries doing nano research and development conform their activities, make sense. Despite their differing value systems, political and legal traditions and systems, all members of the international community should agree at least on a minimum political control and administration in order to ensure that nanoscale research and development is consistent with the ethical principles present in all universal value systems. Defining the perceived benefits and risks of nanoscale science applications, as perceived by the scientific community, industry and business and public, is necessary while policy-makers decide what is best for their community and the nation.

16.5.1 Nano Rules and Regulations

The *laissez-faire* approach of the past has led to a lack of rules and regulations regarding research, development and the deployment of atomically and molecularly engineered produces, services and devices. The absence of appropriate norms at both the international and national levels, reflect what W.F. Ogburn described as the cultural lag (Ref. 13). According to his theory of technological evolutionism there is a gap between the technical development of a society and its moral and legal institutions. It is a social fact that while scientific and technological development in nanomaterials manufacturing is leaping ahead, legal regulations lag behind and leave a gap in terms of trust between the people and public authorities. The present situation has the potential to generate social tensions and problems since the misuse of nanotech for destruction purposes cannot be ruled out explicitly. There is a need for new forms of regulations and international standards to direct the research, development, manufacturing and commercialization of nanotech. However, the mere establishment of a legal framework or standards is not enough since the laws and regulations need to be enforced. A better understanding of the legal implications of nanotech would help the policy-makers in the establishment of a regulatory framework and standards that are consistent with the dominant value system of their society. This is particularly valid and important for developing countries, where new laws and institutions have been introduced, after their independence from colonial rule, without taking into consideration the local culture, institutions and traditions. Since the control and manipulation of the genetic signature imprinted in the DNA of each human being and manipulation of human cells has become possible, including the use of stem cells used in laboratories, a regulatory framework for setting limits to the exploration of human nature has become a necessity. However, none of the regulations should delay or inhibit the growth of knowledge aimed toward the betterment of humankind.

16.5.2 Nano Ethics—A Deontological Code for the Nano Community

Not all that can be done should be done: This is the nucleus of the ethical dimension with regard to research and the technological innovation process, in general, and it applies to nano, in particular. Since science and technology on the scale of the nanometre are not restricted to the domain of materials science but reach out to life sciences, it is important to understand the ethical implications of this new branch. The formalization of a deontological code is needed to prevent nanoscience and technology from derailing and heading into the wrong direction. Ethical guidelines related to research procedures and activities in the realm of atoms and molecules are, in fact, needed to bridge the present gap between science and ethics. One of the core tenets of such a deontological code is the observation, respect and protection of human dignity and non-invasion of the privacy of individuals. From a purely scientific point of view, humans do not enjoy a higher status than other living creatures but from the ethical perspective, humans have a different status because of their highly complex nature resulting from their genetic endowment which equips them with faculties that other living beings do not have. This is where ethics and science clash. Little systematic research into the ethical consequences of nanotech has been undertaken so far. The ethical community needs to be actively involved in the nano debate because the expertise of ethicists is required for softening the confrontation between the two blocks. The big ethical controversies will not focus on the threats to the dignity of the average grown-up adult, but will be concerned with foetuses, children, the terminally sick, elderly and disabled, all of whom belong to the most vulnerable groups of society.

16.6 Nanotech and War—Nano Arms Race

The military apparatus plays a crucial role in the scientific and technological innovation process, and applies to discoveries and development in the nanometer scale as well. In the developed as well as in the developing worlds, large investments are made in national security and defence. In the current international political scenario, the effort to acquire more powerful instruments to secure national integrity and interests, is a common phenomenon throughout the world, regardless of a country's economic status. The use of nano-engineered materials for weaponry and the potential use of atomically manufactured matter and devices for weapons of mass destruction is no longer science fiction and has today become a reality. In view of the increased political instability and feeling of uncertainty in the world today, the need to safeguard a nation's interests has gained importance. It is therefore possible that weapons equipped with nanotech will be deployed in present and future armed conflicts since there is as yet no international or national ban on weapons using nanoparticles and nanotechniques. The direct link between nanotechnology and ballistic studies scares people as also the potential abuse of nano-engineered weaponry (Ref. 12). However, we must state here that research for military use results in innovations that could later be utilized for civil purposes as well. It is still too early to speculate about the possible technological fallout of nano commodities designed for defence but one cannot preclude this possibility.

16.7 Public Perception and Public Involvement in the Nano Discourse

One of the most pressing issues in science is the involvement of the public. The Triple Helix, i.e. academia, government and industry, needs to be extended to a fourth dimension, the civil society, because of the latter's role and relevance in today's world. The public plus the print and electronic media need to be involved from the beginning, since news and information regarding nanoscale science and technology will shape public perception and determine to a large extent, people's knowledge, attitude and behavior toward this new technology. Consumer acceptance is the key when it comes to commercially-developed nanotech products, services and devices because ultimately it is the end-users who will influence the trajectory of nanotech. Therefore, it is of paramount importance to communicate with the public from the very beginning so that people understand rather than misunderstand the novelty of atomically and molecularly engineered products and services. The under-estimation of consumer acceptance and consumer resentment with regard to biotechnologically engineered organisms in the past has revealed the importance of social acceptance since knowledge and perception finally direct people's attitudes *vis-à-vis* new technologies. Studies conducted in the USA have already shown that artificially created matter having the attributes of living creatures, namely the ability to adapt, co-operate, learn and adjust to change occurring in them, or another system scares people (Ref. 12). It is therefore important for such studies to be conducted in the developing world as well, especially in areas where nanoengineered products have to be used. People's apocalyptical predictions with regard to nanotechnology do not find any foundation in rational reasoning. Uncertainty and unpredictability are, however, effective instruments for manipulating the public opinion and for undermining their trust in the development of science and technology.

16.7.1 Nanotech and the Media

In the age of information and an emerging knowledge society in many parts of the world, including the developing countries, mass media play a crucial role and hence should not be neglected. With regard to nano it is even more important to envisage the social function of mass communication tools. The printed and electronic media, including the Internet, known respectively as the fourth and fifth estates, are among the most influential instruments used to shape public perception and people's understanding of nano. Scientific and technological breakthroughs generally reach the public through the media, which is why it is important to inform people in a way they understand. Two possible avenues can be used to assess the trend of nano's presence in the media. If we use the number of scientific publications about nano as a variable indicator over a certain time span, we can assess the trend and trace its development over the past years and forecast future trends. The same can be done with publications in popular science, business media and entertainment. The rapid increase in publications in scientific literature during the past ten years is striking (Fig. 16.1). To put it in perspective, in 1995, 4,372 nano* references were recorded in English scientific publications, while in 2000, there were 11,447 and in the first five months of 2005, alone there were 16,518 titles. These figures, made available by the Institute of Scientific Information (ISI) web of knowledge, do not leave any doubt about the rise in number of scientific publications and this trend will certainly grow progressively. As the analysis here refers only to English language publications traced by ISI, it does not reflect the global scenario. It also includes titles that may not be purely nano-specific but include hybrids.

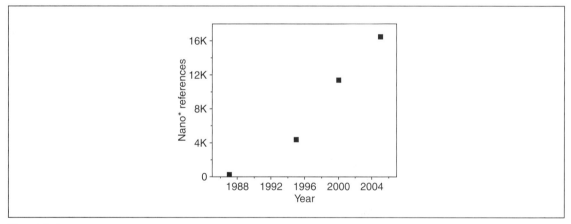

Fig. 16.1: *Increase in nano-related publications in English journals.*

With regard to the popular science literature, business media and entertainment, a similar trend stands out, though it could not be quantified here. Broadcasting, television and film industries as also the computer game industry have already developed a great interest in nano for increasing their clientele and turnovers. Science fiction literature about nano is fascinating not only adults but even the youth, and video games have already paved the way for nano to the children's room. The media hype about new discoveries in the atomic and sub-atomic world gives rise to distortions and reflects the commercial exploitation of an event rather than the aim to popularize the potentials of nanotech or to inform the public of the latest discoveries and achievements in this field. Without any doubt, the media has become an important platform wherein scientists, government officials, pressure and advocacy groups and social activists can voice their opinions and influence the minds of the uninitiated audience.

16.7.2 The Public Eye on Nanotechnology

The exclusion of the civil society in the dialogue on the potential positive and negative implications of nanotechnology and the course on which academia, public and private research funding agencies, and the industry are navigating, will have disturbing consequences and cause a backlash among public opinion. Therefore, the Triple Helix must become aware of the role that the public plays in the advances in science and technology. The participation of the civil society in the current public debate on nano is doubtless a sensitive issue. Public fears, possible negative social response or even rejection of nanotechnology, indicate that the involvement of the civil society, i.e. social movements, non-governmental organizations (NGOs) and community-based organizations (CBOs), is no longer just an option, but a necessity. Resentments and social discontent about nanotechnology have been openly voiced by the call for a moratorium on the deployment of nanomaterials and the opinions of nano-critics and nano-sceptics can no longer be ignored. The mobilization against nanotech by certain social activist groups constitutes a first, but alarming sign that public acceptance cannot be taken for granted. The primary intention of the anti-nano community is to heighten public awareness on the possible negative impacts of nano and to influence the policy-makers to jeopardize technological advancement. These groups are less likely to be influential since in the developing world where the democratic instruments to challenge national policies are limited. Further, social acceptability is more likely where people are willing to accept the risk of new technologies and generally have a positive view of science.

16.8 Harnessing Nanotechnology for Economic and Social Development

16.8.1 Nanotechnology and the Developing World

The manner in which industrialization advanced in the developing world has no parallels in the highly industrialized countries. In the former, various kinds of technologies—from indigenous to state-of-the-art technologies—co-exist and a plurality of different kind of societies live together, ranging from pre-industrial to emerging knowledge societies. The ongoing industrialization and modernization trend in the developing world has generated a range of problems that have culminated in the global phenomena of environmental pollution, widespread diseases and urbanization (Ref. 14). The situation in the developing world has not significantly improved and in certain countries, the condition of the people has deteriorated further. The world's most pressing problems are manifold and relate to a variety of issues. Extreme poverty, lack of education, high rates of mortality and morbidity, widespread epidemics and environmental problems are rampant. Innovative and holistic approaches and strategies thus need to be developed and implemented while addressing these highly complex and intertwined problems, and no technology should be considered irrelevant in this effort. We stated earlier that nanotech techniques can find applications in societies and economies, irrespective of their development or status and that at least theoretically, everyone can benefit from their potential applications. Nanoscale techniques have the potential to be in harmony with both traditions and technological development, which makes them a valid tool in the struggle against poverty. They can indeed make a significant difference in the current scenario and contribute to a more sustainable economic and social development. Several developing countries have recognized nanotech as a catalyst for economic, human, social, technological and environmental development and launched national nanotechnology initiatives. Worldwide, more than one-third of all the nations are promoting research and development, including education and training of nanoscientists and nanotechnologists, and more than seven of these countries belong to the developing world (Ref. 15). To put it in the right perspective, in India for instance, the Department of Science and Technology, has allocated $ 20 million in the present Five-Year Plan for the national Nanomaterials Science and Technology Initiative. Several academic institutions in India already possess the necessary ingredients to compete at a global level and to become research centres of excellence, with highly educated and trained workforce, state-of-the-art research infrastructure and well established links with industry and business. Figure 16.2, shows the investment in nano research in a number of countries, grouped separately. It is clear that the investment in India is still very low in comparison to other countries in the region.

Developing countries are rich in human capital and their brainpower will, in the medium and long-term, reshape the imbalance between the North and South. Nanotech offers a new opportunity to the manufacturing industry in the developing world. The wide range of possible applications of nanoscale technologies suggests that if the industrial sector of developing countries enters the field of manufacturing of nanomaterials, it can enhance its competitiveness in manufacturing at the global level.

16.8.2 Harnessing Nanotechnology for Sustainable Development

For the development and application of technology to be successful, it needs to be designed in consonance with the needs of the target group and to be suited to the socio-economic context. Nanoscale science and its deriving technologies can virtually enhance the lives of nearly every human being, be he rich or poor, because of its pervasive benefits and its suitability in resource-limited settings (Ref. 16). It has the potential to provide the most innovative tools and strategies for fighting poverty-related issues. Five billion people live in the developing world, and their lives and living conditions can be enhanced by the diffusion of applications of discoveries made in this area. In order to substantiate our optimistic views with concrete examples, we have identified seven core areas, wherein nanotechnology can make a significant difference in the developing world. These are detailed below.

1. Economic Development   Nanobiotechnology, involving the biological production and utilisation of nanomaterials, is a promising new field, especially in the developing world with its unparalleled biodiversity. This natural asset of the developing world can be harnessed through nanomaterials synthesis using micro-organisms, including bacteria, viruses, fungi as well as plant and animal-based products. Several examples

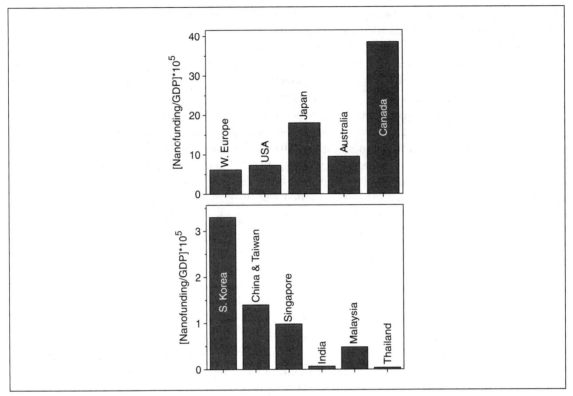

Fig. 16.2: *Investment in nano research for the year 2003 in several countries, given as a fraction of their GDP. Data from http://www.nano.gov/html/res/IntlFundingRoco.htm, www.imf.org/external/pubs/ft/weo/2004/01/pdf/,www.ics.trieste.it/Documents/Downloads/df2625.ppt.*

of nanomaterials synthesis using biology have been reported (Ref. 17). Techniques related to nanobiotechnology do not require large investments and infrastructure, and can therefore be developed on the site of application itself. The green and cost-effective solution to nanomaterials manufacturing is one example which proves that with the requisite knowledge and skills, nanotech can be developed and utilized in the developing world. This approach has been applied to the synthesis of nanoparticles and nanotriangles of gold as well as various other inorganic nanoparticles such as those of CdS and $CaCO_3$. These methodologies can be suitably adapted to facilitate the large-scale synthesis of materials which can be used for applications such as cancer therapy, IR absorbing coatings, etc. The synthesis of nanoparticles in human cells has added a new dimension to this research (Ref. 18). It is also likely that an understanding of the underlying processes may permit us to alter the chemistry to facilitate control of the shape and size of nanomaterials. The biochemical events may be transplanted to other organisms so that processes similar to the bulk production of enzymes become feasible. All these developments could take place in the foreseeable future with an investment that would be much lower than that necessary for chemical or physical routes.

2. Safe Drinking Water Among the numerous possible applications of nanotechnology, the most widespread impact as far as the developing world is concerned may be in the area of water purification. Access to safe drinking water is one of the major concerns in the developing world since almost half of the world population has no access to safe drinking water and basic sanitation. Water purification systems, equipped with nanomaterials and using new kinds of membrane technologies with variable pore sizes as filters, could provide people in any area with safe drinking water. These are easy in application and maintenance, and are already available in the market; the forward-osmosis membrane technology of Hydration Technologies (Ref. 19) is one technique utilising nanotechnology. Thus a combination of nanotechnologies will be useful in providing safe drinking water through cost-effective measures, which will be less dependent on energy resources. Although the product is currently marketed only to provide emergency water supply, large-scale water purification is indeed feasible in the future. To substantiate the validity of these suggestions, we mention the following. Carbon nanotube-based filters can be developed for water purification. The development of a filter which can separate petroleum hydrocarbons from crude oil has been demonstrated (Ref. 20). The filters also remove bacteria from water. With nanotubes, smart sensors can be incorporated into the filter as several nanotube-based sensors are already known. Nanoparticles have been shown to degrade pesticides and pollutants (Ref. 21). Several nanomaterials are known to be anti-bacterial and can be incorporated on various kinds of substrates (Ref. 22). It may be mentioned that we have not listed numerous other discoveries in the area related to this application.

3. Improving Food Security Nutrient deficiency is a widespread phenomenon throughout the developing world, which challenges the physical and mental health of over one billion people. Food starvation can be, but is not always, related to crop failure. Novel techniques using nanotechnology can be applied in agriculture for breeding crops with higher levels of micro-nutrients, enhancing pest detection and control, and improving food processing techniques. The lack of adequate storage facilities besides crop failure is one of the major reasons causing food shortage in the developing world, particularly in the remote areas. In India alone, millions of tons of wheat and rice are rotting in the open because of lack of storage facilities. Food spoils easily, especially in the tropical belt because increased temperatures here favour the growth of micro-organisms in food, which reduce its quality or even render it inedible. Oxygen accelerates the degeneration process because it enables the growth of micro-organisms. It is known that carbon dioxide inhibits the growth of microbes. Carbon nanotubes can be used in food processing and preservation as an oxygen scavenger and can prevent packaged food from deteriorating (Ref. 23). Another application of one of the recent nanoscale innovations involves the use of atomically modified food, which marks the beginning of a radically new paradigm for food production. It has the potential to sidestep the controversial genetically modified food and to increase the yield of agricultural produce. In Thailand, researchers at Chiang Mai University have modified local rice varieties to develop a variant of rice that grows throughout the year by applying nanotechnology (Ref. 24). This particular nanotech technique involves the perforation of the wall of a rice cell through a particle beam for introducing one nitrogen atom into the cell, which triggers the re-arrangement of the rice's DNA. Novel techniques applying nanotechnology could thus contribute towards improving the current food situation, which in several part of the globes, is quite alarming.

4. Health Diagnosis, Monitoring and Screening Nanoscale techniques have the potential to revolutionize the health sector, particularly in the fields of diagnosis, screening and monitoring of diseases and health conditions (Ref. 25). A large spectrum of novel applications using nanoscale techniques in healthcare is possible and this is the beginning of a new paradigm for healthcare. Lack of accurate, affordable and accessible diagnostic tests impedes global health efforts, especially in the remote and inaccessible regions of the world. Many communicable diseases like HIV/AIDS, malaria, tuberculosis and others can be diagnosed with the help of screening devices using nanotechnology. The standard diagnostic tests for these diseases in the developing world are costly, complex, and poorly suited to resource-limited settings. A radically new approach to health diagnosis has been developed in India by the Central Scientific Instruments Organisation (CSIO). Theoretical simulation and design parameters for a micro-diagnostic kit using nano-sized biosensors were completed in 2004 and are ready for clinical trials (Ref. 26). The techniques are based on highly selective and specific biosensors and receptors like antibodies, antigens and DNA, which enable an early and precise diagnosis of various diseases. The diagnostic kit "Bio-MEMS" (micro-electro-mechanical-system) has a size of about 1 cm × 1 cm, costs around Rs 30 per piece is easy to apply. The testing time in this is very fast and only requires a tiny amount of blood. This novel diagnostic tool could also find application in the detection of other diseases and pollutants in the environment, including water and food (Ref. 27). In order to develop therapeutics to combat malaria caused by the parasite plasmodium falciparum, a common disease in many parts of the developing world, Subra Suresh and his team at MIT are using nanotechnology to systematically measure the mechanical properties of biological systems in response to the onset and progression of the disease (Ref. 28). Innovative drug delivery systems, using nanotechnology constitute another area wherein nanotechnology can find application. Cancer is widely prevalent in the developing world as elsewhere and poses a big challenge to human health. The latest results obtained in cancer detection and treatment with nanoscale techniques give the hope that nanotechnology could be heading for a breakthrough in defeating this disease. One way of detecting cancer early, safely and economically, is by the injection of 'molecular beacons' into the body. Britton Chance and his colleagues at the University of Pennsylvania have developed tiny capsules that use the specific biochemical activity associated with a tumor to detect breast cancer (quoted in Ref. 29). As far as cancer treatment is concerned, Jennifer L. West and her team at Rice University, Houston, Texas have developed gold 'nano bullets' that can destroy inoperable human cancers. The nanoshells consist of tiny silica particles plated with gold, which when heated with infrared light, cause the cancer cells to die (Ref. 30). Carbon nanotubes have been transported into the cell nucleus and continuous near infrared radiation absorption of nanotubes causes cell death. This methodology has been used for cancer cell destruction (Ref. 31).

5. Environmental Pollution   Environmental degradation due to unsustainable production techniques and other human activities has exposed the entire world's population to increased risks. Innovative techniques, using nanoengineered materials and devices, can be deployed for the removal of polluting molecules in air, water and soil. Cleaner manufacturing processes and methods, by applying nanoscale techniques, can also contribute to a reduced level of environmental pollution, especially in the developing world where international standards are often not adhered to. For instance the high levels of arsenic in soil and water are posing a major environmental problem in several regions of the developing world. A simple and cheap but effective nanoscale technique to remove arsenic involves the use of TiO_2 nanoparticles (Ref. 32). Nanomaterials have further proved to be very effective in removing metal ion contamination. A wider application of such technologies, which are harnessing discoveries made in nanoscience, could have a positive and wider impact on the health conditions and natural habitat of millions of people.

6. Energy Storage, Production and Conversion   Chronic power shortages and increased need for energy resources because of the rapidly growing population and economies of the developing world, are posing challenges to the energy market. Since almost all sources of energy are not renewable, the world will soon face a global energy supply problem. Solar energy is an interesting and valid alternative, especially in the tropical sunny South. Scientific studies have demonstrated that nanoscale techniques involving nanotubes and nanoparticles lead to increased conversion efficiencies. Semiconducting particles of titanium dioxide, coated with light-absorbing dyes bathed in an electrolyte and embedded in plastic films, are cheap and easy to manufacture and offer a viable alternative to conventional energy production and storage. Because of their low cost-structure, photovoltaics using nanotech constitute a valid alternative for overcoming the problem of power shortage, especially in the developing world. Researchers at Nanosolar, a venture capital start-up based in Palo Alto, California, are developing cheaper methods for producing photovoltaic solar cells by using nanotechnology (Ref. 33). The objective is to boost the power output of nano solar cells and make them easier to deploy by spraying them directly on surfaces. This approach is simple and can be easily replicated in the developing world. These highly efficient solar cells can be made of a mix of alcohol surfactants and titanium compounds sprayed on a metal foil. Within 30 seconds, a block of titanium oxide perforated with holes of nano-meter size rises from the foil. The solar cells form when the holes are filled with conductive polymer and electrodes are added and then covered with a transparent plastic. These are concrete examples wherein nanotechnology can be used for energy storage, production and conversion in the developing world.

7. Global Partnerships   The inclusion of the South in the nano-dialogue has created new platforms and alliances between the North and South and strengthens their ties. The allocation of some of the large public scientific funding of nanoscience and technology could be directed to developing countries in order to foster the development, diffusion and dissemination of nanoscience, engineering and technology in the developing world. Global research networks of excellence create more value for the international scientific community. It is certainly important to encourage international partnerships between the North and South, similar to the first North-South expert group meeting of nanoscientists and nanotechnologists in Trieste (Italy) in February 2005, but there is also a growing need for scientific exchange and alliances among countries of the developing world in view of the ongoing regionalization trends in politics and economics (Ref. 34). Global research networks, including scientific co-operation and collaboration, are needed to find joint solutions for the most pressing problems of the world community. Since partnerships at the regional level are likely to gain importance in the long term, the establishment of South-South nano-networks must be envisaged.

16.8.3 Unexplored Biodiversity of the South—Opportunities for Bio-nanotechnology

The development of nanocomponents that imitate or emulate natural processes can find many applications in many sectors. The protection and preservation of species in the tropical belt has acquired a new dimension because of bio-nanotechnology and its potential applications. The scientific exploration of the nano-bio interface becomes very interesting in view of the developing synthetic life-forms, manufactured organs and bio-nanodevices. Since they have the requisite research infrastructure, developing countries enjoy a strategic advantage vis-à-vis the industrialized world. This is also true because the biodiversity is much larger in the tropical and sub-tropical belt as in other geographical and climatic zones. It is thus up to the developing countries to make use of this distinctive asset and to discover the secrets of this unexplored biodiversity and the particular physical properties of its living organisms. The 'Lotus Effect', discovered by Wilhelm Barthlott and his student Christoph Neinhuis of the University of Bonn, illustrates how knowledge of what happens at the bio-nano interface can be made fruitful. Lotusan, a dirt repellent paint is the commercialized form of the scientific discovery of the Lotus Effect. The developing countries need to discover the richness of their flora and fauna, which will help them optimise the use of bio-nanotechnology. Their rich cultural heritage, including the millennia old traditional knowledge in the areas of homeopathic, Ayurvedic and herbal medicine, along with their large biodiversity provide developing countries with a strategic advantage that the developed countries do not have. It lies in the hands of these countries, especially those that have already adopted nanotechnology initiatives within the bounds of their national technology development policies, to utilize their unique assets for moving ahead in the emerging global nano world.

16.9 Conclusions

The potential societal implications of the scientific and technological innovation in the realm of nano and the future applications of discoveries at the frontiers of science are only partially understood and therefore need to be further explored. The uniqueness and distinctiveness of nanoscience and nanotechnology, especially with regard to its pervasiveness into virtually all spheres of human life, explain why their potential impacts will exceed those of all other conventional technologies hitherto developed. The convergence of the newly emerging technologies of the 21st Century have the potential to revolutionize social and economic development and may offer innovative and viable solutions for the most pressing problems of the world community and its habitat. A better understanding of the potential benefits and hazards of nanoscale science and technology is essential because it will provide policy-makers with better tools to take responsible choices. Nanoscience and its deriving technologies have the potential to improve the state of the developing world, if the applications are designed and tailored to best fit the needs of the people. Unlike other technologies, nanotechnology offer a unique chance to bridge the technological gap between the industrialized and the developing world. But a favourable terrain for the growth of these sprouts needs to be prepared and therefore joint and concerted efforts by all concerned are needed to rule out future factions and new divides.

Review Questions

1. What are the specific implications to society when advanced technologies are implemented?
2. Contrast nanotechnology from other emerging technologies.

3. Summarize the landmarks in nano research.
4. List the specific areas of economic benefit to the developing world particularly from the perspective of biodiversity.
5. How can nanotechnology be used for sustainable development?
6. How can nano research lead to a new political divide?

References

1. Eigler, D.M. and E.K. Schweizer, *Nature*, **344** (1990), pp. 524–26.
2. Snow, C.P., *Two Cultures and a Second Look*, (1963), New American Library, New York.
3. The organic *Weltanschauung* regards all phenomena in the universe as integral parts of an inseparable harmonious entity when exploring the very nature of things and is well established in social sciences and humanities.
4. Christensen, C.M., *The Innovator's Dilemma: When New Technologies Cause Great Firms to Fail*, (2000), Harperbusiness, New York.
5. Roco, M.C., (2003), Government Nanotechnology Funding: An International Outlook, retrieved from http://www.nano.gov/html/res/IntlFundingRoco.htm.
6. Retrieved from http://www.nanobusiness.org/.
7. The projected World GDP is based on the average annual growth rate of 4.2 per cent and 4.1 per cent, respectively for the period from 2006–09 and 2010–15. The data are taken from the *World Economic Outlook*, (2004), published by the International Monetary Fund (IMF). Retrieved from www.imf.org/external/pubs/ft/weo/2004/01/pdf.
8. Herring, R., *Nanotech Patents Proliferate*, (2005), Study Quantifies Huge, Complicated Body of Intellectual Property Generated by Entrepreneurs of Tiny, Complicated Technology, Retrieved from http://www.redherring.com/Article.aspx?a=11866&hed=Nanotech+Patents+Proliferate&sector=Industries&subsector=Biosciences.
9. Nano Science and Technology Institute (NSTI), (2005), *Nanotech Ventures 2005*, Retrieved from http://www.nsti.org/NanotechVentures2005/.
10. Hullmann, A., *Nanotechnology. Europe and the World: The International Dialogue in Nanotechnology*, (2005), Retrieved from www.ics.trieste.it/Documents/Downloads/df2625.ppt.
11. National Nanotechnology Initiative (NNI), *21st Century Nanotechnology Research and Development Act*, Retrieved from http://www.nano.gov/html/about/funding.html.
12. Cobb, M.D. and J. Macoubrie, *Journal Nanoparticle Research*, **6** (2004), pp. 395–405.
13. Ogburn, W.F., *On Culture and Social Change*, (1964), The University of Chicago Press, Chicago.
14. Hinrichsen, D., R. Salem, R. Blackburn, et al., *Population Reports, Meeting the Urban Challenge*, (2002), Baltimore, Maryland: Population Information Program, Center for Communication Programs, The

Johns Hopkins Bloomberg School of Public Health, Vol. XXX, No. 4. Retrieved from http://www.infoforhealth.org/pr/m16/m16.pdf#search='population%20city%20developing%20countries%20half%20of'. This publication gives a comprehensive overview on the urbanization phenomenon in the developing world.

15. Brazil, China, India, Singapore, South Korea, Taiwan, Thailand, the Philippines and others have become actively involved in nanotech.

16. Drexler, K.E., (1986), *Engines of Creation*, Garden City, Anchor Press/Doubleday, New York.

17. See for example, (a) Southam, G. and T.J. Beveridge, *Geochim. Cosmochim, Acta*, **60** (1996), pp. 4369–76. (b) Klaus, T., R. Joerger, E. Olsson and C.G. Granqvist, *Proc. Nat. Acad. Sci.*, **96** (1999), pp. 13611–14. (c) Mukherjee, P., A. Ahmad, D. Mandal, S. Senapati, S.R. Sainkar, M.I. Khan, R. Parishcha, P.V. Ajaykumar, M. Alam, M. Sastry and R. Kumar, 2001, *Angew. Che., Int. Ed.*, **40**, pp. 3585–88. (d). Nair, B. and T. Pradeep, (2002), *Crystal Growth and Design*, **2** (2002), pp. 293–98.

18. Anshup, Sai Venkataraman J., Subramaniam C., Rajeev Kumar R., Suma Priya, Santhosh Kumar T.R., Omkumar R.V., Annie John and Pradeep T., *Langmuir*, **21** (2005), p. 11562.

19. See http://www.hydrationtech.com.

20. Srivastava, A., O.N. Srivastava, S. Talapatra, R. Vajtai and P.M. Ajayan, *Nature Materials*, **3** (2004), pp. 610–614.

21. Sreekumaran Nair, A., R.T. Tom and T. Pradeep, J. Environ. Monitor., **5** (2003), pp. 363–65. Sreekumaran Nair, A. and T. Pradeep, Indian patent No. D-CHE/0561.

22. Prashant, J. and T. Pradeep, *Biotechnology and Bioengineering*, **90** (2005), pp. 59–63.

23. Hechman, M., 2005, *Better Eating through Nanotech*, Retrieved from http://www.extremenano.com.

24. ETC Group, *Jazzing Up Jasmine: Atomically Modified Rice in Asia?*, (2004), Retrieved from http://www.etcgroup.org/article.asp?newsid=444.

25. Frost and Sullivan, *Nanomedicine—Global Technology Developments and Growth Opportunities*, (2004), Farmington, USA.

26. *AZoNanotechnology, Tuberculosis Diagnosis Kit Based on Nanotechnology*, (2004), Retrieved from http://www.azonano.com/details.asp?ArticleID=368.

27. The same organization is developing a diagnostic kit for tuberculosis.

28. Suresh, S., J.P. Spatz, A. Micoulet, M. Dao, C.T. Lim, M. Beil and T. Seufferlein, *Acta Biomaterialia*, **1** (2005), pp. 15–30.

29. Schewe, P.F., B. Stein and J. Riordon, *The American Institute of Physics, Bulletin of Physics News*, **531** (2001), March 22.

30. Hirsch, L.R., R.J. Stafford, J.A. Bankson, S.R. Sershen, B. Rivera, R.E. Price, J.D. Hazle, N.J. Halas and J.L. West, *Proc. Nat. Acad. Sci.*, **100** (2003), pp. 13549–54.

31. Kam, N.W.S., M. O'Connell, J. Wi Westm, J.A. Lsdom and H.J. Dai, *Proc. Nat. Acad. Sci.*, **102** (2005), pp. 11600–11605.

32. See for example, Pena, M.E., G.P. Korfiatis, M. Patel, L. Lippincott and X.G. Meng, *Water Research*, **39** (2005), pp. 2327–37.
33. Future Pundit, (2004), *Nanotech Start-ups Pursuing Cheaper Photovoltaic Solar Power,* Retrieved *from http://www.futurepundit.com*. First published in MIT's Technology Review.
34. The first North-South dialogue on nano was jointly organized by the International Centre for Science and High Technology and the United Nations Industrial Development Organization (UNIOD). For details visit http://www.ics.trieste.it/Nanotechnology/.

Additional Reading

1. *Implications of Emerging Micro and Nanotechnologies*, (2002), National Research Council.
2. Rocco, M.C., *Societal Implications of Nanoscience and Nanotechnology*, (2001), National Science Foundation.

This work was originally published in, *Current Science*, **90** (2006), 645-658.

Appendix

HISTORY OF NANOSCIENCE AND TECHNOLOGY*

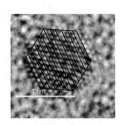

- **December 29, 1959:** *Plenty of Room at the Bottom*

  The lecture of Richard Feynman at the California Institute of Technology. It was titled, "There's Plenty of Room at the Bottom." Here, he proposed the "possibility of maneuvering things atom by atom."

- **Mid-1970s:** *Nanotechnology*

  Idea of molecular nanotechnology, originated in the mind of Eric Drexler, MIT undergraduate. He realized that the biological 'machinery' could be adapted to build non-living products upon command.

- **1974:** *Molecular Devices*

  The first molecular electronic device was patented by Aviram and Seiden of IBM. Professor Norio Taniguchi of Tokyo Science University invented the term *nanotechnology*.

- **1980:** *Molecular Nanotechnology*

  K. Eric Drexler, an MIT student, writes the first paper on advanced nanotechnology.

- **1980s:** *Chemical Synthesis of Nanoparticles*

  Various kinds of nanoparticle systems were made. These include stable, dispersible materials of almost every element and common binary oxides and sulphides.

- **1981:** *Scanning Tunneling Microscopy*

  Binning and Rohrer (IBM) build the scanning probe microscope making it possible to work with atoms and molecules.

  "Surface Studies by Scanning Tunneling Microscopy," Binning G., H. Rohrer, Ch. Gerber and E. Weibel, *Phys. Rev. Lett.*, **49** (1982), p. 57.

- **1983:** *Self-assembled Monolayers*

  Self-assembled monolayers (molecularly thin films) of thiols on gold surfaces were made. These are used for various nanopatterning applications now.

---

* The author is aware of the fact that a selection of this sort reflects personal preferences. As a result, several noteworthy contributions are not listed although the author has no intention to undermine the significance of those works. A brief summary of the work with original reference citation is given so that the reader can explore specific topics further. Such a selection may be useful for classroom presentations.

"Adsorption of Bifunctional Organic Disulfides on Gold Surfaces," Nuzzo, R.G. and D.L. Allara, *J. Am. Chem. Soc.*, **105** (1983), p. 4481.

- **1985:** *Scientists Richard Smalley, Robert Curl, Jr., and Harold Kroto discover spherical cages of 60 carbon atoms called "buckminsterfullerene".*

  "C_{60}: Buckminsterfullerene", Kroto, H.W., J.R. Heath, S.C. O'Brion, R.C. Curl and R.E. Smalley, *Nature*, **318** (1985), p. 162.

- **1986:** *Atomic Force Microscopy*

  Atomic force microscope (AFM) invented.

  "Atomic Force Microscope", Binning, G., C.F. Quate and Ch. Gerber, *Phys. Rev. Lett.*, **56** (1986), p. 930.

- **1986:** *Engines of Creation*

  During the same time period, K. Eric Drexler publishes "Engines of Creation." Drexler presented his provocative ideas on molecular nanotechnology to a general audience.

- **November 9, 1989:** *Writing with Atoms*

  Don Eigler at IBM's Zurich Research Laboratory arranges 35 xenon atoms to write "IBM" using the tip of a scanning tunneling microscope (STM), achieved at temperatures close to absolute zero.

  "Positioning Single Atoms with a Scanning Tunneling Microscope", Eigler, D.M., and E.K. Schweizer, *Nature*, **344** (1990), p. 524.

- **1991:** *Nanotubes Arrive*

  Discovery of carbon nanotubes by Sumio Iijima.

  "Helical Microtubules of Graphitic Carbon", Iijima, S., *Nature*, **354** (1991), p. 56.

- **1992:** *Nanosystems*

  Drexler published *Nanosystems*, a technical work outlining a way to manufacture extremely high-performance machines out of molecular carbon lattice ('diamondoid').

  Drexler, K.E., *Nanosystems: Molecular Machinery, Manufacturing and Computation*, (1992), Wiley/Interscience, New York.

- **1994:** *Gold Particles*

  Stable gold nanoparticles with molecular protection were made in solution.

  "Synthesis of Thiol-Derivatised Gold Nanoparticles in a Two-Phase Liquid–Liquid System", Brust, M., M. Walker, D. Bethell, D.J. Schiffrin and R. Whyman, *J. Chem. Soc., Chem. Commun.*, (1994), pp. 801–802.

- **January 1996:** *Assembling Molecules*

  Scientists at IBM succeed in moving and precisely positioning individual molecules at room temperature.

- **August 1998:** *Aligned Nanotubes*

  Aligned nanotube bundles have been grown on surfaces by the pyrolysis route.

"Large Aligned-Nanotube Bundles from Ferrocene Pyrolysis", Rao C.N.R., R. Sen, B.C. Satishkumar and A. Govindaraj, *Chem. Commn.*, (1998), pp. 1525–1526.

- **July 1999:** *Molecular Logic Gate*

  A team at UCLA and Hewlett-Packard create a molecular 'logic gate'.

  "Electronically Configurable Molecular-Based Logic Gates", Collier, C.P., E.W. Wong, M. Belohradský, F.M. Raymo, J.F. Stoddart, P.J. Kuekes, R.S. Williams and J.R. Heath, *Science*, **285** (1999), p. 391.

- **1999, 2000:** *Y-Junction Nanotubes*

  Using Y-shapes nanochannel alumina, Y-jucntion nanotubes have been synthesized.

  "Y-Junction Carbon Nanotubes and Controlled Growth", Li, J., C. Papadopoulos and J. Xu, *Nature* **402** (1999), p. 253.

  Pyrolysis of a metallocene and thiophene has been used to make such junctions.

  "Y-Junction Carbon Nanotubes", Satishkumar, B.C., P.J. Thomas, A. Govindaraj and C.N.R. Rao, *Appl. Phys. Lett.*, **77** (2000), p. 2530.

- **2001:** *Moore's Law Surpassed*

  In June 2001, Intel Corporation researchers announced that they had created the technology needed to produce the world's smallest and fastest silicon transistor on a mass scale. These switch on and off 1.5 trillion times a second.

- **April 2002:** *Optical Microscopy at Nanometer Scale*

  Scientists used conventional optics to image clumps of bacteria, just 33 nanometres across—much smaller than the wavelength of light used to illuminate them. The study shows that 'far-field' optical microscopes can operate well beyond the diffraction limit ($\sim \lambda/2$). A new kind of optical microscopy is born to look at nano objects.

  "Focal Spots of Size $\lambda/23$ Open Up Far-Field Florescence Microscopy at 33 nm Axial Resolution", Dyba M. and S. Hell, *Phys. Rev. Lett.*, **88** (2002), p. 163901.

- **May 2002:** *Photons can Move Molecules*

  The first conversion of light to mechanical energy by a single-molecule device has been demonstrated.

  "Single-Molecule Optomechanical Cycle", Hugel, T., N.B. Holland, A. Cattani, L. Moroder, M. Seitz and H.E. Gaub, *Science*, **296**, 1103.

- **May 2002:** *Molecular Shuttling Leads to Nanoconstruction*

  Receptor-lipid molecules disperse and regroup across an artificial cell membrane by adding and removing free-floating ligands.

  "Crown Ether Functionalized Lipid Membranes: Lead Ion Recognition and Molecular Reorganization", Sasaki, D.Y., T.A. Waggoner, J.A. Last and T.M. Alam, *Langmuir*, **18** (2002), p. 3714.

- **June 2002:** *Nanofilter-based Chirality Separation*

  A bionanotube membrane that can separate out the left- and right-handed forms of chiral drug molecules has been developed. An antibody which binds preferentially to one mirror image form (or enantiomer) of the chemical is used in this approach.

"Antibody-Based Bio-Nanotube Membranes for Enantiomeric Drug Separations", Lee, S.B., D.T. Mitchell, L. Trofin, T.K. Nevanen, H. Söderlund and C.R. Martin, *Science*, **296** (2002), p. 2198.

- **June 2002:** *Magnetic Spins Store Quantum Information*

Spins of clusters of atoms in a magnetic compound became aligned when a magnetic field was applied. This coherence persisted for up to ten seconds, and the clusters could have information stored on them.

"Coherent Spin Oscillations in a Disordered Magnet", Ghosh, S., R. Parthasarathy, T.F. Rosenbaum and G. Aeppli, *Science*, **296** (2002), p. 2195.

- **June 2002:** *Diffraction Limit and Light Transport*

A large amount of light can pass through a sub-wavelength aperture in a patterned metal film without being diffracted. The diffraction limit could be overcome this way.

"Beaming Light from a Sub-Wavelength Aperture", Lezec, H.J., A. Degiron, E. Devaux, R.A. Linke, L. Martin-Moreno, F.J. Garcia-Vidal and T.W. Ebbesen, *Science*, **297** (2002), p. 820.

- **June 2002:** *Laser Lithography for Cheaper Chips*

Silicon chips could be made more quickly and cheaply by using a new technique which would make it possible to go on with silicon for some more time than previously anticipated.

"Ultrafast and Direct Imprint of Nanostructures in Silicon", Chou, S.Y., C. Keimel and J. Gu, *Nature*, **417** (2002), p. 835.

- **June 2002:** *Magnets Make Logic*

A ferromagnetic NOT gate, a new class of device has been made. It is believed that a full set of logic gates could be developed by using this technique.

"Sub-Micrometer Ferromagnetic NOT Gate and Shift Register", Allwood, D.A., G. Xiong, M.D. Cooke, C.C. Faulkner, D. Atkinson, N. Vernier and R.P. Cowburn, *Science*, **296** (2002), p. 2003.

- **June 2002:** *Nanorod Devices for Photovoltaics*

The plastic solar cells with semiconducting nanorods have been made.

"Hybrid Nanorod-Polymer Solar Cells, Huynh, Wendy U., Janke J. Dittmer and A.P. Alivisatos", *Science*, **295** (2002), p. 2425.

- **June 2002:** *Dip-pen Nanolithography with Biomolecules*

Dip-pen nanolithography has been used to create nanoscale patterns.

"Direct Patterning of Modified Oligonucleotides on Metals and Insulators by Dip-Pen Nanolithography", Demers, L.M., D.S. Ginger, S.J. Park, Z. Li, S.W. Chung, and C.A. Mirkin, *Science*, **296** (2002), p. 1836.

- **July 2002:** *Carbon Nanotube-based Imaging*

Carbon nanotubes have been used as the basis of a cold-cathode X-ray device. It has been used to image a fish and a human hand.

"Generation of Continuous and Pulsed Diagnostic Imaging X-Ray Radiation Using a Carbon-Nanotube-Based Field-Emission Cathode", Yue, G.Z., Q. Qiu, B. Gao, Y. Cheng, J. Zhang, H. Shimoda, S. Chang, J.P. Lu and O. Zhou, *Appl. Phys. Lett.*, **81** (2002), p. 355.

- **August 2002:** *Tagged Nanoparticles for Detection*

Nanoparticle probes have been developed for detecting DNA with unique "fingerprints".

"Nanoparticles with Raman Spectroscopic Fingerprints for DNA and RNA Detection", Cao, Y.W.C., R. Jin and C.A. Mirkin, *Science* **297** (2002), p. 1536.

- **August 2002:** *Converting Alcohol into Carbon Nanofibres*

Carbon Nanofibres have been made from methyl alcohol.

"Carbon nanofibers Synthesized by Decomposition of Alcohol at Atmospheric Pressure", Jiang, N., R. Koie, T. Inaoka, Y. Shintani, K. Nishimura and A. Hiraki, *Appl. Phys. Lett.*, **81** (2002), p. 526.

- **August 2002:** *Liquids Driven by Light*

Liquids have been driven without using mechanical parts. The technique involves shining a light on the surface of the tube and transporting the fluid by a process known as photocapillarity.

"Photon-Modulated Wettability Changes on Spiropyran-Coated Surfaces", Rosario, R., D. Gust, M. Hayes, F. Jahnke, J. Springer and A.A. Garcia, *Langmuir*, **18** (2002), p. 8062.

- **August 2002:** *Nanoscale Patterns to Boost Magnetic Density*

A film-patterning technique that could overcome the problems associated with high-density magnetic recording has been developed.

"Nanoscale Patterning of Magnetic Islands by Imprint Lithography Using a Flexible mold", McClelland, G.M., M.W. Hart, C.T. Rettner, M.E. Best, K.R. Carter and B.D. Terris, *Appl. Phys. Lett.*, **81** (2002), p. 1483.

- **August 2002:** *Nanoparticles to Destroy Bacteria*

Magnesium oxide nanoparticles kill bacteria.

"Metal Oxide Nanoparticles as Bactericidal Agents", Stoimenov, P.K., R.L. Klinger, G.L. Marchin and K.J. Klabunde, *Langmuir*, **18** (2002), p. 6679.

- **August 2002:** *Single Molecule Light Emission*

Electroluminescence from individual molecules of silver has been demonstrated.

"Strongly Enhanced Field-Dependent Single-Molecule Electroluminescence", Lee, Tae-Hee, J.I. Gonzalez and R.M. Dickson, *Proc. Natl. Acad. Sci.*, **99** (2002), p. 10272.

- **September 2002:** *Molecular Electronics Breakthrough*

The highest density electronically addressable memory to date has been developed. The 64-bit memory uses molecular switches. The total area is less than one square micron, giving it a bit density that is more than ten times that of current silicon memory chips.

http://nanotechweb.org/articles/news/1/9/8/1.

- **September 2002:** *Nanografting Makes Tiny DNA Patterns*

  DNA patterns that are one-thousandth the size of those in commercially available microarrays have been developed. This method could help create faster, more powerful devices for DNA sequencing, biological sensors and disease diagnosis.

  "Production of Nanostructures of DNA on Surfaces", Liu, M., N.A. Amro, C.S. Chow and Gang-yu Liu, *Nano Letters*, **2** (2002), p. 863.

- **September 2002:** *GaN Nanowire Laser Emits Light*

  The first GaN nanowire laser has been made.

  "Single Gallium Nitride Nanowire Lasers", Johnson, J.C., Heon-Jin Choi1, K.P. Knutsen, R.D. Schaller, P. Yang and R.J. Saykally, *Nature Materials*, **1** (2002), p. 106.

- **October 2002:** *Atom Lithography for the Future*

  The resolution of 'atom lithography' need not be limited by the wavelength of light.

  "Demonstration of Frequency Encoding in Neutral Atom Lithography", Thywissen, J.H. and M. Prentiss, *New J. Phys.*, **7** (2005), p. 47.

- **October 2002:** *Optical Thermometer with Molecules*

  A common luminescent organic semiconductor based thermometer has been developed. This can possibly do thermal imaging of nanometre-sized devices.

  "A Molecular Thermometer based on Long-Lived Emission from Platinum Octaethyl Porphyrin", Lupton, J.M., *Appl. Phys. Lett.*, **81** (2002), p. 2478.

- **October 2002:** *Molecules to Power Nanoscale Computers*

  A new kind of computing that relies on the motion of molecules rather than the flow of electrons has been demonstrated. Logic gates use cascades of carbon monoxide molecules to transfer data. Devices made in this way have dimensions on the scale of nanometres, which are several orders of magnitude smaller than existing components.

  "Molecule Cascades", Heinrich, A.J., C.P. Lutz, J.A. Gupta and D.M. Eigler, *Science*, **298** (2002), p. 1381.

- **November 2002:** *Carbon Nanotube Makes Transistors Better than Silicon*

  It is claimed that a carbon nanotube transistor has better properties than silicon transistors of an equivalent size.

  "High-κ Dielectrics for Advanced Carbon-Nanotube Transistors and Logic Gates", Javey, A., H. Kim, M. Brink, Q. Wang, A. Ural, J. Guo, P. Mcintyre, P. Mceuen, M. Lundstrom and H. Dai, *Nature Materials*, **1** (2002), p. 241.

- **November 2002:** *Conductance of Hydrogen Molecules*

  The conductance of a single hydrogen molecule has been measured. A single hydrogen molecule has been trapped between two platinum electrodes and its conductance measured. This is a simple test system in which fundamental properties of single-molecule devices can be explored.

"Measurement of the Conductance of a Hydrogen Molecule", Smit, R.H.M., Y. Noat, C. Untiedt, N.D. Lang, M.C. van Hemert and J.M. van Ruitenbeek, *Nature*, **419** (2002), p. 906.

- **December 2002:** *Artificial Nanopores Detect DNA Molecules*

An artificial nanopore has been made by micro-moulding poly (dimethylsiloxane)—(PDMS)—elastomer. The on-chip electronic sensor detected single DNA molecules.

"An Artificial Nanopore for Molecular Sensing", Saleh, O.A. and L.L. Sohn, *Nano Letters*, **3** (2003), p. 37.

- **January 2003:** *Nanoelectronic Devices are Made from Nanowires*

One-dimensional heterostructure electronic devices based on nanowires have been made. Resonant tunnelling diodes by bottom-up assembly of different III/V semiconductor materials have also been made.

"Luminescence Polarization of Ordered GaInP/InP Islands", Håkanson, U., V. Zwiller, M.K.J. Johansson, T. Sass and L. Samuelson, *Appl. Phys. Lett.*, **82** (2003), p. 627.

- **January 2003:** *Functional Devices for Drug Delivery*

A nanoporous material that opens or closes its pores under the influence of light has been created. This hexagonal mesoporous silica may have applications in the controlled release of chemicals such as drugs.

"Photocontrolled Reversible Release of Guest Molecules from Coumarin-Modified Mesoporous Silica", Mal, N.K., M. Fujiwara and Y. Tanaka, *Nature*, **421** (2003), p. 350.

- **January 2003:** *Molecules have been Shown to Store Data*

Data storage devices using rotaxane molecules have been made. The films made by using these molecules have been scanned with an atomic force microscope (AFM) to create a pattern of dots to store the information.

"Information Storage Using Supramolecular Surface Patterns", Cavallini, M., F. Biscarini, S. Léon, F. Zerbetto, G. Bottari and D.A. Leigh, *Science*, **299** (2003), p. 531.

- **January 2003:** *Single Molecule Detection Using Nanoshells*

Single molecules have been detected with the help of surface-enhanced Raman scattering (SERS) by using nanoshells.

"Controlling the Surface-Enhanced Raman Effect via the Nanoshell Geometry", Jackson, J.B., S.L. Westcott, L.R. Hirsch, J.L. West and N.J. Halas, *Appl. Phys. Lett.*, **82** (2003), p. 257.

- **February 2003:** *DNA Computing Device Makes Its Own Energy*

A DNA computing device that provides its own energy has been made. The device uses DNA molecules as both input data and as a fuel source.

"DNA Molecule Provides a Computing Machine with Both Data and Fuel", Benenson, Y., R. Adar, T. Paz-Elizur, Z. Livneh and E. Shapiro, *PNAS*, **100** (2003), p. 2191.

- **February 2003:** *Record-breaking Superconductor Transistor*

  A superconducting amplifier has been built with the highest current and power gains till date. The device can be used in low-temperature applications in the form of read-out elements for quantum computers.

  "Low-Noise Current Amplifier Based on Mesoscopic Josephson Junction", Delahaye, J., J. Hassel, R. Lindell, M. Sillanpää, M. Paalanen, H. Seppä and P. Hakonen, *Science*, **299** (2003), p. 1045.

- **February 2003:** *Nanofibre 'Bandage' to Heal Wounds*

  A nanofibre mat from fibrinogen, a soluble protein that is present in the blood, has been made which can be used as a wound dressing or tissue-engineering scaffold.

  "Electrospinning of Nanofiber Fibrinogen Structures", Wnek, G.E., M.E. Carr, D.G. Simpson and G.L. Bowlin, *Nano Letters*, **3** (2003), p. 213.

- **February 2003:** *Carbon Nanotube Device to Detect Gas Molecules*

  An array of detectors containing single-walled carbon nanotubes, which can sense gases, has been made.

  "Efficient Formation of Iron Nanoparticle Catalysts on Silicon Oxide by Hydroxylamine for Carbon Nanotube Synthesis and Electronics", Choi, H.C., S. Kundaria, D. Wang, A. Javey, Q. Wang, M. Rolandi and H. Dai, *Nano Letters*, **3** (2003), p. 157.

- **February 2003:** *Expanding and Contracting DNA Nanomotor*

  A DNA nanomotor that expands and contracts by up to 5 nm has been made.

  "DNA Duplex-Quadruplex Exchange as the Basis for a Nanomolecular Machine", Alberti, P., and J.L. Mergny, *PNAS*, **100** (2003), p. 1569.

- **February 2003:** *Carbon Nanotube Flow Sensors*

  Carbon nanotubes generate a potential when a liquid flows over it. The potential is sensitive to the flow parameters.

  "Carbon Nanotube Flow Sensors", Ghosh, S., A.K. Sood and N. Kumar, *Science*, **299** (2003), pp. 1042–1044.

- **March 2003:** *Sensing Polymers are Now Nanocrystalline*

  Porous silicon has been used for making nanocrystalline polymer structures. The resulting structures may have applications as sensing devices inside the body.

  "Polymer Replicas of Photonic Porous Silicon for Sensing and Drug Delivery Applications", Li, Y.Y., F. Cunin, J.R. Link, T. Gao, R.E. Betts, S.H. Reiver, V. Chin, S.N. Bhatia and M.J. Sailor, *Science*, **299** (2003), pp. 2045.

- **March 2003:** *Highest Resolution Optical Microscopy*

  'Near-field Raman microscopy' has been used to look at carbon nanotubes revealing 30 nm features.

  "High-Resolution Near-Field Raman Microscopy of Single-Walled Carbon Nanotubes", Hartschuh, A., E.J. Sánchez, X.S. Xie and L. Novotny, *Phys. Rev. Lett.*, **90** (2003), p. 095503.

- **March 2003:** *Carbon Nanotubes Linked Up by Amino Acids*

  Sidewalls of single-walled carbon nanotubes have been functionalized with amino acids.

  "Sidewall Amino-Functionalization of Single-Walled Carbon Nanotubes through Fluorination and Subsequent Reactions with Terminal Diamines", Stevens, J.L., A.Y. Huang, H. Peng, I.W. Chiang, V.N. Khabashesku and J.L. Margrave, *Nano Letters*, **3** (2003), p. 331.

- **April 2003:** *Biomolecular Electronic Devices Show Higher Gain*

  A prototype field-effect transistor using a deoxyguanosine derivative, has been made. The device has a maximum voltage gain of 0.76, which is higher than that reported for other molecular devices.

  "Field-Effect Transistor Based on a Modified DNA Base", Maruccio, G., P. Visconti, V. Arima, S. D'Amico, A. Biasco, E. D'Amone, R. Cingolani R. Rinaldi, S. Masiero, T. Giorgi and G. Gottarelli, *Nano Letters*, **3** (2003), p. 479.

- **April 2003:** *Hydrogen Storage by Nanostructured Graphite*

  Nanostructured graphite has been shown to absorb hydrogen to the tune of 0.20–0.25 per cent by weight.

  "Dense Hydrogen Adsorption on Carbon Sub-Nanopores at 77 K", Kadono, K., H. Kajiura and M. Shiraishi, *Appl. Phys. Lett.*, **83** (2003), p. 3392.

- **April 2003:** *Nanowires Made with Protein Templates*

  On a yeast protein template metallic nanowires have been made by using nanoparticles.

  Scheibel, T., R. Parthasarathy, G. Sawicki, Xiao-Min Lin, H. Jaeger and S. L. Lindquist, *PNAS*, **100** (2003), p. 4527.

- **May 2003:** *DNA Detection by Using Nanotubes*

  A nanotube-based DNA detection method has been developed which uses fluorescence detection.

  "Carbon Nanotube Nanoelectrode Array for Ultrasensitive DNA Detection", Li, J., H.T. Ng, A. Cassell, W. Fan, H. Chen, Q. Ye, J. Koehne, J. Han and M. Meyyappan, *Nano Letters*, **3** (2003), p. 597.

- **May 2003:** *Genes Inserted into Cells by Using Nanotubes*

  Carbon nanofibres have been used to insert plasmid DNA into cells.

  "Intracellular Iintegration of Synthetic Nanostructures with Viable Cells for Controlled Biochemical Manipulation", McKnight, T.E., A.V. Melechko, G.D. Griffin, M.A. Guillorn, V.I. Merkulov, F. Serna, D.K. Hensley, M.J. Doktycz, D.H. Lowndes and M.L. Simpson, *Nanotechnology*, **14** (2003), p. 551.

- **May 2003:** *Large Magnetic Anisotropy*

  Large magnetic anisotropy energy (MAE) of over 9 milli-electron-volts has been found in cobalt atoms placed on a platinum substrates. It is suggested that this discovery would help in the design of new magnetic materials that may be used for data storage.

  "Giant Magnetic Anisotropy of Single Cobalt Atoms and Nanoparticles", Gambardella, P., S. Rusponi, M. Veronese, S.S. Dhesi, C. Grazioli, A. Dallmeyer, I. Cabria, R. Zeller, P.H. Dederichs, K. Kern, C. Carbone and H. Brune, *Science*, **300** (2003), p. 1130.

- **May 2003:** *Enhanced Superconductivity*

  Superconductivity has been shown to be enhanced in a nanoengineered magnetic field.

  "Nanoengineered Magnetic-Field-Induced Superconductivity", Lange, M., M.J.V. Bael, Y. Bruynseraede and V.V. Moshchalkov, *Phys. Rev. Lett.*, **90** (2003), p. 197006.

- **June 2003:** *Nanoparticles Destroy Halocarbons*

  A solution of gold and silver nanoparticles carries out the catalytic destruction of halocarbons in water. The authors claim that this may be used for the removal of halocarbons from drinking water.

  "Halocarbon Mineralization and Catalytic Destruction by Metal Nanoparticles", Nair A.S. and T. Pradeep, *Current Science*, **84** (2003), p. 1560.

- **June 2003:** *Bar-codes with DNA*

  DNA self-assembly around a DNA scaffold has been used to create a bar-code pattern containing information.

  "Directed Nucleation Assembly of DNA Tile Complexes for Bar-Code-Patterned Lattices", Yan, H., T.H. LaBean, L. Feng and J.H. Reif, *PNAS*, **100** (2003), p. 8103.

- **June 2003:** *Gold Nanoparticles Synthesized by Biology*

  A micro-organism, *Rhodococcus* sp., which normally grows on fig trees, has been used to synthesize gold nanoparticles.

  "Intracellular Synthesis of Gold Nanoparticles by a Novel Alkalotolerant Actinomycete, *Rhodococcus* Species", Ahmad, A., S. Senapati, M.I. Khan, R. Kumar, R. Ramani, V. Srinivas and M. Sastry, *Nanotechnology*, **14** (2003), p. 824.

- **July 2003:** *Precision Sensors to Detect Quantum Movement*

  A nanoelectromechanical bridge and a single-electron transistor have been used to detect ultra small displacements.

  "Nanometre-Scale Displacement Sensing Using a Single Electron Transistor", Knobel, R.G., and A.N. Cleland, *Nature*, **424** (2003), p. 291.

- **August 2003:** *Quantum Circuits, Freated and Erased*

  Devices and circuits can be created and changed at the atomic level by using erasable electrostatic lithography.

  "Erasable Electrostatic Lithography for Quantum Components", Crook, R., A.C. Graham, C.G. Smith, I. Farrer, H.E. Beere and D.A. Ritchie, *Nature*, **424** (2003), p. 751.

- **August 2003:** *Wet Nanoparticles Change Structure*

  Certain nanoparticles change their structure by the addition of water.

  "Water-Driven Structure Transformation in Nanoparticles at Room Temperature", Zhang, H., B. Gilbert, F. Huang and J.F. Banfield, *Nature*, **424** (2003), p. 1025.

- **September 2003:** *Nanotubes Sense Helium*

  Multi-walled carbon nanotubes under a positive bias field-ionize passing gas atoms.

"Helium Detection Via Field Ionization from Carbon Nanotubes", Riley, D.J., M. Mann, D.A. MacLaren, P.C. Dastoor, W. Allison, K.B.K. Teo, G.A.J. Amaratunga and W. Milne, *Nano Letters*, **3** (2003), p. 1455.

- **September 2003:** *Biological Cells Imaged with Nanoparticles*

  Proteins have been imaged efficiently and non-destructively by using gold nanoparticles.

  "Single Metallic Nanoparticle Imaging for Protein Detection in Cells", Cognet, L., C. Tardin, D. Boyer, D. Choquet, P. Tamarat and B. Lounis, *PNAS*, **100** (2003), p. 11350.

- **September 2003:** *Nanopores Analyze DNA*

  The properties of single DNA molecules have been studied by using a solid-state nanopore membrane. The technique may be ultimately used for rapid DNA sequencing.

  "DNA Molecules and Configurations in a Solid-State Nanopore Microscope", Li, J., M. Gershow, D. Stein, E. Brandin and J.A. Golovchenko, *Nature Materials*, **2** (2003), p. 611.

- **November 2003:** *Nerve Agent Detector Using Nanotubes*

  A nerve agent detector using single-walled carbon nanotubes has been developed.

  "Nerve Agent Detection Using Networks of Single-Walled Carbon Nanotubes", Novak, J.P., E.S. Snow, E.J. Houser, D. Park, J.L. Stepnowski and R.A. McGill, *Appl. Phys. Lett.*, **83** (2003), p. 4026.

- **December 2003:** *Nanotube Sorter Using DNA*

  Self-assembly of DNA has been used to sort carbon nanotubes according to their diameter and electronic properties.

  "Structure-Based Carbon Nanotube Sorting by Sequence-Dependent DNA Assembly", Zheng, M., A. Jagota, M.S. Strano, A.P. Santos, P. Barone, S.G. Chou, B.A. Diner, M.S. Dresselhaus, R.S. Mclean, G.B. Onoa, G.G. Samsonidze, E.D. Semke, M. Usrey and D.J. Walls, *Science*, **302** (2003), p. 1545.

- **2003:** *Assorted Discoveries*
  - A nanowire has the potential to detect the gene for cystic fibrosis (CF) more efficiently than conventional tests carried out for the disease. CF is the most common fatal genetic disease among people of European origin.
  - A 'nanosensor' that only works when noise is added has been developed. The device uses 'Stochastic Resonance' to enhance sub-threshold signals. In a noise-free environment, the detectors will not receive a signal. If a moderate amount of noise is present, the signal will, as it were, float on top of the noise, while triggering the detectors.
  - Nanoscale fibers that are thinner than the wavelengths of light have been developed.
  - Nanoshells can act as anti-cancer drugs. The light used is near-infrared light which gets absorbed by the shells, thereby heating and destroying the cells which are injected with the shells. Detection of the cancer cells is also done by the shells.
  - Prostate specific antigen (PSA) at extremely low levels in a blood sample has been detected with nanoparticles. This heralds the use of new kinds of detection methods for proteins.
  - Multi-photon microscopy of the blood vessels has been carried out with quantum dots.
  - Nano bones have been implanted in patients.

- A new type of protein chip has been developed on the basis of protein-binding silica-nanoparticles. Chips are analyzed by using MALDI-TOF mass spectrometry.
- It has been shown that an injection of magnetic nanoparticles into the bloodstream can reveal where harmful viruses are located. The particles are coated with antibodies to fight a particular virus, so that they will form clumps that would be visible on conventional body scans.

■ **January 2004:** *World's First Nanotechnology College*

The world's first college of nanotechnology is established at Suny Albany, USA.

■ **February 2004:** *Transition from Quantum to Classical through Decoherence*

Decoherence—The transition from quantum to classical behavior, caused by the thermal emission of radiation was observed in the case of C_{70} molecules at higher temperatures.

"Decoherence of Matter Waves by Thermal Emission of Radiation", Hackermüller, L., K. Hornberger, B. Brezger, A. Zeilinger and M. Arndt, *Nature*, **427** (2004), p. 711.

■ **March 2004:** *Controlling the Rotation of a Molecule*

Simple electron transfer processes and photoexcitation have been used to control the rotary motion of a metallacarborane, resulting in the possible applications for nanovalves and in the modification of surface properties.

"Electrical or Photocontrol of the Rotary Motion of a Metallacarborane", Hawthorne, M.F., J.I. Zink, J.M. Skelton, M.J. Bayer, C. Liu, E. Livshits, R. Baer and D. Neuhauser, *Science*, **303** (2004), p. 1849.

■ **April 2004:** *Nanomehanical Resonator Approaching Quantum Limit*

A tiny nanoelectromechanical arm's vibrations have been measured to probe the limits at which classical physics takes over from quantum behavior.

"Approaching the Quantum Limit of a Nanomechanical Resonator", LaHaye, M.D., O. Buu, B. Camarota and K.C. Schwab, *Science*, **304** (2004), p. 74.

■ **April 2004:** *New Nanocrystal Mesophase Produced by Self-assembly of Nanocrystal Micelles*

The self-assembly of water-soluble nanocrystal micelles leads to the formation of a new nanocrystal mesophase, which is suitable for integration into devices that use standard techniques of microelectronic processing.

"Self-Assembly of Ordered, Robust, Three-Dimensional Gold Nanocrystal/Silica Arrays", Fan, H., K. Yang, D.M. Boye, T. Sigmon, K.J. Malloy, H. Xu, G.P. López and C.J. Brinker, *Science*, **304** (2004), p. 567.

■ **May 2004:** *Single-electron Transistor*

A single-electron transistor has been made which operates by using a nanometer-scale vibrating arm. It was built by using a simple two-step process and unlike previous devices of the kind, it does not require cryogenic temperatures to be operational.

"Silicon Nanopillars for Mechanical Single-Electron Transport", Scheible, D.V. and R.H. Blick, *Appl. Phys. Lett.*, **84** (2004), p. 4632.

- **May 2004:** *Carbon Nanotubes Respond to Magnetic Fields*

  Carbon nanotubes have been found to respond to magnetic fields. In the presence of a magnetic field, semiconducting nanotubes can be made metallic and vice versa.

  "Optical Signatures of the Aharonov-Bohm Phase in Single-Walled Carbon Nanotubes", Zaric, S., G.N. Ostojic, J. Kono, J. Shaver, V.C. Moore, M.S. Strano, R.H. Hauge, R.E. Smalley and X. Wei, *Science*, **304** (2004), p. 1129.

  "h/e Magnetic Flux Modulation of the Energy Gap in Nanotube Quantum Dots", Coskun, U.C., Tzu-Chieh Wei, S. Vishveshwara, P.M. Goldbart and A. Bezryadin, *Science*, **304** (2004), p. 1132.

- **May 2004:** *Nanotube Mixture*

  Rolled-up nanotubes of InAs/GaAs with tube walls containing alternating layers of crystalline and non-crystalline materials have been made. The structure of these radial superlattices was altered by using a laser, which resulted in the production of small regions of β-Ga_2O_3.

  "Radial Superlattices and Single Nanoreactors", Deneke, Ch., N.Y. Jin-Phillipp, I. Loa and O.G. Schmidt, *Appl. Phys. Lett.*, **84** (2004), p. 4475.

- **May 2004:** *DNA Detection Using Bio-bar-code-based Technique*

  DNA detection using the bio-bar-code-based technique achieves a sensitivity similar to the generally used polymerase chain reaction (PCR) method. This technique involves the use of both gold nanoparticles attached to the bar-code DNA as well as the magnetic nanoparticles.

  "Bio-Bar-Code-Based DNA Detection with PCR-Like Sensitivity", Nam, Jwa-Min, S.I. Stoeva and C.A. Mirkin, *J. Am. Chem. Soc.*, **126** (2004), p. 5932.

- **May 2004:** *Nanoscale Conveyors for Nanoassembly*

  An important tool for nanoassembly has been developed, wherein the indium atoms are carried to a precise location by using a multi-walled carbon nanotube.

  "Carbon Nanotubes as Nanoscale Mass Conveyors", Regan, B.C., S. Aloni, R.O. Ritchie, U. Dahmen and A. Zettl, *Nature*, **428** (2004), p. 924.

- **May 2004:** *Solid-state C_{50} Molecules*

  For the first time, solid-state carbon-50 molecules have been prepared by using an arc-discharge technique involving chlorine.

  "Capturing the Labile Fullerene[50] as $C_{50}Cl_{10}$", Xie, Su-Yuan, F. Gao, X. Lu, Rong-Bin Huang, Chun-Ru Wang, X. Zhang, Mai-Li Liu, Shun-Liu Deng and Lan-Sun Zheng, *Science*, **304** (2004), p. 699.

- **June 2004:** *Nanoprticles to Clean Tumours*

  Tumors in mice have been eradicated by photothermal ablation using near infrared-absorbing nanoparticles without affecting the healthy tissues.

  "Photothermal Tumor Ablation in Mice Using Near Infrared-Absorbing Nanoparticles", O'Neal, D.P., L.R. Hirsch, N.J. Halas, J.D. Payne and J.L. West, *Cancer Letters*, **209** (2004), p. 171.

- **June 2004:** *Inkjet Printing at Nanometer Scale*

  Thin layers of molecules have been deposited on cantilever beams using the inkjet-printing technique, which enables the beams to act as chemical or biochemical sensors.

  "Rapid Functionalization of Cantilever Array Sensors by Inkjet Printing", Bietsch, A., J. Zhang, M. Hegner, H.P. Lang and C. Gerber, *Nanotechnology*, **15** (2004), p. 873.

- **June 2004:** *Water Molecules Confined in Nanotube*

  Water molecules confined inside an open-ended single-walled carbon nanotube were studied using neutron-scattering measurements and molecular-dynamics simulations.

  "Anomalously Soft Dynamics of Water in a Nanotube: A Revelation of Nanoscale Confinement", Kolesnikov, A.I., Jean-Marc Zanotti, Chun-Keung Loong, P. Thiyagarajan, A.P. Moravsky, R.O. Loutfy and C.J. Burnham, *Phys. Rev. Lett.*, **93** (2004), p. 035503.

- **June 2004:** *Quantum Dot Infrared Photodetectors*

  Semiconductor nanocrystals (quantum dots) have been used for making infrared detection devices, finding applications in night-vision googles, environmental monitors and military target tracking systems.

  "High Detectivity InAs Quantum Dot Infrared Photodetectors", Kim, Eui-Tae, A. Madhukar, Z. Ye and J.C. Campbell, *Appl. Phys. Lett.*, **84** (2004), p. 3277.

- **June 2004:** *AFM Imaging Less than 1Å Resolution*

  A new higher-harmonic force microscope using a single carbon atom as a probe has been developed, which has a resolution close to three times better than that of the traditional STM.

  "Force Microscopy with Light-Atom Probes", Hembacher, S., F.J. Giessibl and J. Mannhar, *Science*, **305** (2004), p. 380.

- **June 2004:** *Nanoconnections Using Photolithography*

  Interconnects to nanowire devices have been developed by using the photolithography technique which ensures that higher proportions of the nanowires connect to the electrodes.

  "Scalable Interconnection and Integration of Nanowire Devices without Registration", Jin, S., D. Whang, M.C. McAlpine, R.S. Friedman, Y. Wu and C.M. Lieber, *Nano Letters*, **4** (2004), p. 915.

- **June 2004:** *Carbon Nanotubes for Bulb Filaments*

  A light bulb with carbon nanotubes as the filament has been designed, and it is found to have several advantages over the conventional tungsten filament.

  "Carbon Nanotube Filaments in Household Light Bulbs", Wei, J., H. Zhu, D. Wu, B. Wei, *Appl. Phys. Lett.*, **84** (2004), p. 4869.

- **June 2004:** *Exothermic Nanocomposites*

  A nanocompsoite of aluminium and iron oxide that reacts exothermically on igniting has been made. This could have possible applications in explosives, as an energy source in MEMS devices, etc.

"Ignition Studies of Al/Fe$_2$O$_3$ Energetic Nanocomposites", Menon, L., S. Patibandla, K.B. Ram, S.I. Shkuratov, D. Aurongzeb, M. Holtz, J. Berg, J. Yun and H. Temkin, *Appl. Phys. Lett.*, **84** (2004), p. 4735.

- **July 2004:** *STM to Add and Remove Electrons*

  STM has been used for adding and removing single electrons from gold adatoms, thereby controlling the latter's charge state.

  "Controlling the Charge State of Individual Gold Adatoms", Repp, J., G. Meyer, F.E. Olsson and M. Persson, *Science*, **305** (2004), p. 493.

- **July 2004:** *Laser Microstructuring of Gold Nanoparticle Inks*

  A fountain-pen-based laser technique has been developed for depositing gold nanoink stripes as thin as 5 microns, which is simultaneously cured with an argon ion laser.

  "Fountain-Pen-Based Laser Microstructuring with Gold Nanoparticle Inks", Choi, T.Y., D. Poulikakos and C.P. Grigoropoulos, *Appl. Phys. Lett.*, **85** (2004), p. 13.

- **July 2004:** *Imaging the Single Spin of an Electron*

  For the first time, spin of an individual electron has been imaged by combining MRI with AFM. This could eventually lead to the production of three-dimensional images on an atomic scale and as read-out devices for spin-based quantum computers.

  "Single Spin Detection by Magnetic Resonance Force Microscopy", Rugar, D., R. Budakian, H.J. Mamin and B.W. Chui, *Nature*, **430** (2004), p. 329.

- **July 2004:** *Nanotube Forms p-n Junction Diodes*

  A *p-n* junction diode was made from a single-walled carbon nanotube by using electrostatic doping. Unlike the conventional diodes, in this case, applying a voltage altered the nanotube's properties.

  "Carbon Nanotube *p-n* Junction Diodes", Lee, J.U., P.P. Gipp and C.M. Heller, *Appl. Phys. Lett.*, **85** (2004), p. 145.

- **July 2004:** *Looking into a Nanowire Using STM*

  Using STM, the inside of a gallium arsenide nanowire was imaged with atomic resolution. This revealed defects such as planar twin segments and single-atom impurities present in it.

  "Direct Imaging of the Atomic Structure Inside a Nanowire by Scanning Tunnelling Microscopy", Mikkelsen, A., N. Sköld, L. Ouattara, M. Borgström, J.N. Andersen, L. Samuelson, W. Seifert and E. Lundgren, *Nature Materials*, **3** (2004), p. 519.

- **July 2004:** *Growth of Metallic Grains during Recrystallization*

  The changes taking place in the growth of metallic grains during the recrystallization process, have been observed in three spatial dimensions and a one-time dimension by a specially developed X-ray microscope. The growth is found to be less smooth and regular than the theoretical predictions.

  "Watching the Growth of Bulk Grains During Recrystallization of Deformed Metals", Schmidt, S., S.F. Nielsen, C. Gundlach, L. Margulies, X. Huang and D.J. Jensen, *Science*, **305** (2004), p. 229.

- **August 2004:** *Nanoribbons to Channel Light*

  Nanoribbons of crystalline oxide have been used as waveguides for channeling light between devices.
  "Nanoribbon Waveguides for Sub-Wavelength Photonics Integration", Law, M., D.J. Sirbuly, J.C. Johnson, J. Goldberger, R.J. Saykally and P. Yang, *Science*, **305** (2004), p. 1269.

- **August 2004:** *Potential Parts for Nanotechnology*

  The various forms of shapes such as twins, tetramers, triangles, rods and 3D arrays have been made from the packaging RNA (pRNA) molecules.
  "Bottom-Up Assembly of RNA Arrays and Superstructures as Potential Parts in Nanotechnology", Shu, D., Wulf-Dieter Moll, Z. Deng, C. Mao and P. Guo, *Nano Letters*, **4** (2004), p. 1717.

- **August 2004:** *Timed Light Emission*

  Controlling the timing of light emission from Semiconductor Quantum Dots has been achieved by embedding them in a photonic crystal structure.
  "Controlling the Dynamics of Spontaneous Emission from Quantum Dots by Photonic Crystals", Lodahl, P., A.F. van Driel, I.S. Nikolaev, A. Irman, K. Overgaag, D. Vanmaekelbergh and W.L. Vos, *Nature*, **430** (2004), p. 654.

- **August 2004:** *Light Moves Water Droplets by Lotus Effect*

  Water droplets have been found to move on, illuminating with UV light with the aid of a nanowire-coated surface. This phenomenon could have possible applications in the case of microfluidic devices.
  "Lotus Effect Amplifies Light-Induced Contact Angle Switching", Rosario, R., D. Gust, A.A. Garcia, M. Hayes, J.L. Taraci, T. Clement, J.W. Dailey and S.T. Picraux, *J. Phys. Chem. B*, **108** (2004), p. 12640.

- **September 2004:** *NMR at Nanoscale*

  A new NMR method known as better observation of magnetisation, enhanced resolution and no gradient (BOOMERANG) has been developed. Using this, liquids and solids can be imaged, which could finally lead to the development of portable NMR instruments for use at the micron scale and below.
  "Observation of Force-Detected Nuclear Magnetic Resonance in a Homogeneous Field", Madsen, L.A., G.M. Leskowitz and D.P. Weitekamp, *PNAS*, **101** (2004), p. 12804.

- **September 2004:** *Transparent Conducting Nanotubes*

  Transparent electrically conducting films have been created from single-walled carbon nanotubes, which has been used for making an electric field-activated optical modulator.
  "Transparent, Conductive Carbon Nanotube Films", Wu, Z., Z. Chen, X. Du, J.M. Logan, J. Sippel, M. Nikolou, K. Kamaras, J.R. Reynolds, D.B. Tanner, A.F. Hebard and A.G. Rinzler, *Science*, **305** (2004), p. 1273.

- **September 2004:** *Lateral Atom Manipulation Using STM*

  The lateral motion of a "Co" atom on a "Cu" surface was controlled using an STM, i.e. the atom was moved back and forth between neighbouring sites on the crystal's (111) surface. And thereby a novel imaging technique was developed.

"Controlling the Dynamics of a Single Atom in Lateral Atom Manipulation", Stroscio, J.A. and R.J. Celotta, *Science*, **306** (2004), p. 242.

- **September 2004:** *Imaging on sub-Angstrom Scales*

  A crystal was imaged on sub-Angstrom scales by using a technique developed for correcting the aberrations in a scanning transmission electron microscope.

  "Direct Sub-Angstrom Imaging of a Crystal Lattice", Nellist, P.D., M.F. Chisholm, N. Dellby, O.L. Krivanek, M.F. Murfitt, Z.S. Szilagyi, A.R. Lupini, A. Borisevich, W.H. Sides, Jr. and S.J. Pennycook, *Science*, **305** (2004), p. 1741.

- **September 2004:** *Carbon Nanotubes of Record Length*

  A single-walled carbon nanotube, which is four centimeters long, has been prepared and it is believed to be the world's longest one.

  "Ultralong Single-Wall Carbon Nanotubes", Zheng, L.X., M.J. O'connell, S.K. Doorn, X.Z. Liao, Y.H. Zhao, E.A. Akhadov, M.A. Hoffbauer, B.J. Roop, Q.X. Jia, R.C. Dye, D.E. Peterson, S.M. Huang, J. Liu and Y.T. Zhu, *Nature Materials*, **3** (2004), p. 673.

- **September 2004:** *Diamond Films for Biosensor Applications*

  Nanocrystalline diamond has been modified with protein molecules such that the molecules still remain active. Also the enzyme catalase was attached to a diamond film for creating a biosensor for the detection of H_2O_2.

  "Protein-Modified Nanocrystalline Diamond Thin Films for Biosensor Applications", Härtl, A., E. Schmich, J.A. Garrido, J. Hernando, S.C.R. Catharino, S. Walter, P. Feulner, A. Kromka, D. Steinmüller and M. Stutzmann, *Nature Materials*, **3** (2004), p. 736.

- **September 2004:** *Superhard Carbon Phase*

  By cold compression of carbon nanotubes, a quenchable superhard high-pressure carbon phase was synthesized. Its hardness is comparable to that of cubic diamond and it also retained its properties at room temperature.

  "A Quenchable Superhard Carbon Phase Synthesized by Cold Compression of Carbon Nanotubes", Wang, Z., Y. Zhao, K. Tait, X. Liao, D. Schiferl, C. Zha, R.T. Downs, J. Qian, Y. Zhu and T. Shen, *PNAS*, **101** (2004), p. 13699.

- **September 2004:** *Surface Modification of C_{60} Molecules Linked to its Toxicity*

  It has been found that the toxicity of C_{60} molecules to human cells is highly dependent on the molecules attached to the C_{60} molecules surface.

  "The Differential Cytotoxicity of Water-Soluble Fullerenes", Sayes, C.M., J.D. Fortner, W. Guo, D. Lyon, A.M. Boyd, K.D. Ausman, Y.J. Tao, B. Sitharaman, L.J. Wilson, J.B. Hughes, J.L. West and V.L. Colvin, *Nano Letters*, **4** (2004), p. 1881.

- **September 2004:** *Self-assembled Biocidal Nanotubes*

  Biocidal nanotubes have been created which can also self-assemble into a nanocarpet structure. These tubes changed color in the presence of bacteria apart from killing it.

"Self-Assembly of Biocidal Nanotubes from a Single-Chain Diacetylene Amine Salt", Lee, S.B., R. Koepsel, D.B. Stolz, H.E. Warriner and A.J. Russell, *J. Am. Chem. Soc.*, **126** (2004), p. 13400.

- **September 2004:** *Carbon Nanotube Filters*

Multi-functional filters using carbon nanotubes have been developed. These can filter bacteria and viruses from water as well as separate petroleum into its molecular components.

"Carbon nanotube filters", Srivastava, A., O.N. Srivastava, S. Talapatra, R. Vajtai and P.M. Ajayan, *Nature Materials*, **3** (2004), p. 610–614.

- **October 2004:** *Molecular Switches Stabilized by Self-assembled Monolayer*

The electrical switching of single molecules has been stabilized by altering their environment. A self-assembled monolayer has been used for surrounding oligo (phenylene-ethynylene) (OPE) molecules.

"Mediating Stochastic Switching of Single Molecules Using Chemical Functionality", Lewis, P.A., C.E. Inman, Y. Yao, J.M. Tour, J.E. Hutchison and P.S. Weiss, *J. Am. Chem. Soc.*, **126** (2004), p. 12214.

- **October 2004:** *Anti-bacterial Agents to be Released by Silica Nanoparticles*

Mesoporous silica nanoparticles have been produced from room-temperature ionic liquids (RTILs) by using them as a template. For the anti-bacterial ionic liquids, these nanoparticles act as controlled release agents.

"Morphological Control of Room-Temperature Ionic Liquid Templated Mesoporous Silica Nanoparticles for Controlled Release of Antibacterial Agents", Trewyn, B.G., C.M. Whitman and V.S.Y. Lin, *Nano Letters*, **4** (2004), p. 2139.

- **October 2004:** *Nanoparticles for Rapid Bioassay of Food Samples*

A single bacterial cell has been detected within 20 minutes by using bioconjugated nanoparticle-based bioassay. In the case of ground beef samples, this technique has been used to identify *Escherischia coli* bacteria.

"A Rapid Bioassay for Single Bacterial Cell Quantitation Using Bioconjugated Nanoparticles", Zhao, X., L.R. Hilliard, S.J. Mechery, Y. Wang, R.P. Bagwe, S. Jin and W. Tan, *PNAS*, **101** (2004), p. 15027.

- **October 2004:** *Novel Optoelectronic Fibres*

Novel optoelectronic fibres containing metal, insulator and semiconductor layers have been created. These fibres could find applications in photodetectors by weaving them into a spectrometric fabric.

"Metal-Insulator-Semiconductor Optoelectronic Fibres", Bayindir, M., F. Sorin, A.F. Abouraddy, J. Viens, S.D. Hart, J.D. Joannopoulos and Y. Fink, *Nature*, **431** (2004), p. 826.

- **October 2004:** *Dip-pen Nanolithography Boosted by Surfactant*

Using dip-pen nanolithography, it has been found that the surfactant added ink patterned maleimide-linked biotin onto the mercaptosilanized glass. Now the range of ink-substrate combinations used for patterning biotin and other biomolecules such as proteins can be expanded by the use of this technique.

"Surfactant Activated Dip-Pen Nanolithography", Jung, H., C.K. Dalal, S. Kuntz, R. Shah and C.P. Collier, *Nano Letters*, **4** (2004), p. 2171.

- **October 2004:** *Magnetic Nanoparticles to Act as Sensors*

  A new type of magnetic sensor has been developed for detecting biomolecules; this involves measuring the Brownian relaxation of magnetic nanoparticles bound to the target biomolecules.

  "Biological Sensors Based on Brownian Relaxation of Magnetic Nanoparticles", Chung, S.H., A. Hoffmann, S.D. Bader, C. Liu, B. Kay, L. Makowski and L. Chen, *Appl. Phys. Lett.*, **85** (2004), p. 2971.

- **October 2004:** *Atomically Thin Carbon Films*

  A carbon film of only one atom thickness is found to have some possible useful electronic properties. For example, transistors can be made by processing these graphene films for some semiconductor applications.

  "Electric Field Effect in Atomically Thin Carbon Films", Novoselov, K.S., A.K. Geim, S.V. Morozov, D. Jiang, Y. Zhang, S.V. Dubonos, I.V. Grigorieva and A.A. Firsov, *Science*, **306** (2004), p. 666.

- **October 2004:** *Information Transfer along a Carbon Chain*

  Information was transferred along a carbon chain for a distance of more than 2.5 nm by using conformational communication. This could find applications for transmitting and processing data in molecular devices.

  "Ultra-Remote Stereocontrol by Conformational Communication of Information along a Carbon Chain", Lund, A., L. Vallverdú and M. Helliwell, *Nature*, **431** (2004), p. 966.

- **October 2004:** *Nanomechanical Memory Element*

  A high-speed nanomechanical memory element has been made from single-crystal silicon wafers for the first time. This device consists of a vibrating arm that has the ability to switch between two distinct states.

  "A Controllable Nanomechanical Memory Element", Badzey, R.L., G. Zolfagharkhani, Alexei Gaidarzhy and P. Mohanty, *Appl. Phys. Lett.*, **85** (2004), p. 3587.

- **October 2004:** *Nanoscale Water Confinement*

  The closed multi-walled carbon nanotubes of diameter between 2 and 5 nm have been filled with water. This is found to be important for understanding the behavior of a liquid confined at nanoscale.

  "Observation of Water Confined in Nanometer Channels of Closed Carbon Nanotubes", Naguib, N., H. Ye, Y. Gogotsi, A.G. Yazicioglu, C.M. Megaridis and M. Yoshimura, *Nano Letters*, **4** (2004), p. 2237.

- **October 2004:** *Flash Welding of Nanofibres*

  Polyaniline nanofibres have been found to bond together on exposing to a camera flash. This technique could have applications in the preparation of asymmetric nanofibre films, polymer-polymer nanocomposites and photo-patterning polymer nanofiber films.

  "Flash Welding of Conducting Polymer Nanofibres", Huang, J. and R.B. Kaner, *Nature Materials*, **3** (2004), p. 783.

- **November 2004:** *Fluorescent Quantum Dots for Imaging Nanotubes*

  Single-walled carbon nanotubes have been imaged in an optical microscope by using semiconductor nanocrystals. This technique could have applications in the manufacture of carbon nanotube devices.

  "Fluorescence Microscopy Visualization of Single-Walled Carbon Nanotubes Using Semiconductor Nanocrystals", Chaudhary, S., J.H. Kim, K.V. Singh and M. Ozkan, *Nano Letters*, **4** (2004), p. 2415.

- **November 2004:** *InP Nanowires Grown Epitaxially*

  Indium phosphide nanowires have been grown epitaxially onto silicon and germanium substrates for the first time. This could aid in the integration of standard silicon technology with III-V semiconductors, which has good optoelectronic and high-frequency properties.

  "Epitaxial Growth of InP Nanowires on Germanium", Bakkers, E.P.A.M., J.A.V. Dam, S. De Franceschi, L.P. Kouwenhoven, M. Kaiser, M. Verheijen, H. Wondergem and P.V.D. Sluis, *Nature Materials*, **3** (2004), p. 769.

- **November 2004:** *Nanotube Networks Show Percolation*

  Single-walled carbon nanotube networks have been found to be transparent and conducting, while two-dimensional percolation behavior was observed from their direct current conductivity.

  "Percolation in Transparent and Conducting Carbon Nanotube Networks", Hu, L., D.S. Hecht and G. Grüner, *Nano Letters*, **4** (2004), p. 2513.

- **November 2004:** *ZnO Nanowires to Grow on Sapphire*

  Aligned arrays of ZnO nanowires have been horizontally grown on a sapphire substrate. This technique could have applications in the fabrication of devices.

  "Horizontal growth and *in situ* Assembly of oriented Zinc Oxide Nanowires", Nikoobakht, B., C.A. Michaels, S.J. Stranick and Mark D. Vaudin, *Appl. Phys. Lett.*, **85** (2004), p. 3244.

- **November 2004:** *Speedy Doping of Ionic Nanocrystals*

  It has been found that the cations in the ionic nanocrystals can be speedily exchanged with those of another material. This processes is also found to be reversible.

  "Cation Exchange Reactions in Ionic Nanocrystals", Son, D.H., S.M. Hughes, Y. Yin and A.P. Alivisatos, *Science*, **306** (2004), p. 1009.

- **November 2004:** *Multi-functional Nanotube Yarns*

  Multi-walled carbon nanotubes have been dry-spunned into strong twisted yarns having good electrical properties. These fibers could find applications in the case of structural composites, protective clothing, supercapacitors, etc.

  "Multi-Functional Carbon Nanotube Yarns by Downsizing an Ancient Technology", Zhang, M., K.R. Atkinson and R.H. Baughman, *Science*, **306** (2004), p. 1358.

- **November 2004:** *Water-assisted Nanotube Synthesis*

  Water has been found to be a solution for many problems plaguing the synthesis of carbon nanotube currently. During the chemical vapor deposition (CVD) of single-walled carbon nanotubes, it was found to boost catalytic activity.

"Water-Assisted Highly Efficient Synthesis of Impurity-Free Single-Walled Carbon Nanotubes", Hata, K., D.N. Futaba, K. Mizuno, T. Namai, M. Yumura and S. Iijima, *Science*, **306** (2004), p. 1362.

- **November 2004:** *Nanoscale Test Tube to Carry Out Reaction*

  Fullerene epoxide polymer $(C_{60}O)_n$ has been been prepared by polymerization using single-walled carbon nanotubes as a nano test tube. This polymer that is usually seen in a tangled and branched three-dimensional form, was surprisingly found to be linear with unbranched topology.

  "Selective Host–guest Interaction of Single-Walled Carbon Nanotubes with Functionalized Fullerenes", Britz, D.A., A.N. Khlobystov, J. Wang, A.S. O'Neil, M. Poliakoff, A. Ardavan and G.A.D. Briggs, *Chemical Communications*, **2** (2004), p. 176.

- **December 2004:** *STM Probes Lock-and-key Effect in Surface Diffusion*

  The diffusion of some large organic molecules on to a Cu(110) surface was found to depend on their orientation. Here the surface diffusion was observed to follow a lock-and-key model as probed by STM.

  "Lock-and-Key Effect in the Surface Diffusion of Large Organic Molecules Probed by STM", Otero, R., F. Hümmelink, F. Sato, S.B. Legoas, P. Thostrup, E. Lægsgaard, I. Stensgaard, D.S. Galvão and F. Besenbacher, *Nature Materials*, **3** (2004), p. 779.

- **December 2004:** *Nanotube Transistor of sub-20 nm Size*

  A field-effect carbon nanotube transistor device having a channel length of just 18 nm has been created, which is the world's smallest nanotube transistor.

  "Sub-20 nm Short Channel Carbon Nanotube Transistors, Seidel, R.V., A.P. Graham, J. Kretz, B. Rajasekharan, G.S. Duesberg, M. Liebau, E. Unger, F. Kreupl and W. Hoenlein, *Nano Letters*, **5** (2004), p. 147.

- **December 2004:** *Biosensing using Peptide Nanotubes*

  Self-assembled peptide nanotubes have been used for creating a novel electrochemical biosensor. The sensitivity of the biosensor was found to improve several folds in the presence of these nanotubes.

  "Novel Electrochemical Biosensing Platform Using Self-Assembled Peptide Nanotubes", Yemini, M., M. Reches, J. Rishpon and E. Gazit, *Nano Letters*, **5** (2005), p. 183.

- **December 2004:** *Superconductivity Related to Film Thickness*

  It has been found that the superconducting transition temperature for ultrathin lead films varied with the number of atomic layers present in it.

  "Superconductivity Modulated by Quantum Size Effects", Guo, Y., Yan-Feng Zhang, Xin-Yu Bao, Tie-Zhu Han, Z. Tang, Li-Xin Zhang, Wen-Guang Zhu, E.G. Wang, Q. Niu, Z.Q. Qiu, Jin-Feng Jia, Zhong-Xian Zhao and Qi-Kun Xue, *Science*, **306** (2004), p. 1915.

- **December 2004:** *Micro-machined Fountain Pen to Etch Nanopatterns*

  An AFM-based micromachined fountain pen with molecular ink has been used for etching nanopatterns on a surface.

"Micro-Machined Fountain Pen for Atomic Force Microscope-Based Nanopatterning", Deladi, S., N.R. Tas, J.W. Berenschot, G.J.M. Krijnen, M.J. de Boer, J.H. de Boer, M. Peter and M.C. Elwenspoek, *Appl. Phys. Lett.*, **85** (2004), p. 5361.

- **December 2004:** *Fluorescence Microscope Images Nanotubes Inside WBC's*

  Single-walled carbon nanotubes have been imaged inside white blood cells using near-infrared fluorescence microscope.

  "Near-Infrared Fluorescence Microscopy of Single-Walled Carbon Nanotubes in Phagocytic Cells", Cherukuri, P., S.M. Bachilo, S.H. Litovsky and R.B. Weisman, *J. Am. Chem. Soc.*, **126** (2004), p. 15638.

- **December 2004:** *Nanotube-based Optical Sensors for Measuring Blood Sugar Levels*

  Near-infrared optical sensors have been made up of modified single-walled carbon nanotubes for measuring blood sugar levels.

  "Near-Infrared Optical Sensors based on Single-Walled Carbon Nanotubes", Barone, P.W., S. Baik, D.A. Heller and M.S. Strano, *Nature Materials*, **4** (2004), p. 86.

- **January 2005:** *Polymers Synthesized Using DNA Nanomachine*

  A nanomechanical device that synthesizes different products according to their configuration has been made from DNA. This could find applications in the creation designer polymers, encryption of information, etc.

  "Translation of DNA Signals into Polymer Assembly Instructions", Liao, S. and N.C. Seeman, *Science*, **306** (2005), p. 2072.

- **January 2005:** *Nanoscale Switch*

  A nanoscale mechanical switch has been made which can demonstrate basic logic circuits. In future, this could replace semiconductor switches of the electronic devices.

  "Quantized Conductance Atomic Switch", Terabe, K., T. Hasegawa, T. Nakayama and M. Aono, *Nature*, **433** (2005), p. 47.

- **January 2005:** *Nanotubes to Sense Gas Attacks*

  When an atom or a molecule hits carbon nanotubes, there is found to be a change in its electrical resistivity. This could be well exploited by devices that use nanotubes as chemical sensors.

  "Atom Collision-Induced Resistivity of Carbon Nanotubes", Romero, H.E., K. Bolton, A. Rosén and P.C. Eklund, *Science*, **307** (2005), p. 89.

- **January 2005:** *AFM with a Time Resolution of Microseconds*

  An atomic force microscope has been demonstrated to take images of periodic processes with a time resolution of micro-seconds, which is faster as compared to the normal rapid-scan.

  "Atomic Force Microscopy with Time Resolution of Micro-Seconds", Anwar, M. and I. Rousso, *Appl. Phys. Lett.*, **86** (2005), p. 014101.

- **January 2005:** *Nanoneedle Attached to AFM for Nanoscale Operation*

  The nucleus of living cells has been operated upon using by nanoneedles attached to an atomic force microscope.

"Nanoscale Operation of a Living Cell Using an Atomic Force Microscope with a Nanoneedle", Obataya, I., C. Nakamura, S.W. Han, N. Nakamura and J. Miyake, *Nano Letters*, **5** (2005), p. 27.

- **January 2005:** *PbS Quantum Dot Devices Detect IR Wavelengths*

  The world's first solution-processed PbS quantum dot photodetector and photovoltaic devices that respond to infrared have been made. These devices could boost the efficiency of conversion of solar energy into electrical energy.

  "Solution-Processed PbS Quantum Dot Infrared Photodetectors and Photovoltaics", Mcdonald, S.A., G. Konstantatos, S. Zhang, P.W. Cyr, E.J.D. Klem, L. Levina and E.H. Sargent, *Nature Materials*, **4** (2005), p. 138.

- **January 2005:** *Fluorescent Quantum Dots to Label Plant Proteins*

  Fluorescent semiconducting nanoparticles have been employed for labeling plant proteins and this is the first time that they have been used for live imaging in plant systems.

  "Quantum Dots as Bio-Labels for the Localization of a Small Plant Adhesion Protein", Ravindran, S., S. Kim, R. Martin, E.M. Lord and C.S. Ozkan, *Nanotechnology*, **16** (2005), p. 1.

- **January 2005:** *Patterns to Measure Nanoscale Strain*

  For measuring the mechanical strain at sub-micron scale, nanoscale gold networks have been created on polymer surfaces. And a porous alumina stamp coated with gold has been used for making the patterns.

  "Development of Patterns for Nanoscale Strain Measurements: I. Fabrication of Imprinted Au Webs for Polymeric Materials", Collette, S.A., M.A. Sutton, P. Miney, A.P. Reynolds, X. Li, P.E. Colavita, W.A. Scrivens, Y. Luo, T. Sudarshan, P. Muzykov and M.L. Myrick, *Nanotechnology*, **15** (2004), p. 1812.

- **January 2005:** *Assembly of Nanowire Structures Controlled by Biomolecules*

  Protein molecules have been used for controlling the assembly of CdTe nanowire structures and this could have applications in making nanoscale electronic circuits in the future.

  "Biological Assembly of Nanocircuit Prototypes from Protein-Modified CdTe Nanowires", Wang, Y., Z. Tang, S. Tan and N.A. Kotov, *Nano Letters*, **5** (2005), p. 243.

- **January 2005:** *Semiconducting Nanoparticles to Form Aerogels*

  The semiconducting metal chalcogenide nanoparticles have been assembled into aerogels, which have the same optical properties as their constituent nanoparticles.

  "Porous Semiconductor Chalcogenide Aerogels", Mohanan, J.L., I.U. Arachchige and S.L. Brock, *Science*, **307** (2005), p. 397.

- **January 2005:** *Nanocontact AFM to Study Single-electron Effects*

  A nanocontact AFM with a spatial resolution of 50 nm has been used for studying single-electron effects in an individual InAs quantum dot. This method, called 'electrostatic force spectroscopy', doesn't require any leads attached to the system under study.

  "Detection of Single-Electron Charging in an Individual InAs Quantum Dot by Non-Contact Atomic-Force Microscopy", Stomp, R., Y. Miyahara, S. Schaer, Q. Sun, H. Guo, P. Grutter, S. Studenikin, P. Poole, and A. Sachrajda, *Phys. Rev. Lett.*, **94** (2005), p. 056802.

- **January 2005:** *Cytotoxicity of Colloidal Nanoparticles*

  An assay that quantifies the toxicity of colloidal CdSe and CdSe/ZnS nanoparticles in aqueous solution to cells has been developed.

  "Cytotoxicity of Colloidal CdSe and CdSe/ZnS Nanoparticles", Kirchner, C., T. Liedl, S. Kudera, T. Pellegrino, A.M. Javier, H.E. Gaub, S. Stölzle, N. Fertig and W.J. Parak, *Nano Letters*, **5** (2005), p. 331.

- **January 2005:** *Gold Nanoclusters Gets Charged up on a Ceramic Surface*

  Gold nanoclusters on a ceramic surface have been found to gain an electrical charge while acting as a catalyst for the low-temperature oxidation of carbon monoxide.

  "Charging Effects on Bonding and Catalyzed Oxidation of CO on Au_8 Clusters on MgO", Yoon, B., H. Häkkinen, U. Landman, A.S. Wörz, Jean-Marie Antonietti, S. Abbet, K. Judai and U. Heiz, *Science*, **307** (2005), p. 403.

- **January 2005:** *Controlled Deposition of CdSe Nanoparticles on Polymer Templates*

  CdSe nanoparticles have been placed selectively on a diblock copolymer template using electrophoretic deposition technique.

  "Controlled Placement of CdSe Nanoparticles in Diblock Copolymer Templates by Electrophoretic Deposition", Zhang, Q., T. Xu, D. Butterfield, M.J. Misner, D.Y. Ryu, T. Emrick and T.P. Russell, *Nano Letters*, **5** (2005), p. 357.

- **January 2005:** *Raman Laser with a Silicon Nanocrystal*

  Raman lasting on a single chip has been demonstrated. This is based on a low-loss single-mode rib waveguide containing a reverse-biased p-i-n diode structure fabricated on a single silicon chip. This presents an important milestone in the fabricated of optoelectronic devices and could be integrated with CMOS-compatible silicon chips.

  "An All-Silicon Raman Laser", Rong, H.S., A.S. Liu, R. Jones, O. Cohen, D. Hak, R. Nicolaescu, A. Fang and M. Paniccia, *Nature*, **433** (2005), p. 292.

- **January 2005:** *Nanoscale Motors*

  A working example of a Brownian motor is presented. Thermal Brownian motion in combination with non-equilibrium noise has been used to exercise control over the system with nanoscale accuracy. This provides an ideal pathway for fabricating and operating nanoscale devices.

  "Brownian Motors", Hanggi, P., F. Marchesoni and F. Nori, *Annalen Der Physik*, **14** (2005), p. 51.

- **January 2005:** *Gold Nanoantennas to Produce Confined and Enhanced Electric Fields*

  Gold bowtie antennas, fabricated through lithography, have been shown to confine optical fields to spatial scales far below the diffraction limit. This is of vital importance if understanding local electromagnetic enhancements and improved detection capabilities using surface enhances raman spectroscopy (SERS).

  "Improving the Mismatch Between Light and Nanoscale Objects with Gold Bowtie Nanoantennas", *Phys. Rev. Lett.*, **94** (2005), Art. No.017402.

- **January 2005:** *Carbon Nanotubes as Electrodes for Electrochemistry*

  Single walled carbon nanotubes have been used as nanoelectrodes for electrochemistry. These were contact by nanolithography and cyclic voltammetry was performed in aqueous solutions. This study demonstrates the potential of carbon nanotubes as nanoelectrodes.

  "Individual Single-Walled Carbon Nanotubes as Nanoelectrodes for Electrochemistry", Heller I., J. Kong, H.A. Heering, K.A. Williams, S.G. Lemay, and C. Dekker, *Nano Letters*, **5** (2005), p. 137.

- **January 2005:** *Transistors from Carbon Nanotubes*

  Isolated single walled carbon nanotubes are used to construct field-effect transistors of less than 20 nm dimensions. This material could find practical use in the field of nanoelectronics in regulating the current and working of devices.

  "Sub-20 nm Short Channel Carbon Nanotubes Transistors", Seidel, R.V., A.P. Graham, J. Kretz, B. Rajasekharan, G.S. Duesberg, M. Liebau, E. Unger, F. Kreupl and W. Hoenlein, *Nano Letters*, **5** (2005), p. 147.

- **January 2005:** *Aerosol based Approach for Synthesis of Carbon Nanotubes*

  A novel method to synthesize carbon nanotubes based on aerosols is demonstrated. Nanotubes of 0.6–2 nm diameter and 50 nm length was produced on a commercial scale using this method.

  "A Novel Aerosol Method for Single Walled Carbon Nanotubes Synthesis", Nasibulin, A.G., A. Moisala, D.P. Brown, H. Jiang and E.I. Kauppinen, *Chemical Physics Letters*, **402** (2005), p. 227.

- **January 2005:** *Assembling Metallic Nanowires*

  Metallic nanowires have been grown and self-assembled from solution by application of external alternating electric field. The wires showed high electrical conductivity and this array could find use in gas sensors and serve as building blocks for electronic systems.

  "Self-Assembly of Metallic Nanowires from Aqueous Solution", Cheng D., R.K. Gonela, Q. Gu and D.T. Haynie, *Nano Letters*, **5** (2005), p. 175.

- **January 2005:** *Nanorotors that Double up as Catalyst*

  Barcoded gold-nickel nanorods, anchored onto a silicon surface demonstrate rotational motion while catalysing the decomposition of hydrogen peroxide. This discovery presents an important milestone where the catalytic system doubles up as a movable rotor. This will lead to development of nanomechanical actuators and movable parts, useful for fabricating nanodevices.

  "Synthetic Self-Propelled Nanorotors", Fournier-Bodoz, S., A.C. Arsenault, I. Manner and G.A. Ozin, *Chemical Communications*, **4** (2005), p. 441.

- **January 2005:** *Clusters Exhibiting Near-infrared Luminescence*

  Monolayer-protected gold clusters have been shown to exhibit visible-near-infrared luminescence. The luminescence which could be tuned by controlling the core diameter and capping ligands presents interesting development in the field of optoelectronic devices.

  "Near-IR Luminescence of Monolayer-Protected Metal Clusters", Wang, G.L., T. Huang, R.W. Murray, L. Menard and R.G. Nuzzo, *J. Am. Chem. Soc.*, **127** (2005), p. 812.

- **February 2005:** *Nanoparticles to Detect Alzheimer's-related Proteins*

  The concentration of amyloid-β-diffusible ligands (ADDLs) in cerebrospinal fluid has been measured by using a nanoparticle-based bio-barcode assay. This technique could provide a method for the early dignosis of Alzheimer's disease, as ADDLs are likely markers for it.

  "Nanoparticle-Based Detection in Cerebral Spinal Fluid of a Soluble Pathogenic Biomarker for Alzheimer's Disease", Georganopoulou, D.G., L. Chang, Jwa-Min Nam, C.S. Thaxton, E.J. Mufson, W.L. Klein and C.A. Mirkin, *PNAS*, **102** (2005), p. 2273.

- **February 2005:** *Coaxial Nanotubes to Form Buckypaper*

  A high-yield technique for making double-walled carbon nanotubes has been developed, wherein their structures could have superior physical properties as compared to those of single- or multi-walled carbon nanotubes.

  "Nanotechnology: 'Buckypaper' from Coaxial Nanotubes", Endo, M., H. Muramatsu, T. Hayashi, Y.A. Kim, M. Terrones and M.S. Dresselhaus, *Nature*, **433** (2005), p. 476.

- **February 2005:** *End States Viewed in 1D Atom Chains*

  For the first time, end-states in 1D atom chains have been observed. This could have applications in nanoelectronics, as it would improve our understanding of electronic properties in one-dimensional structures.

  "End-States in One-Dimensional Atom Chains", Crain, J.N. and D.T. Pierce, *Science*, **307** (2005), p. 703.

- **February 2005:** *Nanobelts to Detect Nerve Agents*

  For detecting nerve agents, sensors have been made by combining tin oxide 'nanobelts' with low-power micro-heaters. These are ultrastable, highly sensitive and free from the 'Poisoning effect' of metal oxides as found in the previous sensors.

  "Integration of Metal Oxide Nanobelts with Microsystems for Nerve Agent Detection", Yu, C., Q. Hao, S. Saha, L. Shi, X. Kong and Z.L. Wang, *Appl. Phys. Lett.*, **86** (2005), p. 063101.

- **February 2005:** *Nanotube Formation Aided by Liquid Carbon*

  Liquid carbon has been found to be involved in the formation of multi-walled carbon nanotubes. These are believed to form by homogenous nucleation inside the liquid droplets.

  "Liquid Carbon, Carbon-Glass Beads, and the Crystallization of Carbon Nanotubes", de Heer, W.A., P. Poncharal, C. Berger, J. Gezo, Z. Song, J. Bettini and D. Ugarte, *Science*, **307** (2005), p. 907.

- **February 2005:** *Hydrogen Storage Capabilities Improved by Carbon Nanotubes*

  The hydrogen sorption properties of catalyzed sodium alanates were found to improve by the use of single-walled carbon nanotubes. The sorption kinetics of the material also improved by a factor of four by using these nanotubes.

  "The Catalytic Effect of Single-Wall Carbon Nanotubes on the Hydrogen Sorption Properties of Sodium Alanates", Dehouche, Z., L. Lafi, N. Grimard, J. Goyette and R. Chahine, *Nanotechnology*, **16** (2005), p. 402.

- **February 2005:** *Si Nanowire Sensors for Helping Drug Discovery*

  For detecting interactions between small molecules and proteins, a silicon nanowire sensor has been created and this could have applications in drug discovery.

  "Label-Free Detection of Small-Molecule–Protein Interactions by Using Nanowire Nanosensors", Wang, W.U., C. Chen, Keng-hui Lin, Y. Fang and C.M. Lieber, *PNAS*, **102** (2005), p. 3208.

- **February 2005:** *Nanotube Coating for Pyroelectric Detectors*

  The use of single-walled carbon nanotubes as a thermal-absorption coating for pyroelectric detectors has been investigated. This could have applications in measuring the optical power laser of systems.

  "Single-Wall Carbon Nanotube Coating on a Pyroelectric Detector", Lehman, J.H., C. Engtrakul, T. Gennett and A.C. Dillon, *Applied Optics*, **44** (2005), p. 483.

- **February 2005:** *Sonochemical Preparation of Hollow Nanospheres*

  Hollow nanospheres and hollow nanocrystals of molybdenum compounds have been prepared using ultrasound. These could have applications in catalysis, microelectronics and photonics.

  "Sonochemical Preparation of Hollow Nanospheres and Hollow Nanocrystals", Dhas, N.A. and K.S. Suslick, *J. Am. Chem. Soc.*, **127** (2005), p. 2368.

- **February 2005:** *Solar Cells based on Nanocluster Composites Exhibit Unprecedented Efficiency*

  Porphyrin modified gold nanoparticles are complexed with fullerene molecules and assembled into arrays on nanostructured SnO_2 substrate. This electrode exhibits an incident photo-to-photocurrent efficiency of 54% and an efficiency of 1.5%. These numbers are around 45 times higher than cells fabricated with porphyrins or fullerenes as the sole component.

  "Photovoltaic Cells Using Composite Nanoclusters of Porphyrins and Fullerenes with Gold Nanoparticles", Hasobe, T., H. Imahori, P.V. Kamat, T.K. Ahn, S.K. Kim, D. Kim, A. Fujimoto, T. Hirakawa and S. Fukuzumi, *J. Am. Chem. Soc.*, **127** (2005), p. 1216.

- **February 2005:** *Buckypaper from Carbon Nanotubes*

  A high-yield technique for making double-walled carbon nanotubes has been developed, wherein their structures could have superior physical properties as compared to those of single- or multi-walled carbon nanotubes.

  "Buckypaper from Coaxial Nanotubes", Endo, M., H. Muramatsu, T. Hayashi, Y.A. Kim, M. Terrones and N.S. Dresselhaus, *Nature*, **433** (2005), p. 476.

- **February 2005:** *Manipulating and Working with Atoms*

  Highly controlled manipulations of atoms at room temperature have been achieved using atomic force microscopy. This is of importance for fabrication of nanoscale devices with atomic precision.

  "Atom Inlays Performed at Room Temperature Using Atomic Force Microscopy", Sugimoto, Y., M. Abe, S. Hirayam, N. Oyabu, O. Custance and S. Morita, *Nature Materials*, **4** (2005), p. 156.

- **February 2005:** *Nanocrystal Emitting Blue Laser*

  Semiconductor nanocrystals emitting blue laser have been developed. Core-shell systems of CdS/ZnS stabilized in sol-derived silica matrix show lasing properties at room temperature. This could further research for developing tunable emission from such systems.

"Blue Semiconductor Nanocrystal Laser", Chan, Y., J.S. Steckel, P.T. Snee, J.M. Cargue, J.M. Hodgkiss, D.G. Nocera, and M.G. Bawendi, *Appl. Phys. Lett.*, **86** (2005), Art. No. 073102.

- **February 2005:** *Nanoparticle Coated Self-cleaning Surface*

  Coating silica and titania nanoparticles on glass substrates transformed them into self-cleaning surface with good anti-reflective properties. The commercialization of this technology would have a huge impact in semiconductor and coating industry.

  "Self-cleaning Particle Coating with Antireflective Properties", Zhang, X.T., O. Sato, M. Tauchi, Y. Einaga, T. Murakami and A. Fujishima, *Chemistry of Materials*, **17** (2005), p. 696.

- **March 2005:** *Bimetallic Nanoparticle Catalysts to Clean up Trichloroethene*

  The palladium-coated gold nanoparticles have been found to be extremely effective catalysts for breaking down trichloroethene into less harmful products, which has serious implications for human health and environment.

  "Designing Pd-on-Au Bimetallic Nanoparticle Catalysts for Trichloroethene Hydrodechlorination", Nutt, M.O., J.B. Hughes and M.S. Wong, *Environ. Sci. Technol.*, **39** (2005), p. 1346.

- **March 2005:** *Nanoparticle/Copolymer Mixtures to Direct their Self-assembly*

  By adding nanoparticles to diblock copolymers, it has been found that they can redirect their self-assembly. This technique could find applications for chemical sensing, separation, catalysis, etc.

  "Self-Directed Self-Assembly of Nanoparticle/Copolymer Mixtures", Lin, Y., A. Böker, J. He, K. Sill, H. Xiang, C. Abetz, X. Li, J. Wang, T. Emrick, S. Long, Q. Wang, A. Balazs and T.P. Russell, *Nature*, **434** (2005), p. 55.

- **March 2005:** *Molecular Switch Shows Negative Differential Resistance*

  A single molecule attached to two electrodes has been found to show a negative differential resistance. This may be a valuable technique for future research in single-molecule electronics.

  "A Molecular Switch Based on Potential-Induced Changes of Oxidation State", Chen, F., J. He, C. Nuckolls, T. Roberts, J.E. Klare and S. Lindsay, *NanoLetters*, **5** (2005), p. 503.

- **March 2005:** *CdSe Nanoparticle-based Active Tips for NSOM*

  For a near-field scanning optical microscopy, an active optical tip has been created with just a few CdSe nanoparticles at its apex. This tip may even be believed to contain just one nanoparticle.

  "CdSe Single-Nanoparticle-Based Active Tips for Near-Field Optical Microscopy", Chevalier, N., M.J. Nasse, J.C. Woehl, P. Reiss, J. Bleuse, F. Chandezon and S Huant, *Nanotechnology*, **16** (2005), p. 613.

- **March 2005:** *Controlled Evolution of Polymer Crystals on Mica*

  An AFM coated with a polymer has been used for growing crystals of the polymer on a mica substrate. This new technique is capable of starting crystallization from scratch, then controlling and imaging the process as it proceeds in real-time.

  "The Controlled Evolution of a Polymer Single Crystal", Liu, X., Y. Zhang, D.K. Goswami, J.S. Okasinski, K. Salaita, P. Sun, M.J. Bedzyk and C.A. Mirkin, *Science*, **307** (2005), p. 1763.

- **March 2005:** *Bacteria Trapped at Electrode Junctions*

Electric fields have been used to manipulate bacteria into a gap between two electrodes. By measuring changes in the electrical performance of the device, it was possible to detect that the bacteria were in position.

"Manipulation and Real-Time Electrical Detection of Individual Bacterial Cells at Electrode Junctions: A Model for Assembly of Nanoscale Biosystems", Beck, J.D., L. Shang, M.S. Marcus and R.J. Hamers, *Nano Letters*, **5** (2005), p. 777.

- **March 2005:** *Surface Tension to Drive Nanomotors*

A nanoelectromechanical device known as 'relaxation oscillator', which exploits the effects of surface tension, has been made for the first time. This device consists of two droplets of liquid metal on a substrate made of carbon nanotubes, which can be controlled with a small-applied electric field.

"Surface-Tension-Driven Nanoelectromechanical Relaxation Oscillator", Regan, B.C., S. Aloni, K. Jensen and A. Zettl, *Appl. Phys. Lett.*, **86** (2005), p. 123119.

- **March 2005:** *Polymer-coated Au Nanoparticles for Heat Transport*

In the case of polymer-coated gold nanoparticles, it was found that on adding a solvent, the polymer coating swelled leading to a sudden rise in the thermal conductivity of the layer.

"Thermal Transport in Au-Core Polymer-Shell Nanoparticles", Ge, Z., Y. Kang, T.A. Taton, P.V. Braun and D.G. Cahill, *Nano Letters*, **5** (2005), p. 531.

- **March 2005:** *Nanotube Capacitor to Detect Vapors*

Chemical vapors have been detected by using the capacitance changes occurring in a network of single-walled carbon nanotubes. This technique is sensitive to a broad range of gases and also emits a fast response.

"Chemical Detection with a Single-Walled Carbon Nanotube Capacitor", Snow, E.S., F.K. Perkins, E.J. Houser, S.C. Badescu and T.L. Reinecke, *Science*, **307** (2005), p. 1942.

- **March 2005:** *Drawing Sheets of Carbon Nanotubes*

Polymerization of caprolactum in presence of single-walled carbon nanotubes leads to the formation of composite with optimized morphology. This composite could be drawn into sheets with improved mechanical properties, for use in optoelectronics.

"Continuous Spinning of a Single Walled Carbon Nanotube-Nylon Composite Fibre", Gao, B., M.E. Itkis, A.P. Yu, Bekyarova, B. Zhao and R.C. Haddon, *J. Am. Chem. Soc.*, **127** (2005), p. 3847.

- **March 2005:** *Ordered Silicon Nanowire Arrays*

Gold nanoparticles are used as seeds to grow highly ordered silicon nanowire arrays, with the growth taking place perpendicular to the substrate. This materials will find applications in nanoelectronics and easily integrate into nanoscale devices.

"Controlled Growth of Si Nanowire Arrays for Device Integration", Hochbaum, A.I., R. Fan, R.R. He, P.D. Yang, *Nano Letters*, **5** (2005), p. 457.

- **March 2005:** *Amino Acid Detector for MARS Mission*

  A microfabricated capillary electrophoresis instrument for sensitive amino acid biomarker analysis is developed. This has been tested for working in environments similar to that prevailing in Mars and afforded ppb levels of detection.

  "Development and Evaluation of a Microdevice for Amino Acid Biomarker Detection and Analysis on Mars", Skelley, A.M., J.R. Scherer, A.D. Aubrey, W.H. Grover, R.H.C. Ivester, P. Ehrenfreund, F.J. Grunthaner, J.L. Bada and R.A. Mathies, *PNAS*, **102** (2005), p. 1041.

- **March 2005:** *Ultrasensitive Detection of Anthrax Biomarker*

  SERS has been applied for detecting anthrax biomarker up to 10^3 spores, which is less than the critical limit of 10^4 spores. This technique has been successfully transitioned to a field-portable instrument for widespread in anthrax detection.

  "Rapid Detection of an Anthrax Biomarker by Surface-Enhanced Raman Spectroscopy", Zhang, X.Y., M.A. Yong, O. Lyandres and R.P. van, Duyne, *J. Am. Chem. Soc.*, **127** (2005), p. 4484.

- **March 2005:** *Demystifying SERS*

  SERS presents one of the most important tool that provides vibrational information, with a limiting accuracy of a single molecule. The role of nanoparticle surface charge in bringing about SERS has been analysed in great detail. The information presented in critical for choosing and designing substrates for optimized SERS activity.

  "Role of Nanoparticle Surface Charge in Surface-Enhanced Raman Scattering", Alvarez-Pueble R.A., E. Arceo, P.J. G. Goulet, J.J. Garrido and R.F. Aroca, *J. Phys. Chem.*, **109** (2005), p. 3787.

- **March 2005:** *Nanoparticle based Arsenic Removal from Water*

  Iron nanoparticle have been shown to extract and remove arsenic ions from water sources. The removal of highly poisonous arsenic ions from water sources presents a challenge in developing countries and this presents them with a unique method for both *in-situ* and *ex-situ* water purification.

  "Removal of Arsenic (III) from Groundwater by Nanoscale Zero-Valent Iron", Kanel, S.R., B. Manning, L. Chralet, H. Choi, *Environmental Science and Technology*, **39** (2005), p. 1291.

- **March 2005:** *Single Walled Carbon Nanotube for Chemical Detection*

  Single walled carbon nanotubes has been used for fabricating chemicapacitors. These provide highly sensitive detection of vapours and can be used repeatedly.

  "Chemical Detection with a Single-walled Carbon Nanotube Capacitor", Snow, E.S., F.K. Perkins, E.J. Houser, S.C. Badescu, T.L. Reinecke, *Science*, **307** (2005), p. 1942.

- **April 2005:** *Magnetic Nanoparticles Fills Up Nanotubes*

  Carbon nanotubes have been filled up with magnetic nanoparticles. The resulting magnetic nanostructures could have applications in memory devices, medicine and wearable electronics.

  "Carbon Nanotubes Loaded with Magnetic Particles", Korneva, G., H. Ye, Y. Gogotsi, D. Halverson, G. Friedman, Jean-Claude Bradley and K.G. Kornev, *Nano Letters*, **5** (2005), p. 879.

- **April 2005:** *Au Nanoparticles to Monitor Protein Conformational Changes*

  The changes in the folding of a yeast protein were monitored by attaching gold nanoparticles to it. This could be used for sensing faulty protein folding, which plays a role in medical conditions such as Alzheimer's disease, cystic fibrosis and mad cow disease (BSE).

  "Gold Nanoparticles as a Colorimetric Sensor for Protein Conformational Changes", Chah, S., M.R. Hammond and R.N. Zare, *Chemistry and Biology*, **12** (2005), p. 323.

- **April 2005:** *Virus Infection Monitored by Fluorescent Quantum Dots*

  Fluorescent semiconductor nanoparticles have been used for labeling the respiratory syncytial virus (RSV), which shows the increased role of quantum dots in medical imaging.

  "Progression of Respiratory Syncytial Virus Infection Monitored by Fluorescent Quantum Dot Probes", Bentzen, E.L., F. House, T.J. Utley, J.E. Crowe, Jr. and D.W. Wright, *Nano Letters*, **5** (2005), p. 591.

- **April 2005:** *Controlled Formation of Helical Nanobelts*

  Controlling the formation of helical nanobelts from strips of semiconductor has been worked out. It has been found that the design of the spirals could be tailored by altering parameters such as the width of the belt, its crystal direction and the shape of its tip.

  "Controllable Fabrication of SiGe/Si and SiGe/Si/Cr Helical Nanobelts", Zhang, L., E. Deckhardt, A. Weber, C. Schönenberger and D. Grützmacher, *Nanotechnology*, **16** (2005), p. 655.

- **April 2005:** *Fungus Produces Silica Nanoparticles by Bioleaching of Sand*

  The fungus fusarium oxysporum that normally causes disease in plants has been found to create nanoparticles of silica from sand through bioleaching.

  "Bioleaching of Sand by the Fungus *Fusarium oxysporum* as a Means of Producing Extracellular Silica Nanoparticles", Bansal, V., A. Sanyal, D. Rautaray, A. Ahmad and M. Sastry, *Advanced Materials*, **17** (2005), p. 889.

- **April 2005:** *Bent Carbon Nanotubes Grown*

  Carbon nanotubes which change direction along their length have been grown; electric fields have been used to alter the orientation of the tubes.

  "Control of Carbon Capping for Regrowth of Aligned Carbon Nanotubes", AuBuchon, J.F., Li-Han Chen and S. Jin, *J. Phys. Chem. B*, **109** (2005), p. 6044.

- **April 2005:** *Nanotube Transistor Interacts with Biological System*

  The interaction of a nanoelectronic device with an intact biological system has been demonstrated for the first time.

  "Transparent and Flexible Carbon Nanotube Transistors", Artukovic, E., M. Kaempgen, D.S. Hecht, S. Roth and G. Grüner, *Nano Letters*, **5** (2005), p. 757.

- **April 2005:** *Multi-walled Carbon Nanotubes Aligned by Magnetic Bacteria*

  Laterally aligned multi-walled carbon nanotubes have been grown by using magnetic nanoparticles obtained from the bacterium Magnetospirillium magnetotacticum (MS-1).

"Laterally Aligned, Multi-Walled Carbon Nanotube Growth Using Magnetospirillium Magnetotacticum", Kumar, N., W. Curtis and Jong-in Hahm, *Appl. Phys. Lett.*, **86**, p. 173101.

- **April 2005:** *High-speed Nanowire Circuits*

Using nanoscale-building blocks, high-speed integrated nanowire circuits have been made for the first time and these could find applications in lightweight, portable electronics.

"Nanotechnology: High-Speed Integrated Nanowire Circuits", Friedman, R.S., M.C. McAlpine, D.S. Ricketts, D. Ham and C.M. Lieber, *Nature*, **434** (2005), p. 1085.

- **April 2005:** *Diffraction Limit Breached with Silver Superlens*

A silver superlens, which effectively collects evanescent waves, though excitation of surface plasmons has been designed and fabricated. This can be used for imaging structures with a resolution of one-sixth of the wavelength of light, thereby overcoming the diffraction limit. This provides newer capabilities for imaging nanoscale materials.

"Sub-Diffraction-Limited Optical Imaging with Silver Superlens", Fang, N., H. Lee, C. Sun and X. Zhang, *Science*, **308** (2005), p. 534.

- **April 2005:** *Nanoshells for Imaging and Treating Cancer*

Nanoshells, having absorption in the near-infrared region has been used for imaging cancerous cells. Immunotargeted nanoshells allow selective detection and destruction of breast carcinoma cells through photothermal therapy.

"Immunotargeted Nanoshells for Integrated Cancer Imaging and Therapy", Loo, C., A. Lowrey, N. Halas, J. West and R. Drezek, *Nano Letters*, **5** (2005), p. 709.

- **April 2005:** *Generating Triangular Silver Nanoplates on Surface*

Highly ordered triangular silver nanoplates have been generated on a planar surface using a seed-mediated approach. This resulting substrate is a potential candidate for studying metal-enhanced fluorescence and SERS activity.

"Rapid Deposition of Triangular Silver Nanoplates on Planar Surfaces: Application to Metal-Enhanced Fluorescence", Aslan, K., J.R. Lakowicz and C.D. Geddes, *J. Phys. Chem. B*, **109** (2005), p. 6247.

- **April 2005:** *Nanoparticles as Enhancers for Magnetic Resonance Imaging*

Ultrasmall iron oxide nanoparticles have been used as agents to enhance contrast at lymph nodes. Clinical tests have been conducted to successfully image and detect lymph node metastases.

"Diagnostic Performance of Nanoparticle-enhanced Magnetic Resonance Imaging in the Diagnosis of Lymph Node Metastases in Patients with Endometrial and Cervical Cancer", Rockall, A.G., S.A. Sohaib, M.G. Harisinghani, S.A. Babar, N. Singh, A.R. Jeyarajah, D.H. Oram, J.I. Jacobs, J.H. Shpeherd and R.H. Reznek, *Journal of Clinical Oncology*, **23** (2005), p. 2813.

- **April 2005:** *Silver Superlens for Optical Imaging at Sub-diffraction Limits*

An optical superlens having a negative refractive index has been made from a thin layer of silver. This can be used for imaging structures with a resolution of about one-sixth of the wavelength of light, thus overcoming the diffraction limit. This could find applications in imaging nanoscale objects with light.

"Sub-diffraction-limited Optical Imaging with a Silver Superlens", Fang, N., H. Lee, C. Sun and X. Zhang, *Science*, **308** (2005), p. 534.

- **May 2005:** *Biological Systems to be Imaged at Sub-10 nm Resolution*

    By exploiting the piezoelectric effect, the electromechanical imaging of the internal structure of human teeth has been done. This technique called 'piezoresponse force microscopy', could find applications for imaging in a wide range of biomaterials at sub-10 nm resolution.

    "Electromechanical Imaging of Biological Systems with Sub-10 nm Resolution", Kalinin, S.V., B.J. Rodriguez, S. Jesse, T. Thundat and A. Gruverman, *Condensed Matter*, **0504** (2005), p. 0504232.

- **May 2005:** *Nanocrystals for Detecting Genetic Mutations*

    A technique that uses nanocrystals for detecting genetic mutations has been developed, which involves the creation of a bioelectronic coding for point mutations known as single nucleotide polymorphisms (SNPs).

    "Nanocrystal-Based Bioelectronic Coding of Single Nucleotide Polymorphisms", Liu, G., T.M.H. Lee and J. Wang, *J. Am. Chem. Soc.*, **127** (2005), p. 38.

- **May 2005:** *AFM Working in Liquid for Imaging*

    An atomic force microscope that works in liquid has been made which could be used for imaging biological samples, easily oxidizable materials and samples in hazardous environments.

    "Atomic Force Microscope in Liquid with a Specially Designed Probe for Practical Application", Zhang, D., H. Zhang and X. Lin, *Rev. Sci. Instrum.*, **76** (2005), p. 053705.

- **May 2005:** *Attaching Amino Acids to Inorganic Surfaces*

    The adhesion of amino acids to semiconductors, metals and insulators used in electronic devices has been tested. The results have been used for designing an inorganic nanostructure which selectively binds to a specific primary peptide sequence.

    "Differential Adhesion of Amino Acids to Inorganic Surfaces", Willett, R.L., K.W. Baldwin, K.W. West and L.N. Pfeiffer, *PNAS*, **102** (2005), p. 7817.

- **May 2005:** *Conductive Cantilevers for AFM*

    An atomic force microscope with an electrically insulated conductive tip has been tested. The hexagonally packed intermediate layer of the red bacterium *Deinococcus radiodurans* has been imaged.

    "Assessment of Insulated Conductive Cantilevers for Biology and Electrochemistry", Frederix, P.L.T.M., M.R. Gullo, T. Akiyama, A. Tonin, N.F. de Rooij, U. Staufer and A. Engel, *Nanotechnology*, **16** (2005), p. 997.

- **May 2005:** *LED Chips Based on Semiconductor Nanocrystals*

    By replacing external color-converting phosphors with CdSe-based nanocrystals that are incorporated into GaN charge injection layers, more efficient multi-color light-emitting diodes can be made.

"Multi-Color Light-Emitting Diodes Based on Semiconductor Nanocrystals Encapsulated in GaN Charge Injection Layers", Mueller, A.H., M.A. Petruska, M. Achermann, D.J. Werder, E.A. Akhadov, D.D. Koleske, M.A. Hoffbauer and V.I. Klimov, *Nano Letters*, **5** (2005), p. 1039.

- **May 2005:** *Nanotube Substrates Improves Neuronal Electrical Signaling*

Nerve cells from the hippocampus region of the brain have been grown on the substrates containing networks of carbon nanotubes. The neural electrical signaling between cells has been found to improve through the use of these nanotube substrates.

"Carbon Nanotube Substrates Boost Neuronal Electrical Signaling", Lovat, V., D. Pantarotto, L. Lagostena, B. Cacciari, M. Grandolfo, M. Righi, G. Spalluto, M. Prato and L. Ballerini, *Nano Letters*, **5** (2005), p. 1107.

- **June 2005:** *Supramolecular Nanostamping Using DNA*

Mass production of nanodevices could be facilitated by the use of the supramolecular nanostamping printing technique. This method has a resolution of less than 40 nm and uses DNA hybridization for replicating a pattern.

"Supramolecular Nanostamping: Using DNA as a Movable Type", Yu, A.A., T.A. Savas, G.S. Taylor, A. Guiseppe-Elie, H.I. Smith and F. Stellacci, *Nano Letters*, **5** (2005), p. 1061.

- **June 2005:** *Nanotubes Obtained from Grass on Heating*

Multi-walled carbon nanotubes having diameters of 30–50 nm have been prepared by heating grass in the presence of oxygen.

"Obtaining Carbon Nanotubes from Grass", Kang, Z., E. Wang, B. Mao, Z. Su, L. Chen and L. Xu, *Nanotechnology*, **16** (2005), p. 1192.

- **June 2005:** *Functionalized Gold Nanoparticles Mimic Enzyme*

Functionalized gold nanoparticles that can catalyze the formation of silica at low temperatures and neutral pH have been synthesized. This has been achieved by mimicking the polysiloxane-synthesizing enzyme found in the orange puffball sponge Tethya aurantia living in seawater.

"Functionalized Gold Nanoparticles Mimic Catalytic Activity of a Polysiloxane-Synthesizing Enzyme", Kisailus, D., M. Najarian, J.C. Weaver and D.E. Morse, *Advanced Materials*, **17** (2005), p. 1234.

- **June 2005:** *A Single Charged Atom for Field Regulation on a Si Surface*

A single charged atom on a silicon surface has been shown to regulate the conductivity of a molecule nearby for the first time. This could have applications in the development of single molecule transistors.

"Field Regulation of Single-Molecule Conductivity by a Charged Surface Atom", Piva, P.G., G.A. DiLabio, J.L. Pitters, J. Zikovsky, M. Rezeq, S. Dogel, W.A. Hofer and R.A. Wolkow, *Nature*, **435** (2005), p. 658.

- **June 2005:** *Block Copolymers to Form Non-regular Structures for Nanodevices*

A hybrid method for patterning silicon that exploits a mixture of polymers has been designed. This involves the directed assembly of block copolymer blending into non-regular structures for nanoelectronic devices.

"Directed Assembly of Block Copolymer Blends into Nonregular Device-Oriented Structures", Stoykovich, M.P., M. Müller, S.O. Kim, H.H. Solak, E.W. Edwards, J.J. de Pablo and P.F. Nealey, *Science*, **308** (2005), p. 1442.

- **June 2005:** *C_{60} Aggregates in Water for Restricting Bacteria's Growth*

  When C_{60} molecules are exposed to water, the resulting nano-C_{60} aggregates have been found to show a detrimental microbial response. The bacteria's growth and respiration rates have been restricted by these aggregates under certain conditions.

  "C_{60} in Water: Nanocrystal Formation and Microbial Response", Fortner, J.D., D.Y. Lyon, C.M. Sayes, A.M. Boyd, J.C. Falkner, E.M. Hotze, L.B. Alemany, Y.J. Tao, W. Guo, K.D. Ausman, V.L. Colvin and J.B. Hughes, *Environ. Sci. Technol.*, **39** (2005), p. 4307.

- **June 2005:** *ZnO Nanowires for Light-harvesting in Solar Cells*

  Conventional dye sensitized solar cells, modified with crystalline ZnO nanowires have been fabricated. The highly oriented ZnO nanowires provides a greater surface area for light harvesting, using which a light-to-energy conversion efficiency of 1.5% has been demonstrated.

  "Nanowire Dye-Sensitized Solar Cells", Law, M., L.E. Greene, J.C. Johnson, R. Saykally and P.D. Yang, *Nature Materials*, **4** (2005), p. 455.

- **June 2005:** *Measuring Single Molecule Conductivity*

  Results from STM supported by quantum mechanical modeling point to the fact that the electrostatic field emanating from a fixed point charge regulates the conductivity of the nearby molecules. These effects are observable at room temperature and provide vital information for the fabrication of nanoscale devices.

  "Field Regulating of Single-molecule Conductivity by a Charged Surface Atom", Piva, P.G., G.A. DiLabio, J.L. Pitters, J. Zikovsky, M. Rezeq, S. Dogel, W.A. Hofer and R.A. Wolkow, *Nature*, **435** (2005), p. 658.

- **June 2005:** *Demystifying Kinesin Movement*

  Movement of Kinesin and dynein inside a cell has been imaged using fluorescence imaging with one nanometer accuracy. This technique could provide important information about the movement of various organelle and could lead to discovery of newer signaling pathways inside cells.

  "Kinesin and Dynein Move a Peroxiosome in Vivo: A Tug-of-War or Coordinated Movement?", Kural, C., H. Kim, S. Syed, G. Goshima, V.I. Gelfand and P.R. Selvin, *Science*, **308** (2005), p. 1469.

- **June 2005:** *Nanoparticles as Gene-delivery Vehicles*

  Ploy(ethylene glycol) modified gelatin nanoparticles have been tested as gene delivery vehicles for tumor specific cells employing both in-vitro and in-vivo studies. Such a biocompatible system is highly desirable for targeted and controlled delivery of genetic constructs to solid tumors.

  "Tumor-Targeted Gene Delivery Using Poly(Ethylene Glycol)-Modified Gelatin Nanoparticles: In-Vitro and In-Vivo Studies", Kaul, G. M. Amiji, *Pharmaceutical Research*, **22** (2005), p. 951.

- **July 2005:** *Light Driven Nanovalve*

  A nanoscale valve that can be open and shut by shining light has been developed. Based on a channel protein, opens on shining light of wavelength of 266 nm. The pore size is as large as 3 nm. This represents significant progress is designing drug delivery vehicles for automated and targeted drug delivery.

  "A Light-Actuated Nanovalve Derived from a Channel Protein", Kocer, A., M. Walko, W. Meijberg and B.L. Feringa, *Science*, **309** (2005), p. 755.

- **July 2005:** *Reversible Molecular Valve*

  Self-assembly of organic molecules has been used to develop a nanovalve that opens and closes based on the presence or absence of redox species. This is perfect example of self-assembly driven molecular machines and is a major step towards the realization of nanoscale machines.

  "A Reversible Molecular Valve", Nguyen, T.D., H.R. Tseng, P.C. Celestre, A.H. Flood, Y. Liu, J.F. Stoddart and J.I. Zink, *PNAS*, **102** (2005), p. 10029.

- **July 2005:** *Separating Metallic and Semiconducting Carbon Nanotubes*

  A wet-chemical approach for separating metallic and semiconducting carbon nanotubes on a large scale has been achieved. This presents an important development towards usage and fabrication of carbon nanotubes in nanodevices, where the electrical transport has to be predetermined.

  "Large-Scale Separation of Metallic and Semiconducting Single-Walled Carbon Nanotubes", Maeda, Y., S. Kimura, M. Kanda, Y. Hirashima, T. Hasegawa, T. Wakahara, Y.F. Lian, T. Nakahodo, T. Tsuchiya, T. Akasaka, J. Lu, X.W. Zhang, Z.X. Gao, Y.P. Yu, S. Nagase, S. Kazaoui, N. Minami, T. Shizmizu, H. Tokumoto and R. Saito, *J. Am. Chem. Soc.*, **127** (2005), p. 10287.

- **August 2005:** *Carbon Nanotubes Kill Cancer*

  The transporting capabilities of carbon nanotubes, combined with suitable functionalization chemistry and their intrinsic NIR optical absorption properties, can be used for destroying cancer cells.

  "Carbon Nanotubes as Multi-Functional Biological Transporters and Near-Infrared Agents for Selective Cancer Cell Destruction", Kam, N.W.S., M. O'Connell, J.A. Wisdon and H.J. Dai, *PNAS*, **102** (2005), p. 11600.

- **August 2005:** *Nanowires to Study Activities in Brain*

  Silicon nanowires have been used to connect to neurons, based on their size-scale match. The nanowires, connected to axons and dendrons, could prove useful for studying the various signaling pathways occurring inside the brain and could one day lead to the development of newer treatments for brain-related disorders.

  "Detection, Stimulation, and Inhibition of Neuronal Signals with High-Density Nanowire Transistor Arrays", Fernando, P., P.T. Brian, Y. Guihua, F. Ying, B.G. Andrew, Z. Gengfeng and M.L. Charles, *Science*, **313** (2005), p. 1100.

- **August 2005:** *Nanoscale Materials to Prevent Formation of Blood Clots in Arteries*

    Paramagnetic nanoparticles functionalized with fumagillin have been used for delivering drug molecules directly at the site of plaque formation in arteries. This has been images using magnetic resonance imaging and the amount and rate of drug release has been calculated.

    "Endothelial Integrin Targeted Fumagillin Nanoparticles Inhibit Angiogenesis in Atherosclerosis", Patrick, M.W., M.N. Anne, D.C. Shelton, D.H. Thomas, J.D. Robertson, A.W. Todd, H.S. Anne, H. Grace, J.S. Allen, E.K. Lacy, H. Zhang, S.A. Wickline and G.M. Lanza, *Arteriosclerosis, Thrombosis and Vascular Biology*, **26** (2005), p. 2103.

- **August 2005:** *Amazing Ductility of Carbon Nanotube Composites*

    Multiwalled carbon nanotubes have been drawn into sheets several meters long. This discovery provides an avenue for immediate incorporation of nanotubes into light-emitting diodes, electronic displays and conductive substrates.

    "Strong, Transparent, Multifunctional Carbon Nanotube Sheets", Zhang, M., S. Fang, A.A., Zakhidov, S.B. Lee, A.E. Aliev, C.D. Williams, K.R. Atkinson and R.H. Baughman, *Science*, **19** (2005), p. 1215.

- **August 2005:** *Fabrication of Perfect Diode with Carbon Nanotube*

    Carbon nanotubes show perfect diode behaviour with an 'ideality factor' of one, which has been impossible to achieve in semiconductors so far. This points to interesting applications in nanoelectronics and logic circuitry.

    "Photovoltaic Effect in Ideal Carbon Nanotube Diodes", J.U. Lee, *Appl. Phys. Lett.*, **87** (2005), p. 073101.

- **September 2005:** *Y-shaped Carbon Nanotubes Exhibiting Switching Behaviour*

    Y-shaped carbon nanotubes exhibit switching behaviour even without application of an external gate voltage. This means that the current flow could be controlled along the desired pathways, which could prove useful for fabrication and designing nanoscale transistors and logic gates.

    "Novel Electrical Switching Behaviour and Logic in Carbon Nanotube Y-Junctions", Bandaru, P.R., C. Daraio, S. Jin and A.M. Rao, *Nature Materials*, **4** (2005), p. 663.

- **September 2005:** *Nanomotor to Move Objects at Macroscopic Scale*

    A synthetic nanoscale motor has been fabricated to move and transport objects at nanoscale. This motor converts light energy into biased Brownian motion of stimulireponsive rotaxanes and leads to modification of surface tension.

    "Macroscopic Transport by Synthetic Molecular Machines", Berna, J., D.A. Leigh, M. Lubomska, S.M. Mendoza, E.M. Perez, P. Rudolf, G. Teobaldi and F. Zerbetto, *Nature Materials*, **4** (2005), p. 704.

- **September 2005:** *Protein Clusters as Logic Gates*

    Clusters of proteins are potential candidates for performing complex logic operations on a binary scale. Their distinct 'on' and 'off' states makes them ideal candidates. Though postulated on a theoretical basis, researchers are confident of realizing it experimentally.

    "The Logical Repertoire of Ligand-Binding Proteins", Graham, I. and T. Duke, *Physical Biology*, **2** (2005), p. 159.

- **September 2005:** *Cheaper Route to Quantum Dots*

  Synthesis of quantum dots has been achieved at an economically viable industrial scale. The technique which removes the need for costly stabilizing agents and solvents, reduces the cost by 80%. This is crucial for commercialization of the existing technology.

  "The Use of Heat Transfer Fluids in the Synthesis of High-Quality CdSe Quantum Dots, Core-Shell Quantum Dots and Quantum Rods", Asokan, S., K.M. Krueger, A. Alkhawaldeh, A.R. Carreon, Z. Mu, V.L. Colvin, N.V. Mantzaris, and M.S. Wong, *Nanotechnology*, **16** (2005), p. 2000.

- **September 2005:** *Nanoscale Synchronization*

  Two spatially separated nanoscale oscillators (emitting in microwave region) were made to synchronize be carefully tuning their electric field. It was observed that such coordination between the oscillators results in significant narrowing of the emission width and increase in the power. This discovery would result in significant size reduction in transmitters and receivers for wireless communication.

  "Mutual Phase-Locking of Microwave Spin Torque Nano-Oscillators", Kaka, S., M.R. Pufall, W.H. Rippard, T.J. Silva, S.E. Russek and J.A. Katine, *Nature*, **437** (2005), p. 389.

  "Phase-Locking in Double-Point-Contact Spin-Transfer Devices", Mancoff, F.B., N.D. Rizzo, B.N. Engel and S. Tehrani, *Nature*, **437** (2005), p. 393.

- **October 2005:** *Nanoparticle Flow Sensors*

  The flow of liquids has been sensed with an assembly of gold nanoparticles. A flow rate of 500 cc/h on a nanoparticle assembly of a cm^2 area produces a potential of the order of 10–20 millivolts. These kinds of sensors may be used for biological applications.

  "Flow-Induced Transverse Electrical Potential Across an Assembly of Gold Nanoparticles", Subramaniam, C., T. Pradeep and J. Chakrabarti, *Phys. Rev. Lett.*, **95** (2005), p. 164501.

- **October 2005:** *Superconductivity in Graphitic Systems*

  Intercalating metal atoms in graphene sheets makes them superconducting. This discovery could lead to a better understanding of mechanism of superconductivity. This should also trigger further research for high T_c superconductor with carbon nanotubes.

  "Superconductivity in Intercalated Graphite Compounds C_6Yb and C_6Ca", (2005), Weller, T.E., M. Ellerby, S.S. Saxena, R.P. Smith, N.T. Skipper, *Nature Physics*, **1** (2005), p. 39.

- **October 2005:** *Molecular Nanomotor Driven by Chemical Energy*

  A unidirectional molecular motor driven by chemical energy is fabricated. This motor derives its energy from chemical conversions.

  "A Reversible, Unidirectional Molecular Rotary Motor Driven by Chemical Energy", Lectcher, S.P., F. Dumur, M.M. Pollard and B.L. Feringa, *Science*, **310** (2005), p. 80.

- **October 2005:** *Unidirectional Nanomotor on Gold Surface*

  A nanomotor which operates in one direction has been fabricated on the surface of a gold nanoparticle. This motor is capable of turning full 360 degrees with respect to the nanoparticle surface and works on the principle of photo induced cis-trans isomerisation of the molecule. This kind of nanomotors are ideal for use in artificial muscles.

"Unidirectional Molecular Motor on Gold Surface", van Delden, R.A., M.K.J., ter Wiel, M.M. Polard, J. Vacario, N. Koumura and B.L. Feringa, *Nature*, **437** (2005), p. 1337.

- **October 2005:** *Semiconducting Nanowires Show Lasing Action*

   GaN nanowires grown by chemical vapour deposition from their respective organic precursors have been shown to demonstrate lasing action at low thresholds. This phenomenon, observed at room temperature could lead to development of optoelectronic devices.

   "GaN Nanowire Lasers with Low Lasing Thresholds", Gradecak, S., F. Qian, Y. Li, H.G. Park and C.M. Lieber, *Appl. Phys. Lett.*, **87** (2005), Art. No. 173111.

- **October 2005:** *Nanowires have Memory!*

   Theoretical stimulations have shown a psedoelastic behaviour in copper nanowires. High surface stress is thought to bring about crystallographic reorientations, which are reversible to the extent of 50%. The experimental realization of this will provide a vital breakthrough in non-invasive medical diagnosis and treatment.

   "Shape Memory Effect in Cu Nanowires", Liang, W.W., M. Zhou and F.J. Ke, *Nano Letters*, **5** (2005), p. 2039.

- **October 2005:** *Looking below the Surface of Nanomaterials*

   A new technique called scanning near field ultrasound holography (SNFUH) has been developed and implemented to image and study nanostructures existing beneath the surface. This technique can image buried nanostructures with a spatial resolution of 10–100 nm.

   "Nanoscale Imaging of Buried Structures Via Scanning Near Fied Ultrasound Holography", Shekhawat, G.S. and V.P. Dravid, *Science*, **310** (2005), p. 5745.

- **October 2005:** *Patterning Substrates with Virus*

   Genetically modified virus particles have been used to create patterns on gold substrates. This kind of biology-inspired patterning on a conductive substrate could rove very useful for development of biological sensor arrays.

   "Patterned Assembly of Genetically Modified Viral Nanotemplates Via Nucleic Acid Hybridization", Yi, H., S. Nisar, S.Y. Lee, M.A. Powers, W.E. Bentley, G.F. Payne, R. Ghodssi, G.W. Rubloff, M.T. Harris and J.N. Culver, *Nano Letters*, **5** (2005), p. 1931.

- **November 2005:** *Gold Dots used for Negative Permeability*

   Nanofabricated medium consisting of electromagnetically coupled pairs of gold dots shows a strong negative permeability in the visible region. This presents an important discovery in the field of optics and magnetism.

   "Nanofabricated Media with Negative Permeability at Visible Frequencies", Grigorenko, A.N., A.K. Geim, H.F. Gleeson, Y. Zhang, A.A. Firsov, I.Y. Khrushchev and J. Petrovic, *Nature*, **438** (2005), p. 335.

- **November 2005:** *Nanosensor Based on Quantum Dot*

Quantum dots modified with DNA probes have been used to detect DNA targets using Fluorescence Resonance Energy Transfer (FRET). The specificity of this technique is unparalleled, while the detection capabilities of single-point mutations have been reported.

"Single-Quantum-Dot-Based DNA Nanosensor", Zhang, C.Y., H.C.Yeh, M.T. Kuroki and T.H. Wag, *Nature Materials*, **4** (2005), p. 826.

- **November 2005:** *Nanoparticles for Cellular Targeting*

Magnetic, fluorescent nanoparticles functionalized with a variety of biological molecules have been used to differentiate endothelial cells. This material is used for *in-vivo* cancer imaging in pancreas. This could also find application in differentiating cell lines, exploring cellular states and targeting specific cell types.

"Cell-Specific Targeting of Nanoparticles by Multivalent Attachment on Small Molecules", Weissleder, R., K. Kelly, E.Y. Sun, T. Shtatland and L. Josephson, *Nature Biotechnology*, **23** (2005), p. 1418.

- **November 2005:** *Making Strong Nanotube-based Foams*

Arrays of multi-walled carbon nanotubes have been found to possess exceptional compressibility and strength. The foam material withstands up to 15 MPa and could be used as damping layers for high-precision devices.

"Super-Compressible Foamlike Carbon Nanotubes Films", Cao, A., P.L. Dickrell, W.G. Sawyer, M.N. Ghasemi-Nejhad and P.M. Ajayan, *Science*, **310** (2005), p. 1307.

- **December 2005:** *Do Silica Nanotubes Get Wet?*

Capillary action in silica nanotubes have been studied to great detail, employing a variety of solvent systems. The solvent uptake by such nanotubes will provide information about their wettability and their use as potential filters and purifiers.

"Observing Capillarity in Hydrophobic Silica Nanotubes", Jayaraman, K., K. Okamoto, S.J. Son, C. Luckett, A.H. Gopalani, S.B. and D.S. English, *J. Am. Chem. Soc.*, **127** (2005), p. 17385.

- **December 2005:** *Carbon Nanotubes to Absorb Microwaves*

Carbon nanotubes and nanobeads prepared by pyrolysis of camphor have been observed to absorb microwave to the extent of 20 dB. This presents a significant step in studying microwave assisted reactions with nanotubes as templates. Industrial applications of this material are also worth pursuing.

"Application of Carbon Nanomaterial as Microwave Absorber", Sharon, M., D. Pradhan, R. Zacharia and V. Puri, *Journal of Nanoscience and Nanotechnology*, **5** (2005), p. 2117.

- **December 2005:** *Carbon Nanotubes as Fire-retardant*

Carbon nanotube-polymer composites have exhibited excellent fire retardant properties in comparison to traditional halogenated materials. The material behaves like a gel above 200°C and are environmentally more benign than conventional materials.

"Nanoparticle Networks Reduce the Flammability of Polymer Nanocomposites", Kashiwagi. T., F. Du, J.F. Douglas, K.I. Winey, R.H. Harris and J.R. Shields, *Nature Materials*, **4** (2005), p. 928.

References

1. Several nanotechnology websites, especially http://nanotechweb.org
2. Websites of authors
3. www.webofscience.com

Glossary of Nano Terms

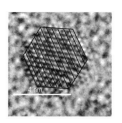

No term can be associated exclusively to an area. However, there are several terms used routinely in nanoscience and technology, just as in any other area. The following constitutes an effort to summarize these terms.

Ab initio: From first principles, generally used in the case of theoretical calculations.

Acoustic microscope: An AFM enabling one to image the topography of a sample, and simultaneously study the ultrasonic surface vibrations.

AFM: see Atomic Force Microscope.

AMFM: Attractive-mode force microscope. Used for biological imaging.

aNFOM: Apertureless NFOM. By measuring the modulation of the scattered electric field from the end of a sharp silicon tip which is scanned close to the surface, an image is constructed. Resolutions of the order of a few nm can be achieved.

Ångstrom (Å): Unit of length used to measure atoms and molecules, 10-10 meter (m) or 0.1 nanometer (nm).

ARTM AFM: Attractive Regime Tapping-Mode AFM.

Assembler: An all-purpose device for guiding chemical reactions by positioning molecules. A programmable molecular machine, which can build any molecular structure or device from simpler chemical entities. This will be similar to an assembly machine in a factory. Originally introduced by Eric Drexler.

Atomic force microscope (AFM): A device used to get topography of a surface at atomic resolution. This is a scanning probe microscopic (SPM) device. It involves the scanning of a probe over a surface to be investigated in atomic steps.

Atomic manipulation: Manipulating atoms using SPM techniques.

Atomistic simulations: Computations of the properties of nanoscale systems using sophisticated models.

Bacteriophage (phage): Any virus that infects bacteria. Organisms used for the study of molecular genetics and are widely used as cloning vectors.

BEEM: Ballistic electron emission microscopy. Electron tunneling from the tip to the base electrode injects ballistic electrons. The current due to these electrons provides information on the interface electronic structure.

BEES: Ballistic Electron Emission Spectroscopy. See BEEM.

Beanstalk: A tower stretching from planetary surface to synchronous orbit. Proposal to generate electricity by putting solar panels in space and transferring the energy using carbon nanotubes has fascinated scientists.

Bit: A binary digit. The fundamental unit of information in Information Theory.

Bioborg: A biological cyborg without non-organic components.

Bio-assemblies: Biomolecular assemblies. Assemblies containing several biologically relevant molecules.

Biocompatibility: The ability of the biological system to accept foreign objects.

Bioinformatics: The study and compilation of the information content of biological objects, particularly in relation to genetic materials.

Biomedical nanotechnology: Referring to the subject area of nanomedicine.

BioMEMS: MEMS used in medicine.

Biomimetic: Imitating, copying, or learning from nature. Nanotechnology is practiced by nature and adaptation of these principles can result in improved technology. See wet nanotechnology.

Biomimetics: The branch of study dealing with the structures and functions of biological substances of artificial products mimicking biology.

Biomimetic chemistry: Chemistry of new molecules, molecular assemblies and materials with biomimetic functions. These materials (such as synthetic cell membranes) are designed for specific applications.

Biomimetic materials: Materials that imitate, copy or learn from nature.

BioNEMS: Biofunctionalized nanoelectromechanical systems. NEMS which can interact with biology.

Biopolymeroptoelectromechanical Systems (BioPOEMS): Combining optics and microelectromechanical systems which can be applied for biology.

Biotin: A low molecular weight compound used as a co-enzyme. As they bind tightly to the egg protein avidin, this is used as a marker in a number of nanotechnology applications.

Blotting: A technique in which macromolecules separated on a gel are transferred to a sheet of paper, which immobilizes them. The molecules can be analyzed by other techniques subsequently.

Blue Goo: Opposite of Grey Goo. Beneficial nanobots. See other terms, Grey Goo, Red Goo, Green Goo.

Bot: A hardware or software robot or program.

Bottom-up: Building larger objects starting from smaller building blocks. Starting from molecules and atoms; common approach in chemistry.

Brownian motion: Inherent motion of particles in a fluid owing to thermal agitation, named after Robert Brown (1827), its discoverer. This motion is central to many of the properties of natural systems.

Bulk technology: Technology which is based on the manipulation of atoms and molecules in bulk (referring to micrometers).

Buckminsterfullerene: The spherical cage structures of carbon containing 60 atoms, resembling a football. Named after the architect Buckminster Fuller, who is famous for the geodesic domes. *See* also Fullerenes.

Bucky balls: Short name for Buckminsterfullerene molecules.

C-AFM[1]: Contact AFM. The scanning probe tip is in contact with the sample surface being investigated.

C-AFM[2]: Conducting AFM. Contact mode AFM with conductive tip. The tip-surface current is monitored.

C-AFM/STM: Combined AFM/STM.

CAM: Constant-amplitude mode of non-contact AFM. Here the tip is oscillating with a constant amplitude of 1–10 nm at a given frequency. The interactions with the surface changes the frequency. The amplitude of oscillation is kept constant. When the cantilever oscillation is damped as a result of interaction, the excitation voltage to keep the amplitude constant, Vexc, increases. The shift in the frequency and the shift in the excitation voltage can be used for imaging the surface and/or studying the strength of interaction.

Carbon nanotubes: See Nanotubes.

CFM: Chemical Force Microscopy or Contact Force Microscopy.

Cavitation: The formation of bubbles in a fluid during high-power sonication (physics); formation of a cavity by either normal or pathological biological processes (medicine).

Chemical Vapour Deposition (CVD): A technique used to prepare thin films, widely used in materials synthesis. Carbon nanotubes are largely made by this route.

Chemical force microscopy: An AFM technique with molecularly functionalized tip, used to probe the chemical characteristics of surface-bound molecules. Used to study properties such as strength of single molecules and bonds.

Chemomechanical conversion: Conversion of chemical energy into mechanical energy.

Chiral: The property of an asymmetric molecule, the structure of which cannot be superimposed on its mirror image. The molecule exists in two forms called enantiomorphs, which are mirror images of each other. The solutions of these rotate the plane of polarized light in different directions, left (levo) or right (dextro), making the enantiomers levorotatory or dextrorotatory. There is also a racemic mixture, which has an equal fraction of both and therefore, will not be optically active.

Cognotechnology: Convergence of nanotechnology, biotechnology and IT for control of mind.

Colloid: A state of matter in while particle dimension changes from that of true molecules and bulk suspensions. The dimensions vary from 1 nm to 100 nm or larger. Such solutions possess a range of properties. The particles may have charge and can be separated by electrophoresis, except at their isoelectric point (when no net charge exists). Originally used to refer to amorphous state, in contrast to crystalloids.

Computational nanotechnology: Computational studies of nanomaterials and structures. Almost any property of a material can be simulated.

CNT: Carbon nanotube.

Constant-current mode for STM: In constant-current mode the feedback keeps the tip-surface current constant and the distance between the sample and tip is monitored.

Constant-force mode for AFM: The feedback keeps the force between the cantilever and the sample constant.

Constant-height mode for SPM: Similar to the constant-current or constant-force modes but the feedback mechanism is turned off and so the technique is faster.

Contact mode: A technique used in AFM.

CP-AFM: Conducting probe AFM. This is used to measure I–V characteristics of a nanoscale sample while doing AFM.

Dalton: Unit of molecular mass. Approximately equal to the mass of a hydrogen atom (1.66×10^{-24} g).

Dendrimers: Large molecules. From the Greek word *dendra*, tree, a dendrimer is a branching polymer. The diversity of the branches vary depending on the generation of its growth: several of them are of nanometer dimensions. Dendrimers are used in nanoparticle synthesis.

Dendrite: An extension of a nerve cell, it receives stimuli from other nerve cells.

DFM: Dynamic Force Microscopy or Dissipation Force Microscopy.

DFS: Dynamic Force Spectroscopy. Here, the resonance frequency of the cantilever as a function of the resonance amplitude is measured.

Diamondoid: Structures resembling diamond with three dimensional network of covalent bonds. Structures of this kind are formed mostly with second row atoms. Most useful diamondoid structures are carbon rich. Structures have some of the properties of diamond such as high strength.

Dimer: A combination of two similar or dissimilar molecules, the starting point of making larger structures. Homo and hetero dimers refer to similar and different constituents, respectively.

Dip pen nanolithography: An AFM-based soft-lithography technique.

Disassembler: An instrument able to take apart structures a few atoms at a time. See assembler.

DNA chip: A chip built to identify mutations or alterations in the DNA code.

DNA Microchip: See DNA chip.

Dry nanotechnology: Fabrication of nanostructures without biology or wet chemistry. A carbon nanotube based technology will be dry as against a technology with DNA which will be 'wet'. See wet nanotechnology. These two terms were originally introduced by Prof. R.E. Smalley.

EC SPM: Electrochemical SPM.

EFM: Electrostatic Force Microscopy. Electrostatic force between the DC biased conductive tip and the surface is measured by measuring the change in the resonance frequency of the cantilever. Also used to refer to another technique, Electric Force Microscopy.

Electrophoresis: The movement of charged colloidal particles through a medium due to applied electrical potential. The method is used to separate macromolecules such as proteins on the basis of number of charges carried by the molecule. Several sub-nano particles of atomic specificity are separated using this technique.

Entanglement: From quantum mechanics, entanglement is a relationship between two objects in which they both exhibit superposition but once the state of one object is measured, the state of the other is also known. It is a hotly researched topic today, both in theory and experiment.

Escherichia coli (E. coli): Rodlike bacterium normally found in the colon of humans and other mammals and widely used in biomedical research, as an indicator organism. In biophysics several fundamental aspects of biology are investigated with this.

Eucaryote (Eukaryote): Living organism composed of one or more cells with a distinct nucleus and cytoplasm. This class includes all forms of life except viruses and bacteria, which are procaryotes.

Femto: A prefix meaning Femtometer, 10^{-15} m. Unit suitable to express the size of atomic nuclei as against Angstrom (10^{-10} m), used for measuring atomic dimensions.

Femtosecond: 10^{-15} s.

Femtotechnology: Manipulating materials on the scale of elementary particles (leptons, hadrons, and quarks). A technology after picotechnology, which is below the length scale of nanotechnology. Also used as Femtotech.

FFM: Frictional force microscopy.

FMM: Force modulation mode, Force modulation microscopy.

Flagellum (plural flagella): Long, whiplike protrusion whose undulations drive a cell through a fluid medium.

Fluorescein: Fluorescent dye emitting light when irradiated at the right frequency. Several biologically friendly fluorescent molecules, especially fluorescein isothiocyanate (FITC) are used for probing biological processes in-situ.

Fractal: A mathematical object having fractional dimension.

Fullerenes: Fullerenes are a molecular form of pure carbon discovered in 1985 and rewarded with a Nobel prize for the discoverers (Kroto, Smalley and Curl) in 1996. They are cage-like structures of carbon atoms, the most abundant form produced is buckminsterfullerene (C_{60}), with 60 carbon atoms arranged in a spherical structure. There are larger fullerenes containing from 70 to 500 carbon atoms. Various other forms of fulleneres are now known, even with other elements.

Functionalized/functionalization: The attachment of a chemically active moieties to an inert molecule/entity.

Gene Chip: See DNA chip.

Genome: The genetic information carried by an organism.

Gray Goo or Grey Goo: Destructive nanobots. opposite of Blue Goo. Several other Goo's have been proposed.

GTP (guanosine 5'-triphosphate): Major nucleoside triphosphate used in the synthesis of RNA and energy-transfer processes. Cell stores energy in this and its hydrolysis releases it.

Heisenberg uncertainty principle or uncertainty principle: A quantum-mechanical principle, according to which the position and momentum of an object cannot be precisely determined. "The more precisely the POSITION is determined, the less precisely the MOMENTUM is known" (Werner Heisenberg).

HeLa cell: A human epithelial cell line that grows vigorously in culture, used for biological investigations. Derived from a human cervical carcinoma.

Hemoglobin: The major protein in red blood cells that is involved in oxygen binding.

HOPG: Highly Oriented Pyrolytic Graphite. HOPG is widely used in scanning tunneling and force microscopies as the substrate material because of its atomic flatness and cleanness.

Hydrophilicity: Tendency to mix with water; wettable. Hydrophilic groups interact with water, so that hydrophilic regions of protein or biomolecule tends to reside in water.

Hydrophobicity: Tendency of hating, not to mix with water; nonwetting.

IC-AFM: Intermittent contact AFM, also called the high-amplitude resonance mode or tapping mode.

IET: Inelastic electron tunneling.

IFM: Interfacial force microscopy.

Immunoglobulin (Ig): An antibody molecule. Higher vertebrates have five classes of immunoglobulin—IgA, IgD, IgE, IgG, and IgM—each with a different role in the immune response.

In vitro: A process taking place outside the cell, as opposed to in an organism (*in vivo*) (Latin for "in glass").

In vivo: In the cell or organism (Latin for "in life").

Ion beam lithography (IBL): Lithography using ions rather than light beams.

Ion channel: A water-filled channel across the lipid bilayer through which specific inorganic ions can diffuse. In cells, voltage responsive proteins make the channels open or close.

Isoelectric point: The pH at which a charged molecule has no net electric charge. The molecule does not move at this pH in an electric field.

KPFM: Kelvin probe force microscopy, In the Kelvin method (or vibrating capacitor method) two conductors are arranged as a parallel plate capacitor, and the spacing is kept small. The contact potential difference (CPD) between the two materials is sensitive to the changes in the surface electronic structure of the plates. There are various ways by which measurements can be done which can be used to understand the structure of interfaces.

KPM: Kelvin probe microscopy.

Langmuir-Blodgett (LB) method: A technique used to make ultrathin films, first a Langmuir film is made (at air-liquid interface) which is then compressed and transferred to a substrate.

LCD: Liquid crystal display, most important technology used in flat panel displays. Alignment of liquid crystals can be altered with electric current. Assume that light is transmitted through a crystal in one alignment and in another alignment caused by the electric filed, the transmission is affected. This can be used to cerate a display if alignment at individual locations can be altered. By packing red, blue and green emitting crystals appropriately, a full colour display can be developed.

LEDs: Light Emitting Diodes, Traditionally LEDs are created from two semiconductors, although there are several organic analogues now. LEDs are durable, with low power requirement and come in various colours. Arranging red, blue and green LEDs next to each other can create a display. As LEDs are larger, they will not get the resolution required for most indoor displays.

LFM: Lateral Force Microscopy.

Low-dimensional structures: Example: quantum wells, quantum wires and quantum dots.

Ligand: A binding agent, which links on a metal ion or atom or group of atoms.

Liposome: Phospholipid bilayer vesicle formed from an aqueous suspension of phospholipid molecules.

Lysis: Rupture of a cell's plasma membrane, leading to the release of cytoplasm and the death of the cell.

Lysosome: Membrane-bounded organelle in eucaryotic cells containing digestive enzymes.

Mechatronics: Electronics and mechanical engineering together, so as to make intelligent machines.

Membrane channel: A protein complex that allows inorganic ions or small molecules to diffuse passively across the lipid bilayer.

Membrane potential: Voltage difference across a membrane due to a slight excess of positive ions on one side and of negative ions on the other. A typical membrane potential for an animal cell plasma membrane is 60 mV (with inside negative relative, in comparison to the surrounding fluid).

Membrane protein: Protein that is normally closely associated with a cell membrane.

Membrane transport: Movement of molecules across a membrane assisted by a membrane transport protein.

MEMS: Microelectromechanical systems: generic term to describe micron scale electrical/mechanical devices.

Mesoscale: A device or structure larger than the nanoscale (10^{-9} m) and smaller than the megascale. Typically in the range of 10^{-7} m. Used also as mesoscopic.

MFM: Magnetic force microscopy or Manipulation force microscopy. The manipulation force microscope is an AFM adapted to measuring the force necessary to displace micron-size samples adhering to surfaces.

Microtubule: Long, cylindrical structure composed of the protein, tubulin. It is one of the three major classes of filaments of the cytoskeleton.

Molecular assembler: Also known as an assembler.

Molecular beam epitaxy: (MBE) Process used to make compound (multi-layer) semiconductors involving deposition of layers one by one.

Molecular electronics (ME): Electronic devices with molecular parts rather than the continuous materials as in Si.

Molecular integrated microsystems (MIMS): Microsystems in which functions found in biological and nanoscale systems are integrated with materials.

Molecular machine: Machine with molecular parts which converts mechanical energy to motion.

Molecular mechanics: A method by which the molecular potential energy function is computed to understand and predict various properties of the system.

Molecular manipulator: A device with atomically precise positioning, and can be used to make complex structures.

Molecular recognition: Molecules adhere in a highly specific way recognizing the other, in a structured fashion.

Molecular wire: A molecular wire, a quasi-one-dimensional molecule, which can transport charge carriers, from one end to the other.

Monomer: Primary unit from which a polymer is constructed.

Moore's law: Coined by Gordon Moore originally in 1965, future chairman and chief executive of Intel. The of number transistors packed into an integrated circuit had doubled every year since the technology's inception four years earlier. In 1975, the time period was revised to every two years. People use a time scale of 18 months, which is obeyed approximately all through the semiconductor revolution.

MSTM: Magnetic STM.

Nanite: Nanorobot or nanobot. Machines with atomic-scale components.

Nanoarray: An ultra-sensitve array used for biomolecular analysis, much more dense and compact than a microarray.

Nanoassembler: A nanoscale device which combines individual molecules into the structures required.

Nanobalance: A balance small enough to weigh viruses and other sub-micron scale particles. Several approaches for nanobalances have been reported based on carbon nanotubes. Some are mentioned in the chapter on experimental methods.

Nanobiotechnology: Application of nanotechnology to learn biological systems and use of this knowledge for devices and methods.

Nanobot: Nanorobot or Nanomachine.

Nanobubbles: Nanosized bubbles with or without molecules or objects. We have a chapter on such objects in this volume.

Nanochips: Next generation chips with higher storage density.

Nanoclusters: Aggregates of atoms, molecules or clusters in the nanoscale.

Nanocones: Objects made with carbon and carbonaceous materials, used for liquid delivery and light transport applications. These objects may be of use to NSOM and in nanoscale chemistry/biology.

Nanocrystals: Crystals of elements or compounds in the nanoscale, with which they manifest properties different from their bulk counterparts. The crystal structure may be the same or different from their bulk analogues.

Nanofabrication: Fabrication of materials and devices using assemblers starting from molecules, same as nanoscale engineering.

Nanofilters: Objects used for selective filtration of molecules depending on their sizes.

Nanofluidics: The study of manipulation of nanoscale quantities of fluids.

Nanoimprinting: Also called soft lithography. A technique developed by Whitesides and coworkers using mould based printing, the moulds have nanoscale features. The creation of features sometimes utilize self assembled monolayers (SAMs).

Nanoindentation: Force exerted by a diamond indenter is measured as it makes an indent of nanometer dimensions in a material. The stiffness and hardness of the material are extracted. Mechanical properties of the material are understood from such studies.

Nanolithography: Nanoscale writing. This uses a variety of techniques some of them are covered in this book.

Nanomachine: A molecular machine made to perform mechanical functions.

Nanomachining: Machining of matter by modifying materials in the nanoscale.

Nanomanipulator: A virtual reality device connected to a scanning probe microscope, allowing manipulation of atoms. Here individual molecules, DNA and atomic objects can be manipulated. Nanomanipulation refers to the process.

Nanomanufacturing: Molecular manufacturing.

Nanomaterials: Materials in which one of the dimensions of the constituent objects is under 100 nm.

Nanomechanical: Dealing with mechanical properties of nano objects.

Nanometer (nm): 10^{-9} m.

Nanooptics: Investigation of light-matter interaction at the nanoscale.

Nanopens and Nanopencils: Objects used in dip pen lithography allowing drawing of objects for nanoelectronics.

Nanopharmaceuticals: Use of nanoscale objects for the use of delivery, transport, uptake and delivery of pharmaceuticals.

Nanopipettes: Objects similar to pipettes with which controlled delivery of chemicals, reagents and light at nanometer scale areas. An area of research used intensely in biology for manipulating living organisms.

Nanopores: Channels with dimension as small as single molecules or DNA which can be used for molecular separation. This pores can be in membranes, ceramic objects such as zeolites. This kind of objects can simplify biomolecular separation.

Nanoprobe: Nanoscale devices used to understand the properties of materials and systems.

Nanoreplicators: A nanobot capable of replication.

Nanorobot: Nanobot.

Nanorods: Objects with diameter in the nanodimension, length can be from nanometers to microns. Several nanorods are known, with metals and semiconductors.

Nanoropes: Nano objects such as particles, rods and tubes connected to form a rope.

Nanoscale: The dimension of 1-100 nm.

Nanoscopic: Dealing with nanoscale.

Nanosensor: Sensor for chemical or physical properties made with nanomaterials.

Nanoshells: Shells with wall thickness or dimension in the nanoscale. Nanoshells of metals, semiconductors, insulators and polymers are known. We have a chapter on this in the volume. Also used as nanobubbles.

Nanosieving: A molecular sorter, used in the separation of biomolecules.

Nanosprings: A nanowire turned to form a spring.

Nanosurgery: Molecular repair and cell surgery.

Nanosystem: An assemblage of nanoscale components designed for a specific function.

Nanoterrorism: Antisocial activities with nanobots.

Nanotube: A cylindrical tube of carbon with diameter in the range of nanometers and length of the order of microns, discovered by Sumio Iijima, 1991. See the chapter on carbon nanotubes.

Nanowires: Similar to nanorods, but of longer length than rods.

Nanowetting: Wetting behavior studied at the nanoscale.

NEMS: Nanoelectromechanical systems: Nanoscale electrical/mechanical devices.

NSOM: Near field scanning optical microscopy, also SNOM.

OLED or Organic LED: LEDs based on carbon-based molecules.

Oligomer: Short polymer, between monomer and polymer, from Greek oligos, few, little.

Passivation: Covering the reactive surface with molecules or atoms so as to eliminate dangling bonds. A method used to protect nanoparticles.

PCR (polymerase chain reaction): Technique used for amplifying specific regions of DNA by cycles of DNA polymerization, each followed by a brief heat treatment, used to separate complementary strands.

Phase imaging: Imaging by mapping the phase of the cantilever oscillation in the Tapping Mode.

Organelle: Subcellular compartment, located in the cytoplasm, surrounded by a membrane (e.g., lysosome, mitochondrion).

Pico technology: Next smaller step after Nanotechnology.

Proteomics: The study of proteins and their functions expressed by a genome.

Protoplasm: The total contents of a living cell.

Proximal probes: Devices capable of fine positional control of atoms and molecules, enabling nanotechnology.

PSD: Position sensitive detector or post source decay, in mass spectrometry.

QFM: Quasinoncontact force microscopy.

QPM: Quantum proximity microscopy.

Quantum: A discrete quantity of electromagnetic radiation or amount of energy associated with a process.

Quantum computer: Computer utilizing quantum properties of atomic or subatomic matter.

Quantum confinement: Confinement of electrons when the size of matter reduces below the exciton radius of the semiconductor. At the confined state, quantum effects are exhibited by the material.

Quantum cryptography: A cryptographic system utilising quantum properties. Unsolicited visitors alter the quantum state of the system and so are detected.

Quantum dot lasers: Lasers utilizing energy states of nanomaterials.

Quantum dots: Nanocrystals or nanoparticles. Refer to confined electrons. Electrons in them occupy discrete states as in the case of atoms and therefore, quantum dots are referred to as artificial atoms.

Quantum Hall effect: Quantised resistance observed in some semiconductors at low temperatures.

Quantum mechanics: The mechanics of subatomic, atomic and molecular objects. Motion of subatomic particles is described by this mechanics, the central aspect of it is the wavefunction which describes the system under investigation completely.

Quantum mirage: The possibility of transporting information utilizing the unique properties of quantum confined electrons.

Quantum tunneling: Electronic transport across a barrier without medium.

Quantum well: A thin layer of narrow band gap semiconductor sandwiched between insulators. The electrical transport is confined to the thin layer.

Quantum wire: A thin line of conducting material in which electrical transport is confined within one dimension.

Qubit: The analogue of a bit in quantum computing.

RC SFM: Resonance contact scanning force microscope, a method used for high speed imaging for real time applications in which noise arising due to tip-surface interaction is reduced.

Red Goo: Refers to designed nanotechnological objects for destruction, similar to grey goo, which is accidentally created.

SAP: Scanning atom probe.

SCM: Scanning capacitance microscopy, used to map the local capacitance of a surface.

Self-assembly: Molecular organization without external stimulus.

Self-repair: Ability to heal itself without external influence.

Self-replication: An ability of an assembler to replicate itself to a definite amount within a fixed period.

SAM: Scanning acoustic microscope or Self-assembled monolayer.

SEM: Scanning electron microscopy, imaging technique in which a focused beam of electrons is used to scan the sample and the generated secondary electrons and ions as well as deflected primary electrons are used to image the sample.

SFM: Scanning force microscopy (atomic force microscopy), Shear force microscopy.

SFS: Scanning force spectroscopy, Static force spectroscopy.

Shape memory alloys: Alloys which "remember" their shape and return to their shape after being bent. The commonly used one is Nitinol, which gets its name from the constituent elements and the discovery team (Nickel/Titanium/Naval Ordinance Laboratory). Typical composition is 55%–56% nickel and 44%–45% titanium.

SICM: Scanning ion-conductance microscopy.

Smart materials: Products with ability to respond to the environment, such as shape transformation.

SNOM: Scanning near field optical microscopy: same as NSOM.

SPE: Scanning probe electrochemistry.

Spintronics: Electronic devices exploiting the spin of electrons, in addition to charge.

SPM: Scanning probe microscope (SPM), including AFMs and STMs, in which effect of interaction of a sharp probe with the sample is measured to infer atomic structure of the material.

SP STM: Spin-polarized STM.

SSPM: Scanning surface potential microscopy.

SThM: Scanning thermal microscope.

STM: Scanning tunneling microscopy, also Scanning thermal microscopy, Imaging technqie using local temperature variations of the sample.

STS: Scanning tunneling spectroscopy.

Superlattices: Materials which exhibit double periodicity. A nanoparticle can be crystallized into a superlattice in which there will be a periodicity within the nanoparticle (gold lattice for example) and a periodicity of the particles themselves. Such superlattices are known in conventional materials also, in addition to multilayered films.

SVET: Scanning vibrating electrode technique.

Tapping mode AFM: The probe tip is oscillated at a fixed frequency, close to the resonance frequency and scanned. Amplitude of the oscillation changes with the topography of the sample, which is used for imaging.

Technocyte: A proposed nanoscale device in the bloodstream for repair or protection.

Technofobes (or Technofobits): Those who have a phobia to technology.

Top-down approach: Making nanosystems, materials and devices starting from bulk materials.

Tribology: Study of friction, wear and lubrication of interacting surfaces.

Tunneling: A truly quantum mechanical effect due to the wave nature of matter by which a particle can cross over a barrier if its wavefunction has a finite probability of existence.

UAFM: Ultrasonic AFM.

UFM: Ultrasonic force microscopy.

UHV: Ultra high vacuum, a requirement of most of the high quality scientific instrumentation, considered as vacuum better than 10^{-8} torr.

van der Waals force: Weak intermolecular attractive forces.

Vesicle: Small, membrane-bounded, spherical organelle in cells.

VT SPM: Variable temperature SPM.

Wet nanotechnology: Bionanotechnology, where nanosystems function in the presence of water. The nanomaterials are membranes, enzymes and such biological entities. All life is based on such technologies.

Young's modulus: A modulus relating tensile (or compressive) stress to strain.

Zeptosecond: 10^{-21} s.

Zygote: A fertilized egg.

References

Websites

1. http://www.nanotech-now.com/nanotechnology-glossary-N.htm
2. http://www.foresight.org/UTF/Unbound_LBW/Glossary.html
3. http://www.nano.org.uk/vocab_terms.htm
4. http://www.jpk.com/index2.htm
5. http://www.aleph.se/Trans/Global/Omega/omeg_term.html
6. http://www.orionsarm.com/glossary.html
7. http://www.nanoworld.org/spmglossary/glossindex.htm
8. http://www.nanomedicine.com/NMIIA/Glossary.htm
9. http://www.aleph.se/Trans/Tech/Nanotech/index.html

Books

1. H.S. Nalwa (Ed.), *Encyclopedia of Nanoscience and Nanotechnology*, Vol. 1–10, American Scientific Publishers, 2004.
2. Kenneth J. Klabunde (Ed), *Nanoscale Materials in Chemistry*, Wiley, New York, 2001.
3. C.N.R. Rao and A. Govindaraj, *Nanotubes and Nanowires*, Royal Society of Chemistry, London, 2005.
4. C.J. Chen, *Introduction to Scanning Tunneling Microscopy*, Oxford University Press, 1993.

INDEX

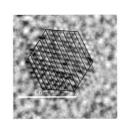

$(\sqrt{3} \times \sqrt{3})\ R30°$ 131
$(\sqrt{7} \times \sqrt{7})\ R10.9°$ 133
Abbe criterion   22
Absorption Spectroscopy   190, 201, 228
Adhesive force   335
Affinity biosensors   146
Ag (111)   133
Agrarian era   8
'All-trans'   133
Amphiprotic species   265
Antibody capped nanoparticles   273
Antibody–nanoparticle hybrid systems   271
Anti-epidermal growth factor receptor   274
Aristotle   158
Armchair tubes   115
Astigmatism   27
Atomic Force Microscopy, AFM   48, 334
Atomic manipulations   47
Attenuation length   35
Au (111)   131
Au-Fe$_3$O$_4$ 276
Aüger electron spectroscopy, AES   62, 69
Ayurveda   9
Azobenzenes   321

Binning, Gerhards   350
Biodiversity   367
Bohr radius   180
Boyle, Robert   158
Bragg's law   77
Bright field   34
BSA-capped gold nanoparticles   274
Buckministerfullerence, C$_{60}$ 89
    Chemistry   96
    Critical point   104
    Endohedral chemistry   99
    Ferromagnetism   103
    Orientational ordering   100
    Solubility   104
    Superconductivity   102

Capacitive charging   138
Capping agents   181
Carbon onions   253
Carbon tetrachloride, reaction with   256
Catenanes   323
Charge coupled device, CCD   39
Chemical sensors   149
Chemical vapour deposition, CVD   61
Chinese hamster ovary cells   277
Chiral tube   115
Chiral vector   115
Chitosan   307
Chromatic abberration   27
Cluster size equations, CSEs   172
Clusters   156
    Formation   158
    Gas phase clusters   156
    Ionic clusters   172
    Metal clusters   167
    Noble gas clusters   171
    Nuclearity   157
    Semiconductor clusters   168
Cold field emitter   24
Complimentary DNA strands   287
Confinement   180
    1D confinement   181
    2D confinement   180
Confocal microscopy   55
Contrast transfer function   35
Core-shell nanoparticles   215
    Application   206, 234

Index

Bimetallic 221
Biological applications 234
Catalysis 235
Characterization 200, 225
Chemical reactivity 237
Magnetism 235
Metal-metal oxide 216
Polymer coated 222
Properties 227
Semiconductor 222
Coulomb blockade 291
Coulomb gap 212
Coulomb staircase 197
Covalent approach 317
Cox 91
Cyanoplyynes 91

Dark field 34
Debye-Scherrer method 78
Dendrimers 272
Diagnosis 365
Diagnostic applications 274
Diffraction 27
Digital imaging 39
Digital revolution 350
Dispersion diagram 230
DNA-modified nanoparticles 264
Drexler, Eric 4, 350
Drinking water 365
Dry nanotechnology 5
Dynamic light scattering, DLS 74

Eigler 47
Einstein, Albert 62
Electrochemical sensors 293
Electrochemistry, core-shell 227
Electromagnetic lenses 23
Electron detector 29
Electron energy loss spectroscopy, EELS 38, 62
EELS at the nanometer scale 41
Electron gun 23
Electron lenses 25
Electron microscopies 20
Electronic noses 292
Electrons 180

Electrostatic attraction 265
Ellipsometry 136
Emission spectroscopy 190
Energy dispersive spectrometry, EDS 32
Energy production 367
Energy storage 367
Environmental Transmission Electron Microscopy, ETEM 40
Environmental pollution 366
Epidermal growth factor receptor, EGFR 274
Euler's theorem 118
Everhart-Thornley (E-T) detector 29
Excited states
Dynamics 191
Life time 191
Exciton radius 180, 290

Faraday 9
Fast atom bombardment, FAB 60
Feynman, Richard 10, 301, 317, 349
Field emission 23, 124
Flow sensor 124
Fluorescein isothiocynate@SiO_2 254
Fluorescence line narrowing, FLN 189, 191
Fluorescence resonance energy transfer, FRET 58, 272
Focused ion beam, FIB 61
Food security 365
Fourier transform ion-cyclotron resonance, FT-ICR 92
Frank-Condon rule 66
Friction force microscope, FFM 336
Fuller, Buckminster 89
Functionalized metal nanoparticles 204
Applications 206, 234

Gas proportional counter 34
Geiger counter 80
Global market 356
Glossary 413
Glucose oxidase, GOD 145
Gold nanorods 298
Gold nanoshells for blood immunoassay 258
Gramicidine 147
Green nanotechnology 353

Index

Hawkins 100
Head disk capacity 338
HEK 308
Helicenes 321
HER2 258
High resolution electron energy loss spectroscopy, HREELS 73
Human papilloma virus, HPV 309
Hydrogen 124

Immunoassays 254
Immunoglobuline G-capped gold nanoparticles 273
Immuno-localization 272
In vivo targeting 272
Industrial age 8
Industrial revolution 349
Information age 8
Inorganic fullerenes 125
In-situ Nano measurements 42
Intracellular SERS probes 277
Ion cyclotron resonance, ICR 163, 165
Ion gate 165
Ion selective films 257
Ion sensing 149
Ionization potential 175
Isoelectric points, pI 265

Jellium model 176

Kaldor 91
Kikuchi patterns 34
Knudsen cell 160
Koopmans' approximation 71
Kratschmer 93

LaB_6 24
Lactobacillus 10
Langmuir, Irving 129
Langmuir-Blodgett Technique 129
Laser vaporization 159
Layer by layer, LbL 224
Leeuwenhoek, Antony van 19
Lens aberrations 26
Liquid drop model 176
Liquid metal ion source 161
Lithium-drifted silicon 33

Lithography 288
Lotus effect 368
Low energy electron diffraction, EELD 139
Low voltage SEM 31
Lycurgus cup 9

Mackay icosahedra 172
Magic numbers 171
Magnetic nanoparticles 270, 312
Magnetosperillum magentotactium 10
Matrix assisted laser desorption mass spectrometry, MALDI MS 61
Mesoporous silica 218
Metcars 169
Metallofullerenes 105
Microanalysis 32
Micro-electro mechanical systems, MEMS 340
Mie's theory 230
Molecular computing 319
Molecular devices 319
Molecular electronics 151
Molecular machines 318
Molecular motors 4, 318
Molecular nanomachines 316
Molecular nanotechnology 4, 350
Molecular ratchet 321
Molecular shuttles 323
Molecular transistor 123
Molecules inside the nanoshells 253
Monolayer protected metal nanoparticles 200
    Characterization 200
    Preparation 200
Monolayers 128
    Composition 143
    Decomposition 142
    Patterning 142
Moore, Gordon 11
Moore's Law 11
MOSFET 327
Motor proteins 319
Multi-walled nanotubes, MWNTs 119

Nano arms race 360
Nano divide 355
Nano economy 355
Nano policies 358

Nano rules   358
Nano, historical   349
Nano-bio hybrid systems   271
Nanobiology   263, 264
Nanobiotechnology   278
Nanobubbles   244
Nanobusiness   355
Nanocapsules   244
Nanocrystal   179
    Chemical properties   196
    Electronic structure   120, 187
Nano-electro-mechanical systems, NEMS   294
Nano-ethics   359
Nanofiber probes   297
Nanofuture   352
Nanolubrication   339
Nanomachine   147, 313
Nanomedicines   301
Nanopores   304
Nanorod-biomolecule hybrid systems   268
Nanosensors based on quantum size effects   291
Nanosensors   285
Nanoshaving   143
Nanoshells   244
    Metal nanoshells   249
        Gold   249
        Silver   250
    Nanoshells from liposomes   251
    Oxide nanoshells   246
        Hollow silica   246
        Titania   248
        Zinc oxide   248
        Zirconia   247
Nanotechnology and the developing world   362
Nanotechnology   350, 360, 363
    History   372
    Investments   356
    Media   361
    Milestones   350
    Reliability   357
    Safety   357
    Risks   357
    Sustainable development   363
Nanotribology   330
Nano-tribometer   331
Nanotube based filters   125

Nanotubes   114
    Electronic structure   120, 187
    Filling   119
    Helicity   115
    Synthesis   117
    Transport properties   122
Nasal administration   307
Near field SERS   297
Nipkow disk   56
Noble metal materials   269
Non-covalent approach   317

Ocular administration   307
Olfactory receptors   3
Optical limiting   233, 239
Optical non-linearity   233
Optical properties   175, 228
Optical spectroscopy, core-shell   225
Oral administration   305

P16   309
Paracelsus   9
Particle engineering   216
Peptide coated quantum dots   272
Perception   289
Photo ionization efficiency, PIE   174
Photoelectron microscopy   71
Photoelectron spectroscopy, PES   62
Photoemission electron microscopy, PEEM   72
Photographic films   39
Photoluminescence excitation, PLE   189
Photosynthesis   3
Photovoltaic devices   4
Piezoelectric drives   45
Piranha solution   130
Plasmon absorption band   201
Polypyrrole coated SiO_2 223
Promyelocytic leukemia   273
Prostate cancer cell   273
Prostrate-specific antigen, PSA   277
Pulsed arc cluster ion source   160

Quadrupole mass filter, QMF   163, 164
Quantum corral   47
Quantum dots   179, 311
    Chemical synthesis   185

Index

Properties with size   194
Synthesis   117
Quartz crystal microbalance, QCM   136, 146, 296, 333
Quasi-elastic light scattering, QELS   74
Quasi-static approximation   229

Raman spectroscopy   74
Rayleigh criterion   22
Recognition   287
Redhead equation   137
Reflection-absorption infrated spectroscopy, RAIRS   136
Resolution   20
Resolving power   20, 22, 34
Reverse saturable absorption, RSA   233
Rohrer, Heinrich   350
Rotaxanes   323

Scan coils   28
Scanning Aüger microscopy, SAM   72
Scanning Electron Microscope, SEM   20
Scanning Nearfield Optical Microscopy, SNOM   56
Scanning Probe Lithography, SPL   51
Scanning probe microscopies, SPM   43
Scanning thermal microscopy, SThM   49
Scanning tunneling microscopy, STM   44
Scanning tunneling spectroscopy, STS   47
Scanning tunneling optical microscopy, STOM   58
Scherrer formula   78, 81
Schottky field emission   24
Screening   365
Secondary ion mass spectrometry, SIMS   59
Self-assembled monolayers, SAMs   129
   Self-assembly   128
   Thermal stability   140
Semiconductor nanocrystals   270
Sensing   236
Sensors   144
Shake-down   71
Shake-off   71
Shake-up   71
Siegbahn, Kai   62
SiHa   274, 309
Single electron devices   196

Single molecule devices   320
Single-walled nanotubes, SWNTs   119
Smalley, Richard E   91
Smart dust   298
Smoke sources   160
Smoluchowski's formula   226
SNMS   59
Societal implications   348, 352
Spherical aberration   26
Spherical quantum well   188
Spiropyran   321
Steam engine   349
STEM   37
Storage   124
Sum frequency generation, SFG   139
Superlattices   208
Superlubricity   337
Supersonic (free jet) nozzle sources   160
Supramolecular systems   321
Surface Enhanced Raman scattering, SERS   297
Surface enhancement   70
Surface force apparatus, SFA   332
Surface plasmon resonance spectroscopy, SPRS   290
Surface plasmon resonance, SPR   267
Switches   320
Switching speed   327

Tectodendrimers   305
Templates   151
Therapeutic applications   310
   Gold nanoparticles   310
Time of flight, TOF   163, 164
Transistor   8
Transition dipole moment   73
Transmission Electron Microscopy, TEM   34, 192
Triptycene   320
Tunneling current   44
Turbostratic constraint   116

Ultraviolet photoelectron spectroscopy, UPS   62, 68

van der Walls   133
Vancomycin   267
Vancomycin-capped gold nanoparticles   275
Vibrational spectroscopies   72

Watt, James   349
Wavelength dispersive spectrometry, WDS   32
Wehnelt cap   23
Wet nanotechnology   5
Wetting control   150
Wien filter   163
Wudl, Fred   97

X-ray diffraction, XRD   75, 192, 225
X-ray photoelectron spectroscopy, XPS   62

Y junction   123
Young's modulus   42, 336

Zeta potential   226
'zigzag' tubes   115